W9-BWY-109

Female Control:
Sexual Selection by
Cryptic Female Choice

MONOGRAPHS IN BEHAVIOR AND ECOLOGY

Edited by John R. Krebs and Tim Clutton-Brock

Five New World Primates: A Study in Comparative Ecology, *by John Terborgh*

Reproductive Decisions: An Economic Analysis of Gelada Baboon Social Strategies, *by R.I.M. Dunbar*

Honeybee Ecology: A Study of Adaptation in Social Life, *by Thomas D. Seeley*

Foraging Theory, *by David W. Stephens and John R. Krebs*

Helping and Communal Breeding in Birds: Ecology and Evolution, *by Jerram L. Brown*

Dynamic Modeling in Behavioral Ecology, *by Marc Mangel and Colin W. Clark*

The Biology of the Naked Mole-Rat, *edited by Paul W. Sherman, Jennifer U. M. Jarvis, and Richard D. Alexander*

The Evolution of Parental Care, *by T. H. Clutton-Brock*

The Ostrich Communal Nesting System, *by Brian C. R. Bertram*

Parasitoids: Behavioral and Evolutionary Ecology, *by H.C.J. Godfray*

Sexual Selection, *by Malte Andersson*

Polygyny and Sexual Selection in Red-winged Blackbirds, *by William A. Searcy and Ken Yasukawa*

Leks, *by Jacob Höglund and Rauno V. Alatalo*

Female Control: Sexual Selection by Cryptic Female Choice, *by William G. Eberhard*

Female Control: Sexual Selection by Cryptic Female Choice

WILLIAM G. EBERHARD

Princeton University Press
Princeton, New Jersey

Published by Princeton University Press,
41 William Street,
Princeton, New Jersey 08540
In the United Kingdom: Princeton University Press,
Chichester, West Sussex

Library of Congress Cataloging-in-Publication Data

Eberhard, William G.
Female control : sexual selection by cryptic female choice / William G. Eberhard.
p. cm. — (Monographs in behavior and ecology)
Includes bibliographical references (p.) and indexes.
ISBN 0-691-01085-4 (alk. paper). — ISBN 0-691-01084-6 (pbk.: alk. paper)
1. Sexual selection in animals. 2. Reproduction—Regulation. I. Title. II. Series.
QL761.E23 1996
591.56—dc20 95-25639

This book has been composed in Adobe Times Roman

Princeton University Press books are printed on acid-free paper and meet the guidelines for permanence and durability of the Committee on Production Guidelines for Book Longevity of the Council on Library Resources

Printed in the United States of America

10 9 8 7 6 5 4 3 2 1

10 9 8 7 6 5 4 3 2 1 (pbk.)

To the memory of George C. Eickwort

Contents

Preface and Acknowledgments

My effort to review this subject completely was ultimately defeated by the size and complexity of the literature on animal sexual behavior and physiology. I attempted to consult original literature rather than reviews, especially after finding that the latter often gave overly simplified renditions of earlier studies, sometimes with a serious male-active, female-passive bias. But I was not completely successful, as can be seen from the references.

I also came to realize that there was no sense in trying to achieve a complete, synthetic overview of cryptic female choice. The field is too young for a useful synthesis. Thus, this book aims to function more as an alarm and a rallying point, to awaken researchers to important but heretofore relatively neglected and only partially appreciated types of female participation in sexual interactions. I regret the probable omission of some studies which could be cited as examples of cryptic female choice or the other phenomena discussed, and apologize to the authors of works I could have cited but did not. In searching for general, wide-ranging trends, I opted, when possible, in favor of taxonomic diversity rather than exhaustive reviews of particular groups. The scope of the examples cited should be taken with these qualifications in mind.

I am grateful to a large number of people for help of different sorts. The following gave permission to use figures: Tom Battin, P. S. Chen, Joshua Ginsberg, Enrique Gonzalez, Darryl Gwynne, Robert Jackson, John Linley, Tomohiro Ono, Merja Otronen, Michael Robinson, Rafael Lucas Rodriguez, Nick Smythe, and Paul Watson.

The following persons kindly provided references: Tim Birkhead, Fred Bercovitch, John Christy, Francesco Dessi-Fulgheri, Peter Dunn, Jessica Eberhard, Kyle Harms, Yosiaki Itô, Scott Pitnick, Stan Rand, H. Robertson, Richard Rowe, Neal Smith, Gabriele Uhl, Rob Voss, Bill Wcislo, Mary Jane West-Eberhard, Masaaki Yamagishi, and David Zeh.

The following allowed me access to unpublished manuscripts and data: Göran Arnqvist, Robin Baker, Tom Battin, Mark Bellis, Andreas Berghammer, William Brown, John Christy, Carlos Cordero, Pauly Eady, Enrique Gonzalez, Pamela Gregory, Sandy Harcourt, Dan Howard, Bernhard Huber, Craig LaMunyon, Scott Pitnick, Viterbo Rodriguez, Rafael Lucas Rodriguez, Nick Smythe, Randy Thornhill, Don Windsor, and Jeanne and David Zeh.

The following people kindly checked over and corrected my descriptions of their work: Göran Arnqvist, William Brown, Terry Christensen, Enrique Gonzalez, Darryl Gwynne, Dan Howard, Bernhard Huber, Robert Jackson, Jan Lifjeld, John Linley, Merja Otronen, Alfredo Peretti, David Queller, Leigh Simmons, John Sivinski, Bob Suter, Gabiele Uhl, Otto von Helversen, Paul Ward, Paul Watson, and Jeanne and David Zeh. I also benefited from useful discussions with Carlos Cordero, Gustavo Hormiga, Egbert Leigh, Pedro Leon, Oscar Rocha, Rafael Lucas Rodriguez, and Mary Jane West-Eberhard.

Molly Denton performed yeoman labors in typing cramped, cryptic script, and in organizing and checking references.

Thanks are also due Chris Spain, for inspiring the imaginary conversation which got this project off the ground. The gift of a graphics program from Paul and Carolina Hanson made the preparation of the figures a much more enjoyable project.

The staff of both the Smithsonian Tropical Research Institute library, especially Angel Aguirre and Vielka Chang, and the Smithsonian Institution library in Washington, D.C., helped locate, copy, and send many of the references in the approximately two meters of xerox copies through which I have burrowed.

I am grateful to the director of STRI, Ira Rubinoff, for his considerate, tolerant, and sustained support over the years. My research has been supported by the Smithsonian Tropical Research Institute, and the Vicerrectoría de Investigación of the Universidad de Costa Rica.

Darryl Gwynne, John Christy, and Jeanne and David Zeh deserve special thanks for reading over and making detailed and thoughtful comments on major portions of the manuscript.

I am particularly grateful to Rafael Lucas Rodriguez, Bernhard Huber, Scott Pitnick, and Bill Wcislo, who read and commented on the entire manuscript, and especially to Mary Jane West-Eberhard for extensive and helpful suggestions.

Female Control:
Sexual Selection by
Cryptic Female Choice

1 What Is Cryptic Female Choice?

1.1 Sexual Selection Results from Competition for Female Gametes, Not for Females

This book is about a neglected aspect of sexual selection: female processes that affect male reproductive success and occur after the male has succeeded in coupling his genitalia with those of a female.

In sexually reproducing animals with internal fertilization, reproduction typically begins with precopulatory courtship—a series of more or less complex interactions between the male and female that culminate with the male introducing his genitalia (or genitalic products such as a spermatophore) into the female's body. These interactions are relatively easily observed, and both theoretical and empirical studies of sexual selection have concentrated on the "preliminaries" which result in intromission. Events that happen later, after genitalic coupling, can also be crucial in determining a male's reproductive success, but they have been largely ignored.

Darwin, for instance, in his chapter on "Principles of Sexual Selection," refers repeatedly to competition between male animals in terms of attempts to "possess," "obtain," or pair with females (Darwin 1871). He states, with reference to humans, that "as far as selection is concerned, all that is required is that choice should be exerted before the parents unite" (p. 894). The implication is that sexual selection results from competition for sexual access to females. Subsequent studies of sexual selection have generally maintained this precopulatory emphasis (e.g., summaries in Bateson 1983; Bradbury and Andersson 1987; Andersson 1994). Indeed, the common use of the term courtship has a similar focus: "Courtship. Overtures to mating. . . . in a narrow sense all patterns of behavior that initiate mating or aim to do so. . . . more broadly conceived, encompassing all behavior patterns of precopulation, pair formation, and pair bonding" (Immelmann and Beer 1989, p. 62).

Inherent in this emphasis on the events preceding copulation is the assumption that copulation leads directly and inevitably to fertilization of the female's eggs. Examination of the male and female morphologies of animals with internal fertilization strongly suggests, however, that the opposite may generally be true: copulation, in fact, seldom leads directly and

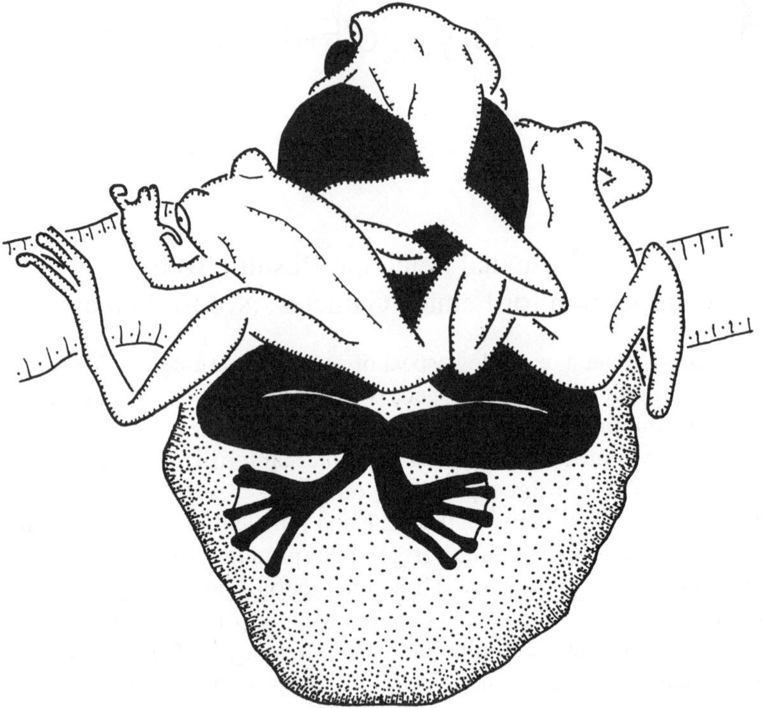

Figure 1.1 The "active male" aspect of sperm competition is exemplified by males of the frog *Chiromantis xerampelina*. The males (white) are jostling for position on a female (black), which is ovipositing into a foam nest (stippled). Due to his small size, a male in amplexus (central male) cannot position his cloaca as close to the female's cloaca as can a peripheral male (male on left). (After Jennions et al. 1992)

inevitably to fertilization[1] (Eberhard 1985). Instead, even after copulation has begun, processes that are sometimes or always under direct control of the female (Thornhill 1983; Eberhard 1985; Birkhead et al. 1993a), as well as direct competitive interactions between males at the level of their ejaculates ("sperm competition" of Parker 1970c), can influence the chances that a given copulation will result in fertilization.

Study of male-male interactions via their ejaculates (fig. 1.1) has increased recently, stimulated by Parker's pioneering paper (1970c) and by the discovery of sperm removal and displacement devices on the male genitalia of some insects (e.g., Waage 1979, 1984; Ono et al. 1989; Yokoi 1990; Gage 1992; Gack and Peschke 1994). Additional male-male competitive devices include the ability to swamp other males' ejaculates with large numbers of sperm (e.g., Kenagy and Trombulak 1986, and Ginsburg and Huck 1989 on mammals; Birkhead and Møller 1992 on birds; Dickinson 1988 on a beetle), to seal off the entrance to the female with a plug (e.g., Aiken 1992 on a beetle), to push previous sperm to sites where they will not be used (Waage 1984 on odonates), to seal them off from the exit duct with a hard, gel-like substance (Diesel 1990 on a crab), and, perhaps, to lure previously deposited sperm out of hiding in the female to sites where they can be diluted or otherwise eliminated (Scott and Williams 1993 on *Drosophila*).

The female side of the postintromission story, which is the theme of this book, has been much less studied, despite the much greater opportunities that females have to influence events. Various female-controlled processes and activities occur after intromission has occurred, and they have been shown to vary in ways that can modify the chances that a given copulation will result in offspring. A possibly incomplete list includes the following: premature interruption of copulation; denial of deeper genitalic access to internal sites where the male's sperm will have a better chance of being used; lack of sperm transport to storage and/or fertilization sites within the female; discharge or digestion of the current male's sperm or those of previous or subsequent males; lack of ovulation; lack of preparation of the uterus for implantation of embryos; abortion; lack of oviposition; rejection or removal of mating plugs; prevention of removal of plugs by subsequent males; removal of sperm-injecting structures (spermatophores) before they have transferred their contents; selective use of stored sperm; failure to trigger sperm-injecting mechanisms of spermatophores; failure to modify insemination ducts, making remating more difficult; selective fusion with sperm that have reached the egg; failure to invest maximally in the mate's offspring; and lack of rejection of subsequent advances of other males (see chaps. 3 and 4 for examples of species in which these processes occur). Additional possibilities include lack of sperm activation, lack of sperm nourishment, failure to seize and open spermatophores in the bursa, and changes in sperm leakage and the efficiency of sperm usage. If females do not respond equally to all conspecific males in any one of these processes, and thereby prejudice paternity in favor of certain types of male, then "cryptic" female choice (Thornhill 1984) occurs.

This distinction between pre- and postcoupling female choice is parallel to that which has been made on the male side between classic, precopulatory sexual selection by direct competitive interactions between males and postintromission sexual selection by competition between their ejaculates ("sperm competition"; Parker 1970c). Just as the idea of sperm competition extends the possiblity of male-male competition past the moment when copulation begins, so the idea of cryptic female choice extends the possibility of female choice past initiation of copulation.

Biases in female responses like those just listed can fit Darwin's original criteria for sexual selection.[2] A given male's reproductive output can be reduced by the attempts of other males to reproduce with the same female.

Such discrimination is "cryptic" in the sense that it is a hidden, or internal, decision made by a female after the more obvious decision to copulate. Even though a male was accepted as a partner in copulation, he may be rejected as a father of the female's offspring. Copulation has been commonly used as the criterion for final acceptance in female choice; but if cryptic female choice occurs, a male that copulates with a female may nevertheless fail to sire her offspring. Cryptic female choice is possible because genitalic coupling does not necessarily result in successful intromission, nor does successful intromission necessarily result in successful insemination, nor does successful insemination necessarily result in the sperm arriving at appropriate storage or fertilization sites, nor does successful storage of sperm necessarily result in fertilization of the eggs.

I will also differ from Darwin's original presentation (1871) in using the phrase "female choice" in a wide sense that is not limited only to processes that involve her nervous system (as, indeed, Darwin himself did later; see Darwin 1882 in Andersson 1994). For example, suppose that the female has a structure with which the male must mesh in order to successfully achieve copulation and insemination, that females mate with more than one male, and that males within the population differ with respect to their abilities to mesh with the female, so that some are able to transfer more sperm than others. If a male's chances of paternity increase when he transfers more sperm, then there will be sexual selection favoring those males able to mesh best with the female. This is similar to the way sexual selection has been defined for plants (Stephenson and Bertin 1983). By changing the structure (either evolutionarily, over time, or facultatively from one male to the next), a female can change the criteria of this selection.

A concrete example may help clarify this idea. The female of the cassidine plant beetle *Chelymorpha alternans* has a long, coiled spermathecal duct with frequent reversals in the direction of coiling (fig. 1.2). The male has an even longer genitalic sclerite, which is threaded up this duct and through which sperm flow during copulation (V. Rodriguez 1994a; V. Rodriguez et al., subm.). When different males mate with the same female, males with a longer sclerite sire significantly more offspring; and when a male's sclerite is artificially shortened, the female stores less of his sperm (see chap. 7.2). The exact mechanism in the female that causes this bias is not known. It could be as simple as the length of the spermathecal duct, with males possessing a longer sclerite more often threading it all the way to the spermatheca. In other words, by imposing a mechanical barrier (in this case, a long, complexly coiled duct), the female can discriminate among potential sires of her offspring with respect to the mechanical properties of their bodies (in this case, a long, flexible sclerite). No female receptors or effectors need be involved (see Huber 1993b for an application of this reasoning to spiders).

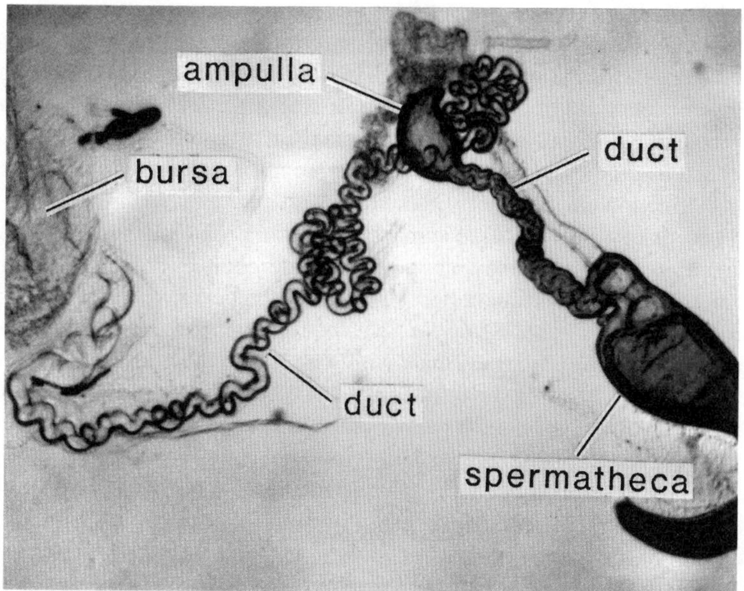

Figure 1.2 The "playing field" on which male *Chelymorpha alternans* tortoise beetles compete, with the advantage going to males with longer genitalia. The male's genitalic "internal sac" enters the thin-walled bursa of the female, and from there his long, thin flagellum (which is about three times longer than his entire body) is threaded up the complexly coiled duct leading to the female's spermatheca (note reversals of direction of coiling; see also fig. 7.8). Usually the flagellum reaches only the ampulla (a "flagellum trap"), where it becomes coiled in a tight tangle.

If sexual selection does indeed involve both sperm competition and cryptic female choice, then it is better understood as being driven by competition among males for access to females' gametes rather than just access to the females themselves. This book, then, represents an attempt to make this small extension to Darwin's theory of female choice, complementing the extension already made regarding male-male (sperm) competition (Parker 1970c). The basic objective of this book is to illustrate the many ways in which cryptic female choice occurs in nature, and to present data indicating that it is probably much more common than previously suspected.

1.2 How to Recognize Cryptic Female Choice

Sexual selection by cryptic female choice can result from a female-controlled process or structure that selectively favors paternity by conspecific males with a particular trait over that of others that lack the trait when the female has copulated with both types. The selection is "cryptic" in the sense that the traditional methods of determining male reproductive success by counting the number of copulations with potentially fertile females will not detect it. In some cases the distinction between cryptic female choice and classic, overt female choice is difficult or perhaps impossible. I will discuss some difficult cases in an attempt to clarify the basic concepts and distinctions involved.

Perhaps first it is useful to start with an example of what cryptic female choice is *not*. A female may bias a given male's chances of paternity simply by altering the proportion of her copulations that occur with that male. The female of the chaffinch *Fringilla coelebs*, for example, may bias competition between sperm from the male with which she is paired and from extra-pair males by changing the frequency with which she leaves the vicinity of her mate, solicits copulations from other males, and allows such copulations to occur (Sheldon 1994). These female behavior patterns may result in sexual selection by female choice. But it will not be *cryptic* female choice if changes in a male's paternity success can be completely accounted for by the percentage of copulations that were with him. If, on the other hand, frequencies of paternity are *not* the same as frequencies of copulations (as, in fact, is the case in this species, where extra-pair copulations have a greater probability of resulting in offspring; Sheldon and Burke 1994; see fig. 4.16 and chap. 4.13 for additional examples), then it is possible that cryptic female choice is also occurring.

Male participation in a female's internal reproductive processes is generally only indirect and subject to female cooperation. There is one possible exception, however—that of nourishment of the female. Nourishment of the female by the male (either orally or via seminal products introduced into her reproductive tract) can benefit the male in two nonmutually exclusive ways: (1) as "paternal effort," the investment of nutrients or defensive compounds, which increases the number and/or quality of the female's offspring; or (2) as "mating effort," which increases the male's fertilization success with respect to that of other males that mate with the same female (e.g., Gwynne 1984c). If the male can count on his sperm being used to fertilize at least some of the extra eggs or higher quality offspring that result from his paternal investment, selection may favor such donations (see summary in Thornhill and Alcock 1983). This kind of investment may evolve under natural selection rather than sexual selection if it does not exclude other males from mating with the female or gaining access to her eggs. Sexual selection by cryptic female choice could occur, however, if females receiving larger investments delay longer in returning to sexual receptivity (e.g., Oberhauser 1989, 1992 on butterflies; Simmons and Gwynne 1991 on a katydid). It could also occur if some males were better able to induce the female to use the nutrients they donate to make more eggs or larger eggs, rather than use them for other functions (e.g., increase female longevity).

In many discussions of male nutrient contributions, females are seen as passive offspring-generating machines whose outputs are directly manipulated by males through their contributions. Determining the relative importance of male contributions versus female choosiness (i.e., how unbiased the female actually is) may be difficult in particular cases. In addition, a difficult distinction must be made between male contributions of nutrients versus male manipulation of the female via hormonelike substances that trigger physiological or behavioral processes in her body. See chapter 6.6 for a detailed discussion of this topic.

A second type of difficulty in distinguishing cryptic female choice can occur in species with maternal care. If a male has a greater ability to induce his mate to make larger maternal investments in the offspring he has sired than she would have made when paired with other males (say, make larger eggs; see chap. 4.10), cryptic female choice could occur. Some possible consequences of manifestations of this type of male ability would not be "cryptic," however, if standard methods of determining lifetime male reproductive success were used (e.g., Clutton-Brock 1988).

Just as in precopulatory female choice (e.g., Andersson 1994), a female exercising cryptic choice can bias paternity by passively allowing other processes, such as direct male-male aggression or sperm competition, to dictate which males will fertilize her eggs (e.g., Dickinson, in press). The line between male and female control can be extremely difficult to determine, and this, added to the usual supposition of male control (see sec. 1.9), means that possible female roles are often neglected. A concrete example (from McVey and Smittle 1984; McVey 1988) is useful to illustrate how cryptic female choice might not be recognized as something distinct from precopulatory female choice.

Some males of the dragonfly *Erythemis simplicicollis* defend territories at ponds with floating vegetation on or near which females oviposit; others are nonterritorial satellites. Females visit ponds daily to oviposit. Following copulation with a territorial male, the female begins a bout of oviposition while the male guards her. During oviposition the female "wanders" over the surface of the water (searching for appropriate sites? dispersing her eggs?), sometimes leaving the territory of the male with whom she has just mated, and moving into that of a neighboring male (who promptly mates with her). The female thus tends to lay more eggs in larger territories before "wandering" into another territory (McVey 1988). The last male to mate with a female fertilizes essentially all of the eggs laid immediately after copulation, but only about 65% of eggs laid 24 hours later, when sperm mixing has occurred (McVey and Smittle 1984).

Females prefer some possible oviposition sites with particular types of water plants over others. They also "refuse" to oviposit after some copulations (27% of copulations with satellite males; 7% of copulations with resident territorial males); following a refusal, the female flies away from the pond, then returns within a few minutes and is generally captured and mated by another male. In addition, females lay an average of 20% of their eggs on returning the next day without further copulation (McVey 1988).

Direct male-male combat certainly influences territory ownership and size, and may also influence pond choice (competition for territories was stronger at one pond with more attractive oviposition sites; McVey 1988). So male-male battles certainly filter the males with respect to their abilities to hold large territories. Females are therefore more likely to interact (and mate) with some males than others just prior to ovipositing. But females "refuse" to oviposit about four times more often after copulating with satellite males than with territorial males, thereby exercising further, cryptic selection against satellites. In addition, if after copulating with some males and not others females increase rates of oviposition, or wander less while ovipositing, or are more likely to behave inconspicuously on arriving at the pond the next day and thus reduce their chances of a subsequent mating before at least some of their eggs are laid, cryptic female choice may further influence male reproductive success.

I believe that many evolutionary biologists have underestimated the possibilities for cryptic female choice. Males who lose in a second, postcopulatory phase of female choice will fail to sire offspring just as certainly as those who lose out in the precopulatory phase. Likewise, any general treatment of sexual selection that omits the cryptic aspects of choice risks serious error because the rules of the internal game may be drastically different from those in force prior to copulation as regards such criteria as male vigor, size, or strength. Yet cryptic choice does not even appear in the indices of recent major books on sexual selection (Bateson 1983; Bradbury and Andersson 1987; Andersson 1994).

1.3 Female "Rules of the Game"

Female roles in postintromission discriminations among males can be subtle, and are easily underestimated. Consider, for instance, how male seminal products transferred along with sperm stimulate oviposition in the cricket *Acheta domesticus* (Destephano and Brady 1977). The seminal fluid contains an enzyme (or complex of enzymes) that catalyzes the production of prostaglandin, a substance which induces oviposition by the female (Destephano and Brady 1977). The male enzyme produces prostaglandin in the female by acting on a substrate, arachidonic acid, which the female accumulates in her spermatheca. Attention has generally been focused on the active role of the male in producing and transferring such products (e.g., Chen 1984; Gillott 1988) and their adaptive significance to the male. What is often overlooked is that both the female's responses to such products and the thresholds at which these responses occur may also be evolved characters that have important evolutionary consequences for both males and females.

In the first place, female production of arachidonic acid and its inclusion in the spermatheca prior to mating establish a critical condition that influences male reproduction. Without the arachidonic acid, the selective consequences for the male of transferring or not transferring the enzyme would be entirely different.

Variations in female responses to different quantities of the enzyme suggest a further, selective effect on male reproduction. An evolved female response to a seminal substance (say, a lubricant or a compound that nourished the sperm) might originate when, for example, the substance became used by the female as a cue that insemination had occurred and she could proceed to oviposit. In such a case, one would expect that the female's threshold for a maximum response to this substance would be fixed at or just below the lower limit of the amount transferred during normal copulations. This would be an appropriate threshold if females were to lay the maximum number of eggs in the shortest time possible, probably often an adaptive trait. Oviposition would be triggered equally by all successful copulations, and cryptic female choice would not occur via this mechanism.

The actual data on *A. domesticus* contradict this prediction. The threshold for the female's maximum oviposition response is substantially *above* the lower limit of the amounts that males normally transfer in nature (Murtaugh and Denlinger 1982, 1985). Thus the reduced number of eggs laid after those matings in which smaller amounts of this substance are transferred is a variable that constitutes a mechanism of cryptic female choice with respect to the amount of this accessory gland product. If males differ genetically in their abilities to produce enough substance to elicit a maximum female response, the result can be changes in gene frequencies. One advantage to the female of imposing such a "rule of the game" could come from producing sons better able to elicit maximum oviposition responses from their mates. Similar examples of female thresholds for maximum response that are above the amounts of material transferred by some males also occur in oviposition in the locust *Schistocerca gregaria* (Leahy 1973), and for remating in *D. funebris* (fig. 1.13). Characteristics of this sort, which have not traditionally been considered to be under sexual selection by female choice, are discussed in chapter 6.

Apparently minor details of female morphology can also have large but subtle selective consequences for males, so that females may exercise

Figure 1.3 Male control, but on female sufferance. A male *Truljalia hibinonis* tree cricket cleans his genitalia just after mating. He is ingesting the sperm from another male, which he removed from the female. Even though, as in this species, males sometimes appear to be in control of sperm precedence patterns, detailed consideration shows that this control is dependent on characteristics of the female, such as the length of the spermathecal duct and the form and extensibility of the spermatheca. (From Ono et al. 1989; photo courtesy of T. Ono)

strong influence even in cases where the male seems to exercise conspicuous, active control. For example, Ono et al. (1989) showed that males of the tree cricket *Truljalia hibinonis* (fig. 1.3) are able to remove a substantial fraction of the sperm deposited by previous males from the female's sperm storage organ (spermatheca). When a female has mated with two males, 87.5 ± 12.3% of the sperm in the female (by volume) are from the second male. Sperm from previous males are apparently removed by a "flushing" mechanism. The second male inserts his genitalia all the way to the anterior end of the female's spermatheca; as his viscous ejaculate emerges it pushes much of the previously deposited sperm posteriorly, eventually forcing them out the entrance of the spermatheca and onto the shaft of his penis, where they adhere. They are then removed when the male withdraws his penis from the female.

The second male's behavior and morphology clearly result in removal of the first male's sperm. But several aspects of the morphology of the female's reproductive tract could (and probably do) influence the success of the male flushing mechanism:

1. A folded rather than a smooth spermathecal wall could protect sperm from being flushed out. Even fine irregularities in the inner lining of the spermatheca could affect the likelihood that previous sperm will be flushed out. Fine striae and scalelike projections occur in the spermathecae of several insects (Davey 1958; Klostermeyer and Anderson 1976; Camacho 1989) and spiders (Huber 1994b). In the fly *Drosophila wassermani* sperm cells are normally found wrapped around internal recurved hooks near the entrance of the spermatheca (Pitnick and Markow 1994).

2. Increased elasticity of the walls of the spermatheca (or relaxation of muscles in its walls) could allow the spermatheca to expand to accommodate a larger proportion of both sperm masses and thus reduce the amount of sperm flushed out. Substances within the spermatheca could also increase or decrease the viscosity and stickiness of the first male's ejaculate and (perhaps to a lesser extent because of the presence of the male himself) the viscosity of that of the second male and thus influence the effectiveness of the flushing action. Cryptic female choice could occur if any of these factors under female control resulted in some males being favored reproductively over other conspecifics because of their abilities to deal with the female's design.

3. The very short spermathecal duct (apparently essentially non-existent in *T. hibinonis* according to the figure in Ono et al. 1989) and the large opening from the vagina to the spermatheca are both important in allowing a male to insert his genitalia all the way into the anterior end of the spermatheca. If access were more difficult, as in other orthopterans (e.g., Gregory 1965) and in many other animal groups (see chap. 7.1), flushing would be more difficult or physically impossible, unless the male developed a long, eversible structure and the ability to thread it up the long duct. This has occurred in some grasshoppers but not in others. In *Locusta migratoria* the last male apparently flushes out previous sperm (Gregory 1965; Parker 1984); in catontopine, cyrtacanthacridine, and romaleine grasshoppers the male's spermatophore does not reach the end of the long female duct where previously deposited sperm are stored (Pickford and Gillott 1971; Whitman and Loher 1984). Even when the male reaches deep into the female, the possibility of strong back currents created by insertion of male structures into thin, rigid, blind-ended female ducts (the fluid contents would be pushed out the open end) can make female participation via active absorption of liquids through the walls of her duct or the spermatheca important (Linley 1981; Linley and Simmons 1981).

One might object that such sperm-retaining characteristics of the female cannot be selective and therefore cannot exercise "choice." But in morphologically facilitating or impeding the attempts by subsequent males to override her earlier decision to allow insemination by a previous male, the female can selectively influence male paternity success. By morphologically or physiologically biasing the outcome in favor of males with certain traits, she can increase the chances that her sons will have such traits and thus improve their chances of reproduction.

A second example of apparent male control is sperm removal with hooks and spines on the male genitalia in odonates (e.g., Waage 1979, 1984). Again, the structure of some female storage organs can influence male success. For example, in *Mnais pruinosa* the spermathecal duct is so narrow that it completely prevents access of the male genitalia (Siva-Jothy and Tsubaki 1989a); in *Argia* the enlarged distal area of the spermatheca is also probably inaccessible for the male (Waage 1986). The arrangement of muscles in the walls of the bursa and the spermatheca (Siva-Jothy 1987a) indicates that a female can modify the size and shape of these structures (Waage 1984), and their size and shape could be important determinants of quantities of sperm removed. For instance, contraction of the muscles in the wall of the bursa or the spermatheca, or of the muscles that bring the bursa closer to the external genital opening (Siva-Jothy 1987a), would presumably facilitate sperm removal.

Certainly the degree to which even those sperm in the most accessible portion of the female are cleaned out appears to vary within a species. For example, the coefficient of variation of the volume of sperm remaining after long copulations in *Mnais pruinosa* was 71% (Siva-Jothy and Tsubaki 1989a).

In addition, successful sperm removal depends, to some extent, on the tangled mass commonly formed by odonate sperm (Waage 1984), which is not typical of many other animals. This characteristic seems paradoxical in the context of sperm competition by removal: a male whose sperm were less clumped and tangled and thus more difficult for subsequent males to remove would lose fewer offspring to subsequent males. Presumably odonate sperm form tangled masses because of some other advantage, perhaps associated with traits of the female or her eggs, which more than compensates for these disadvantages.

More direct avenues of influence are also open to female odonates. The females of *Erythemis simplicicollis* and *Mnais pruinosa* modify the number of eggs laid during the burst of oviposition immediately following copulation with some types of males, and they sometimes avoid subsequent copulation prior to subsequent bursts of oviposition (McVey 1988; Siva-Jothy and Tsubaki 1989a,b). An even more dramatic indication of possible female influence on paternity in this order famous for supposed male control comes from the recent discovery that female odonates sometimes emit droplets of sperm during or immediately after some copulations but not others (fig. 1.4; E. Gonzalez, pers. comm.; see chap. 3.3). This behavior has been observed in three genera, including *Coenagrion*; in this genus penis morphology suggests that males also use their genitalia to remove previ-

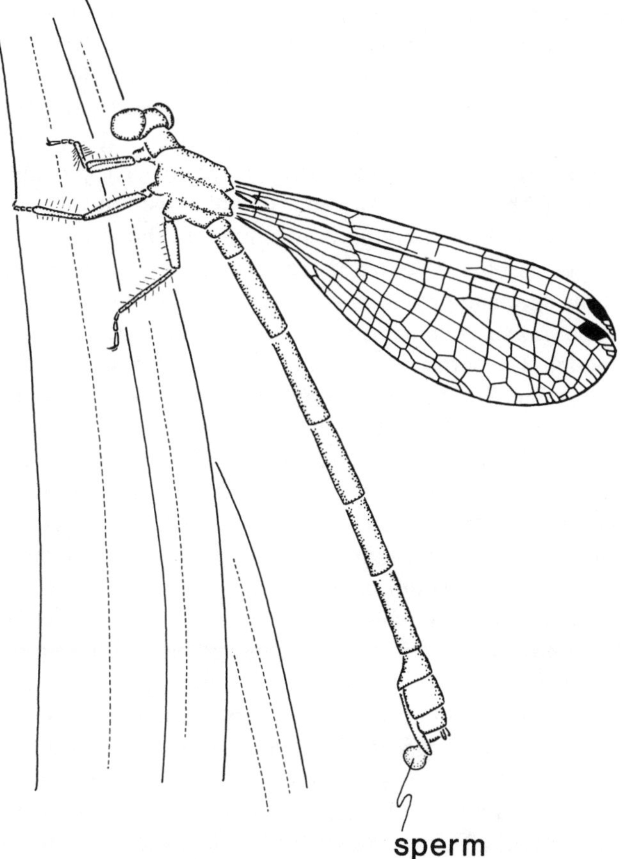

Figure 1.4 Possible female participation in a supposedly male-controlled process. This female *Paraphlebia quinta* damselfly is emitting a viscous mass containing sperm from her genitalia. More than one-third of the copulations in this species are accompanied by sperm emission, and emission is more frequent in copulations with males of one of the two different wing morphs. (Drawn from a photo by E. Gonzalez)

ously deposited sperm or push them away from fertilization sites (Waage 1986; Cordero et al. 1995).

A female could use her "rules of the game" to obtain sons with particular traits by favoring males with those traits. A particular sperm removal device on the male genitalia could be favored by morphologically or behaviorally making it particularly easy for males with such devices to remove her stored sperm (or by making it more difficult for males lacking such devices). She could favor other male traits which were not directly associated with sperm removal (e.g., dark color, large size) by altering the difficulty of sperm removal according to whether the male had such traits.

Examples from other animals show that female rules can also be purely behavioral. The males of some katydids attach packets of sperm to the female which only slowly transfer their sperm to the female's internal reproductive tract. Females eat these packets, however, and some males produce an additional mass, the spermatophylax (fig. 1.5), which the female eats first. This can give the sperm in the first packet (the ampulla) more time to enter her body. The order in which the female consumes the spermatophylax and the ampulla is not imposed, however, by the male; the male tactic is only effective if the female adheres to the rule of the game "eat the spermatophylax first" (see chap. 4.4). In fact, the female of the cricket *Gryllodes supplicans* sometimes "breaks the rule" by dropping the spermatophylax before she has finished it and removing the ampulla long before it has emptied its sperm into her (Sakaluk 1984; see also fig. 4.7).

Even simple female inactivity can be crucial for the success of male manipulations. For example, in the beetle *Tenebrio molitor* studied by Gage (1992), the male's genitalia apparently reach the female's bursa but do not enter her spermathecal duct. Spines on the male's penis shaft trap and remove at least some of the sperm from previous males which are in the bursa. At longer time intervals after previous matings, fewer sperm are removed, presumably because they have moved or have been moved by the female into her spermatheca. If, as is the general rule in insects (Omura 1938; Callahan and Cascio 1963; Davey 1965b; Engelmann 1970; Thibout 1977; Okelo 1979; Linley and Simmons 1981; Heming-van Battum and Heming 1986; Gillott 1988; LaMunyon and Eisner 1993), *T. molitor* females are largely in control of sperm migration or transport in their bodies, then a female could select against a male by moving his sperm more slowly out of the bursa and into the spermathecal duct, where they are presumably less susceptible to removal by subsequent males. A transport bias may in fact occur in *T. molitor* since, in some cases in which the second mating occurs very soon after the first, almost no sperm are removed by the second male (fig. 2 of Gage 1992). This result could also be due to first matings that resulted in relatively little sperm transfer. However, the standard errors in the numbers of sperm transferred in copulations of this species (Gage and Baker 1991) suggest that this is unlikely. In any case, the general point remains that variations in the rates of sperm transport by the female can sometimes affect the chances of sperm removal by subsequent males.

Similarly, the rate at which sperm are used to fertilize eggs can affect the relative importance of sperm removal and other factors that determine sperm precedence values. If, for instance, the proportion of offspring sired by the second of two males to mate with a female (the P2 value) is influenced by the relative number of his sperm in the female, then remating with a female that expresses a given volume of the sperm-containing fluid from

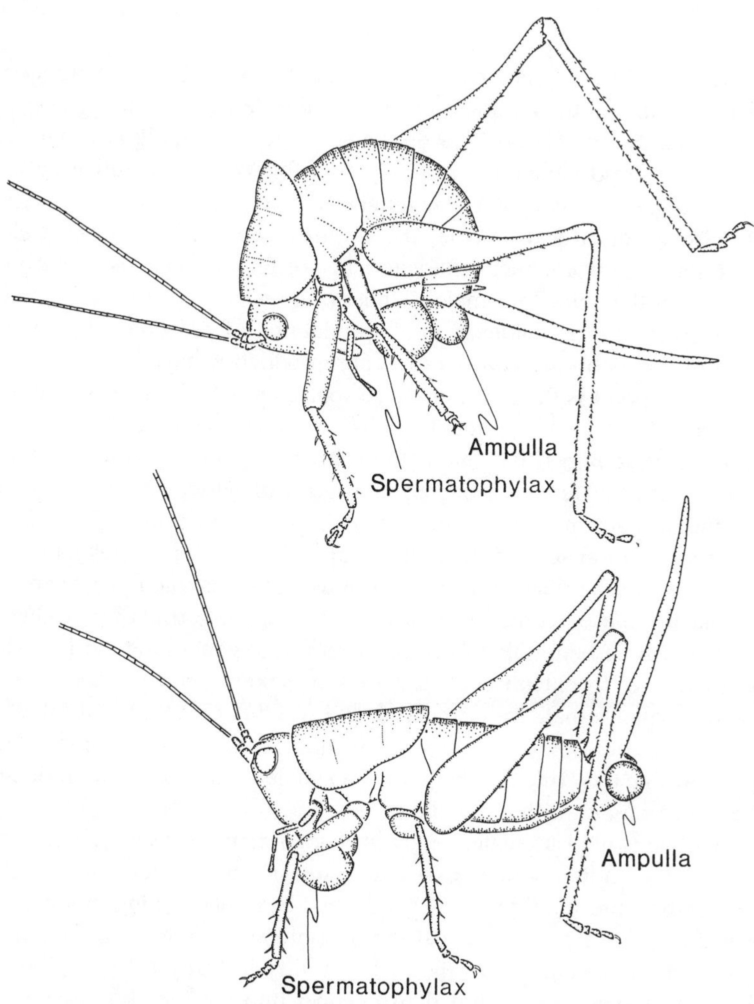

Figure 1.5 A female *Requena verticalis* katydid sets a behavioral "rule of the game." The male of this species attaches a spermatophore to the female which consists of a pair of sperm-containing structures (ampullae), and a pair of larger, spermless masses (spermatophylaxes). Sperm is gradually transferred to the female from the ampullae over the course of about 120 min after copulation ends. Soon after the male releases her, the female curves her abdomen forward and reaches back to grasp the spermatophylax with her mouthparts (top). The spermatophylax breaks free, and she spends on average the next 5 hr feeding on it (bottom). She then pulls the ampullae free of her genitalia and devours them also. This arbitrary sequence of female behavior represents a "rule of the game" for males. The female could as easily break off the spermatophylax, set it aside, and eat the ampullae with most of their sperm still inside (as indeed sometimes occurs in the cricket *Gryllodes supplicans*; Sakaluk 1984, 1985). By consistently imposing this and other, additional rules, the female *R. verticalis* favors males which make a larger spermatophylax by laying larger eggs that he has fertilized. An increase in the tendency to discard the spermatophylax or to consume it faster could change the venue, imposing different selective pressures on males. (After Thornhill and Gwynne 1986)

her spermatheca onto each egg, then replaces it with fluid lacking sperm with the result that the number of sperm used to fertilize each egg declines logarithmically (see Harbo 1979 on the honeybee *Apis mellifera*) will represent a greater possible reproductive payoff for a male than remating with a female that uses a constant, low number of sperm for each egg.

As already discussed, female *Acheta domesticus* crickets offer a clear illustration of how chemical characteristics of females can also determine the "rules of the game" for males. Under other conditions (e.g., no arachidonic acid in the spermatheca), male production of prostaglandin synthetase would probably have very different reproductive payoffs.

The basic point is that, even in those species in which males seem to be in control of sperm usage, female behavior, morphology, and physiology can influence competitive interactions involving male gamete success, conferring on the female a degree of control of paternity. In an analogy with human sporting events, the female's body constitutes the field on which males compete, and her behavioral and physiological responses set at least some of the rules by which they must play. And the female can gain from biasing the outcome in favor of particular male traits by producing sons that will be better able to compete in future generations. It is important, therefore, to extend considerations of sexual selection to ask how these important female traits evolved, and whether or not sexual selection has played a role in their evolution. Changes in the female characteristics that determine the "rules of the game" for male competition can influence the effectiveness of different male tactics.

As will be argued at greater length in chapter 2, many female characteristics that evolved in other contexts may result in sexual selection on males, and later become modified as a result. For example, a long oviduct that evolved to permit gradual deposition of a protective covering around the egg would incidentally favor the male's ability to transport his sperm (or induce the female to transport them) deeper into the female. Once established by natural selection, a female trait of this sort could subsequently be modified by sexual selection. For instance, the oviduct might become further elongated due to the advantage of biasing paternity in favor of males that were especially good at getting their sperm deeper into the female. A similar argument has been made in connection with classic, precopulatory sexual selection by female choice in water striders (Rowe et al. 1994; Weigensberg and Fairbairn 1994; Arnqvist, in press). The female water strider's active resistance to being mounted by a male is favored by natural selection because solitary females can often forage more effectively and avoid predators more successfully; it has the incidental consequence of favoring, via sexual selection, traits that increase the male's ability to resist the female's attempts to dislodge him.

Of course, neither the characteristics of the male nor those of the female are likely to be static and unchanging. Males are expected to respond evo-

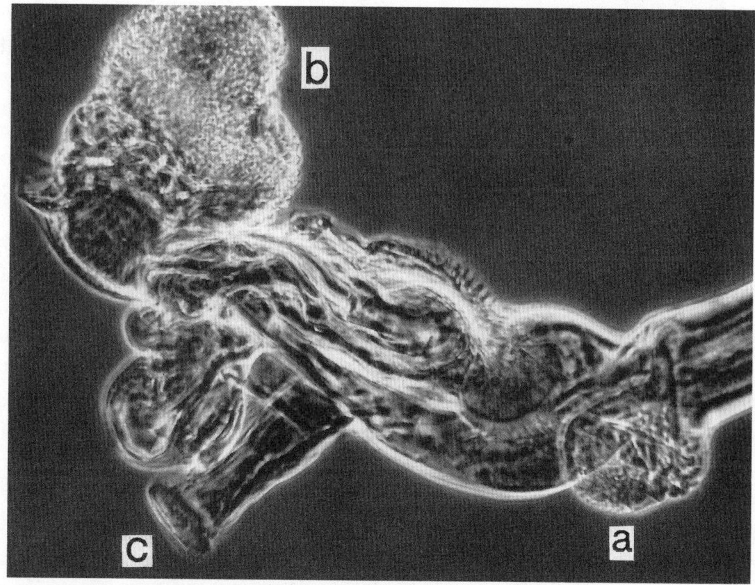

Figure 1.6 The intricate and beautiful genitalia of a male *Ceratitis capitata* medfly exemplify the unpredictability and potential complexity of characters under sexual selection. Male medflies must go through a veritable odyssey to fertilize a female's eggs. A male starts by defending a leaf in a lek, where he repels other males and attempts to attract a female using a pheromone. If a female arrives on his leaf, the male courts her using visual, chemical, and probably auditory cues to induce her to allow him to mount (she often leaves instead, or knocks him off). Once mounted, he must induce her (with tactile cues) to extend her ovipositor so he can grasp its tip with his genitalia and intromit (the female sometimes fails to extend her ovipositor). Once his genitalia are inside her, the male must unfold his long intromittent organ (about 40% the length of his body), and inch it farther inward, probably using a small inflatable sac (a) to push against the walls of her long vagina. When the tip of his genitalia finally reaches the upper end of her vagina, it must hit a still-deeper target. A second inflatable sac (b) apparently drives a tubular structure (c) into a cone in the wall of the vagina, apparently thereby aligning the male's gonopore with the openings of the ducts to the female's sperm storage organs. Even then the male's work is probably not done, as he courts the female during copulation by rubbing her abdomen with his legs. Transport of his sperm probably depends on contractions of muscles in the walls of her spermathecal ducts. Female criteria for accepting a male, his genitalia, and his sperm may include many different male stimuli. Such an array of characters suggests that an argument in the form, "If a female chooses among males on the basis of one particular character, then she will probably not use other characters for further selection," is incorrect. (Photo courtesy of M. Vargas and H. Camacho; from Eberhard 1990b)

lutionarily to female-imposed selection by "solving" the problems posed by the female's reproductive morphology, physiology, and behavior; and females are expected to evolve new mechanisms to discriminate (fig. 1.6) (e.g., West-Eberhard 1983 on "piling on" characters; Eberhard 1985 on genitalic design). To evaluate the hypothesis of cryptic female choice at any particular moment in evolution (i.e., the present), one can only check on the state of this presumed "arms race" (see secs. 1.6 and 1.9 of this chapter for critical discussions of this concept).

1.4 Taxonomic and Conceptual Biases of This Book

Sexual selection by cryptic female choice can theoretically occur in any sexually reproducing group of animals. Ideally questions about the variety of mechanisms and the limits of the taxonomic distribution of cryptic female choice should be answered with data from all animal groups that have sexual reproduction. Unfortunately, there are very few studies in groups other than insects and mammals that provide the kinds of data needed for most of the questions discussed in this book. Although scattered examples are drawn from a wide variety of groups, including nematodes, spiders, scorpions, ticks, copepods, crabs, frogs, fish, birds, and snakes, by force of circumstance the majority of the examples involve insects and mammals. In order to generalize, one must assume that these groups are not exceptional with respect to the rest of the animal kingdom. Fortunately, insects and mammals represent two extremes with respect to the length of time sperm remain viable within the female (Parker 1970c). At least in this respect, the combination of the two groups may give a fair reading of general trends. In addition, the majority of animals *are* insects, and insects are extremely diverse in many aspects of their reproductive biology. I know of no a priori reason to expect that mammals or insects should be atypical with respect to the likelihood of cryptic female choice.

I have also not included plants in any of the discussions. Because of their immobility, female plants can often only effectively exercise choice among males by receiving pollen promiscuously and then choosing afterward. There is a substantial literature on this topic (Willson 1979, 1990, 1994; Willson and Burley 1983; Stephenson and Bertin 1983; Queller 1987, 1994). The mechanisms that seem to be used frequently in plants are very similar to a few of the animal mechanisms: biased release and promotion of pollen tube growth (biased sperm transport and sperm usage in animals); and biased endosperm and embryo growth (biased abortion and priming the uterus for implantation). Plants may differ from animals in that avoidance of endogamy is more often an important advantage of cryptic choice mechanisms (e.g., Cruzan and Barrett 1993). Inbreeding avoidance per se would not necessarily result in any overall advantage for any given male trait. However, inbreeding avoidance is less likely to be selectively important for male than for female plants, and a male trait that diminished the bias of related females against his pollen and their resulting offspring could gain a sexually selected advantage of the sort discussed in this book.

As suggested by Queller (1987, 1994), studies of sexual selection in animals would do well to include ideas from plant evolution to widen the contexts considered. For instance, the emphasis botanists have placed on male and female physiological processes as mechanisms for making and influencing choices in plants (Stephenson and Bertin 1983; Queller 1987, 1994) presaged much of the discussion of similar phenomena in this book (especially chap. 6). Queller's discussion of female interests, and of active and passive female mechanisms that can result in differential male reproductive success, is more subtle than any I have found on similar topics in animals.

A further bias concerns the ideas discussed. Possible female roles in the control of paternity will be emphasized, even though it is obvious that males are also involved and that to some extent they can adjust (both facultatively on the short term, and on the long term) to the female's traits. A bias of this sort is appropriate at the moment. Cryptic female choice involves a relatively neglected way of looking at phenomena in a rapidly expanding area of research. The most important function of this book is to orient future thinking and research, rather than to attempt to give a "final" synthesis, which in any case is not even remotely possible at this stage. For instance, it is not possible to assess the relative importance of cryptic female choice with respect to classic, overt, precopulatory female choice, either in terms of the number of species involved or its relative importance in given species. The overall balance in the relative degrees of control achieved by males and females over female reproductive processes is also generally unknown.

As an empirical biologist, however, I have developed antibodies to presentations of intellectually titillating theoretical possibilities made without any serious attempt to check provocative ideas against the hard realities of nature. I firmly believe that such filtering of theory is crucial to increased understanding of biology. I have thus tried hard to impose a further bias and discuss only feasible types of interactions that flow from "reasonable" interpretations of hard data from real animals. Only in chapter 8, where possible implications of cryptic female choice for other phenomena are explored, is this filter partially relaxed.

1.5 Relationship with Genitalic Evolution

Female responses that result in cryptic female choice can be cued on any one of a variety of male characteristics. One set of such characters is male genitalia (fig. 1.6), and I have argued elsewhere that male genitalia are often under sexual selection by cryptic female choice (Eberhard 1985). The hypothesis concerning genitalia is dependent on cryptic female choice being a common phenomenon. In contrast, the hypothesis of cryptic female choice does not depend on the validity of these ideas concerning genitalia. In fact, I think one can make a strong case for the general importance of cryptic female choice without ever referring to the possibility of cryptic female choice on male genitalia.

To make the case for sexual selection on male genitalia, I listed several possible mechanisms of cryptic female choice and presented a few cases in which stimuli from male genitalia appear to trigger female responses (Eberhard 1985). Other possible, nongenitalic triggering stimuli were not considered. In fact, the apparent *lack* of courtship during and following copulation (which further data now suggest is very widespread; see chapter 5) was cited as evidence contradicting the idea that male genitalia evolved under sexual selection by cryptic female choice (Eberhard 1985, p. 155). So cryptic female choice was left more as a hypothetical possibility than a demonstrated phenomenon of general importance. The present book, then, represents a "next step" by examining a more complete set of male cues (both genitalic and nongenitalic) that can be used by females, giving a more complete list of possible mechanisms by which cryptic female choice can be exercised, citing more evidence that female choice actually occurs, and discussing why and how it evolves and some of its possible consequences.

1.6 Relationship with Male-Female Conflict

Cryptic female choice inevitably results in a conflict between the reproductive interests of males and females: a male will benefit most if every female with which he mates uses only his sperm to fertilize all her eggs for the rest of her life; and females exercising cryptic choice will fail to do this for at least some of their mates. Different kinds of selection can cause male-female conflicts to arise, and conflicts can be played out in different ways (Parker 1979; Alexander et al., in press). Morphological evidence suggests that certain types of selection and conflict may be particularly common with cryptic female choice.

In some situations a male's interactions with a female can directly reduce her production of offspring. For instance, precopulatory harassment of an unreceptive female by a male can reduce her ability to forage (Thornhill and Alcock 1983); prolonged mounting or copulation can increase her vulnerability to predation (e.g., Arnqvist, in press); or an overly powerful hormonal induction of oviposition can cause her to invest so heavily in offspring as to shorten her life (e.g., Kasule 1986; Chapman et al. 1995) (see Alexander et al., in press). Under such conditions, female ability to resist males indiscriminately (to be able to fail to cooperate with any male, no matter what his phenotype) could be favored by natural selection. Indiscriminate resistance would seem especially advantageous once a female has mated and is directly engaged in producing offspring. Some female barriers that evolved in such contexts could be insuperable and not result in sexual selection on males to overcome them. Others might be only partially successful and could result in sexual selection, favoring (by either overt or cryptic choice) those males best able to overcome them. Female resistance can occur externally (e.g., a female's struggles to free herself from the male's grasp) or internally (e.g., prevention of deep genitalic penetration).

A second type of female benefit that can result in male-female conflicts is the production of superior offspring. In this case, selection seems more likely to favor selective rather than indiscriminate female resistance, and sexual selection (both precopulatory and cryptic) on male abilities to overcome female resistance can ensue. Restated in terms of female cooperation with males and the events that lead to fertilization, direct benefits to females via natural selection can lead to mechanisms for categorical lack of cooperation; indirect benefits may often lead to *selective cooperation* with some males and not others.

Female morphology indicates that selective cooperation, rather than indiscriminate resistance, is common in some male-female conflict situations resulting from cryptic female choice. Males of many animals have specialized grasping organs and genitalia that can potentially manipulate the female. In many species the area of the female's body that is grasped is not modified in any way for such contact (e.g., front legs of diplurid spiders, caudal filaments of calanoid copepods, antennae of sminthurid collembolans; Coyle 1985b, 1986; Blades 1977; Blades and Youngbluth 1979; Massoud and Betsch 1972). In other species, such areas are modified for contact, and the form of the modification is revealing. The female generally has grooves, cavities, and slits into which the male structures fit, thus mechanically *aiding* the male (e.g., the genital cavity of some millipedes: fig. 1.17, Tadler 1993; the abdomen of some spiders: fig. 1.7, Huber 1995; the abdomen of some female bees and wasps: Toro and de la Hoz 1976, Toro 1985, Toro and Carvajal 1989; the thorax of some odonates and cicindellid beetles: Freitag 1974; see summary in Eberhard 1985). This cooperation by the female is morphologically selective, however; male grasping or holding structures with inappropriate designs will be unable to mesh with these "cooperative" structures of her body.

In contrast, "aggressive" female resistance structures in such areas, such as spines, inflatable sacs, and shields that might offer indiscriminate resistance to male grasping structures, are virtually unknown. The only excep-

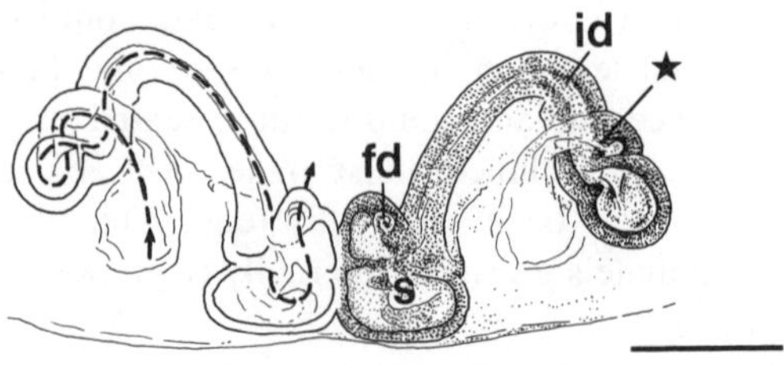

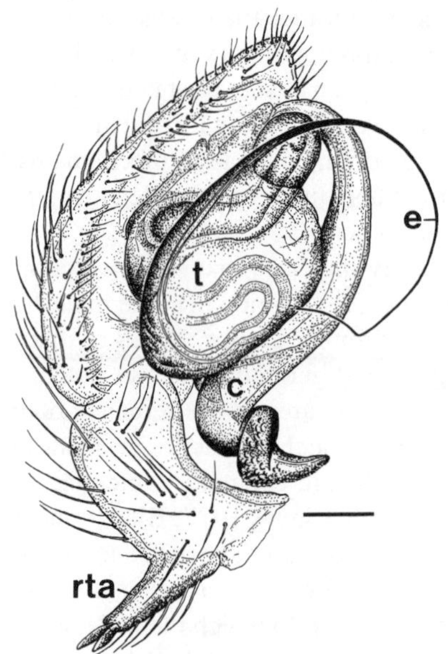

Figure 1.7 The genitalia of a spider illustrate selective cooperation by females with male attempts to intromit. The male genitalia (palps) (*bottom*) of the spider *Dictyna uncinata* include a spurlike prominence (rta) which is used to lock the male genitalia into position on the surface of the female's genital plate (epigynum) (*top*). This allows the male to introduce his intromittent organ (e) deep into the female's insemination duct (id) and transfer his sperm to her spermathecae (s). At the start of a copulation the male scrapes each palp repeatedly across the surface of the female's abdomen until the tips of the "ctinidia" at the tip of the rta are inserted into a pocket (fovea) in the female's abdominal cuticle (fd). In one species group of *Dictyna*, the male rta is especially long, and the female fovea is especially distant from the epigynum. In another species group the rta is absent, and the female abdomen lacks foveae. Far from opposing the hooking action of the male's rta, the female's fovea *facilitates* engagement; but her cooperation is discriminating, as it is available only for those males with appropriate morphology. (From Huber 1995)

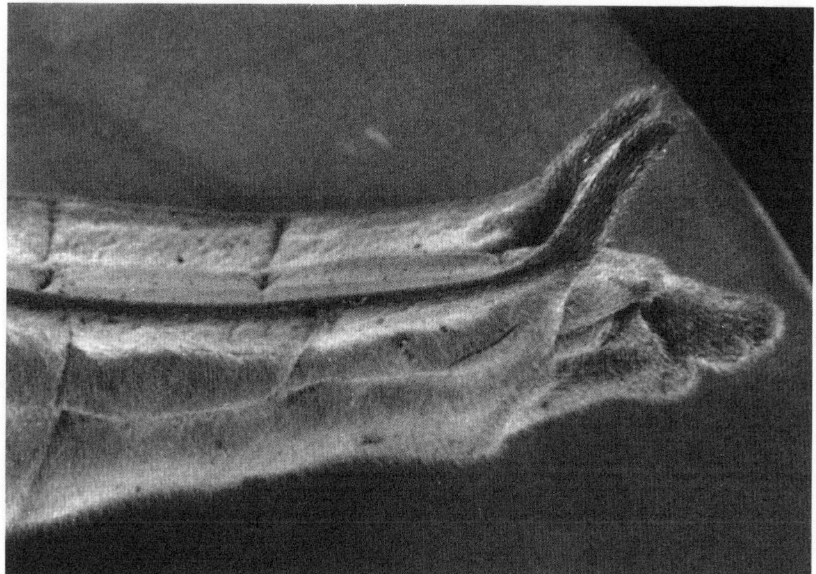

Figure 1.8 This pair of spines near the genitalia of a female *Gerris incognitus* water strider is a rarity—a female structure whose apparent function is to *repel* males. Much more commonly, females have structures such as grooves, pits, and slits (e.g., figs. 1.7, 1.17) whose apparent function is to *accept* male structures (but to accept them selectively, favoring those with appropriate designs). Female waterstriders suffer sharply increased predation when mounted by males, and the spines may be a naturally selected character to reject males indiscriminately. (From Arnqvist and Rowe 1995)

tion of which I know is a pair of spines near the genital opening of the female water strider *Gerris incognitus* (fig. 1.8) (Arnqvist and Rowe, 1995). Female water striders often suffer sharply increased predation when mounted by males (Weigensberg and Fairbairn 1994; Arnqvist, in press), so the spines may be the result of natural selection on females to reject males. (It is also possible that the spines in this species are selective, favoring some male morphologies or behavior patterns over others.) In addition, the often seemingly arbitrary copulatory courtship movements (e.g., stroking the female's dorsum) and their relatively gentle effects on the female (Eberhard 1991, 1994; chap. 5) do not have the physically forceful nature one would expect of direct male-female conflict. There are, of course, many cases in which cryptic female choice is influenced by male traits that do not depend on their physical mesh with her body, such as chemical stimuli in the semen (chap. 6). The degree of "cooperation" of the female characteristics that interact with these male traits cannot be inferred as easily, and female costs and benefits in nature are not yet known (Chapman

Figure 1.9 The typical copulatory posture of the crab-eating macaque *Macaca fascicularis* showing "cooperative" female behavior that facilitates reception of stimuli from the male. The female looks back at the male's face as he grimaces and opens his mouth. She also reaches back to grasp his hind leg (which trembles during ejaculation). In one species in this genus the male does not change his facial expression before ejaculation, and the female apparently does not look back; in several species the female reaches back just at the moment of the male's ejaculation (see table 5.1). (After Kanagawa et al. 1972)

et al. 1995). There are also some behavioral intimations of female cooperation in copulation (e.g., fig. 1.9). It remains to be seen if selective cooperation by females is also the general rule in these contexts.

Seen from the point of view of a male attempting to overcome a female's resistance to his advances, cryptic choice situations may be less likely to involve forceful male-female conflict than precopulatory interactions. A male engaged in attempts to copulate with a female can gauge her resistance and harass her or otherwise increase his efforts until he succeeds. But a male attempting to elicit female responses such as sperm transport, ovulation, or rejection of further males is in a poor position to judge the female's compliance (see chap. 2.4) and thus cannot utilize the efficient "escalate until the female yields" tactic (e.g., Clutton-Brock and Parker 1995).

In sum, male-female conflicts of interest are an inevitable consequence of cryptic female choice. Female morphology suggests that in cryptic female choice contexts, females may tend to cooperate selectively with males rather than reject them indiscriminately, and to derive sexually selected benefits rather than only naturally selected payoffs from doing so. Facultative escalation of male efforts to overcome female resistance is probably often less feasible in cryptic female choice situations than prior to copulation.

1.7 Previous Biases: Male-Female Cooperation and "the Good of the Species"

This book includes many new "why" questions applied to the functional morphology and physiology of reproductive systems. Functional morphology and physiology are old, well-worked, and very successful fields of biology. The power of their techniques and logic to improve understanding of biological structures and processes was obvious long before Darwin. Increased understanding of evolutionary processes has led to many advances in understanding morphological details.

The field of reproductive morphology and physiology is particularly ripe for reexamination. Advances in evolutionary analysis have led to research showing both the generally overriding importance of individual reproduction in evolution and the existence of widespread conflict of reproductive interests between males and females. Many studies of reproductive processes and morphology (past and present) still assume, however, that during copulation males and females are cooperating in order to achieve fertilization, thereby promoting the interests of their species. This assumption is now known to be frequently misleading, and our improved understanding can lead to new interpretations.

Thus, in the framework of modern evolutionary biology, the discovery of a valve in the spermathecal duct (fig. 1.10) thought to function as the female's "cooperative" contribution—to keep sperm *in* a storage organ—is likely to raise the possibility that the valve may instead function to keep some males' sperm *out* (see chap. 7.1). Similar reinterpretations have resulted in important advances in several other fields of biology, such as ecology, animal behavior, molecular biology, epidemiology, and intraorganismic interactions (Williams and Nesse 1991; Williams 1992).

An additional insight stemming from modern selection theory is that, just as a plasmid within a bacterium or an organelle within an eukaryotic cell must be seen as an "organism" with its own reproductive interests (Eberhard 1980, 1990a; Cosmides and Tooby 1981; Saumiton-Laprade et al. 1994), so individual sperm cells are reproductive units. Their interactions with other sperm in the same ejaculate, with sperm from other ejaculates, and with the female and her gametes can have important evolutionary consequences, and analysis of both male and female structures must take them into account (see chap. 2).

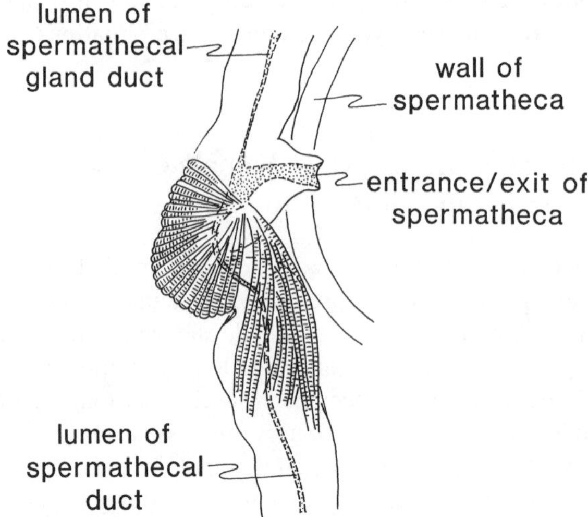

Figure 1.10 The muscles at the junction of the spermathecal duct with the spermatheca of the honeybee *Apis mellifera* probably function as a valve that controls sperm movement. The muscles have been called a "sperm pump" (Laidlaw 1944), but they seem designed, rather, to close the duct by squeezing and bending it sharply when they contract. Whether this closure functions to keep sperm in the spermatheca, or to prevent them from entering, or both is not known. (After Wigglesworth 1965)

1.8 Previous Biases: Overly Strict Categorizations and "Fertilization Myopia"

Morphology

Studies of reproductive processes have also traditionally been plagued by the age-old bias favoring typological or normative descriptions that ignore or minimize variability. This type of focus has impeded progress in biology throughout history (Mayr 1982). An example in sexual selection was previous resistance to ideas related to alternative mating tactics (Andersson 1994). Morphologists have generally described "the" process of copulation and insemination in a species, giving the impression that the sequence of events is always the same and neglecting events that occur in unsuccessful pairings (for a useful, early exception, see Norris 1933). A narrow focus of this sort can, of course, be helpful when attempting to elucidate how male and female gametes eventually get together, at least if there is only one such "successful" sequence. Even if there is only one such sequence, however, a restrictive emphasis on successful copulations can fail to give a complete picture of all important events during mating, and can be espe-

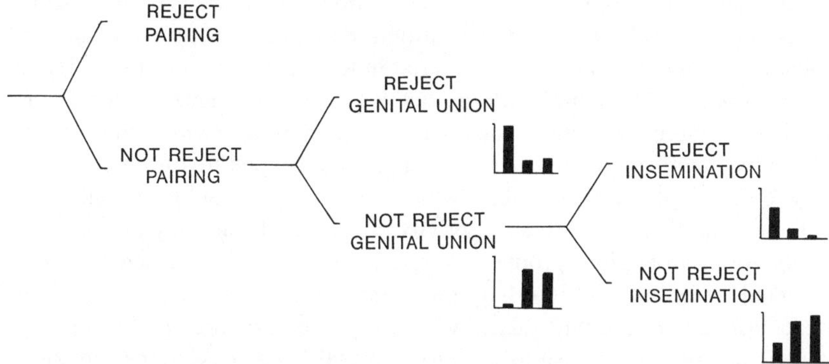

Figure 1.11 Decisions in the sex life of a female *Culex pipiens pipiens* mosquito, illustrating variance in postcoupling female responses that could be of selective importance to males. The percentages of the time particular decisions were made when the female was 6–12 hr old (*N* = 94), 24 hr old (*N* = 39), and 48–120 hr old (*N* = 57) are given (bars, left to right) in the graphs. Young females tend not to be inseminated when pairing occurs, and older females tend to be inseminated, but there are frequent exceptions. The reproductive payoffs for males able to induce the more "difficult" young females to accept insemination could result in strong selection. (Data from Lea and Evans 1972)

cially misleading regarding selection on male traits. Detailed studies that have been alert to variation and to events associated with *failures* in sperm transfer, fertilization, or offspring production have revealed a wealth of ideas about function (e.g., Overstreet and Katz 1977 on sperm transport in rabbits; Watson 1991 on preinsemination copulations in spiders; Baker and Bellis 1993a on orgasms and sperm rejection in human females; V. Rodriguez 1994a, V. Rodriguez et al., subm., Eberhard and Kariko, in press, on intromissions with and without sperm transfer in beetles).

Variations in reproductive processes such as transport of sperm to storage areas or discharge of sperm during or following a copulation can radically change a male's chances of fertilizing the female's eggs (fig. 1.11). Since fertilization success is so crucial to determining male fitness, it is very likely that some male reproductive structures have evolved to function in ways that reduce the probabilities of "failed" copulations or the severity of failures. Thus, until variations in reproductive process are documented, it is likely that the functional significance of some male structures and behavior patterns will remain unintelligible. Similarly, if there is more than one way a male can maneuver his gametes into the vicinity of fertilizable eggs and have the nuclei of his sperm accepted for fertilization, selection on his reproductive behavior, morphology, and physiology may be affected. Several concrete examples of such alternatives are discussed in chapter 2.2.

Another manifestation of "fertilization myopia" is the frequent use of females in optimal reproductive condition in studies of copulation and fertilization. Often they are virgins, reared in laboratory conditions designed to maximize survivorship and reproductive prosperity. This facilitates the work of researchers, as their study animals are more likely to perform complete reproductive responses. But it can give a false picture of the problems confronting real males in nature, which must often encounter females not at their peak of reproductive receptivity. They may be already mated, relatively young to begin reproduction, relatively old, malnourished, recently depleted by investment in other young, or already pair-bonded to another male. For instance, female rats which are older require more preejaculatory and ejaculatory intromissions both to trigger the luteal cycle that prepares the uterus for implantation of embryos, and to increase the number of developing young (Davis et al. 1977; for similar data on deer mice, voles, and hamsters, see Yamanaka and Soderwall 1960; Huck and Lisk 1985 and references therein). A male *Peromyscus maniculatus* mouse must execute more intromissions preceding ejaculation if he is to sire the offspring of a female that has recently mated with another male, and the female is less likely to conceive if the second male is unfamiliar to the female (the males make appropriate behavioral adjustment for these differences in female responsiveness; see chap. 5.2) (Dewsbury and Baumgardner 1981). A male's success with such marginally responsive females could have a substantial effect on his overall reproductive success. By studying only interactions with "easy" females, researchers run the risk of misunderstanding important aspects of male and female behavior (Davis et al. 1977).

If one can use classic precopulatory courtship behavior as a guide, the bias toward using "easy" females in studies of reproductive morphology and physiology may lead to substantial underestimates of the importance of both male courtship and female coyness. Thus males often completely omit substantial portions of precopulatory courtship behavior when courting especially receptive females (e.g., Jackson 1981; Andersson 1994).

A related problem stems from the frequent assumption that the morphological details of the process of insemination can be adequately understood by studying copulations with virgin females. Such females are convenient study subjects because all sperm can be safely attributed to the copulating male. But it is abundantly clear that many copulations in nature are with nonvirgin females (e.g., chap. 9.4). Selection in such copulations may be quite different on some male abilities, such as removal of copulatory plugs (chaps. 4.2, 4.3), induction of emission or repositioning of previous sperm (chaps. 3.3, 4.7), induction of ovulation and oviposition (chaps. 3.7, 3.9), or physical removal or displacement of previous sperm or resistance to displacement (e.g., Waage 1979, 1984; Harshman and Prout 1994).

This means that, just as in behavioral and ecological studies where large sample sizes are used to take variation into account, so studies of the morphology and physiology of insemination and fertilization should employ numerous replicates. This is apparently often not the case. Judging by the references in the recent book by Birkhead and Møller (1992), for instance, it appears that there are few if any studies of this kind with adequate replications in birds. Some published descriptions of copulation are based on samples of one (e.g., Diesel 1990). And many (most?) morphological studies of sperm transfer do not even give sample sizes. Although a single observation is better than none for descriptive purposes, understanding the evolution of reproductive characters demands more.

Behavioral Evolution and Physiology

The same tendency to ignore variation occurs in interpretations of sperm precedence patterns that use only average percentages of offspring (e.g., Gwadz et al. 1971; Walker 1980; Gwynne 1984c; Parker 1984; Ridley 1988, 1989a,b). Females are usually allowed to mate with only two males, and usually only under a particular set of conditions (e.g., only one length of time between first and second copulations; usually either with or without intervening oviposition).

Even under such standardized conditions, however, sperm precedence studies typically reveal high levels of variance around these average values (fig. 1.12) (see Lewis and Austad 1990 on insects; table 9.2 for other groups). A few males characteristically have much greater fertilization success in these studies. If such biases are consistent (a possibility which is seldom checked), they could result in strong selection on males. Changes in experimental conditions, such as the length of time between successive matings or the number of males involved, also reveal substantial variability (Birkhead and Møller 1992 on birds; Ginsberg and Huck 1989, Schwagmeyer and Folz 1990 on mammals; Parker and Smith 1975, Gromko et al. 1984, Dickinson 1988, Yamagishi et al. 1992, Otronen 1994a,b on insects; Zeh and Bermingham, in prep., on pseudoscorpions). Which sets of experimental conditions are most similar to conditions in nature is generally unknown and often not even discussed.

A tight focus on average values can be misleading. For instance, in the orthopteran *Decticus verrucivorus* the average paternity for the second of two males to mate with a female (the P2 value) was 49.6%, and it was concluded that paternity was probably determined by random mixing of sperm (Wedell 1991). But the variation in P2 values was extreme (from 2.6% to 100% of the offspring of only fifteen females), a much larger range than would be expected from random mixing of even only approximately similar numbers of sperm. An even more dramatic spread of P2 values occurred in the moth *Choristoneura fumiferana* (100% paternity for the

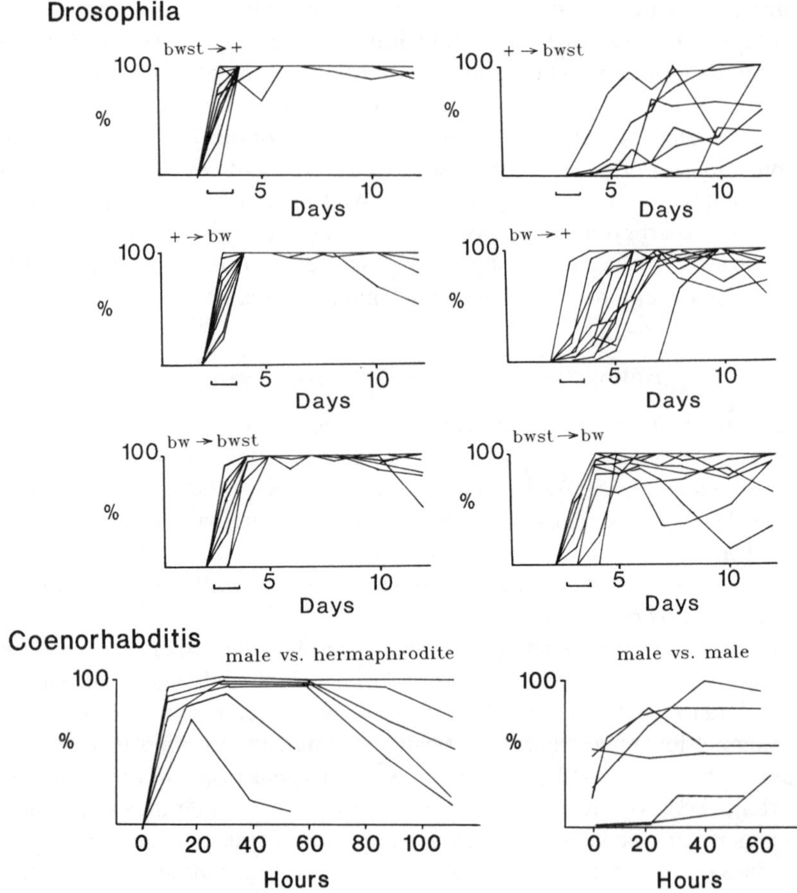

Figure 1.12 Variation in patterns of sperm precedence in the fly *Drosophila melanogaster* (top) and the nematode *Coenorhabditis elegans* (bottom). The common practice of characterizing a species with a single sperm precedence (P2) value can hide a great deal of variation within and among individuals. Each line represents the proportion of a single female's offspring which were fathered by the second of two males (P2 values), plotted over time since the second mating. Values vary both with the length of time since mating with the second male, and with the genetic strains of the parents (+, bw, bwst in the flies). In addition, there is great variation among individual females. Each point in the *D. melanogaster* graphs represents about twenty-five offspring in the early portions and about ten in the later portions of the female's life. (After Prout and Bundgaard 1977) The *C. elegans* graph on the left represents male vs. hermaphrodite sperm competition; a hermaphrodite first stored some of its own sperm and then mated with a male. The graph on the right represents male vs. male competition; two males mated with a single female. (After Ward and Carrel 1979)

first male in six pairs, 100% for the second in six other pairs, and a mix in three others; Retnakaran 1974).

Some authors have combined data including only average sperm precedence values with black-and-white characterizations of the design of sperm storage organs (elongate vs. spherical; cul-de-sac vs. conduit; Walker 1980; Austad 1984; Thomas and Zeh 1984). Even accepting the P2 values and morphological categorizations at face value, early apparent trends (more elongate spermathecae with higher P2 values, conduit design with especially low P2 values) have not proven to be consistent (see Ridley 1989b and Dickinson, in press, on elongate spermathecae in insects; Radwan and Witalinski 1991, Shuster 1991, and Masumoto 1993 on conduit storage organs in mites, isopods, and spiders).

Another important and common way in which variation is neglected concerns studies of the effects of male seminal products on females. The literature on male accessory gland substances in insects, for example, is replete with declarations that when these substances are transferred to the female during copulation, they render her unreceptive to all further copulation attempts. When one examines the original data, however, they usually indicate that the effect was not so definitive: only some fraction of the females refused remating; or the refractory effect began only some hours after copulation ended; or the effect gradually wore off; etc. (e.g., fig. 1.13). The possible reproductive payoffs to males that are able to induce refractory behavior in a larger proportion of their mates, or induce such behavior sooner or for longer, are obvious. This kind of variation, which may be extremely informative, is hidden by overly categorical summaries.

Again, concentrating on average values is useful in some contexts. Large scale trends (e.g., shape of spermatheca vs. P2 values) are more easily tested and understood in comparative studies when categorical descriptions are used. But variation is characteristic of this and many other phenomena related to animal reproduction. Highlighting the variation and searching for reasons why it occurs, instead of sweeping it under the rug of average values and categorical statements, promises to yield important advances in understanding both male and female reproductive processes. Not only may the variants be adaptive, but they indicate the potential for future evolution.

Theoretical analyses of the benefits thought to accrue to females that discriminate between possible mates have also suffered from overly sharp categorizations. When a particular benefit, such as the female gaining access to higher-quality feeding territories, is found to be involved in a given species, the tendency has been to characterize the species by this type of payoff and neglect the possibility that other female benefits occur in the same species (e.g., Maynard Smith 1991). There is no reason, however, to expect a priori that females should obtain only one type of gain (material

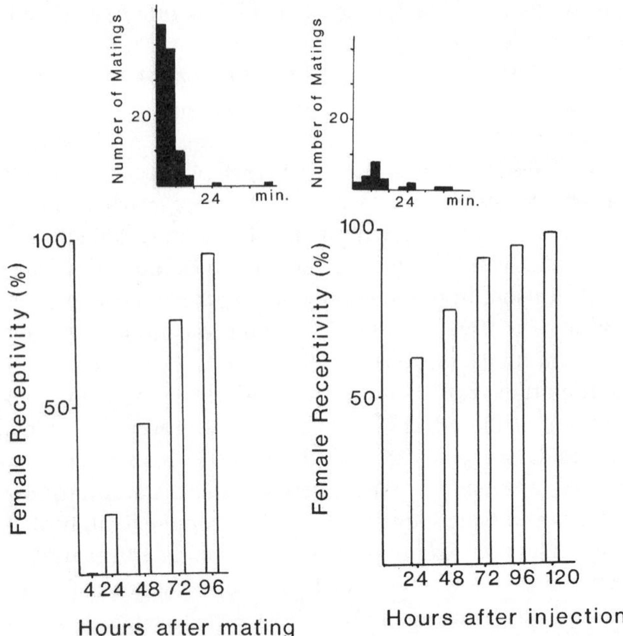

Figure 1.13 Variable effects on female receptivity of both natural mating and standardized injections in *Drosophila funebris*. *Above*: Delay until mating by 100 virgin females (*left*) and 100 females mated 24 hrs earlier (*right*). *Below*: Gradual increase in receptivity after a normal mating (*left*) and after injections of 2–3 micrograms of purified male accessory gland substance PS-1. Female responses varied substantially (e.g., at 24 hrs after the injection and 48 hrs after natural matings). (After Baumann 1974b)

resources, good viability genes, good attractiveness genes; fig. 1.14). There are, in fact, data which suggest different combinations of payoffs to choosy females. Many female songbirds may use territory quality as a criterion in forming a pair bond with one male, and subsequently apply other criteria in selecting males for extra-pair copulations (Birkhead and Møller 1992) (and perhaps subsequently exercise cryptic female choice to bias fertilizations in favor of some males over others; see chap. 4.13).

1.9 Previous Biases: Male Control and Female Passivity

A final problem (fig. 1.15) is a selective emphasis of those aspects of reproductive processes that are under male control, due to the perhaps unconscious assumption that female roles are passive. This is probably part of a general male-centered tradition in biology (Wasser and Barash 1983; Eber-

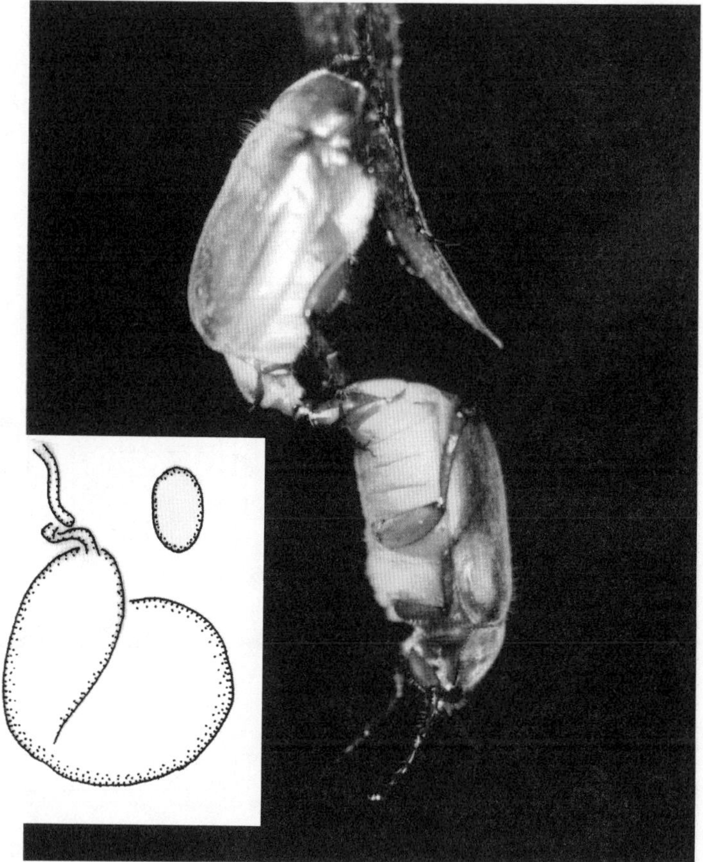

Figure 1.14 Male donation of resources to the female does not preclude sexual selection by female choice on other male characters. During copulation in *Phyllophaga setifera menetreisi* beetles (the male is below, hanging by his genitalia), the male transfers a large quantity of nonproteinaceous substance to the female, filling her large bursa sac (a mature egg is drawn alongside the full bursa sac at the right for scale). Despite this material contribution, *Phyllophaga* males show several signs of being under sexual selection by cryptic female choice: their genitalia are extremely elaborate and differ dramatically among species; male abdominal sclerites are species-specific in form and rub on females during copulation; males of at least four species perform rhythmic genitalic thrusting within the female; and copulatory courtship occurs in 6 of 7 species studied and varies among species (Morón 1986; Eberhard 1993b and unpub.). (From Eberhard 1993b)

hard 1990c; Batten 1992). Although there are important exceptions (e.g., Walker 1980; Wasser and Barash 1983; Thornhill and Alcock 1983; Thornhill 1983; Knowlton and Greenwell 1984; Constanz 1984; Gromko et al. 1984; Simmons 1986, 1987a; Dickinson 1988 and in press; Ward 1993; Birkhead et al. 1993a; Birkhead 1994; Gowaty 1994; Keller and

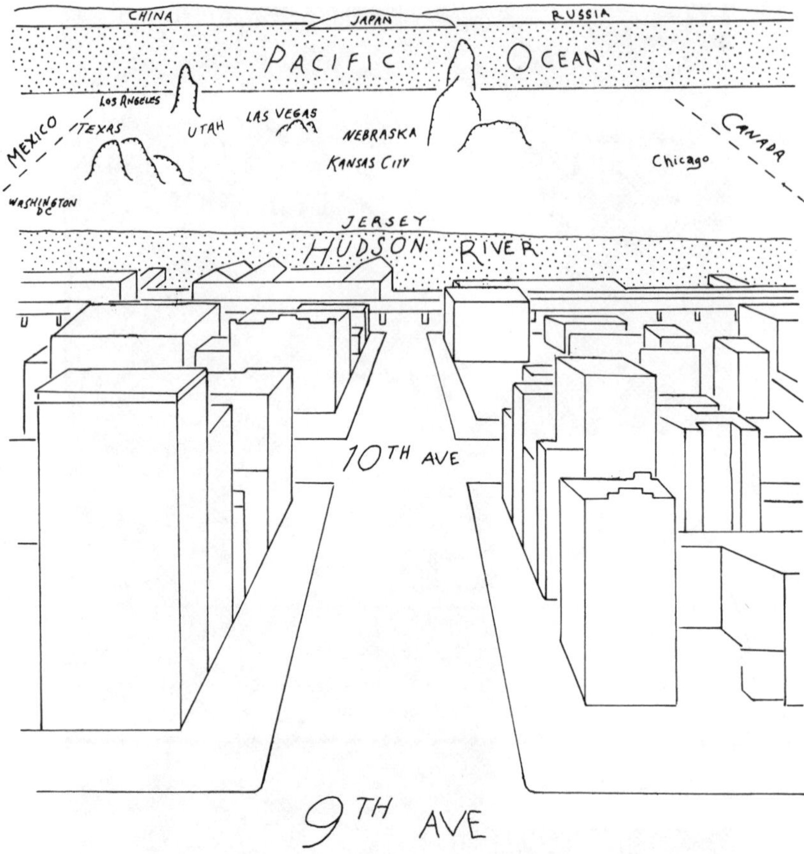

Figure 1.15 Biased reporting—in this case information from an inhabitant of New York regarding the geography of the United States and surrounding areas. Biologists have suffered from a similar "gender provincialism" in reporting on behavioral, evolutionary, and physiological interactions between males and females. (Adapted by W. Eberhard from a *New Yorker* cover by S. Steinberg)

Reeve 1995; L. Brown et al. 1995), recent studies still routinely emphasize the male side of male-female sexual interactions. The following are some illustrative examples, chosen to include work of the leaders in the field and to show especially common biases.

Theoretical models of the use of stored sperm to fertilize eggs typically do not consider the possibility of differential release or usage by females (Gilbert et al. 1981a; Parker 1984, 1990; P. Smith et al. 1988; Parker et al. 1990; Parker and Simmons 1991; Simmons and Parker 1992). When sperm precedence was discussed in terms of the effects of female morphology and behavior (e.g., Walker 1980 on insects), sperm movements were nevertheless discussed in terms of the male's abilities to move sperm, without mentioning the possibility (or rather, the near certainty) that the female also participates in moving them, and in most cases is probably the major force in their transport (Omura 1938; Callahan and Cascio 1963; Davey 1965b,c; Wigglesworth 1965; Engelmann 1970; Thibout 1977; Okelo 1979; Linley and Simmons 1981; Heming-van Battum and Heming 1986; Gillott 1988; Tschudi-Rein and Benz 1990; Sugawara 1993; LaMunyon and Eisner 1993). Some listings of mechanisms that could bias the number of sperm arriving in storage, or the use of one male's sperm over that of another, did not even include the possibility of active female participation, even though sperm transport by females is well documented (Dickinson 1986; Rubenstein 1989; Schwagmeyer and Folz 1990; Parker and Simmons 1991; Gage 1991; Simmons and Parker 1992; Madsen et al. 1992; Inoue et al. 1992; Birkhead and Møller 1993b; Olsson et al. 1994). Systematic listings of mechanisms of female control of fertilization (Eberhard 1985; Birkhead et al. 1993a) included only a small fraction of the twenty or more possible mechanisms discussed in chapters 3 and 4. Even the "father" of cryptic female choice, Randy Thornhill, has discussed the function of a male clamping structure, the notal organ of the scorpionfly *Panorpa vulgaris*, in terms of only the female's escape from the male's hold, without considering the many other possible cryptic female choice mechanisms (oviposition, sperm transport, etc.) that may be involved (Thornhill and Sauer 1991).

The very name often used for this field of study, "sperm competition," (fig. 1.16) emphasizes active male roles over possible female participation (e.g., "sperm screening"; see also Birkhead 1994). This bias in the name can have serious effects on the field itself. For instance, all of the types of adaptation expected under "sperm competition" listed by Andersson (1994) (male plugs female genitalia, male removes sperm from previous males, etc.) are clearly in the male-active, female-passive tradition (see Gack and Peschke 1994 for a similarly limited list).

The common characterization of some aspects of male-female conflicts of interest as "arms races" (e.g., "males invest in armament to achieve matings and females invest in means to prevent extra matings" in Parker 1984, p. 24; see also Cronin 1991) is another case in point. This is a reasonable analogy from the male point of view. Continued adjustments by males that improve their abilities to overcome all changing types of female resistance are favored. But from the female's vantage, the interaction is probably often more subtle. Rather than being in an all-out arms race, the female is in a race to be able to *surrender or accede selectively* (figs. 1.7, 1.17). After all, it is not in the female's interest to successfully resist all male

Types of Male - Male Reproductive Competition

	Intrasexual interactions	Intersexual interactions
Prior to initiation of copulation	male-male battles	"classic" female choice
After initiation of copulation	sperm competition (sensu strictu)	cryptic female choice

Figure 1.16 Labels for the types of sexual selection that result from reproductive competition among males that occur before and after copulation begins. The label "sperm competition" has generally been used in an inclusive sense, to designate all male-female interactions that occur after copulation has begun. As is clear from the analogies in this table, however, this male-centered terminology can result in an underestimation of the roles of females. Just as it is useful to distinguish male competition that involves direct male-male aggression from that involving female biases in favor of certain males prior to copulation, so also a similar differentiation should be made later, distinguishing direct male-male battles ("sperm competition *sensu strictu*") from female-imposed biases. Because the crucial interactions after copulation has begun are played out on or within the female's body, the likelihood that females will have important effects on competition among males is greater after the initiation of copulation than before.

adaptations—by doing so her eggs would remain unfertilized. The relevant female morphology usually consists of selective *acceptance* structures such as pits and grooves rather than active rejection structures such as erectable spines (summary in Eberhard 1985; see also Toro and del la Hoz 1976; Coyle 1985b, 1986; Toro 1985; Toro and Carvajal 1989; Belk 1991; Huber 1994b, 1995; for an unusual exception, see Arnqvist and Rowe, 1995).

Even the pivotal event for sexual selection, the fusion of a small, mobile sperm cell with a large, resource-laden egg cell, is typically misrepresented with the phrase "sperm penetration" of the egg. In fact, the egg is much more active than the sperm (see chap. 4.11), and the process more nearly resembles "swallowing and then gently coddling" the sperm.

The presumption in early studies of extrapair copulations in "monogamous" birds that males were forcing copulations on females (e.g., McKinney et al. 1984), rather than that females were actively soliciting and promoting them (which has proven to be common; Heg et al. 1993; Burley et al. 1994) is another manifestation of a bias toward male control and female passivity. Studies of differences in fertilization success between conspe-

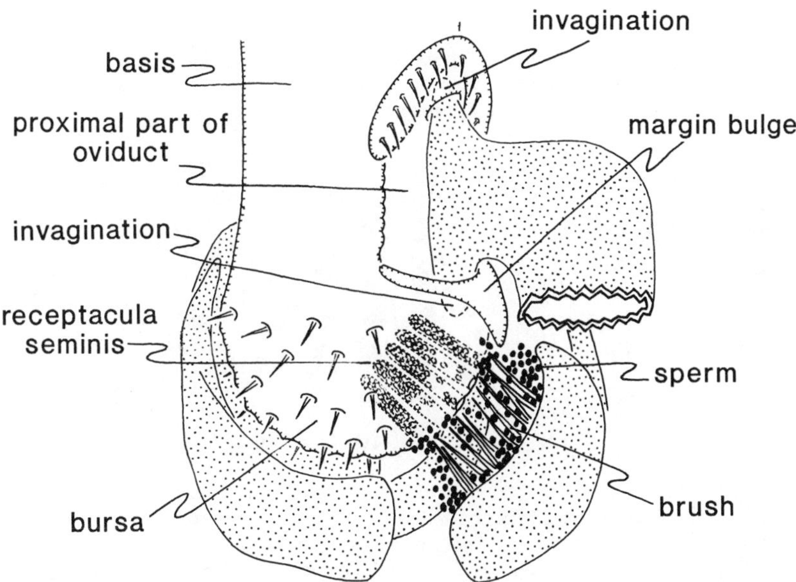

Figure 1.17 The female genitalia of the millipede *Craspedosoma transsilvanicum* illustrate how the "selective surrender" metaphore for male-female interactions is more appropriate than that of an "arms race." The female has special pockets (labeled "invagination" and "margin bulge" in this schematic figure) whose only apparent purpose is to receive processes of the male genitalia (stippled). The resulting close fit locks the animals together, and the male then utilizes brushes to introduce his sperm into the female's multiple seminal receptacles ("receptacula seminis"). The "cooperative" female pockets in this millipede are especially dramatic, because the female's behavior prior to copulation suggests resistance. All copulations began with a "wrestling match" in which the female rolled up her body and the male attempted forcefully to uncoil her; females were less able to escape from males once the genitalia were locked together. (After Tadler 1993)

cific genetic strains routinely address variation among male strains but seldom among female strains (table 9.3). Failures of copulations to result in sperm transfer are attributed to male failures without even considering the possibility of female exclusion or rejection of male genitalia or sperm (e.g., O. Taylor 1967; Somers et al. 1977).

Male control is sometimes assumed even when male and female morphology indicate female influence is far more likely. For example, mating-induced redistribution of sperm already in the female's spermatheca in species whose spermathecae are inaccessible to the male's genitalia was nevertheless attributed to the male with no mention of a possible female role (Radwan and Witalinski 1991 on a mite; Birkhead and Hunter 1990 on several insects). The timing of copulations in birds to coincide with ovulation time both with respect to the time of day and the number of days until the last egg is ovulated has often been assumed to result in an increased

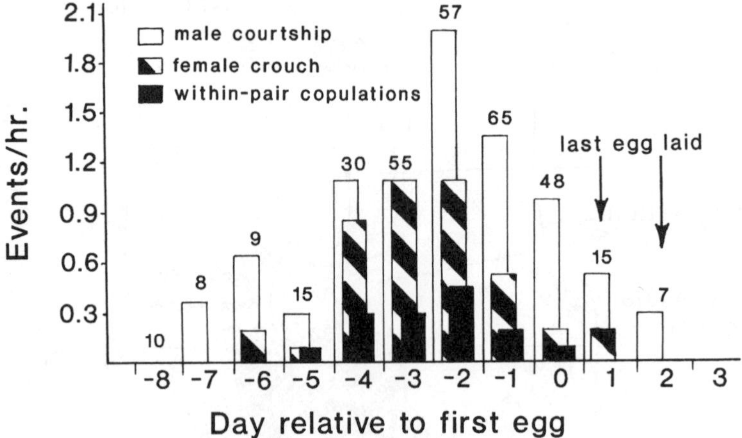

Figure 1.18 "Miscoordination" of male mating behavior and the female fertile period in pairs of the red-winged blackbird *Agalaeus phoeniceus*; it is possible (though as yet untested) that active female control of sperm transfer or subsequent transport (or ejection) may occur. Each egg is fertilized about 24 hr before it is laid, so the first egg was fertilized on day "−1," and the last, depending on the size of the clutch, on day "1" or "2." Since in birds the last male to copulate with a female before the egg is laid is generally more likely to fertilize the egg, the sharp *decrease* in male courtship and copulation just as fertilizations are beginning is paradoxical, at least under the supposition that females participate only as passive receivers of sperm. Numbers over the bars are number of observations. (After Westneat 1993)

likelihood of fertilization (e.g., Westneat 1993). This despite the fact that transportation of sperm by the female is necessary if fertilization is to occur, and despite the lack of correlation of male and female behavior with the usual assumptions regarding fertilization probabilities in some species (e.g., Westneat 1993 on the red-winged blackbird *Agelaius phoeniceus*; fig. 1.18) (see Birkhead and Møller 1993b; chap. 8.5). The possibility that some female crickets sometimes actively discard a portion of the male spermatophore so as to reduce sperm transfer rather than "losing" it was originally ignored (Sakaluk 1984, p. 609; see Sakaluk 1985 for reinterpretation of this behavior as cryptic female choice). Some studies of the functional significance of spines on the penes of mammals concentrated on their effects on the male rather than the female (Johnson et al. 1986; Dixson 1991a), ignoring the near certainty that the spines stimulate females during copulation (Gruelich 1934; Zarrow and Clark 1968; O'Hanlon and Sachs 1986; Dixson 1987).

Temporary postcopulatory associations between males and females are often described as "mate guarding," even when the times involved are quite short, as for example in the bug *Triatoma mazzottii* (about 100 sec; Rojas et al. 1990) and in the wasp *Aphytis melinus*, in which the male performs courtship behavior for only about 2 min, then leaves (Allen et al. 1994). Several authors have argued that a tendency for copulations themselves to be longer when more males are in the vicinity is evidence that they function as guarding (summary in Alcock 1994). But the females are not necessarily such passive pawns for males. For instance, chemical or mechanical stimuli from copulation could induce females to become unreceptive, or to oviposit more quickly (see chaps. 3 and 6 for data on many species). It is possible that a longer copulation is advantageous for the male when other males are nearby because he could thus reduce the risk that the female would accept another male after they parted, not by physically excluding the other male, but by eliciting increased female cooperation (e.g., Dickinson, in press).

The very observational data that are reported in publications have probably been seriously biased by lack of emphasis on female roles (Eberhard 1994 on male and female courtship during copulation; see chap. 5.1), so this can become a vicious cycle of bias in both concepts and data. Discussions of data demonstrating that females sometimes discard sperm from previous matings when stimulated appropriately by males sometimes include phrases such as "sperm removal" or "indirect sperm removal" by the male (Birkhead and Hunter 1990; von Helversen and von Helversen 1991; Birkhead and Møller 1992) when the male has no direct access to the sperm and must rely on female cooperation.

Lifjeld et al. (1993) and Dunn et al. (1994b) give telling critiques of previous male-centered attempts to understand patterns of extra-pair copulations and paternity in birds that failed to fit with observations of species in which females control copulation and fertilization. L. Brown et al. (1995) make similar criticisms of attempts to link mating systems to sizes of male gonads and genitalia without taking into account the morphology of female genitalia (see also secs. 1.2–1.3 above).

It has been argued that male-centered views have a theoretical justification. Selection on males to influence sperm usage patterns is probably generally much more intense than that on females (Parker 1984). A male who fails to win can suffer complete loss of offspring; a female who fails to win loses quality of offspring, but not the offspring themselves. Because males are under more intense selection, it has been supposed that they will generally prevail in such conflicts (Parker 1984). This prediction, however, hinges on the likelihood that mechanisms that influence sperm usage will be equally likely to arise in males and females (Parker 1984). As has already been noted by some authors (e.g., Ward 1993), and as I will show in the next chapter, females far outstrip males in this respect. Thus the expectation of male triumph in these evolutionary battles is called into doubt.

1.10 Summary

Females in animals with internal fertilization have partial or complete control over many processes that can influence a male's chances of fathering their offspring. A probably incomplete list includes prevention of ejaculation and deep penetration, premature interruption of copulation, lack of sperm transport, emission of the sperm of the current male, retention of sperm from previous males, digestion of sperm, failure to nourish stored sperm, biased use of stored sperm, selective fusion with sperm which have reached the egg, failure to transfer sperm from previous males to a site where the current male can remove or deactivate them, lack of ovulation, lack of preparation of uterus for implantation, biased abortion, lack of oviposition, reduction of investment in offspring, rejection of mating plugs, prevention of removal of previous plugs, removal of sperm-injecting spermatophores before sperm transfer is complete, prevention of injection of sperm from spermatophores, and remating with other males (see chaps. 3 and 4 for examples of species in which females perform these processes).

If these processes bias the probability that some particular type of male will sire the offspring of females who copulate with more than a single male, they are exercising cryptic female choice. The male features that are under selection are those that cause the female to differentiate among males. The female choice is "cryptic" in the sense that it will be missed using classical, Darwinian criteria for sexual selection by female choice: a male can succeed in being accepted as a partner in copulation and achieve genitalic coupling but nevertheless be rejected as a sire. The female may perform, or fail to perform, crucial postintromission processes that modify his chances of fertilizing her eggs. These considerations, along with evidence for direct male-male competitive interactions at the level of their ejaculates, lead to the conclusion that sexual selection is better understood as a result of competition among males for access to female gametes rather than for access to the females themselves.

Since several aspects of competition between males for female gametes are often played out within the bodies of females, female traits can influence male access to their gametes in numerous and often subtle ways. Previous studies have tended to underemphasize female roles. The importance of female morphology and behavior can be illustrated even in extreme cases, in which males seem to be in control of competitive outcomes. In an analogy with sporting events, both the characteristics of the playing field on which males are competing and many of the rules by which they must abide are set by the female. Females can (and often do, as I will show in later chapters) easily tilt either the field or the rules in favor of males with particular characteristics.

Previous studies of the reproductive processes that occur following genitalic coupling have often suffered from several biases. Studies of reproductive morphology and physiology often include the mistaken notion that males and females consistently cooperate during mating. Variation in morphological and physiological processes during insemination and the subsequent sperm usage is often ignored. Females are often assumed to be passive, and there is an overemphasis on male rather than female perspectives in many evolutionary interpretations.

Notes

1. I will use the definitions of Page (1986): *copulation*—the act of pairing and intromission (at least partial insertion of the male genitalia into the female's body) which may, but does not necessarily, result in insemination; *insemination*—deposition of sperm in the female; *fertilization*—entry by a sperm cell into an egg, and subsequent fusion of their nuclei to form a zygote.

2. Arnold (1994) has argued that most selection in such a postintromission context should not be considered sexual selection, but rather "mate fecundity selection." I follow a number of other authors (e.g., West-Eberhard 1983, 1991; Willson 1994; Andersson 1994) in disagreeing, for the following reasons. One crucial property of sexual selection, which makes its results differ from those of natural selection, is that sexual selection continues to favor innovation more consistently than natural selection (Darwin 1871; West-Eberhard 1991). This is because the advantage of having a given trait is largely determined by whether other members of the species have the same trait. For example, once an adjustment to an environmental change has been made (say a change in an enzyme makes it able to function at peak efficiency at a new environmental temperature), there may be little or no selective advantage to further innovation. In contrast, a new male plumage that is more effective at intimidating other males (or charming females) will only give a male greater reproductive success as long as other conspecific males lack the device. As soon as the plumage design becomes established in the population, the male will need an additional change in this character (or others) if he is to gain an advantage over other males. As Darwin (1871, p. 583) wrote: "In regards to structures acquired through ordinary or natural selection, there is in most cases, as long as the conditions of life remain the same, a limit to the amount of advantageous modifications . . .; but in regard to structures adapted to make one male victorious over another, either in fighting or in charming the female, there is no definite limit to the amount of advantageous modification; so that as long as the proper variations arise the work of sexual selection will go on." Improvements in a male's ability to induce females to perform cryptic female choice processes can result in this sort of competitive advantage in species in which females copulate with more than a single male. They are thus appropriately considered to be under sexual selection.

2 Selection on Cryptic Female Choice

Little previous theoretical work has been directly related to sexual selection by cryptic female choice. What are probably the two most important papers (Parker 1984; Knowlton and Greenwell 1984) suggest it is unlikely to be especially widespread. As mentioned in the last chapter, Parker (1984) noted that males will be selected to overcome mechanisms of cryptic female choice, and that selection in such conflict situations is expected to be more intense on males than on females. A male who fails to overcome female resistance will lose his entire genetic representation in her offspring; a female who fails to resist a male may suffer some loss in quality of her offspring, but generally will not lose the offspring themselves. Thus Parker (1984) predicted that, other things being equal, males will generally prevail. I will argue in this chapter that this is not the expected outcome, because the "other things" are usually not equal.

Knowlton and Greenwell (1984) used game theory to model male-female conflict with respect to sperm competition. They took into account only one possible female mechanism to manipulate the results of a mating (premature termination of copulation), and assumed that the female could reject a male who attempted to co-opt her chances of choosing other males as fathers. They found that female control is likely to evolve under some conditions (when male manipulative behavior begins late in copulation, or is not especially effective, or when males are able to remate quickly and do not invest heavily in the female and her offspring). They concluded that males are expected to win in other situations. Both Knowlton and Greenwell and Parker added, however, that it is difficult to predict the resolution of sexual conflict regarding sperm usage. Although Parker suggested that the stronger selection on males implies that they will often triumph, he noted (p. 21) that "contrary cases abound."

I believe that these pioneering discussions seriously underestimated the likelihood that selective control of fertilization by females will evolve for three reasons:

1. Natural rather than sexual selection on several processes under female control is likely to result in the evolution of mechanisms that can later be used to discriminate between males, thus predisposing females to make such discriminations.

2. The number of processes by which a female can influence paternity was seriously underestimated—instead of one mechanism, there are more than twenty.

3. Genetic variance in males is not necessarily rapidly exhausted by sexual selection, so female selectiveness is more likely to be favored via acquisition of advantageous genes from males than previously thought.

2.1 Female Control Mechanisms and Natural Selection

Many possible mechanisms of cryptic female choice are useful to females in the context of natural selection, and may well have originally evolved in species in which sexual selection by cryptic female choice did not occur. For instance, stimuli from the male or his ejaculate would be appropriate cues to trigger female behavioral and physiological processes necessary for successful insemination and fertilization, including immobility, lordosis (or other postures to allow intromission), sperm transport, sperm maintenance, ovulation, oviposition, embryo implantation (as in mammals), and rejection of further males. Many female structures that evolved under natural selection in other contexts could also be employed to counter male mating attempts. For instance, the bursal muscles of female *Macrohaltica jamaicensis* beetles, which probably help move eggs down the duct during oviposition, can also apparently deny the male genitalic access necessary for sperm transfer (see fig. 3.6; Eberhard and Kariko, in press). The very presence of an egg part way down the oviduct can effectively block access to insemination or fertilization sites in the female (e.g., fig. 2.1 on a fly; see also Birkhead and Møller 1992 on birds). Such structures would thus have no "cost" associated with the male-denying usage. Extended female "memory" of having mated (either as hormonal changes or as classic neuronal memory) would be especially useful under natural selection in the context of females deciding whether to begin oviposition or to remate. It could then be used to compare copulations with different males.

The fact that gamete fusion involves so much active cooperation on the part of the egg (see chaps. 1.9, 4.11) suggests that cryptic female choice may be extremely ancient. Sperm cells have stipped-down bodies and compacted DNA that is incapable of transcription or replication, which are presumably adaptations to increase sperm mobility and numbers and perhaps also to avoid damaging competitive intraejaculate interactions (Sivinski 1984). A sperm is thus essentially helpless when it meets the egg. Egg responses and resources are crucial to pull the sperm into the egg and exclude others, to decondense the sperm nucleus and reactivate its DNA, and to bring about subsequent intracellular movements that result in fusion of

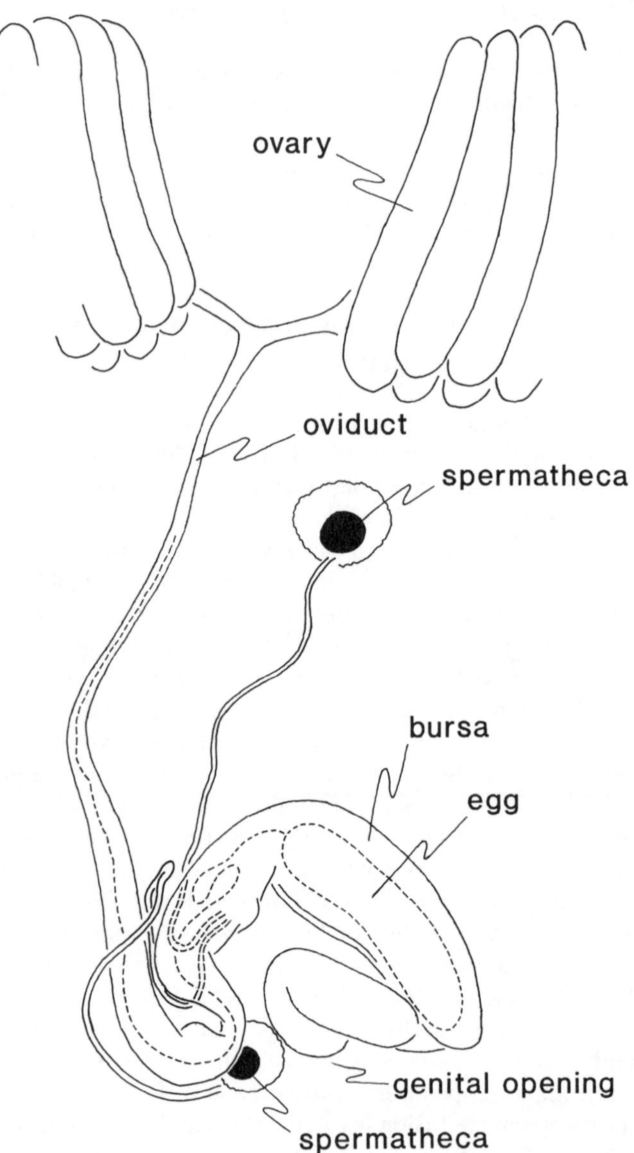

Figure 2.1 A female's control of male access can be simple and inexpensive, both energetically and in terms of special structures. During the extended process of oviposition of her load of about one hundred eggs in a cowpat, the female of the sepsid fly *Achisepsis* sp. nr. *diversiformis* usually carries an egg in the bursa, near the distal end of her oviduct. This is the same site where the male's genitalia are inserted during copulation, and where he deposits a large mass of sperm. The blockage resulting from this detail of female oviposition behavior may be responsible for the very unusual mating system in this and the related species, *Sepsis cynipsea* (Parker 1972): the male finds the female on or near an oviposition site, mounts and courts her, but does not copulate with her until she has laid her entire set of eggs.

egg and sperm nuclei (e.g., Deuchar 1975). Any sperm better able to trigger the egg responses that lead to the fusion of their nuclei would be favored (a possible example of selection at this level is described in chap. 4.11).

The preexistence of such mechanisms would make it evolutionarily "easy," in two different senses, for females to evolve abilities to choose among males cryptically. Few mutations would be needed for choice to begin to occur, so the original appearence of the trait would be likely. Few resources would have to be allocated specifically to this function, so the trait would have little or no material or energetic cost.

The "cost" to the female in material and energy of exercising cryptic choice against certain males may even be negative in some cases. For instance, in some, and probably most, insects, birds, and mammals (e.g., Walton 1960; Davey 1965b; Linley and Simmons 1981; Birkhead and Møller 1992), energy-consuming movements of structures in the female reproductive tract such as peristaltic contractions, beating of cilia, and absorption of liquids are apparently necessary to move the male's gametes to storage and/or fertilization sites. Sperm attraction, maintenance, and activation, which can result from products of female glands (e.g., Villavaso 1975b on the boll weevil *Anthonomus grandis*; Koeniger 1986 on the honeybee *Apis mellifera*), undoubtedly also often involve certain metabolic costs (e.g., fig. 2.2). A female could gain a net increase in energy reserves by failing to transport gametes to her spermatheca, degrading them before they get there, allowing them to leak out, or using them inefficiently to fertilize eggs, especially those eggs laid soon after copulation.

Another potentially important factor is natural selection on female mechanisms to filter the sperm in a given ejaculate. *Intra*ejaculate competition between sperm is perhaps close to universal, since sperm numbers are virtually always greater than egg numbers. Even in groups in which the female mates with a single male, and uses all or most of his sperm to fertilize eggs (as may happen in some nemotodes; Ward and Carrel 1979; Hass 1990), within-ejaculate competition would still be likely, because those sperm achieving early rather than late fertilizations would be less susceptible to death due to female mortality.

There is reason to expect that females are often selected to manipulate intraejaculate competition. Even if the female mates with only a single male, selection could favor female features that reduce sperm access to eggs because it reduces the likelihood of fertilization by substandard sperm that would produce embryos that are less likely to survive, as well as the likelihood of having her eggs exposed to too many sperm at once (see Cohen 1991; Birkhead et al. 1993a; and chap. 7.1 for evidence on sperm quality and polyspermy). The existence of female characteristics (e.g., the corona radiata and cumulus surrounding mammalian eggs) that impede the access of conspecific male sperm, but not that of the sperm of other, very

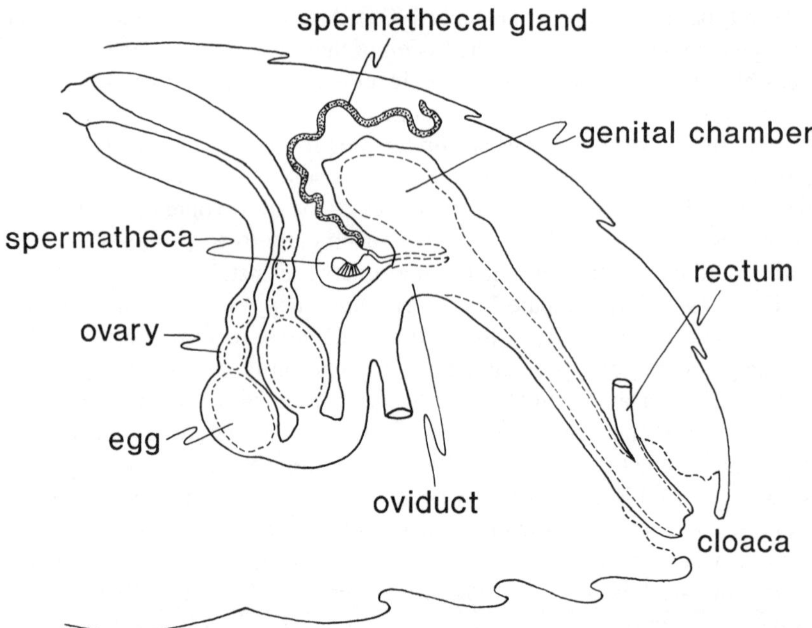

Figure 2.2 The spermathecal gland (stippled) of the boll weevil *Anthonomus grandis* offers a reminder that females can incur net *benefits* when they act against a copulating male's reproductive interests. Products of this gland are necessary for sperm to reach the spermatheca (Villavaso 1975b), and probably also contribute to their maintenance once they arrive there. By failing to accept or transport a male's sperm, the female can reduce her need to synthesize spermathecal gland products. (After Burke 1959)

distantly related species and even of unicellular algae (Talbot et al. 1985) suggests the probable importance of female-imposed bias in access (Cohen 1991). Mechanisms that reduce sperm access (for instance to avoid polyspermy) could later function to selectively favor some males over others (e.g., Birkhead et al. 1993a; Keller and Reeve 1995); if females of the species mated with multiple males they could gain the additional advantage via interejaculate competition of producing sons with more competitive sperm (Devine 1984).

Other potentially important kinds of adaptation under natural selection are mechanisms that originally evolved to protect the female from infection via her reproductive tract (e.g., low pH, presence of mucus and potentially spermicidal macrophage cells in mammals; Roldan et al. 1992). Such traits could make female reproductive tracts hostile for sperm motility and survival. This could thus both reduce sperm numbers and filter out certain classes of sperm cells. Lengthening and narrowing the ducts could have similar defensive and filtering effects (see Birkhead et al. 1993a; see chap. 7.1).

These kinds of female traits set important conditions ("rules of the game") for sperm survival and success. A given type of sperm may be inferior in one female milieu but not in another. For instance, Zimmering and Fowler (1968) concluded from genetic studies of *Drosophila melanogaster* that "a sperm is determined as functional or non-functional depending at least in part on the genotype of the female into which the sperm is transferred" (p. 362). Such filtering may sometimes result in more vigorous offspring, although this is certainly not a necessary correlation (see Madsen et al. 1992 for a possible case involving sperm from different males). In such cases, counteradaptations by males to overcome intraejaculate filtering would be unlikely, as long as the female mates with only a single male. Improved quality of offspring would also be to the male's advantage since all his sperm are equally related to him.

By simply supposing that there are differences in the average abilities of sperm from different males to adapt to hostile conditions within the female, and that the female mates more than once (e.g., Gromko et al. 1984 on *Drosophila*), one has conditions sufficient for cryptic female choice to occur. If there were also variation among females in the hostility of their reproductive tracts, females could gain from accentuated hostility by biasing paternity in favor of males with better-adapted ejaculates. These considerations suggest that the evolution of cryptic female choice may often be virtually inevitable, because it is so unlikely that these kinds of fitness-affecting variations would not occur.

Male-female conflict of interest (e.g., Parker 1979; Alexander et al., in press) is still another context in which female mechanisms could evolve under natural selection that would enable the female to discriminate among males. Consider, for instance, the common male ability to induce females to increase rates of oviposition via behavioral and chemical stimuli (see chaps. 3.7, 3.9, 3.10, 6.2, 6.3 for many examples). Male reproductive interests could often favor more rapid, massive oviposition before the female remates than would be advantageous for the female. Natural selection favoring female abilities to resist powerful male stimuli could produce female traits that would favor those males with enhanced abilities to induce oviposition. Such a female trait could then confer, incidentally at first, the additional benefit of an increase in the abilities of her male offspring to induce oviposition.

In sum, females are substantially more "predisposed" to exercise control in male-female conflict situations within their own bodies than has been generally appreciated. Since one of the factors influencing the outcome of such conflicts is the ability of one sex to manipulate the other, based on existing adaptations (Parker 1984), females are more likely to prevail than previously supposed.[1] Just as with characters involved in classic, precopulatory sexual selection (Andersson 1994), both natural and sexual selection may act on particular traits.

2.2 Categorical Descriptions and the Multiplicity of Female Sperm Storage Mechanisms

As already discussed in chapter 1, descriptions of sperm transfer, storage, and fertilization, and the physiological processes leading to offspring production usually recognize little if any variation. Categorical descriptions can give a false impression of a lack of variation in processes and outcomes during and following copulation, and this can result in an underestimate of the likelihood that variations in female morphology, physiology, and behavior can influence male reproductive success.

Perhaps the best-studied aspect of copulation and fertilization, at least from the point of view of morphology, is sperm storage. In this section I will discuss sperm storage as an example, to show that many species may have alternative mechanisms or sites. Both the existence and the use of different storage mechanisms are probably under at least partial control of the female. This multiplicity implies that females have more possible ways of influencing a male's paternity than is suggested by many descriptions of sperm storage.

Multiple Specialized Sperm Storage Organs

The females of some groups have multiple sperm storage sites. Multiple spermathecae are widespread in Diptera (which usually have two or three; e.g., fig. 2.1) and to a lesser extent other insect orders (Davey 1958; Imms 1964), many crustaceans, millipedes (e.g., fig. 1.17), solifuges, scorpions, spiders, and some mites, pseudoscorpions, amblypygids (usually two, but in some many more; Comstock 1948; Thomas and Zeh 1984; Hjelle 1990; Bauer and Martin 1991; Huys and Boxshall 1991). Birds have both multiple storage sites of a single type (moderate to large numbers of sperm storage tubules) and two additional, alternative sites—the cloaca, which may hold pools of sperm for several hours, and the infundibulum, near the ovaries (Lake 1975; Davies 1992; Birkhead and Møller 1992). At least some snakes have a similar array (multiple folds and tubules in two different portions of the oviduct; Halpert et al. 1982; fig. 2.3), and multiple tubules also occur in several lizards (W. Fox 1963; Cuellar 1966; Connor and Crews 1980). Several different sites in the mammalian female reproductive tract may serve for temporary storage of sperm (Austin 1969; Overstreet

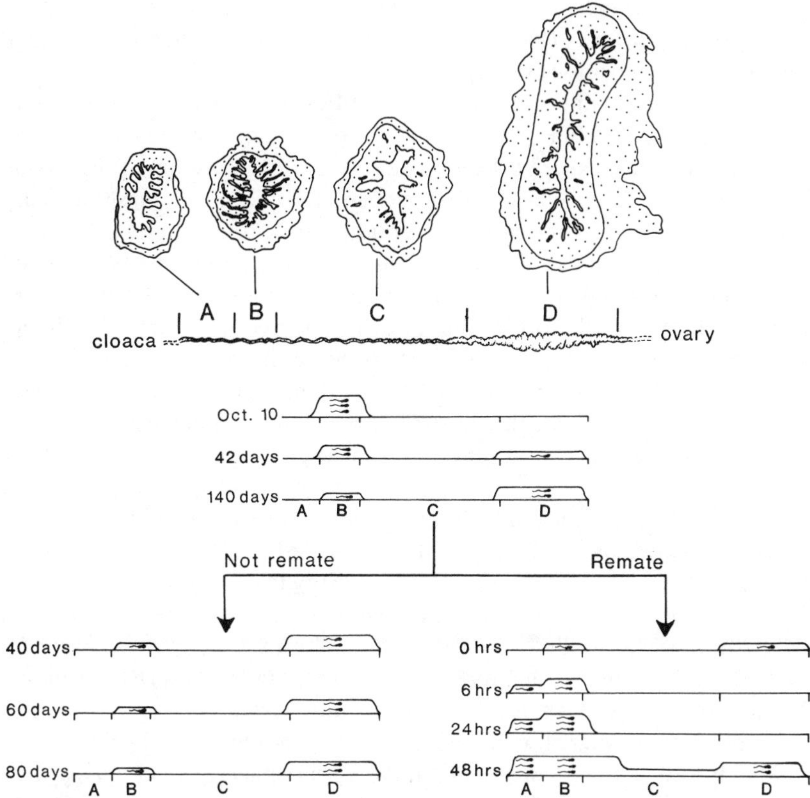

Figure 2.3 Two storage sites with deep fingerlike folds and tubes (B and D) in the oviduct of the red-sided garter snake *Thamnophis sirtalis*, and the complex sequences of sperm movement, storage, and elimination preceding ovulation and fertilization. Snakes collected in the fall (Oct. 10) harbored recently introduced sperm in region B. During hibernation some of this sperm moved to region D (or, more likely, was moved by the female, since sperm which moved was generally found embedded in sloughed-off portions of the oviduct wall). Some females did not remate in the spring (left sequence; time in days after female emerged from hibernation), and their sperm remained stored in both regions until ovulation and fertilization. Other females remated in the spring (*right*; time in hours after copulation), which resulted in almost immediate degradation of sperm in region D, and degradation "to a lesser extent" of that in region B. Many but not all sperm from spring copulations moved (or, again more likely, were moved) relatively quickly (compared with fall matings) to region D. Presumably sperm in D are more likely to fertilize eggs, because they are nearer the ovaries, but no direct data are available. (After Halpert et al. 1982)

and Katz 1977; Fenton 1984). The diversity of storage sites in species of the same genus of bats (Fenton 1984) suggests possible alternative storage (and perhaps deposition) strategies.

The existence of multiple storage sites offers the female the opportunity to bias paternity via biased transport to different storage sites, biased survival at different sites, or biased transport from different storage sites to the fertilization site (e.g., Ward 1993; see chap. 3.5). That such biases actually occur is suggested by data from several groups. For example, a given female dunnock (*Prunella modularis*) can have some storage tubules packed with sperm while others nearby are completely empty (see fig. 8.2; Birkhead et al. 1991). Sharp differences in the degree of filling of different storage tubules in the same female also occur in the Japanese quail (*Coturnix japonica*; Birkhead and Møller 1992) and the lizard *Anolis carolinensis* (Connor and Crews 1980). Similarly, some queen honeybees returning from a nuptial flight (two of twelve) had a sperm mass in only one of their two oviducts (short-term storage sites; Laidlaw 1944). Some differences in filling with apparently nonclumped sperm are so dramatic that biased female transport or uptake seems to be the only reasonable explanation (figs. 2.3, 8.2; see also chap. 3.5). Female ability to control sperm entry into storage in birds, for instance, is suggested by the fact that the accumulation of sperm in storage tubules is both faster and greater in preoviposition female Bengalese finches and turkeys than in laying females (Birkhead and Møller 1992), although this could conceivably be due to variations in male ejaculate size associated with the reproductive condition of the female.

The dynamics of sperm storage in the garter snake *Thamnophis sirtalis* suggests both the possibility of alternative male strategies and of female control over their effectiveness (fig. 2.3; Halpert et al. 1982). Sperm from fall copulations that end up in region D (the infundibulum) instead of B nearer the cloaca are often (always?) degraded if the female remates in the spring, while some of those in B may survive. Infundibulum (D) sperm may also have advantages, however, since they presumably have better access to unfertilized eggs than those in B if the female does not remate. Details of the histological and physiological events associated with sperm migration (formation of a matrix of sloughed cells from the wall of the female duct within which the sperm are embedded, and a surge of estrogen that may cause oviductal contractions) suggest an active role for the female in determining sperm survival at different sites (Halpert et al. 1982).

Another female-mediated process—sperm movement from different sperm storage organs to fertilization sites—has been explored in two genera of flies (*Drosophila* spp. and *Scathophaga stercoraria*). In both groups there is a bias, with sperm in one type of organ being more likely to be used for fertilization than that from the other (Fowler 1973; Gromko et al. 1984; Markow 1985; Ward 1993). In *D. melanogaster*, sperm in the two lateral spermathecae are used less readily than sperm from the central "ventral receptacle." From the female's point of view, sperm in the spermathecae may represent a "fail-safe" reserve, to be used mainly if sperm from the ventral receptacle run out and she fails to remate (Fowler 1973).

Since females in these groups probably influence sperm transport into storage (male genitalia do not apparently enter the relatively long spermathecal ducts), those males better able to prejudice the entry of their sperm into favored storage organs (by direct positioning of their genitalia or their sperm, or by inducing appropriate female responses) may have a reproductive advantage if females mate more than once. A biasing ability of this sort has apparently not been investigated in *Drosophila*, but it may occur in *Scathophaga stercoraria*. Larger males, probably by inducing appropriate responses in females, are better able to have their sperm stored in the spermathecae where they are more likely to be used for fertilization (Ward 1993; see also chap. 3.5).

Sperm at Other Sites in the Female Reproductive Tract

Although multiple spermathecae or other specialized sperm storage organs are probably the exception rather than the rule, there is reason to suspect that the use of alternative sites within the female for sperm storage may nevertheless be relatively common. This is because in many groups (e.g., most insects; Imms 1964) the egg is fertilized at a site (the oviduct) that is *more* accessible to the male than is the sperm storage site. For instance, the male genitalia of the bruchid beetle *Callosobruchus maculatus* fill the female's bursa, where fertilization probably occurs; but they do not extend even part way up the long thin spermathecal duct (Eady 1994b). Alternative storage sites can be as simple as the lumen of the oviduct versus the folds of its lining (e.g., Constanz 1984 on a poeciliid fish).

For long-term storage, a male probably usually needs to get his sperm into "the" sperm storage organs of the female, where special conditions presumably favor their survival (see, however, the end of this section). On the short term, however, sperm in the bursa may even have an *advantage* over spermathecal sperm. For instance, rates of egg hatch in female boll weevils (*Anthonomus grandis*) from which the spermatheca had been removed were not even slightly reduced during the first five days after mating (89.3% vs. 88.5% in operated and unoperated females; Nilakhe and Villavaso 1979). Living sperm capable of fertilizing eggs were present in the bursa and/or oviduct for up to at least fourteen days after mating (Villavaso 1975a). The majority of sperm retained in the female of this species after mating are in the bursa (about 200,000, vs. about 15,000 in the spermatheca; Nilakhe and Villavaso 1979). If bursal sperm are sometimes able to fertilize eggs before the micropyle of the egg reaches the opening of the spermathecal duct, then bursal or oviductal sperm would have a reproductive *advantage* over spermathecal sperm. Sperm may also survive for some time in the female's vagina in another beetle, *Tribolium castaneum* (Schlager 1960), in the bursa of the mosquito *Psorophora howardii* (for as long as 24 hr after mating; Lum 1961), and the bursae of several odonates (McVey and Smittle 1984). In the damselfly *Mnais pruinosa*, bursal sperm have essentially complete precedence over spermathecal sperm soon after copulation (Siva-Jothy and Tsubaki 1989a,b).

The probable importance of bursal sperm in fertilizing eggs is also indicated in some groups by the existence of multiple micropyles (sperm entry points) (fig. 2.4) and male sperm-removing devices that only reach to the bursa and not the spermatheca (e.g., Waage 1984 on odonates; Yokoi 1990 and Gage 1992 on beetles). This, in turn, means that a categorical statement of my own (Eberhard 1985) must be modified. Even though male genitalia fail to reach "the" site of sperm storage in particular groups, genitalic evolution cannot be assumed not to have been affected by selection for sperm removal in these groups.

In other species of insects, bursal sperm are at more of a disadvantage, and the role of the female in determining advantages is relatively clear. In the honeybee *Apis mellifera*, sperm are eliminated from all portions of the female's reproductive tract other than the spermatheca soon after the queen's nuptial flight (Page 1986). The plug in the micropyle of the egg of the housefly *Musca domestica*, which is removed only after the egg is in position at the mouth of the spermathecal duct to be fertilized by spermathecal sperm (fig. 7.6; Leopold et al. 1978), would seem to preclude the possibility of fertilization by sperm in the bursa.

As in many other insects (Davey 1965b), the egg of the silkworm *Bombyx mori* is positioned so that its micropyle lies just at the mouth of the fertilization canal from the spermatheca, and sperm have been observed entering the egg at this site. In other species, however, the positioning is presumably less precise. For instance, the egg of the roach *Periplaneta* has up to one hundred micropyles at one end (Davey 1965b). The positioning of eggs and their micropyles in this and many other groups could bias the likelihood of fertilization by sperm stored at different sites in the female. Any variations in the female morphology and behavior controlling such positioning of eggs (I know of no studies that have even addressed this question) could result in changes in fertilization biases in the many species in which standard accounts suggest that such female alternatives are not feasible. The possible importance of alternative sperm storage sites is also in accord with changes in storage sites over evolutionary time. For instance, in the true bugs the spermathecae have been replaced several times as sperm storage organs by possible alternative sites, nearer the ovaries (Tingidae, Reduviidae, perhaps also Pachynomidae), or the bursa itself

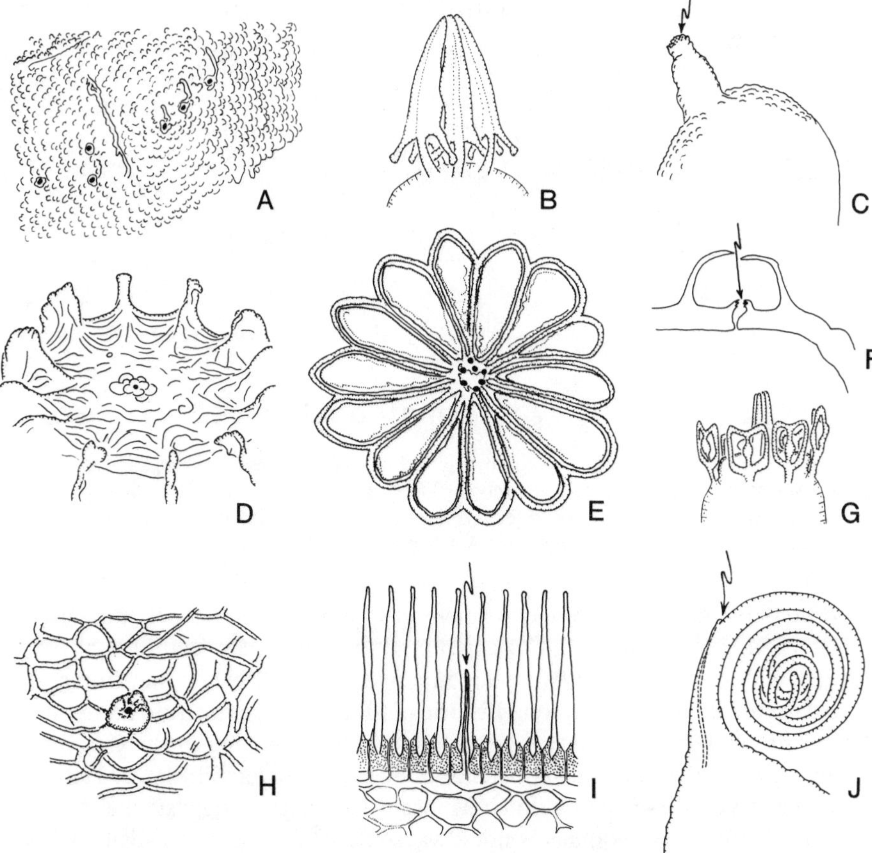

Figure 2.4 A sampler of micropyles (sites of sperm entry, indicated with black dots or arrows) on the eggs of insects, showing a spectacular diversity of complex structures that are presumably involved in the process of sperm entry into the egg. Ignorance regarding the functional morphology and behavior of sperm is such that it is not even possible to present reasonable speculations regarding the possible significance of this diversity. A: *Perlodes microcephala* (Plecoptera); B: *Ortholomus* sp. (Hemiptera); C: *Sialis lutaria* (Neuroptera); D: *Hypena proboscidalis* (Lepidoptera); E: *Habrosyne pyritoides* (Lepidoptera); F: *Columbicola columbae* (Phthiraptera); G: *Kleidocerys resedae* (Hemiptera); H: *Fannia canicularis* (Diptera); I: *Cylapofulvius* sp. (Hemiptera); J: *Acanalonia bonducellae* (Homoptera). (Drawn to different scales; after Hinton 1981a,b)

(Miridae; Schuh and Slater 1995). Females may sometimes abandon former storage sites after males begin to exploit others (see Eberhard 1985 for other possible examples).

Even the importance of storage in specialized female organs for long-term sperm survival of sperm may not in fact be as crucial as it might seem. Long-term storage of viable sperm occurs in relatively unspecialized

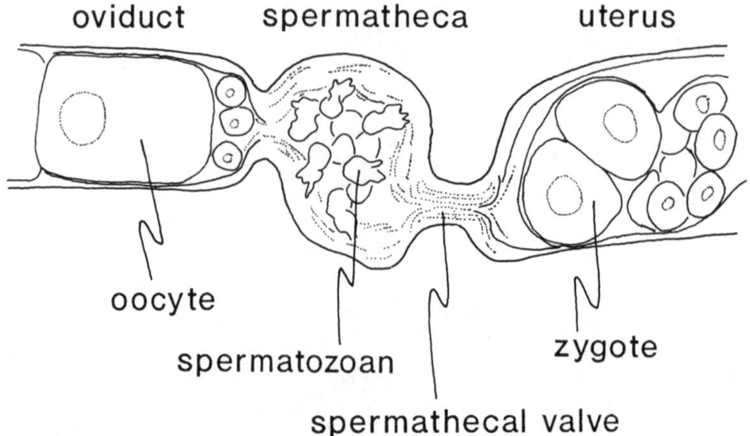

Figure 2.5 Extreme simplicity in a spermatheca. Sperm are lodged in a small segment of the oviduct in the nematode *Coenorhabditis elegans*. Despite this simplicity, live sperm can be stored in nematodes for at least 5 days and are sometimes used with perfect efficiency (one sperm for each egg). (After Ward and Carrel 1979)

portions of female reproductive tracts in groups with little evolutionary history of sperm storage (up to about eight months in a poeciliid fish: E. Clark and Aronson 1951; two years in a frog: Metter 1964). Long storage also occurs in groups with long histories of internal insemination but with only simple female storage organs such as just a widening of the oviduct (fig. 2.5) (up to a year and six batches of eggs in a spider: Uhl 1992, 1994; up to two years in ticks: Feldman-Muhsam 1986). This suggests that specialization in storage organs is not always crucial for sperm survival, and, consequently, that too much emphasis on sperm transfer to specialized storage organs may be misleading.

Sperm Outside the Female Reproductive Tract

An additional, as yet relatively unexplored possible source of variation in sperm storage is the possibility that sperm may leave the female's reproductive ducts and move through her body cavity to reach sperm storage organs and/or fertilization sites. The possibility of female control, or lack of control, of such processes is mostly unstudied, though the morphology of the lining of her reproductive tract would seemingly have important effects. Artificial intraperitoneal or hemocoelic injections of sperm produces pregnancies in a wide variety of animals, including rhesus monkeys, cows, guinea pigs, pigeons, chickens (Rowlands 1958), and such "traumatic insemination" occurs naturally in leeches, onychophorans, and bedbugs (Austin 1975; Hendricky et al. 1978; Carayon 1966, 1975). Rowlands notes that when one horn of the uterus of a sheep is tied off near the cervix, pregnancies nevertheless occur in both sides of the uterus. Living sperm were found in the peritoneal fluids of five of fourteen women sampled 24 hr after copulation (Blandau 1969). Sperm of the nabid bug *Aleorhynchus plebejus* migrate from the bursa to the ovary by moving between the muscle fibers of the oviducts (Davey 1985). Sperm can pass through the wall of the spermatheca in the worm *Stuhlmannia variabilis* (Adiyodi 1988). Perhaps the striking and otherwise puzzling tendency for sperm storage organs, such as spermathecae in many insect groups, to be heavily sclerotized (e.g., figs. 1.2, 2.1) is related to prevention of sperm escape (see Linley and Simmons 1981 for another, not necessarily exclusive explanation that is based on sperm transport mechanisms).

Sperm penetration of female tissues, and fertilization of eggs that have not yet left the ovary, is apparently the rule in ticks and mites (Alberti 1991). In gamasid mites it is possible to trace the probable evolutionary history of the ability of sperm to invade the female this way (Alberti 1991). The first step was apparently a reorganization of the female's ovary, favored by natural selection because it accelerated egg development, but which made sperm access more difficult via the ancestral route in the female. Sperm migration through the hemocoel became common, and the sperm acquired a derived, "ribbon" morphology. An additional possible consequence was the subsequent development of alternative insemination sites, commonly on the legs of females (Alberti 1991). This seems to be an especially dramatic case of a female-imposed "rule of the game" that originated due to natural selection and resulted in selection on male sperm transfer abilities (without information on female remating patterns it is not clear, however, whether sexual selection on males was involved; see chaps. 1.2, 2.7). A similar evolutionary pattern of repeated separation of insemination sites from the gonopore (eight separate evolutionary origins), and subsequent divergent migrations of insemination sites occurs in copepods (Huys and Boxshall 1991). It is apparently not known whether these animals also have invasive sperm.

Female Variations and Male Dimorphisms

In general, the common alternatives just discussed seem to be storage sites with good access to fertilizable eggs but poor conditions for sperm survival, versus sites with higher long-term survival probabilities but poorer access to eggs. The selective importance to males of this kind of tradeoff is evidenced by a number of male dimorphisms. For instance, long-term versus short-term storage adaptations occur in the copepod genus *Euchaeta* (Hopkins and Machin 1977; Hopkins 1978; Ferrari 1978; Ferrari and Dojiri 1987) and snowcrabs (*Chionoecetes opilio*: Beninger et al. 1993). Males

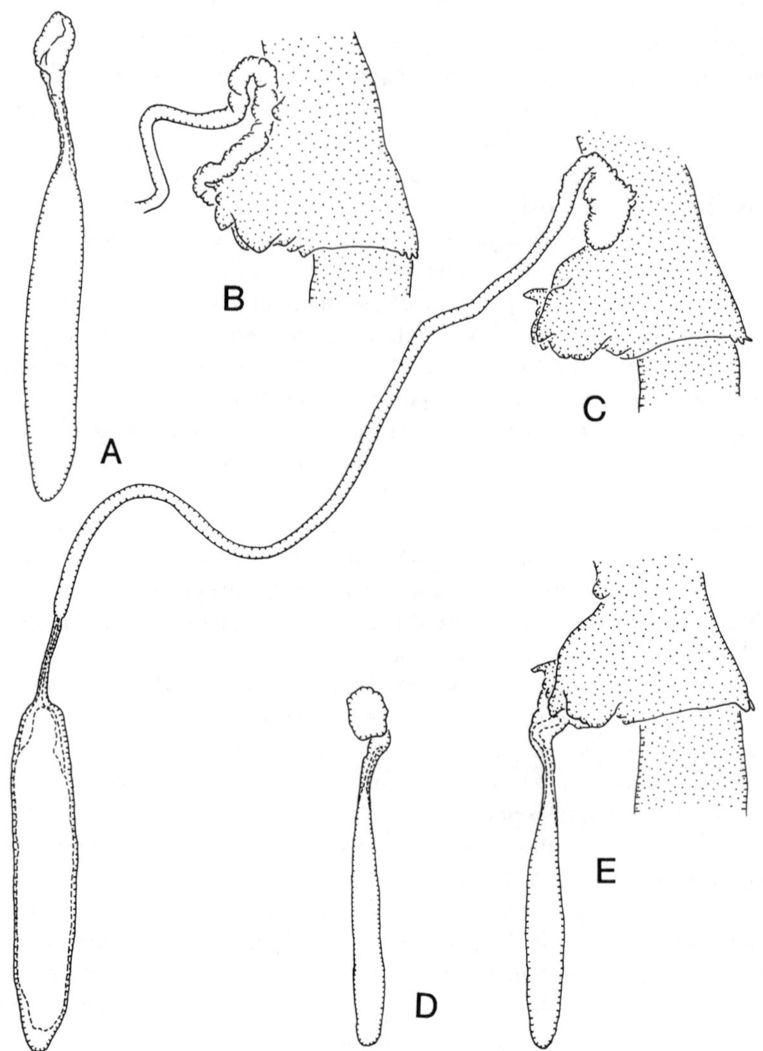

Figure 2.6 Dimorphic spermatophores (A–C and D, E) of the copepod *Euchaeta antarctica* exemplify strategies of long-term and short-term sperm delivery, thus emphasizing the importance to males of variations in sperm storage in females. Indirect evidence indicates that they function as follows. Smaller spermatophores (D,E) are attached directly to the female's gonopore (female stippled), transfer their sperm rapidly, and then fall off quickly. Larger spermatophores are longer (A from male, C on female) and are attached more firmly due to large attachment plates and the rugose female cuticle at attachment anterior to the gonopore. A large spermatophore stays attached until after the smaller spermatophore has fallen off, and later extrudes a process that enters the female's gonopore (B). The female may actively influence the time before a given spermatophore falls off (Blades and Youngbluth 1979). Males may often attach both types of spermatophore during a single mating. (After Ferrari and Dojiri 1987)

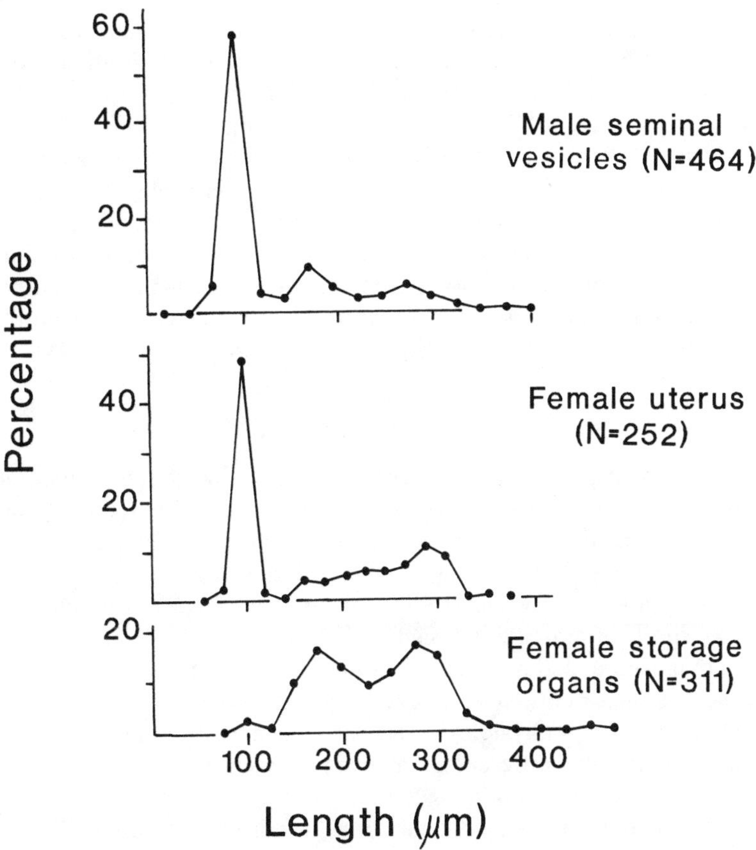

Figure 2.7 The polymorphic sperm of *Drosophila obscura* may have different functions. The proportions of shorter sperm in the male and his ejaculate are much higher than in the female storage organs, suggesting that different sperm morphs may function in different ways. (After Beatty and Sidhu 1969)

make two different types of spermatophore, one apparently specialized for long-term fertilization opportunities and the other for the short-term (fig. 2.6). Male *Loligo paelii* squid place their spermatophores either at sites within the female that offer good access to eggs but poor survival (near the distal end of the oviduct), or poor access but better survival (near the mouth). Spermatophores are placed in the oviduct only if the female is about to oviposit (Drew 1911).

In several groups (e.g., rotifers, molluscs, lepidopterans; see Baccetti and Afzelius 1976) sperm are polymorphic in form (e.g., fig. 2.7). In some cases alternative sperm forms may be specialized for other functions in addition to fertilization per se (e.g., penetration of the female in a rotifer:

Koehler 1965; perhaps inactivation of sperm from other males in lepidopterans: Silberglied et al. 1984; perhaps formation of a plug in the rat: Baker and Bellis 1988, 1989a). In others, different sperm forms may all be fertile, but specialized for alternative tactics for gaining access to eggs. The males of the *obscura* group of *Drosophila* have within-ejaculate dimorphisms in sperm length (Beatty and Sidhu 1969; Joly et al. 1991). The significance of such dimorphisms is unclear. In one species sperm lengths in the female's storage organs were quite different from those in the male's ejaculate (fig. 2.7) (Beatty and Sidhu 1969). In *D. teissieri* different geographic races have monomorphic or polymorphic sperm. Males with monomorphic sperm sire larger proportions of offspring than males with polymorphic sperm when both mated with at least some females (Joly et al. 1991).

Consequences

If accounts of copulation and fertilization tend to be oversimplified and incomplete, as I have just argued is probably the case for sperm storage, there may be multiple unreported ways by which sperm can reach storage and fertilization sites and bring about fertilization and the production of offspring in a given species. Multiple alternatives may exist in less well-studied aspects of reproduction such as induction of sperm transport, ovulation, oviposition, and rejection of further males (chaps. 6.2 and 6.3 give evidence of such multiplicity in chemical cues from the male's semen). Female morphology, physiology, and behavior may often affect the relative probabilities that different alternative processes will result in fertilization and offspring production.

In addition to the more subtle variations just discussed, twenty different possible cryptic female choice mechanisms with concrete examples are discussed in chapters 3 and 4, several of which are actually combinations of several mechanisms, and there are still others that are feasible. This multiplicity of female mechanisms reduces the likelihood that males will be able to evolve overall control of female reproductive processes. Consider, for instance, the possibility that the males in a given evolutionary lineage were able to seize complete control of one particular female process (as, for example, was previously thought to be the case in the mosquito *Aedes aegypti*, where a male transfers enough antiaphrodisiac substance to inhibit remating by >60 females: Craig 1967; see, however, Young and Downe 1982). Female *A. aegypti* can nevertheless exercise cryptic female choice by employing different mechanisms, such as sometimes failing to allow proper coupling for successful transfer of sperm (and antiaphrodisiac) by some males (Gwadz et al. 1971).

Multiple female control devices could lead to cycles of selection on different male abilities, with ability or lack of ability to influence one female mechanism altering the payoffs of additional abilities to influence other mechanisms. One expected result would be that a variety of different male and female control mechanisms may accumulate in given species. Another would be that the costs and benefits of different male and female tactics would change over time, making ESS analyses such as those of Parker (1984) and Knowlton and Greenwell (1984) more complex, and the endpoints of male games less stable.

Some female responses may exercise less intense selection on males than others. For example, in a species in which females remate frequently and males trigger vitellogenesis, a given male might benefit from the triggering effects of other males that had copulated with the same female. Other possible "transferable" effects include induction of oviposition, ovulation, and preparation of the uterus for implantation. On the other hand, other effects (chaps. 3 and 4) are probably "nontransferable." For instance, triggering of sperm transport mechanisms probably usually only benefits the triggering male.

2.3 Genetic Variance among Males

One way females may benefit from being selective in their choices of sires for their offspring is by obtaining better genes for their offspring—either genes that increase offspring viability ("good viability genes," often called "good genes"), or genes that increase the male offsprings' abilities to win out in sexual competition with other males (greater fighting abilities, or greater attractiveness to females—"good attractiveness genes," often called the "arbitrary traits" model; e.g., Maynard Smith 1991). Parker (1984) argued against the importance of genetic benefits to female selectiveness (in this case of overt precopulatory choice) because of the theoretical likelihood that such good genes will go to fixation. He argued instead that other, naturally selected advantages, such as avoidance of harassment by courting males, are what explain the tendency for females to remate. Neither idea (lack of genetic variability or importance of harassment) seems likely to be generally true.

Recent empirical studies have demonstrated that heritable variation occurs in sexually selected characters in nature (e.g., summaries in Cade 1984; Andersson 1994; see also Charalambous et al. 1994; Moore 1994; fig. 2.8). Andersson (1994) concluded that "with reservations for the problems of obtaining unbiased estimates . . . , sexually selected traits seem to have substantial heritabilities, as large as those of other morphological or behavioral traits" (p. 49). Most studies involved male characters under classic, precopulatory sexual selection, but a focused search for genetic variation for both the ability to displace resident sperm in the female and to

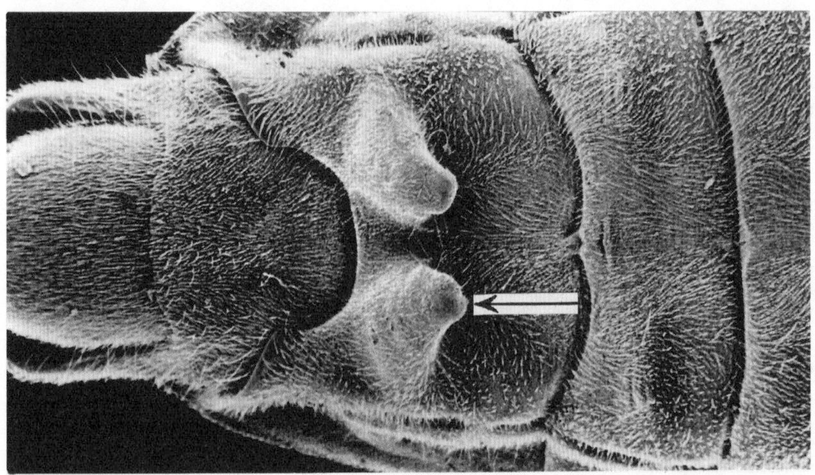

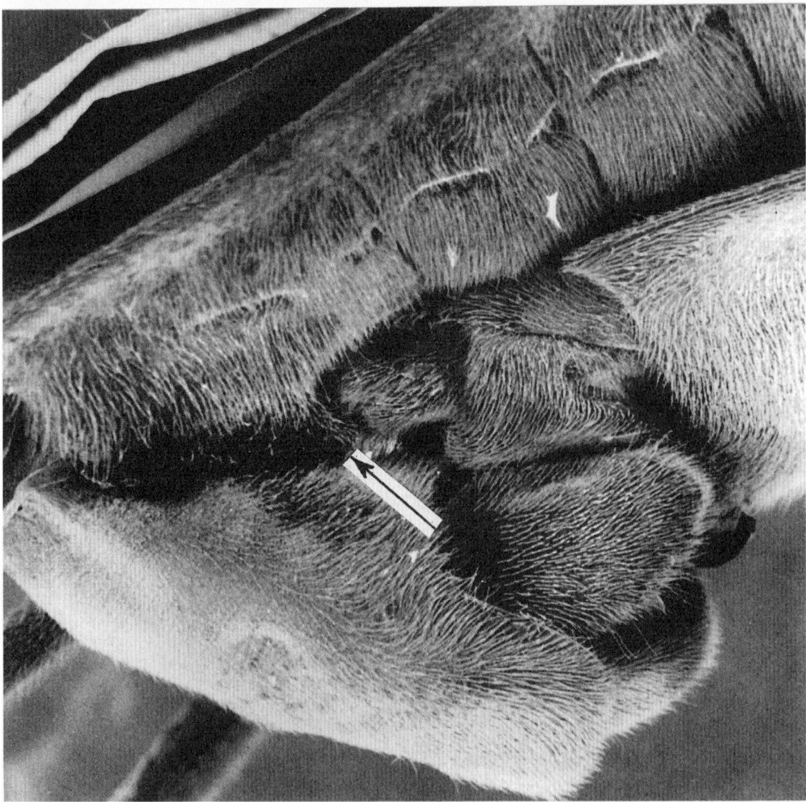

Figure 2.8 A male clasping device in the water strider *Gerris odontogaster* that is apparently under sexual selection by female choice, and in which there is genetic variation among males. When a male mounts a female and clamps her abdomen with his (*bottom*), the female always resists violently. She eventually ceases struggling if she is unable to

resist displacement also revealed substantial variation in natural populations of *D. melanogaster* (Clark et al. 1995). The existence of such genetic variation removes one of the reasons for doubting genetic payoffs to female choice (via either good viability or good attractiveness genes; Parker 1984; Williams 1992).

A second theoretical sticking point, the presumed improbability of a linkage disequilibrium between male display characters and female preference traits, has also been eroded by the recent discovery of correlations of this sort in several animal groups (Gilburn et al. 1993; Andersson 1994; Pomiankowski and Sheridan 1994; Charalambous et al. 1994).

Furthermore, it is important to realize that cryptic female choice mechanisms are expected often to persist even during periods in which genetic variability among the males is lacking. This is because females can come to depend on male signals to trigger their own reproductive processes. For instance, females of the bug *Rhodnius prolixus* are apparently unable to transport sperm from the bursa to the spermathecae unless a product of the male accessory glands is present in the semen (which induces muscle contractions in the female ducts; Davey 1958). Natural selection could favor female dependence of this sort because of reduced costs to the female, and/or reduced errors in triggering). Such dependence could maintain the female response to male signals during periods in which males did not vary genetically. Whenever new male variants arose, they could be subject to sexual selection on the basis of their differential abilities to trigger the preexisting mechanism of the female.

The same reasoning applies to periods during which cryptic female choice ceases to act because females copulate only once. Female responses to male cues could be maintained by natural selection while females were monogamous, and these preexisting responses would be available for competitive exploitation by males if females began to mate multiply. This persistence of female dependence on stimuli from males is equivalent to the continued need for sperm to trigger development (pseudogamy) in many parthenogenetic groups that were derived from sexually reproducing ancestors (e.g., Stenseth et al. 1985; Kirkendall 1990).

---

displace the male, and he then mates. Observations in captivity show that males with longer clasping structures (which are not significantly correlated with male size) are better able to resist female struggles; observations in the field confirm that the males found riding females have longer claspers than solitary males (but are not significantly larger). The female-imposed behavioral "rule of the game" (all mountings are followed by violent opposition), which probably results from natural selection on females to reduce predation risks (females are more vulnerable when carrying a male), results in some types of males having more opportunities to mate. It is not known whether this advantage results in greater rates of egg fertilization for males. (Photos courtesy of G. Arnqvist; bottom photo from Arnqvist 1989)

2.4 Conditions Favoring the Evolution of Cryptic Female Choice

William Brown (in prep. a) gives a general discussion of the conditions under which cryptic female choice is more or less likely to occur. He argues that sometimes precopulatory choice is difficult or impossible for females if males are able to force females to copulate ("convenience polyandry" of Thornhill and Alcock 1983). In such cases, he argues, cryptic choice will be the only mechanisms of discrimination available to the female, and thus be more likely to evolve. Birkhead et al. (1993a) make a similar argument regarding female traits that impede access of sperm to eggs. A second type of situation in which relatively indiscriminate mating by females (and thus cryptic choice) might be favored is when males donate physical resources to the female when they copulate (the "material benefit polyandry" of Thornhill and Alcock 1983) (W. Brown, in prep. a).

While these arguments are logically consistent, I believe they can give only rough, imprecise predictions. This is because, in the end, sexual selection by females is not logical. There is simply no "rule" that says that a female will use only a certain maximum or a certain minimum number of criteria in choosing mates. Male signals can be "piled on" each other (fig. 1.6) (West-Eberhard 1983; Iwasa and Pomiankowski 1994). For instance, male *Oecanthus nigricornis* crickets court females before, during, and after copulation using stridulation, several types of substrate vibration, pheromones, and gland products that the female ingests (Bell 1980), and females exercise both overt, precopulatory choice on the basis of some of these signals and cryptic female choice by ovipositing more rapidly after mating with larger males (W. Brown, in prep. a,b; see also chap. 3.7).

Male signals can also fall into disuse in the face of new ones. Some signals may even be ignored in some contexts but used in others (Zuk et al. 1992). The fact that precopulatory female choice appears to occur in many of the species in which cryptic female choice also occurs (chaps. 3, 4, and 5.1; Eberhard 1994) is evidence that female choice at one stage of the mating process does not preclude the potential for choice at a later stage. Similarly, in precopulatory courtship there is no necessary correlation (positive or negative) between the complexity of male display behavior and of male display morphology (Andersson 1994). In sum, the presence or absence of sexual selection (of whatever type) prior to copulation is probably at best a weak predictor of whether cryptic female choice is likely to occur.

It is worth noting in this context that the kind of male coercion that can lead to forced female "choices" prior to copulation (e.g., Smuts 1992; Small 1993 on male coercion in primates) is much less feasible for many cryptic female choice mechanisms. A male may well be able to force a female to hold still and allow him to copulate, but *forcing* ovulation, sperm transport, or preparation of the uterus for implantation is another story. In general, the male probably does not even know whether the female has cooperated with him in these processes. The failure of male birds to reduce their paternal investment when their copulations with her were not successful in siring all her offspring (e.g., Westneat 1995) is a possible illustration of this problem for males (see chap. 1.6). Thus cryptic female choice seems more likely to be free of the effects of males who harass females until they comply. In other words, females are probably usually less limited in their discriminations.

2.5 A Test Case: Bedbugs

The expectation that males will tend to end up "in control" can be tested in groups in which males have achieved "breakthroughs" in their interactions with females. The bedbugs (cimicids) and their relatives are an instructive example, since at one point in this evolutionary line males achieved unusually extensive control of fertilization. By developing hypodermic insemination (the ability to inject sperm into the female's body cavity at any one of many different sites), and sperm capable of negotiating their way through the female's body cavity to her ovaries, the males performed an "end run" around intermediate female structures. This insemination system is quite effective, as recently mated females typically have the bases of their ovaries swollen with large numbers of waiting sperm (Carayon 1966, 1975; possible variations are not mentioned, however).

Female cimicids nevertheless then apparently evolved a new set of structures in their oviducts and elsewhere to receive male genitalia and sperm. These tissues are made up of cells derived from cell types usually associated with combating infections, and their common effect is to *kill* sperm. In what are thought to be highly derived species, the flow of sperm reaching the base of the ovaries is reduced to a trickle, with most sperm left dead along the way (Carayon 1966). In these derived forms, females appear to have regained a great deal of control over the use of sperm.

Hypodermic insemination occurs in one other major group of insects, the small parasitic order Strepsiptera (Engelmann 1970) and several other smaller groups (summary in Eberhard 1985). Unfortunately, other detailed studies like those of Carayon are not available, so it is difficult to determine if female "countermeasures" have evolved. The available fragmentary evidence suggests perhaps not. In the strepsipteran *Acroschismus wheeleri* the spermatozoa pass along the brood canal and its ducts into the hemocoel, where free-floating eggs develop. Sperm persist in the coelom for up to

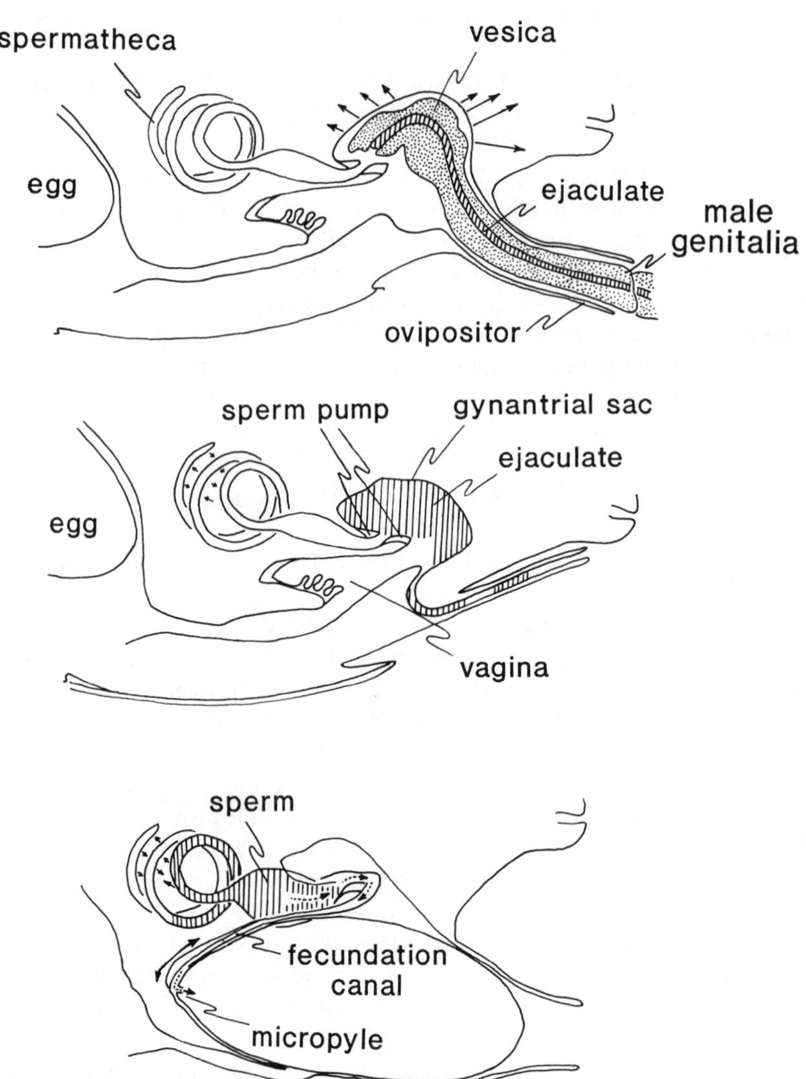

Figure 2.9 The complex internal workings of the "gynantrial complex" in the bug *Hebrus pusillus*, a derived character of gerromorph bugs that results in female control of both insemination and fertilization. During copulation the male genitalia (stippled in upper drawing) enter through the female's ovipositor, and the male's vesica reaches her gynantrial sac. The walls of the gynantrial sac are provided with numerous dilator muscles, whose contraction (arrows in upper drawing) apparently suck the ejaculate (cross-hatched) from the male genitalia as he withdraws after about 12 min of copulation (males lack a sperm pump). Movement of sperm to the spermatheca is then accomplished by the combination of the contraction and relaxation (small arrows in middle drawing) of muscles that compress and extend the coils of the spermatheca (indicated only diagrammatically here; the average number of loops was thirty-nine, and the spermatheca was

several weeks after mating (Engelmann 1970). The lack of detailed studies precludes the drawing of any confident conclusions in this case.

Another probable case in which increased female control has evolved is the female gynantrial complex, a derived characteristic of gerromorph bugs. Detailed morphological study of the hebrid bug *Hebrus pusillus* showed that "two primary functions of the gynantrial complex are to decrease access of males to eggs and to increase the role of the female in fertilizing them" (Heming-van Battum and Heming 1986, p. 160). The muscles of the female are arranged so as to draw the sperm from the male's genitalia with a pumping action, transport them to her spermatheca with another pump, and then stretch and release the canal through which sperm are pushed for fertilization (fig. 2.9). The gynantrial complex is thought to have originated to bring about fertilization of the proportionately large eggs of a gerromorph ancestor, and to have later acquired characteristics that increased female control (Heming-van Battum and Heming 1986).

2.6 Relationship between Cryptic Female Choice and "Sensory Traps"

One way in which male signals[2] can evolve under sexual selection by female choice is by mimicking stimuli that females have evolved, under natural selection, to use in other contexts. Thus, for instance, a male bee could increase his chances of attracting a female, or at least of getting her attention, by emitting an odor similar to that of flowers that the female visits to obtain pollen and nectar. Successful use of such a "sensory trap" (West-Eberhard 1983; Ryan 1990; Christy 1995) depends both on the male's ability to mimic the stimulus convincingly, and on the appropriateness of the female's preexisting response for promoting the male's reproductive interests (fig. 2.10). In both respects, evolution of male traits involved in triggering cryptic female choice processes seem particularly likely to involve the use of sensory traps.

---

substantially longer than the female herself!), the sperm pump at the entrance to the spermatheca, and perhaps movements of the sperm themselves. Fertilization occurs when an egg enters the vagina (bottom), and the female uses still other muscles to stretch the fecundation canal so that its opening is positioned opposite the micropyle of the egg. Contractions of muscles compressing the coils of the spermatheca (small arrows), and extensions (produced by muscles) and retractions (produced by elastic cuticle) of the fecundation canal, probably combined with movements of the sperm themselves, result in sperm moving to the micropyle and fertilizing the egg. Female control over insemination may explain why ejaculates in this species are "undersized" (only about one-third the amount needed to fill the spermatheca; see chap. 8.1). (Simplified after Heming-van Battum and Heming 1986)

Figure 2.10 A possible sensory trap used during copulation. A male lion bites the female on the nape of the neck during copulation. Lion cubs, like the young of many carnivores, cease moving and hang limp when their mothers grab them by the nape to carry them to another site. Ewer (1973) argues that the male may expoit a "sensory trap" when he grasps the female this way, increasing the likelihood that the female will not struggle and resist his advances. (After Ewer 1973)

As already noted in section 2.1, natural selection would often favor females' use of stimuli from the male himself. His sexual products, courtship behavior, and/or stimuli from the act of copulation itself probably often trigger many female reproductive processes. To "trap" the female, the male need only evolve to accentuate those aspects of his own behavior or morphology which females of his species are already using as cues. Such evolution is "easy," in the sense that few changes in the male would be necessary, so mutations and genetic recombinations to produce them are relatively likely to occur and may be relatively inexpensive in terms of reduced viability.

With respect to the appropriateness of female responses for favoring the male's reproduction, the processes involving cryptic female choice such as sperm transport, ovulation, oviposition, preparation of the uterus for implantation, and lack of sexual receptivity to other males can all be crucial determinants of male reproductive success. In addition, the direction of change in the female responses that are expected to evolve under natural selection is in each case the same as that which favors the male's reproductive interests (*increase* oviposition, *increase* sperm transport, *increase* ovulation, *decrease* receptivity to other males, etc.). The likely reproductive payoff to a male able to induce an increase in any of these female responses is obvious.

A related question concerns the persistence of sensory trap responses in females after males begin to exploit them. Evolutionarily sustained female responses in the face of changing male signals seem especially likely with cryptic female choice mechanisms. This is because one probable message communicated to the female ("You have been mated") by the male signals that trigger cryptic choice mechanisms will remain true, even when the particular signals used by the male change. In the previous example of a sensory trap, the male bee is "telling a lie" when he emits the floral odor: he is not really a flower. His chemical display is thus constrained to maintain a resemblance to the scent of flowers, if he is to continue to take advantage of the female response that evolved under natural selection. In contrast, a male copulatory signal informing the female "you have been mated" is seldom likely to be a lie in the context of natural selection. There is thus almost no restriction on the type of male signal; the male can convey this same message by tapping the female on the head with a leg, rubbing her genitalia with his, transferring a particular chemical in his semen, biting her on the leg, etc.

This combination of natural and sexual selection may often lead to evolutionary changes in female criteria responses. Continued association of male stimuli with reproductive responses would be favored by natural selection—for instance, by favoring inhibition of oviposition until copulation has occurred, and thus avoid laying unfertilized eggs. In addition, sexual selection could favor changes in the same female criteria due to the advantage of favoring preferential fertilization with sperm from those males best able to induce oviposition responses. The female would obtain sons better able to induce similar favorable responses in the next generation of females. Elaboration of male signals would not seem to have negative effects on females, as long as the males are not telling lies and continue to "mimic themselves" and male signaling per se does not reduce female reproduction. In some cases male signaling might reduce female reproduction if, for example, by lengthening copulation time to deliver more stimulation the male might expose the female to greater predation, or if male induction of oogenesis and oviposition were so powerful that some eggs were laid in suboptimum habitats or the female's reproductive lifetime were reduced (see Chapman et al. 1995; see also chap. 4.10).

In some cases (see chap. 6) male signals have evolved a second, more spectacular mimicry, in which seminal products mimic the female's own messenger molecules. This second type of mimicry may in effect short-circuit other portions of the female's response mechanisms, such as her sense organs and her central nervous system, by acting directly on the target organs such as the ovary, or on the ganglia controlling oviposition (see fig. 6.3). In these more direct, "ultimate" sensory traps, the male is mimicking the female's own internal signals, and the variation in the signal he sends is sharply limited by female sensitivity, which is presumably determined by receptor molecules in the membranes of her target organs. As I will show in chapter 6, this limitation is probably quite strict, because messenger molecules such as hormones are evolutionarily quite conservative. In contrast to the usual, indirect manipulation of female responses, the male is in this case attempting to "lie" ("this hormone molecule was sent by your own body because oviposition is the best thing to do now"). Natural selection on females exposed to such male substances may favor the evolution of traits that reduce her likelihood of being deceived, but such escape may be limited due to the importance of other responses to these same molecules.

2.7 Do All Female-imposed "Rules of the Game" Result in Sexual Selection on Males?

Female morphology, physiology, and behavior determine the playing field on which males must compete, and many of the rules by which they must play. It might be argued that this is of little import to understanding sexual selection. A great many female traits that determine the "playing field" may have never been involved in sexual selection. For example, the pH of the external portions of the female reproductive tract of some mammals is quite low, and the opening to the outside is relatively small. These characteristics mean that only males with relatively thin intromittent organs and strongly buffered semen can fertilize their eggs. But these female traits probably evolved in the context of preventing microbial invasion of their reproductive tracts (e.g., Birkhead et al. 1993a). Surely it is a mistake to discuss the evolution of such traits as vaginal pH in the context of sexual selection?

This may often be true from the female's point of view. From a male's point of view, however, any female trait that produces a bias favoring the reproduction of some males over that of others can potentially result in selection favoring those male characters that increase the male's chances of fertilizing eggs. Whatever the current function of such a female trait, if it has the effect of filtering or favoring some males over others (e.g., Birkhead et al. 1993a), it can be a factor in selection on males. Only if there is no variation in male reproduction due to variation in the ability to cope with female rules will the rules fail to result in selection.

When such filtering occurs, does it result in natural selection or sexual selection on males? Darwin's basic insight in distinguishing sexual selection from natural selection was that it involved *competition* between males. This is a crucial distinction, because intraspecific adaptations to competition (sexual selection) can give only transient evolutionary advantages. As a given trait that gives an advantage spreads through a population, any additional male trait that happens to give an additional advantage will be favored more strongly, and so on. The result is that sexually selected characters tend to evolve especially rapidly, and assume relatively elaborate forms (e.g., West-Eberhard 1983).

Thus, for instance, if the females of a species with a low vaginal pH mate with only a single male (i.e., only natural selection acts on males), any male able to buffer his semen just sufficiently to allow *enough* sperm to survive so that all of the female's gametes are fertilized by healthy sperm will have adapted to this natural selection. If, however, the female mates with more than one male, and if the differences in numbers of sperm from the two males that survive passage through the ducts with a low pH affect a male's chances of paternity (for example, if those sperm of his which survive the low pH are diluted by those of other males), then males will be under sexual selection. "Enough" sperm at the fertilization site may be much higher, as it will be influenced by the number of sperm from competing males, not just by the number of eggs to be fertilized. This selection is likely to result in more elaborate and effective abilities to promote sperm survival. This distinction is not made in some previous discussions of the evolution of male abilities to promote sperm survival in hostile female environments (e.g., Birkhead et al. 1993a).

Thus the female traits that mediate cryptic female choice may evolve due to advantages conferred by natural selection, or by sexual selection, or both. Although such distinctions are of obvious interest, I will generally not attempt to make them in this book. They may often depend on subtle, variable, and as yet little-studied characters such as the number of times females copulate, the time between copulations, and rates of sperm usage. My goal here is only to establish the simpler propositions that a variety of mechanisms that can produce cryptic female choice exist, and that they are common.

A related point is that the distinction between natural and sexual selection on males must be made differently for individual sperm cells than for ejaculates. Even if a female mates only once, the sperm in a single ejaculate nearly always outnumber the eggs. Thus, competition among sperm is almost inevitable, in the sense that one sperm's success in fertilizing an egg probably results in the failure of others. Whether or not it is an important force in nature depends on whether or not the genes in a sperm cell can manifest themselves in the sperm's phenotype. If they are completely repressed, as they are often said to be, with their nuclei highly condensed and containing DNA-binding proteins (e.g., Mulcahy 1975; Sivinski 1984; Poccia 1989; Cummins 1990), then the sperm's genes have their "hands tied" with respect to gaining any reproductive advantage over other sperm from the same male, and no selection on competitive abilities at this level is expected.

There are, however, some indications that sperm phenotypes are not always independent of their genotypes (see chap. 4.6; also Bellve et al. 1987). In addition, female discriminations may also occur after the sperm has entered the egg and its chromatin has decondensed, so that transcription from its DNA is possible (see chap. 4.11). It is thus of interest to sketch out the expected selective regimes, although no concrete examples are available. If sperm phenotypes do differ according to the genes they carry, then intraejaculate sexual selection could occur. If those sperm which are better at gaining access to eggs are also more likely to produce superior (or inferior) zygotes, this competition could have consequences for both the male and the female. There are apparently no data available on this point (Birkhead et al. 1993a). There is no correlation between visibly abnormal morphology and chromosomal abnormalities in human sperm (Cummins 1990). The possibility that competitively superior male reproductive cells give rise to superior zygotes has played a large role in discussions of similar phenomena in plants (see chap. 1.4). If this association occurs in animals, the reproductive interests of both male and female would be likely to be identical: both parents would be favored by production of offspring with those sperm producing superior zygotes. Both male and female will, for example, be favored by female traits that exclude from the vicinity of fertilizable eggs any sperm likely to produce a defective offspring (i.e., all sperm that carry nuclear damage from chiasmas). Male traits to overcome or counteract female characters that function to filter sperm in this way are not expected to evolve. In fact, the striking abundance of nucleolytic enzymes in the seminal plasma of mammals may be a male adaptation to help dispose of imperfect sperm (Mann and Lutwak-Mann 1981).

A second possibility is that some sperm are superior in achieving fertilizations but produce zygotes that are no better than those from other sperm. If females are strictly monandrous, there will be no advantage to either male or female in having these sperm win out over others. No female characters are expected to evolve to filter the sperm so as to accentuate differences.

If, on the other hand, females are polyandrous and sperm from more than one male compete to fertilize the same eggs, then sperm-level competitive adaptations could be favorable to both males and females. Males that produced larger proportions of more competitive sperm in their ejaculates would stand to obtain more fertilizations. Selection on females, through their sons' abilities to make superior sperm, could also favor female characteristics that increased the probability that only competitively superior sperm will achieve fertilizations. This, in turn, would increase selection on males to overcome such female-imposed filters. In sum, males are not to be expected to evolve to "overcome" female *intra*ejaculate filtering mechanisms, because their reproductive interests are the same as those of the female. But male abilities to overcome *inter*ejaculate filters would be favored, via sexual selection.

2.8 Good Genes, Runaway, or Endless Race?

There are several possible benefits to females of discriminating among prospective fathers of their offspring (Maynard Smith 1991; Williams 1992; Andersson 1994). In some cases females can obtain direct advantages, such as access to resources controlled by some males but not others, avoiding reproductive mistakes with asymmetric fitness consequences such as mating with incompletely fertile conspecific males or males of other species (Williams 1992), or avoiding male manipulations such as long pairings that result in increased predation or decreased foraging and have direct negative effects on female reproduction (e.g., Parker 1979; Arnqvist, in press; Alexander et al., in press).

Other females may obtain indirect benefits such as male genes that will give their offspring greater viability (for instance, greater resistance to parasites; "good viability genes"). Another type of indirect benefit is genes that will give their male descendants better abilities to induce females of future generations to allow them to father their offspring ("good attractiveness genes"). The relative importance of these factors is currently under debate.

A few possible direct benefits seem less likely to accrue to females exercising cryptic (as opposed to overt, precopulatory) choice. Thus males are less likely to be able to sense cryptic biases, and so would be less likely to reward such biases by granting increased access to resources they control. In general, however, there is no theoretical reason to expect that cryptic female choice will result in any one of these different possible female payoffs. There is, however, indirect empirical evidence that cryptic female benefits often include good "attractiveness genes." Male stimuli involved in cases of probable cryptic female choice often seem likely to be relatively poor indicators of male "vigor" or overall quality. They may differ in this respect with many classic, precopulatory male signals (Halliday 1987; see, however, Christy and Eberhard in prep.). Many probable triggers of cryptic female choice, such as genitalic thrusting, kicking, rubbing, tapping, licking, and rocking during copulation (see chap. 5), or transfer of hormonelike substances in the semen (chap. 6) involve relatively small movements or quantities of materials that apparently require little effort by the male, and are not typically accompanied by any female resistance (fig. 2.11). Such

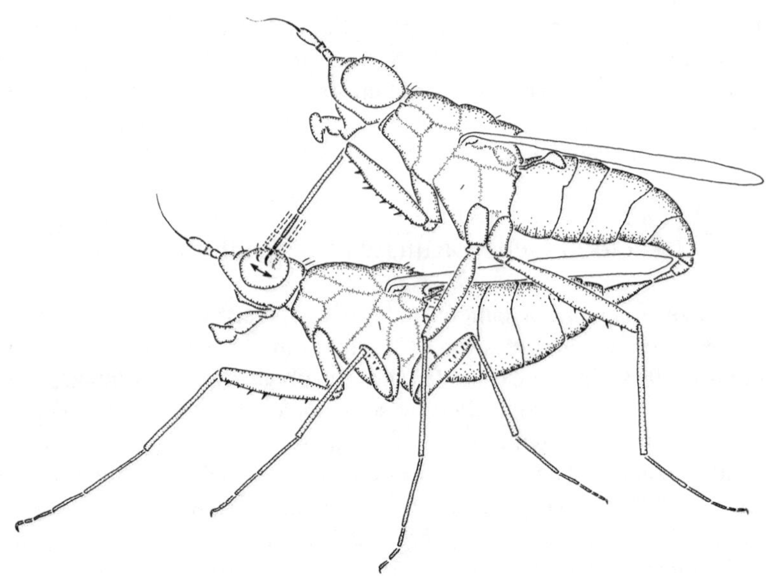

Figure 2.11 The relatively minor movements of copulatory courtship in the nereid fly *Glyphidops* sp. illustrate the improbability, typical of copulatory courtship, that male stimuli provide the female with reliable information regarding the male's vigor. During the approximately thirty-second copulations, the male alternately rubs one front tarsus and then the other with a jittery motion (dotted lines) on the motionless female's eye or her prothorax. The male then withdraws his genitalia, dismounts, and leaves. (Drawn from photographs and field notes; only the legs on the near side are shown)

traits seem unlikely to have any relation to the male's possession of superior "viability genes."

In addition, the differences in copulatory courtship behavior between closely related species (Eberhard 1991a, 1994) do not make sense in the context of measures of male "vigor." Take, for instance, the moths *Micropteryx* studied long ago by O. W. Richards (1924). During the approximately 3 min mating, the male of *M. calthella* raises his wings repeatedly, each time vibrating them for about a second. In *M. thunbergella* the male beats his wings briefly prior to copulation, but during the approximately 30 sec copulation he instead vibrates his abdomen. Why would brief wing vibrations be a good indicator of male viability in one species and abdominal vibrations in another? Unless one invokes additional selection on male communicatory abilities (i.e. "attractiveness genes"; e.g., Hill 1994) one must make the ad hoc argument that the male needs a strong abdomen in one habitat and strong wings in another. Even if this were true, the movements appear too trivial to really distinguish especially vigorous males.

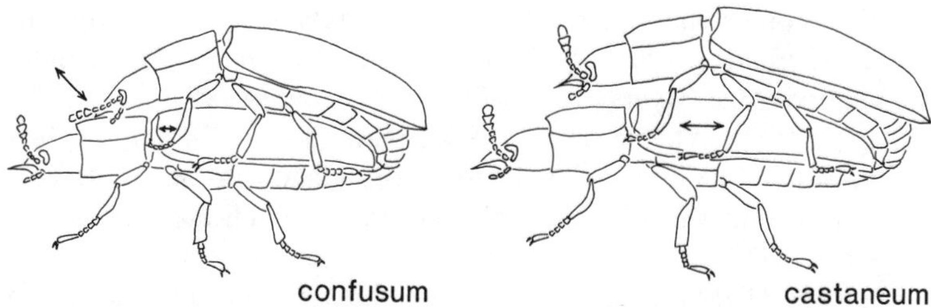

Figure 2.12 The differences in the copulatory courtship behavior of males of two species of *Tribolium* beetles are difficult to understand using good viability genes arguments. Neither the mouthpart tapping by *confusum* nor the two-legged elytral rubbing by *castaneum* (as opposed to one-legged rubbing in *confusum*) seems likely to convey reliable information to the female regarding male viability or vigor. And even if they did, there is no obvious reason, barring further discoveries such as a greater importance of simultaneous two-legged movements in the natural habitat of *T. castaneum*, why one type of movement would be a good indicator in one species and another in the other. (After Wojcik 1969)

The idea of correlations with different needs in different habitats becomes almost silly when the differences seen in other congeneric species are listed: the site on the female's body (tips of the elytra vs. elytra, abdomen, and legs) that is rubbed by the male's legs (R. Rodriguez and Eberhard, in prep., on two species of the beetle *Pseudoxychila*); the rhythm with which the male taps her (postcopulatory displays in *Nasonia* wasps: van der Assem and Werren 1994); rubbing the female's elytra with leg I instead of tapping them with legs II and III in *Omaspides* beetles (V. Rodriguez 1994c); side-to-side rocking instead of forward-backward movement (several species of *Phyllophaga* beetles: Eberhard 1993b); vibrations that accompany rubbing in one species but not in another (R. Rodriguez and Eberhard 1994 on two species of *Xyonysius* bugs); body lifting instead of antennal vibrations in two species of *Brachypnoea* beetles (Eberhard 1994); tapping instead of vibrating the legs on the elytra in both the ladybird beetles *Epilachna* and the scarabs *Strigoderma* (Eberhard 1994; see this reference for further examples) (fig. 2.12).

Intraspecific differences in copulatory courtship that have been documented seem equally arbitrary: wing buzzing in one population of the bee *Nomia triangulifera*, antennal tapping in another (Wcislo et al. 1992); shuddering the entire body in one population of the weevil *Nicentrus lineicollis* and rubbing the female's abdomen with the middle and hind legs in another (Eberhard 1994); vibrating the abdomen rapidly in one population of the stinkbug *Mormidea notulata* and shuddering the entire body in another (Eberhard 1994). Similarly trivial intrageneric differences occur in genitalic thrusting behavior (e.g., see fig. 5.12; table 5.1), which also probably often functions as courtship (Dewsbury 1988; see also chap. 5.2).

Male genitalic characters, which are probably also under cryptic female choice (Eberhard 1985; chap. 7.2), are generally relatively poor indicators of male body size (Byers 1976, 1983, 1990; Coyle 1985a; Eberhard 1985; McAlpine 1988; Wheeler et al. 1993; Eberhard et al., in prep.). The often relatively small amounts of male accessory gland substances affecting female reproductive processes also seem unlikely indicators of male vigor (chap. 6.7; see, however, Cordero 1995 for other views on accessory gland substances). There are, admittedly, exceptions. For instance, the up to 6 hr preinsemination copulations of the spider *Neriene litigiosa* are energetically expensive (Watson and Lighton 1994), and in several other insects and arachnids the males make significant investments in seminal products and are unable to remate or make similar investments for substantial periods until these products are replenished (e.g., Gwynne 1984a, 1988a; Wedell and Arak 1989; Simmons and Bailey 1990; Gwynne and Simmons 1990; Gwynne and Brown 1994 on orthopterans; Drummond 1984; Oberhauser 1989 on lepidopterans; Polis and Sissom 1990 on scorpions). But in general neither the behavior patterns nor the structures and chemicals involved in copulation seem usually designed to be reliable indicators of superior viability genes in the male.

It is necessary to acknowledge that, at least in the best-studied species, *Drosophila melanogaster*, genes affecting male courtship characters may be especially prone to have pleiotropic effects (Hall 1994). The pleiotropic traits often have little or no obvious relation (e.g., the song mutants *cac* and *diss* also produce abnormalities in the visual system; the *yellow* and *ebony* body color mutants are associated with abnormal courtship; Hall 1994). It is thus possible that I am mistaken in deducing that there is a general lack of association between species differences and male viability. However, most behavioral traits whose genetic control has been studied in *D. melanogaster* involve relatively crude behavioral differences as compared with documented differences between species in copulatory courtship. It is thus not clear whether the *Drosophila* studies are applicable in this discussion.

In sum, the supposition that these male behavior patterns are correlated with overall male vigor cannot be eliminated definitively, but it seems doubtful.

It is thus difficult to explain why such male displays so often diverge in closely related species using just the good viability genes argument, unless periods of selection favoring good attractiveness genes also occur (Hill 1994). One alternative possibility would be to invoke genetic drift. Genetic drift explanations seem unlikely, given the potentially huge effects on male fitness of the female reproductive processes involved, and the consistency with which these (as opposed to many other characters) diverge intragenerically (chap. 5).

A further possibility is the formerly popular but more recently strongly criticized idea that the male displays function to allow the female to determine the male's species identity (for arguments on this point with reference to classic, precopulatory courtship displays, see Paterson 1978, 1982; West-Eberhard 1983; Andersson 1994). Species isolation arguments also seem unlikely in the case of copulatory courtship, however. Species discrimination would be much more profitable, and thus more likely to evolve in both males and females, when carried out early in sexual interactions (e.g., Alexander 1964; Eberhard 1985; Shapiro and Porter 1989).

It is important to remember that what females gain from being selective is a separate issue from the main argument being advanced in this book, which is primarily concerned with showing that there are cryptic *mechanisms* by which females may exercise their selectivity, and that such selectivity is common. Female discriminations via the cryptic choice mechanisms listed above could be made on the basis of male resources, male vigor, fertility, species identity, or arbitrary male stimuli that have no relationship to other selectively important characters. The theme of this book concerns the mechanisms by which females choose and the fact that they do so, not the ultimate benefits they obtain in making such choices.

2.9 Summary

Several important discussions of evolutionary resolutions of male-female conflict over sperm usage patterns underestimated the probability that females would retain at least partial control. Females are probably often preadapted in several ways to influence events associated with male-male ejaculate competition, due to natural selection in other contexts such as efficiency of sperm usage, avoidance of infections of their reproductive tracts, avoidance of loss of eggs through lack of fertilization, and protection of their eggs from defective or overly abundant sperm. In fact, some ancient adaptations that presumably increase sperm numbers and mobility make them relatively helpless and dependent after they reach the egg, suggesting that female ability to bias crucial reproductive processes may be extemely old. Utilization of such potential discriminatory mechanisms in the context of selectively favoring some males over others probably has little or no energetic cost for the female, and may even sometimes result in energetic benefits.

Overly simplified descriptions and fertilization myopia in many accounts of copulation and sperm transfer probably obscure a number of additional contexts (such as alternative pathways for sperm to reach eggs) in which female-controlled morphology and behavior can bias a male's chances of paternity. In addition, the multiplicity of different female mechanisms affecting sperm usage makes total male control, or even control of a single process, unlikely.

The sensory trap model of sexual selection by female choice is likely to apply in many cases of cryptic female choice because several reproductively important female responses to male stimuli associated with copulation and insemination (e.g., transport of sperm, initiation or acceleration of oogenesis, ovulation, initiation of oviposition, reduction of receptivity to future mating attempts, preparation of uterine lining for implantation) are very likely to evolve under natural selection. Evolutionary modification (accentuation) of these stimuli is especially feasible for males, since (1) the males themselves are the sources of the original stimuli; and (2) the polarities of female responses likely to be favored by natural selection (e.g., *increase* oviposition, *reduce* receptivity to other males) are in all cases those that would favor the male's reproductive interests. In addition, male signals triggering cryptic female choice may be unusually free to evolve in diverse directions, because they may often be free of the need to resemble the "trap" stimuli to which females originally responded. This is because the message to the female that is being "mimicked" and elaborated upon by the male ("you have been mated") can continue to be communicated even though the male signal changes radically.

A female could benefit from exercising cryptic choice in a variety of ways, and thus the current debate over the relative importance of different hypothesized advantages to females is independent of the general argument that cryptic female choice is widespread. In fact, mechanisms of cryptic female choice may be maintained by natural selection, due to the evolution of female dependence on males for signals to trigger reproductive processes, even during periods in which sexual selection is not possible (lack of genetic variation in males, strict female monandry). The details of the male stimuli that probably trigger cryptic female choice mechanisms do suggest, however, that benefits from good "attractiveness genes" are frequently involved. Both the male signals themselves, which presumably function to influence cryptic female choice (see chaps. 5–7), and the differences between them in closely related species generally seem likely to be poor indicators of overall male vigor.

Notes

1. It is perhaps also useful to note in passing that there are certain conditions that might be thought to favor cryptic female choice, but which will not have this effect. Some authors have proposed relatively indirect payoffs for female control which, as noted by W. Brown (in prep., a), are unlikely. Drummond (1984) and Gwynne

(1984c) argue that in species in which males make nutrient donations to females during copulation, female spermathecal morphology that results in disproportionate use of sperm from her latest mate would evolve as a reward for males. It is argued that females with such spermathecal morphology would accrue nutritional benefits, because selection would favor increased tendencies in males to mate with nonvirgin females. While male donations would indeed be expected to be more selectively advantageous to males if second male sperm precedence (P2) values are high, this would be a secondary effect of spermathecal morphology resulting in high P2 values, not the immediate reproductive payoff needed to explain why the morphology arose in the first place.

2. Strictly speaking, a "cue" (a male trait that has not evolved under selection to increase its stimulatory properties that release female responses but is used nevertheless by the female to trigger a response) can evolve into a "signal" (a trait that has been selected to increase its response-triggering properties) (see Williams 1992 for a discussion of this distinction). The line between the two is easily crossed, and the distinction may be subtle. I will generally sidestep this issue by using the more general term "stimulus."

3 Principal Mechanisms of Cryptic Female Choice

3.1 Criteria

Previous chapters listed possible mechanisms with which a female could theoretically influence a male's chances of fertilizing her eggs after allowing him to begin copulation. The question remains whether any of these mechanisms are actually used in ways that would result in cryptic female choice. This chapter and the next examine specific cases that document female discriminatory behavior, the male traits to which females respond, and variations in reproductive payoffs to different males.

The chapters are organized by possible mechanisms of cryptic female choice, with those which appear to be more general mechanisms included in this chapter. Other, possibly less common mechanisms, which are documented in fewer species or which depend on less ubiquitous female traits, are presented in chapter 4, which also includes a discussion of both chapters. One or more species is discussed for each mechanism. These species were chosen either because they are especially well studied or offer particularly dramatic illustrations of certain points. Species that may use the different mechanisms are listed in tables 3.1–4.8.

Keep in mind that making a strong case that a given female characteristic results in sexual selection by female choice (cryptic or not) is necessarily complex. Evidence should show at least the following:

1. Female responses to some conspecific males differ from those to others (if females respond equally to all males, no selection can occur).
2. Such discrimination occurs under natural conditions (if females in nature mate only once, for example, cryptic choice among males in captivity would be biologically irrelevant).
3. The discrimination results in differences in reproductive success for the males involved (if, for instance, the first male to mate with a female always obtains all fertilizations, female discrimination among subsequent males would have no reproductive significance).
4. Female biases are associated with particular male characteristics (if female favoritism is bestowed randomly on different males, it will have no selective effect).
5. Variation among males in characters used by females to discriminate is associated with genetic differences (otherwise female discrimination

will have no evolutionary effect on males). This last criterion is "optional" in the sense that it need not be fulfilled to demonstrate that female choice has occurred in the past.[1]

Most of the examples in the text and in tables 3.1–4.8 are tentative, in the sense that not all conditions for a strong demonstration are fulfilled. The lack of positive evidence on some points is in nearly all cases due not to data which suggest other conclusions, but to lack of appropriate studies. Similarly, even the best-studied and most widely cited cases of probable precopulatory sexual selection by female choice, such as the male tail feathers in widowbirds and swallows (Andersson 1982; Møller 1988) or the songs of túngara frogs (Ryan 1985), do not include demonstrations of all these points (see Andersson 1994 for a summary).

Also keep in mind that in none of the examples is the male character that is thought to be involved in cryptic female choice necessarily the only factor involved in triggering reproductive decisions by the female. In many cases, for example, the male also performs courtship before the character thought to be involved in cryptic female choice is brought into play. For example, a female *Metaplastes ornatus* katydid responds (or fails to respond) to male calls prior to copulation by approaching from a distance, before subsequently responding to movements of his subgenital plate in her oviduct during copulation by emptying her spermatheca of sperm from previous males (von Helversen and von Helversen 1991). Many processes in animal behavior, as in embryology, involve a sequence or hierarchy of decisions, and female reproductive behavior is no exception.

3.2 Sometimes Discard Sperm of Current Male

In many animals only a fraction of the sperm transferred by the male to the female is actually taken up and stored in the female's reproductive tract. In some species a portion of the male's ejaculate is emitted from the female either during or immediately following copulation (fig. 3.1; table 3.1). Close study of some species has shown that the proportion of different males' ejaculates that is dumped in this way is not constant, and that this variation can have reproductive consequences for males.

Perhaps the best-studied species in which the female discards sperm from the current male during or immediately following copulation is the fly *Dryomyza anilis*. Mating behavior typically occurs as follows (Otronen 1984a,b, 1989, 1990). The male mounts the female near an oviposition site, and the pair then moves a short distance away, where copulation occurs. The male begins copulation with an intromission that lasts just under a minute, during which he transfers sperm to the female. He then withdraws his genitalia and taps rhythmically on the female's external genitalia with

Table 3.1
Possible Cases of Cryptic Female Choice in Which the Female Sometimes Discards the Sperm of the Current Male

	Criteria						
Species	*I*	*II*	*III*	*IV*	*V*	*VI*	*References*
Rhabditis 2 spp. (Nematoda)	Y	Y	Y?[a]	?	?[b]	?	Rehfeld and Sudhaus 1985
*Coenorhabditis elegans** (Nematoda)	Y	?	Y?[c]	?	?	?	Barker 1994; Ward and Carrel 1979
*Chelymorpha alternans** (Coleoptera)	Y	Y	Y[d]	Y	Y (genital size)	?	Rodriguez 1994a; V. Rodriguez et al., subm.
Macrohaltica jamaicensis (Coleoptera)	Y	Y	Y[e]	Y	?[b]	?	Eberhard and Kariko, in prep.
Culicoides melleus (Diptera)	Y	Y	Y[e]	Y	N?[f]	?	
*Dryomyza anilis** (Diptera)	Y	Y	Y[g]	Y	Y (No.genit.taps)	?	Otronen 1990; Otronen and Siva-Jothy 1991
Physocyclus globosus (Araneae)	Y	Y?	Y[g]	Y?[h]	?[b]	?	Eberhard et al. 1993; W. Eberhard, unpub.
Pholcus phalangioides (Araneae)	N(?)	Y	?	?	?	?	Uhl et al. 1993
Gallus gallus (Aves)	Y	Y	?	?	Y (soc. status)	?	Thornhill in Birkhead and Møller 1992
Equus spp. (Perissodactyla)	?	?	?	?[i]	?	?	Tyler 1972; Ginsberg and Rubenstein 1990
Oedocoileus virginianus (Artiodactyla)	Y	Y?	?	Y?[j]	?[b]	?	Warren et al. 1978
Macaca mulatta (Primates)	?	?	?	Y	?[b]	?	Kaufmann 1965
*Homo sapiens** (Primates)	Y	Y	Y?[a]	Y	Y (female orgasm)	?	Baker and Bellis 1993a,b

Notes: Those marked with (*) are discussed in the text.

The criteria are as follows: I: Do females treat some males differently from others? II: Are differences in female treatment of males likely to occur in nature? III: Do differences in female treatment of males result in differences in male reproduction? IV: Do females mate with more than one male in nature? V: Do females use male characteristics to determine how to treat males? VI: Are variations in the male character used by females determined genetically? (Y = yes; Y? = probably yes but no direct data; N = no; ? = no data available)

[a] Female response alters number of sperm transferred, but sperm precedence patterns are not known.

[b] Possible cues for discriminating among males were not checked.

[c] So much sperm was dumped, or so few sperm were found in the female afterward, that reduction in male paternity probabilities seems inevitable. Doubly mated females produced offspring of both males.

[d] Since sperm numbers in the female are strongly modified by the female response, and since precedence is thought to be influenced by relative numbers of sperm from different males, sperm emission presumably has an effect on male reproductive success.

[e] Females sometimes eject spermatophores partially or completely full of sperm. The pattern of sperm precedence is unknown.

[f] Female behavior varies in accord with whether the male has mated recently. If there is variability among males in the ability to recover more quickly from mating, this female criterion could produce discrimination among males.

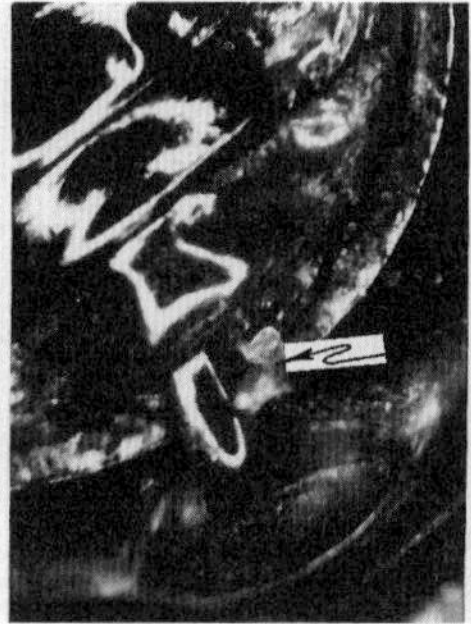

Figure 3.1 Love's labor lost. *Left*: A female Grevy's zebra (*Equus grevyi*) ejects a large volume of liquid (arrow), presumably semen, while the male is still dismounting after copulating. Females of this species sometimes mate with more than a single male during one estrous period, so sperm dumping could influence a male's chances of fathering the female's offspring (the volume emitted in one case was estimated at 300 ml). In the probable sister species, *E. burchelli*, females are monogamous and do not eject semen following mating. (From Ginsberg and Huck 1989)

Center: Sperm ejection in a pair of copulating *Chelymorpha alternans* beetles (ventral view of male and female abdomens). The arrow indicates a mass of sperm that emerged from the female during copulation. Sperm ejection is associated with a nearly twofold reduction in the number of sperm that are eventually stored in the female's spermatheca. Ejection sometimes occurs during copulations with virgin females, and since the female's spermatheca holds about two times the amount of sperm in a male's ejaculate, ejection is not due to lack of space. Ejection in matings with virgin females is more frequent if one portion of the male's genitalia, the flagellum, is shorter. (V. Rodriguez 1994a; V. Rodriguez et al., subm.)

Right: Sperm ejection in the pholcid spider *Physocyclus globosus* (posterior view of the abdomen of a female just after copulation). The encapsulated sperm, which emerged during the latter portion of copulation, are visible as a white mass (arrow) on the abdomen of the female. In at least some cases the sperm belong to the copulating male, since sperm ejection sometimes occurs in copulations with previously unmated females. Similar masses of sperm are also ejected from the female in a related spider, *Pholcus phalangioides*. (Uhl et al. 1994)

---

[g] Direct paternity analysis was made with the sterile male technique.

[h] Females are known to associate (and probably mate) with several different males in the field (Eberhard 1992b), but mating was not checked directly. Captive females remated readily. Overall sperm precedence pattern with two matings suggested sperm mixing in the female, but showed very large variance.

[i] Large amounts of liquid, apparently semen, emerged immediately after the male withdrew his penis. Free-ranging females sometimes mate with different males during a given estrus.

[j] A viscous liquid (presumably semen) dripped from the vagina immediately after at least six of forty-four copulations, and three does urinated immediately after mating. Liquid "usually" dripped from the female after the second or third copulation on the same day. Data on remating were only for captive females.

his genitalic claspers, and simultaneously presses her abdomen with his hind legs in the same rhythm (the number of taps averages 22 ± 6, mean ± SD). The male performs several tapping sequences (8.7 ± 5.2 in the first copulation bout, and lower numbers later, with a range of 8–31 in an entire mating). Each tapping sequence ends with a longer clasper contact (6 ± 2 sec), unless the female is successfully resisting tapping (see below).

Following a copulation bout, the female moves back to the oviposition site and begins to lay eggs, during which time she is usually guarded by the mounted male. Before oviposition begins, the female discharges a droplet—about the size of a single ejaculate—which contains sperm. After the female has laid some eggs, the pair moves away again for another copulation bout. The male is apparently responsible for the termination of oviposition bouts that often end with renewed copulation attempts and the male fluttering his wing as if to fly (M. Otronen, pers. comm.). Often the pair again moves away from the oviposition site. Pairs repeat this cycle of copulation and oviposition up to six times. Fewer eggs are laid in the first oviposition sequence than in later sequences.

There were several indications of male-female conflict during copulation bouts. The female sometimes resisted both intromission and tapping by attempting to walk and by lowering her abdomen to make genitalic contact difficult or impossible; struggling pairs sometimes shook vigorously and rolled on the ground. The male occasionally dismounted after several unsuccessful attempts to tap. The strength of female resistance varied, and sometimes resistance only began after several tapping sequences had already occurred. Larger females resisted more often and spent less time in copulation bouts. Larger males succeeded in performing more tapping sequences than smaller males.

By pairing females with two different males, one of which had been exposed to a sterilizing dose of radiation, Otronen tested the effect of male postcopulatory behavior. If the second male was removed after he had performed a lower number of postcopulatory tapping sequences, his sperm fertilized a lower percentage of the eggs that the female laid following this copulation. Average paternity ranged from 18% (no tapping sequences) to >70% (25 tapping sequences) (Otronen 1990; fig. 3.2). Within a given oviposition sequence the percentage of fertilization by the second male gradually declined. This probably explains why males interrupted oviposition repeatedly to copulate and then tap the female again (Otronen 1994).

Female *D. anilis* thus exercised cryptic choice in at least two different ways after being inseminated; both mechanisms may favor larger males, and thus reinforce precopulatory selection favoring large males (Otronen 1993a). They biased sperm usage (perhaps by putting a larger or smaller proportion of the present males' sperm in the drop they discarded just before ovipositing; Otronen and Siva-Jothy 1991) in accord with whether or

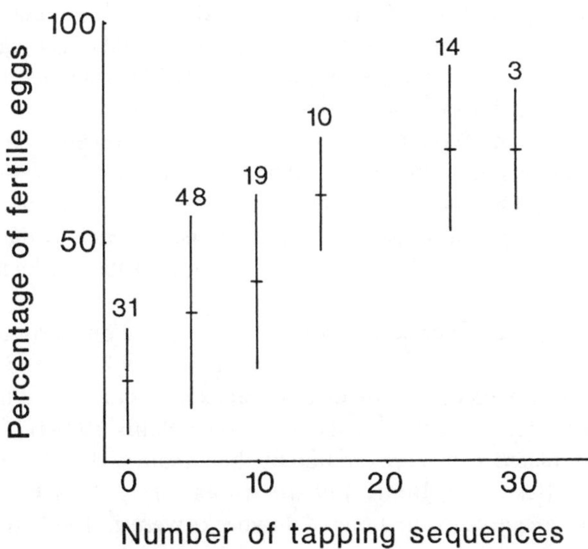

Figure 3.2 Fertilization rates (averages, with standard deviations and sample sizes) give experimental proof of cryptic female choice in the fly *Dryomyza anilis*. A fertile male was allowed to mate with a female that had previously mated with a sterile male. The fertile male's success was indicated by the percentage of fertile eggs laid by the female. In this species the male taps rhythmically on the tip of the female's abdomen with his genitalic claspers following copulation. The female then emits a droplet of sperm just before oviposition, containing sperm from both the current male and previous males. Fertile males were allowed to perform predetermined numbers of postcopulatory tapping bouts before being removed, and the female was then allowed to oviposit. Females that received more tapping bouts laid more fertile eggs, apparently having discarded a larger proportion of sperm from the previous male before beginning to oviposit. Females resist male tapping by bending their abdomens ventrally, making it difficult for the male to tap, and males usually dismount after several unsuccessful attempts to tap, so the numbers of males that succeeded in performing larger numbers of tapping sequences were smaller. (After Otronen 1990)

not postcopulatory tapping sequences occurred. They also determined, at least to some extent, the number of tapping sequences. By resisting more or less strongly, a female could influence the number of tapping sequences a male was able to perform and presumably thus bias the level of sperm precedence he achieved. Larger males also achieved higher fertilization success when the number of tapping sequences was artificially held constant (Otronen, 1994b), but it is not yet certain whether different amounts of sperm from the males or different female treatment of the sperm was responsible.

Otronen and Siva-Jothy (1991) studied the mechanism within the female by which postcopulatory tapping increased fertilization success, using males whose sperm was radioactively labeled. They found that a large fraction of the sperm discarded in a droplet belonged to the last male, and that a greater proportion of the sperm in the last male's ejaculate was retained in the female's bursa when the male was allowed to tap the female (12.8 ± 6.1 tapping sequences) than when he was prevented from tapping her. The relationship of these data to the biases in egg fertilization is uncertain, however, since it is not certain where eggs are fertilized in this species. Recently mated females had sperm in the bursa, in each of the three spermathecae, and (apparently in smaller numbers) in the oviduct just above the bursa.

Humans offer another particularly intriguing case of sperm dumping, in which it appears that the male's behavior during copulation and perhaps his morphology influences how much of his ejaculate is discarded in the "flowback" soon after copulation ends (Baker and Bellis 1993a). On average, 35% of ejaculated sperm are emitted by the female within 30 minutes after copulation (Baker and Bellis 1993a). A total of 127 flowbacks and 109 additional subjective evaluations of flowbacks were collected by thirty-two different women. The number of sperm in the flowback was compared with the anticipated total number in the male's ejaculate, which was calculated using a formula taking into account the midpoint of the number of sperm in observed ejaculates of that male, the time elapsed since his last ejaculation (corrected for whether ejaculation resulted from masturbation or copulation), the percentage of time the pair had spent together since their last copulation, and the female's weight (these factors were found to influence sperm numbers in ejaculates in a separate study in which ejaculates were collected in condoms; Baker and Bellis 1989b) (if ejaculates are smaller when condoms are used, these data will give overestimates of the fraction represented in flowbacks).

The estimated proportion of the ejaculate discarded in the flowback was lower when the woman experienced an orgasm during or following ejaculation than when she did not (Baker and Bellis 1993a). The smallest relative flowbacks occurred when the woman's orgasm took place soon after the man's ejaculation. Orgasm in women is known to involve contractions of the walls of the outer third of the vagina, which create a higher pressure in the vagina than that in the uterus (Masters and Johnson 1966; Jones 1984). The temporary coagulation of human semen that normally occurs about 1 min after ejaculation is thought to be a male adaptation to prevent or minimize flowback (R. Jones 1984).

The data on flowbacks cannot be related directly to the chances of fertilization (and in fact, the only conception that occurred during the Baker and Bellis study apparently resulted from a copulation in which the woman did not experience orgasm and in which flowback was relatively large). In addition, other details of copulation, such as when the male genitalia are withdrawn from the female, can also influence flowback (Masters and Johnson 1966). Nevertheless, it seems probable that at least some flowbacks, which ranged up to nearly 100% of estimated ejaculations, reduce the probability of conception. Thus female responses to male stimulation during copulation probably play a role in determining the chances of conception in humans, as had been predicted (Eberhard 1991b).

An additional survey of human copulation behavior showed that the probability of female orgasm, and in particular those orgasms most likely to retain sperm (those which occur simultaneously with or just following ejaculation), was greater if the man's fluctuating asymmetry index was lower (the symmetry of the two sides of his body was greater; Thornhill et al., 1995). Such symmetry has been associated with increased male mating success in other species (Watson and Thornhill 1994), and may be associated with greater precision in developmental designs (i.e., perhaps superior viability genes; Polak and Trivers 1994). Other male traits, including dominance status and stimulation during copulation, correlate with the probability of female orgasm in the monkey *Macaca fuscata* (Troisi and Carosi 1994; see chap. 5.2).

Flowback of semen in other mammals may also be under female control. Thus in artificial insemination studies of the house mouse, *Mus musculus*, high cervical tension was correlated with reduced amounts of the insemination fluid leaking from the vagina (Leckie et al. 1973). Similarly, in the domestic pig *Sus scrofa*, use of an artificial penis with a helical tip similar to that of the male penis reduced the amount of flowback (Evans and McKenna 1986). Flowbacks occur in several other mammals (e.g., Morton and Glover 1974a,b on rabbits; Ginsberg and Rubenstein 1990 on zebras [fig. 3.1]; Birkhead et al. 1993a also list monkeys, rats, pigs, and sheep), and female "orgasms" that involve energetic peristaltic contractions of the vagina or other portions of the female reproductive tract moving their contents deeper into the female also occur in other nonhuman mammals (VanDemark and Hays 1952 and references therein on cattle and rabbits; Evans 1933 on the dog *Canis familiaris*).

Still another dramatic case of sperm ejection occurs in the soil nematode *Coenorhabditis elegans*. Direct observations under the microscope of copulations involving virgin females showed that in 42% of 102 copulations the female apparently rejected some or all of the recently deposited semen from the uterus: "The vulva would open and the entire mass of seminal fluid would appear to be blown out of the uterus under pressure. This usually resulted in the spicules [intromittent organs] of the male also being blown out of the vulva" (Barker 1994; see also Ward and Carrel 1979 on this species). While the criteria responsible for rejection by females are yet to be established, the high frequency of rejections and the often sweeping nature of the rejection leave little doubt that this female behavior has significant effects on a male's chances of paternity.

Sperm ejection is particularly dramatic in this species because the number of progeny per worm is more than doubled when substantial numbers of sperm are transferred (compared with matings with no transfer), and the reproduction of hermaphrodites is sometimes limited by lack of sperm (Ward and Carrel 1979). In fact, hermaphrodites that were mated by males but did not receive sperm had *reduced* numbers of progeny, perhaps because their own sperm were disabled or became less fertile due to some product from the male (Ward and Carrel 1979). Possibly the advantage to hermaphrodites of avoiding fertilization by sperm from other individuals is to avoid reduction of their genetic representation in the progeny. Males are often quite rare (0.1% in one population; Maupas 1900 in Ward and Carrel 1979). Sperm "leakage" from recently mated females also occurs in several other nematode species in the genus *Rhabditis* (Rehfeld and Sudhaus 1985; Hass 1990; fig. 3.3).

Sperm ejection soon after copulation (presumably sperm of the current male) has also been seen in several other groups. It occurs in three bird species (Birkhead and Møller 1992). In the jungle fowl, it may tend to follow copulations with subordinate as opposed to dominant males (R. Thornhill, pers. comm.). What appeared to be sperm was also ejected by the female immediately following two of three copulations by the Mexican milk snake (*Lampropeltis triangulum sinaloae*; Gillingham et al. 1977) and may also occur in rattlesnakes (*Crotalus*; Stille et al. 1986) and the parrot *Coracopsis vasa* (Wilkinson and Birkhead 1995). Sperm is ejected, along with other material soon after copulation, by females of *Drosophila mercatorum* (Ikeda 1974) and many other *Drosophila* species (T. Markow, pers. comm.), and in several other insects (Eberhard 1994). One particularly well-studied case of dumping sperm, in the chrysomelid beetle *Chelymorpha alternans*, is probably a cryptic female choice mechanism favoring males with longer intromittent genitalia and is discussed in chapter 7.2 (V. Rodriguez 1994a; Rodriguez et al., subm.). As argued at the end of the next section, sperm emission is probably much more common than has been previously appreciated.

3.3 Sometimes Discard Sperm of Previous Males

In species in which females store sperm and in which they at least sometimes mate with a second male while they have viable sperm stored from the previous mating, the second male's chances of fathering the female's offspring can be increased if sperm from previous matings are removed.

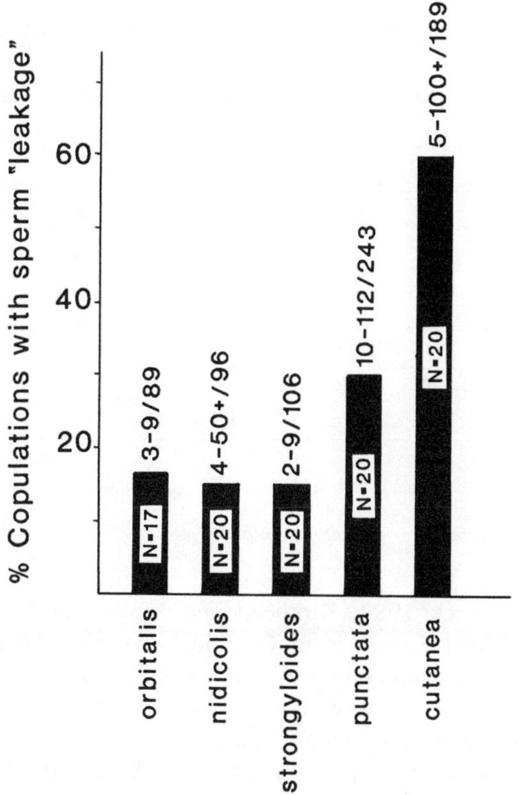

Figure 3.3 Frequencies of sperm "leakage" from female *Rhabditis* nematodes just after copulation ends. Above each bar is the range of numbers of sperm that emerged from the female/the average number of sperm transferred. (After Hass 1990)

Physical displacement of sperm by the male, using his genitalia to either scoop or snag the previous male's sperm and pull them out, or to push them into sites where they are less likely to be used to fertilize the female's eggs, occurs in a variety of odonates (Waage 1979, 1984, 1986; Miller 1984, 1990; Siva-Jothy 1987a; Siva-Jothy and Tsubaki 1989a). Males of two beetle species in different families can also remove sperm which are still in the bursa and have not been transferred to the spermatheca (Gage 1992; Yokoi 1990). Flushing previous sperm, using the male's own ejaculate, occurs in a cricket (Ono et al. 1989), and flushing with sprays of seawater may occur in some sharks and rays (Leigh-Sharpe 1920, 1922 discussed in Eberhard 1985).

In most animals, however, the male's genitalia do not reach deep enough into the female to manipulate stored sperm (R. Smith 1984; Eberhard 1985;

Table 3.2
Possible Cases of Cryptic Female Choice in Which the Female Sometimes Discards Sperm of a Previous Male

	Criteria						
Species	*I*	*II*	*III*	*IV*	*V*	*VI*	*References*
Paraphlebia quinta (Odonata)	Y	Y	Y?	?	Y (male morph)	?	E. Gonzalez, in prep.
Spodoptera litura (Lepidoptera)	N?	?	?	?	?[a]	?	Etman and Hooper 1979
Thamnophis sirtalis (Reptilia)	Y	Y	Y	Y	?[a]	?	Halpert et al. 1982; Schwartz et al. 1989
Prunella modularis (Aves)	?	Y	Y?	Y	Y (cloacal pecks)	?	Davies 1992

Notes: All are discussed in the text.
Each example is classified with respect to criteria for proof that cryptic female choice occurs. See table 3.1 for the criteria and key.
[a] Possible cues for discriminating among males were not checked.

Ginsberg and Huck 1989; Birkhead and Møller 1992; see, however, chap. 2.2), so direct removal is not feasible. In contrast, the ultimate fate of stored sperm in these and probably in most other groups can be influenced directly by the female. In most groups the storage site is not the fertilization site (Eberhard 1985), and female mechanisms such as cilia and muscles move stored sperm to fertilization sites (e.g., Villavaso 1975a; V. Rodriguez 1994b on insects; Gomendio and Roldan 1991 on mammals; Birkhead and Møller 1992 on birds), and could conceivably also move them to bring about removal (table 3.2).

The females of a bird, the dunnock *Prunella modularis*, often discard sperm from previous males before remating when stimulated during precopulatory interactions (Davies 1983, 1985). Males pecked at the female's cloaca during precopulatory displays, and as the male pecked, the female's cloaca became pink and distended, from time to time making strong pumping movements. A small droplet of sperm and/or a fecal pellet was sometimes ejected during pumping movements. The small sizes of the sperm masses made it impossible to see whether or not females always responded to cloaca pecking by dumping the contents of their cloaca. Davies (1985) notes that "there is little doubt that this [sperm ejection] is what the male is waiting for during the pecking display, because he looks at the ejected droplets carefully and, as soon as they are produced, he copulates" (p. 636).

Variations in the numbers of cloacal pecks also suggest that they are adaptations to induce the female to discard previous mates' sperm. The more time the female (which is often polyandrous) had just spent close to another male (less than 10 m away), the more cloaca pecks the male performed (Davies 1983). Direct data on the possible effects on paternity are not available, but it seems likely that sperm ejection influences the probability of fertilization. Sperm usage patterns in birds are commonly based on dilution, or what has been called the "raffle" competition (a male's chances of paternity are directly proportional to the fraction of the sperm stored in the female that belong to him; Birkhead and Møller 1992), so lowering the number of sperm from previous mates could be reproductively advantageous. The male's apparent close observations of ejected droplets and his insistence on cloacal pecking when the female had been with another male point to the same conclusion. There are indications that cloaca pecking in another bird, *Passer domesticus*, may also serve to induce the female to eject sperm from other males (Møller 1987c).

What may turn out to be the most dramatic case of discriminatory ejection of prior sperm occurs in the damselfly *Paraphlebia quinta* (E. Gonzales, pers. comm; see fig. 1.4). As is typical in damselflies, males seize and mate with females near oviposition sites, and females often oviposit immediately after copulation. Copulation is also typical in that initiation and termination are under female control (she must bend her abdomen forward to the "wheel position" to permit intromission), while males initiate and terminate tandem clasping. Also typical is the "pumping" phase of copulation, involving active movement of the male's genitalia and abdomen. Pumping is associated in other species with repositioning or removal of sperm within the female (e.g., Waage 1979, 1984). An "inactive" phase, which is associated in other species with transfer of the sperm of the present male to the female (Waage 1979, 1984), occurs at the end of copulation.

Copulation in *P. quinta* differed in that genitalic contact was often broken repeatedly before the final, inactive phase, despite the fact that the pair remained in tandem. After some but not all such separations, the female emitted a small viscous mass (see fig. 1.4) which, when examined under a microscope, proved to contain sperm (E. Gonzalez, pers. comm.).

Sperm emission did not occur in all matings. The males of *P. quinta* are dimorphic; some, which are on average larger and more dominant in aggressive interactions, have dark wing tips, while others have clear wings. Copulations with clear-winged males averaged more than twice as long (av. 40.9 ± 13.9 min vs. 16.6 ± 5.4 min, $P > .001$), they included more interruptions (av. 2.6 ± 2.1 min vs. 1.0 ± 1.0 min; $P > .008$ with Mann-Whitney U-test), and more often included sperm emission (50% vs. 29.2%). The difference in sperm emission is not significant, however (chi-squared = 1.57, df = 1, $P < 0.2$). In fact, the likelihood that sperm emission would occur following a male withdrawal was no different for hyaline-winged males than for dark-winged males (21% of 34 withdrawals vs. 21%

of 33 withdrawals; only the first emission during a copulation could be distinguished, however; withdrawals at the end of copulations that were not preceded by previous emissions were counted).

These data are still preliminary. The association of the final inactive phase with transfer of sperm by the current male needs to be confirmed before the emitted sperm can be confidently assigned to previous males. The degree of male participation in sperm emission is also not yet clear, as he may initiate the removal of sperm that are eventually discarded from the spermatheca or bursa. If this is the case, the questions of whether or not the discarded sperm would in any case have been out of position to fertilize the eggs, whether the female is capable of moving them back into storage, and why sperm are emitted in only a fraction of the copulations need to be answered. In *Coenagrion scitulum*, however, sperm are not removed from the female by the male during copulation (Cordero et al. 1995), and video tapes of mating reveal apparent sperm droplets emitted from the female during pauses between the repeated intromissions (E. Gonzalez, pers. comm.). It seems very likely that female behavior (at the very least the periodic breaking of genitalic connections, which is crucial if sperm are to be discarded during copulation) has differential impact on the reproduction of different males.

Dumping of previous males' sperm may also occur in some of the species mentioned in the previous section. Sperm masses were discarded in 13.1% of matings with virgin female *C. alternans* beetles (fig. 3.1), but more than six times as often (83.6%) in matings with nonvirgins (V. Rodriguez 1994a). Similar figures for the spider *Physocyclus globosus* (fig. 3.1) were 0% of eight matings with virgins and 100% of eleven matings with nonvirgins (W. Eberhard, unpub.). In neither species is it known whether the sperm emitted by nonvirgins were from the first or second male.

Sperm dumping in general (including sperm from previous and former males) may be quite common. Emission of sperm from the female during or immediately following copulation occurred in 25% of fifty-three species of insects and spiders surveyed with this question in mind (Eberhard 1994). Sperm dumping is probably seriously underreported in published accounts, simply because it is not often looked for and is relatively inconspicuous and easily confused with defecation. For instance, my student R. L. Rodriguez and I both firmly believed that sperm ejection did not occur in the tiger beetle *Pseudoxychila tarsalis*, which we had observed extensively, until a reexamination of videotapes of matings that we had both watched firsthand showed females unmistakably depositing spermatophores on the substrate. In addition, careful observations of females are usually made only during and immediately following copulation, so later ejections would be missed (ejection occurs, for example, in female *Macrohaltica*

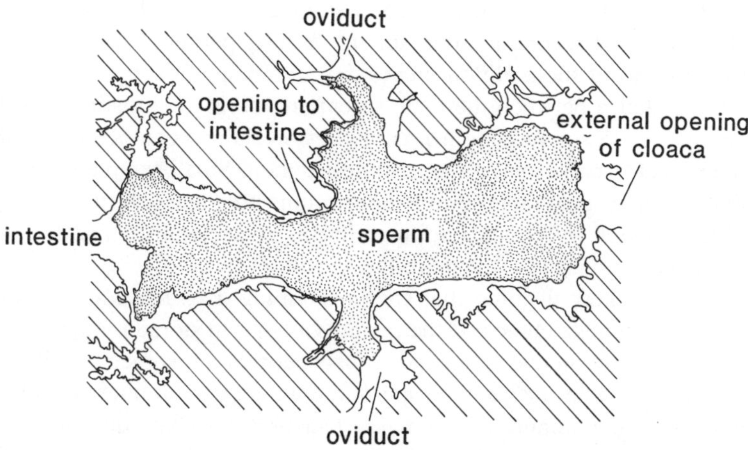

Figure 3.4 Sperm rejection by the female can be subtle and difficult to observe. In this section of the lower reproductive tract of a female *Anolis carolinensis* lizard 4 hr after copulation, a major portion of the sperm has entered the intestine rather than the oviducts, despite the fact that males deposit their sperm at the entrances to the oviducts. Such internal misrouting (and subsequent death) of sperm could easily be missed when checking for possible sperm rejection. (After Conner and Crews 1980)

beetles up to more than a day after copulation has ended; Eberhard and Kariko, in press).

Even in the relatively well-studied odonates, it appears that previous authors may have missed sperm dumping by females. Enrique Gonzalez (pers. comm.) has noted small drops with sperm in one species in each of three different genera (*Paraphlebia*, *Cora*, and *Palaemnema*), and a similar drop in a research video of mating *Coenagrion scitulum* (which was not noticed by the colleague who made the video). The apparently incongruous high frequency of sperm dumping ("leakage") in nematodes, an otherwise very poorly studied group, may be explained by the fact that observations of nematode mating were necessarily made under the microscope, where individual sperm cells can often be easily observed. Observations like those of Poulton (1886, cited in Richards 1927) of "faeces of a special character" being voided by the female grasshopper *Podisma pedestris* just before copulating, and of the female *Anolis* lizards, which wipe the cloaca by dragging it on the ground immediately after copulation (Stamps 1975; Marquez 1994 on *A. aquaticus*) merit further study.

In addition, sperm ejected from storage organs but not from the female's body, or which was degraded in the storage organs, would also be missed (e.g., fig. 3.4). For instance, females of the garter snake *Thamnophis sirtalis* often mate in the autumn before entering hibernation (Whittier et al. 1987). If the female remates in the spring, most of the sperm from the previous fall mating are extruded from storage sites and degraded in the 6 hr following the second mating (see fig. 2.3; Halpert et al. 1982). Direct physical effects of the male on previously stored sperm are precluded by the large distance between the male's genitalia during copulation and the storage sites (>25 cm, judging from drawings in Halpert et al. 1982), so the female probably participates in sperm destruction. Small sample sizes and qualitative rather than quantitative descriptions preclude determination of possible variation in female responses in this species, other than the fact that they only sometimes remate in the spring.

It should be kept in mind that sperm dumping does not automatically signal cryptic female choice. If the quantity of sperm stored by the female is less than that in an ejaculate, and if this sperm is not subject to dilution by the sperm of other males, then sperm dumping by the female would not be of any selective importance to males. I suspect, however, that situations in which sperm dumping does not have reproductive consequences for males may not be common, since such "overly large" numbers of sperm appear not to be "mistakes" by males, but instead are probably often adaptations to dilute the sperm of other males (e.g., Gilbert et al. 1981a,b on *Drosophila melanogaster*; Eady 1995 on a beetle; Kenagy and Trombulak 1986 on mammals; Birkhead and Møller 1992 on birds; see also chap. 8.1).

The timing of sperm ejection can also be crucial. Ejection of the current male's sperm before mixing with previous sperm has occurred could severely reduce the current male's chances of paternity; ejection after sperm mixing might have little or no effect if the female does not mate again. Conversely, ejection of a given volume of sperm would be most damaging to the previous male if it occurred before his sperm mixed with the sperm of the current male. The timing of ejection with respect to possible mixing of ejaculates is unknown in most cases.

3.4 Sometimes Prevent Complete Intromission and Ejaculation

The words *intromission* and *copulation* are used more or less interchangeably throughout this book. In fact, however, copulation in some species is a complex process, involving a sequence of different stages or degrees of intromission. In contrast to mammals, genitalic coupling between male and female in many other animals is not equivalent to intromission (fig. 3.5). A male can succeed in introducing his genitalia into the female, but nevertheless fail to position them in the way needed to bring about successful sperm transfer (e.g., figs. 1.11, 3.6; table 3.3).

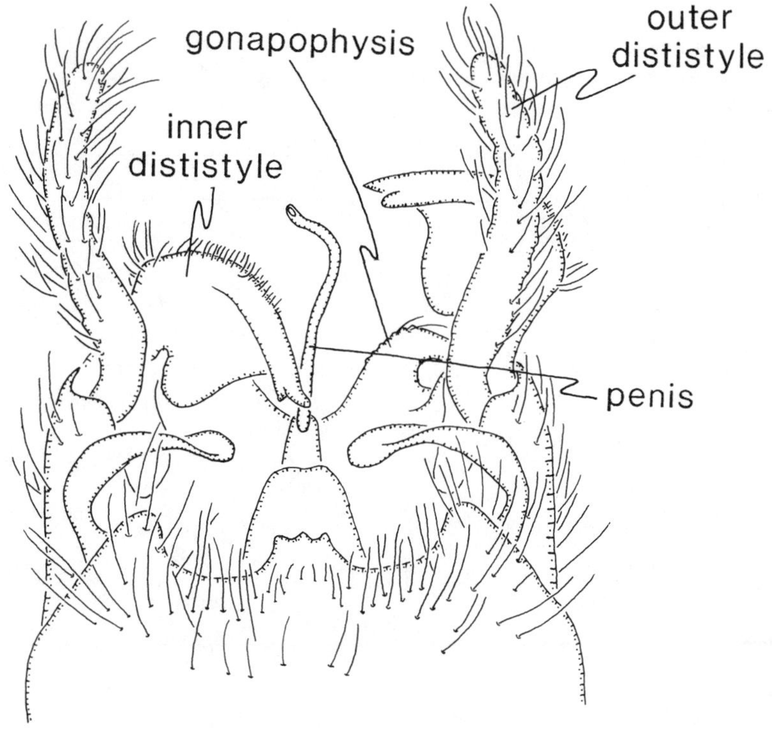

Figure 3.5 The male genitalia of the crane fly *Dolichopeza tridenticulata* illustrate the inadequacy of oversimplified phrases such as "genitalic coupling" or "insertion" to describe copulation in many animals. "Coupling" in these insects is not a single event, but rather a complex, multistage process. It starts with loose seizure of the tip of the female's abdomen with the outer dististyles, followed, after some maneuvering, by a tight clamp with the inner dististyles. Following this the male uses his gonapophyses to pry apart female structures to expose the opening to her bursa, into which he finally inserts his penis. (After Byers 1961)

Not all subprocesses in intromission automatically follow each other. In the grasshopper *Locusta migratoria*, for example, the male first couples his genitalia to those of the female by hooking her subgenital plate with his epiphallus. He then attempts to pull the female's plate downward and press it against his lowered epiproct where, if he is successful, it is held firmly. The male cerci also grip the female's abdomen on either side of the plate. Only after having succeeded in this maneuver does the male attempt to insert his aedeagus between the valves of the female's ovipositor and initiate sperm transfer per se (Gregory 1965). Failures at both stages are common. Of sixty-four cases of probable attempts to hook the female

Table 3.3
Possible Cases in Which Females Exercise Cryptic Female Choice by Sometimes Preventing Complete Intromisssion and Ejaculation

	Criteria						
Species	*I*	*II*	*III*	*IV*	*V*	*VI*	*References*
Chorthippus curtipennis (Orthoptera)	Y	Y?	Y[a]	Y?[b]	?[d]	?	Hartmann and Loher 1974
Locusta migratoria (Orthoptera)	Y	?	?	Y[c]	?[d]	?	Gregory 1965, Parker and Smith 1975
*Macrohaltica jamaicensis** (Coleoptera)	Y	Y	Y	Y[c]	?[d] (cop. court)	?	Eberhard and Kariko, in prep.
Macrodactylus spp. (Coleoptera)	Y	Y	?	Y	?[d] (cop. court)	?	Eberhard 1993a
Panorpa latipennis[e] (Mecoptera)	Y	Y	Y[f]	Y	Y (rapists)	?	Thornhill 1980a, 1984
Aedes taeniorhynchus[g] (Diptera)	Y	?[h]	Y	?	?[d]	?	Lea and Evans 1972
Aedes aegypti (Diptera)	Y	?[i]	Y	?	?[d]	?	Spielman et al. 1967, 1969
Culex pipiens (Diptera)	Y	?[h]	Y	?[j]	?[d]	?	Lea and Evans 1972
*Anastrepha suspensa** (Diptera)	Y	Y	Y[k]	?	Y (song intensity)	?	Webb et al. 1984
Glossina pallidipes (Diptera)	Y	Y[l]	Y	Y	Y? (length jerks[m])	?	Jaenson 1979, 1980

Notes: Those marked with (*) are discussed in the text.

Each example is classified with respect to criteria for proof that cryptic female choice occurs. See table 3.1 for the criteria and key.

[a] Females that allowed only brief intromissions were not attracted by courtship songs, so it is uncertain how often they would meet males in nature. Rapid termination of copulation by males resulted when the spermathecal duct of the female was artificially shortened, suggesting that female closure of the duct may be involved.

[b] Remating was relatively common and occurred readily in captivity, suggesting it occurs in nature.

[c] Direct paternity analysis was made with the sterile male technique.

[d] Possible cues for discriminating among males were not checked.

[e] Mechanism by which sperm arrival in female storage organ is prevented was not determined.

[f] Males executing forced copulations are discriminated against with respect to number of sperm transferred to the female spermatheca; this presumably has an effect on male reproductive success.

[g] Semen was deposited but apparently never entered the female in 25–50% of those virgin females that accepted genitalic coupling but were not inseminated. The authors state that the semen reached only the cloacal hollow but did not eliminate the possibility that it was ejected from deeper within the female. On drying, this semen apparently prevented subsequent males from inseminating the female and the female's eggs remained unfertilized, so such rejections may be unnatural artifacts (perhaps such females never encounter males in nature).

[h] Female probably must visit a male swarm to mate, and this behavior was not studied in detail. Lea and Evans (1972) and Lea and Edman (1972) demonstrated that female avoidance behavior both before and after she is seized by the male effectively deters male in *Culex pipiens* both in the lab and in the field (see fig. 1.11).

[i] The effects of age and mating status were not absolute; even recently emerged females (25–29 hr old) were occasionally inseminated when they mated (1 of 12; Spielman et al. 1969); nine of thirty-eight females were inseminated when remated within 5 hr after the first mating (Spielman et al. 1967); and oviposition increased the likelihood of

subgenital plate ("rubbing with genital armature extended") seen in fifty pairs, only 54.7% led to an attempt to insert the aedeagus; and in only 9.8% of forty-one such attempts did the male succeed in complete coupling of genitalia (Parker and Smith 1975). In another grasshopper, *Chorthippus curtipennis*, females sometimes appear to prevent spermatophore transmission after having allowed the male to insert his aedeagus, and he rapidly withdraws (Hartmann and Loher 1974). When the female's spermathecal duct was artificially truncated to a length shorter than that of the male genitalia, males made similar rapid withdrawals. If females are able to close this portion of the spermathecal duct, this may be the mechanism by which spermatophore transfer is prevented.

Another species in which coupling without sperm transfer sometimes occurs, and in which failures at different stages are probably sometimes due to lack of female cooperation, is the yellow fever mosquito *Aedes aegypti* (Gwadz et al. 1971). Successful copulation is again a multistage process, in which the male first clamps the female cerci with his genitalic claspers, then props open the female's genitalic lips with his paraprocts and everts his aedeagus to deposit sperm at the mouth of the entrance to her bursa with a nonintromittent "kiss" (Spielman 1964). When free-flying males couple with unreceptive females, the male genitalic claspers do not seize the female in the usual copulatory position, and as a result the male fails to emit sperm.

Experimental manipulations showed that active female participation is apparently involved both in failures to achieve appropriate genitalic clasping of refractory females, and in achievement of appropriate positioning with receptive females. When the female's genitalia were denervated by removing the terminal ganglion in her abdomen, insemination frequency fell in normally receptive virgin females, even when the male succeeded in coupling his genitalia (from 95% of 19 sham-operated mature virgins to 52% of 138 operated mature virgins). In complementary experiments, insemination frequency rose when males coupled with normally unreceptive females who had been denervated. Insemination frequencies with teneral

---

remating (Young and Downe 1982). At intermediate ages (40–50 hr) about half were inseminated when they copulated (30 of 54; Spielman et al. 1969). It was not certain that rejection involved the same mechanism in the female in both cases (age, mating status). Depending on the behavior of females in the wild (at what age do they visit male swarms? How long after a first copulation does a female encounter a male for a second copulation?), there may or may not be opportunity for discrimination among males.

[j] Multiple paternity sometimes occurs when the female remates (Bullini et al. 1976).

[k] Inability to reach the female bursa with genitalia presumably precludes ejaculation or places sperm in a disadvantagous position in the female (see Solinas and Nuzzaci 1984 on a related species).

[l] Some females mate more than once in the field before producing offspring (Jaenson 1980).

[m] The length of the final phase of copulation, which involves the male making a high-pitched whining sound, trembling his legs, and jerking his body rhythmically with increasing amplitude, was significantly shorter ($x = 5.8 \pm 1.7$ min, $N = 15$) when insemination did not occur than when it did ($x = 7.7 \pm 1.9$ min, $N = 156$).

virgins went from 0% ($N = 165$) with intact females, to 33% ($N = 171$) with denervated females; with matrone-injected females, they went from 0% ($N = 85$) with intact females to 42% ($N = 148$) with denervated females (Gwadz et al. 1971; Gwadz 1972). Thus even though males successfully clasped the genitalia of refractory females, appropriate positioning of their genitalia was prevented by a mechanism whose operation was related to activity of the female nervous system. Males who had achieved an inappropriate genitalic union with a denervated female were sometimes able, through persistent attempts, to reposition their genitalia appropriately, if the female did not resist (Gwadz 1972).

The behavior of another mosquito, *Sabethes cyaneus*, gives a particularly dramatic illustration of the probable role of sexual selection in a multistage intromission process (Hancock et al. 1990). After courting the female by waving the colorful and conspicuous paddles on his middle legs and inducing her to lower her abdomen, the male achieved a "superficial" genitalic coupling with his gonostyli. He then resumed waving, in a different pattern, for an average of sixteen more waves before attempting to press his genitalia more tightly against those of the female (successful in twenty-three of thirty-two pairs). Sperm transfer occurred only after this genitalic shift (none of thirty-one females that were dissected after superficial coupling and before the shift were inseminated). Hancock et al. (1990) interpret the second bout of waving as courtship to induce the female to allow the genitalic shift to occur. After achieving the shift, the male performed other behaviors (waving, side-to-side gyration of his body) that may represent even further courtship (see chap. 5.1 on copulatory courtship).

Female responses necessary for successful intromission sometimes involve inconspicuous, internal events (fig. 3.6). For example, male *Culicoides melleus* flies copulate readily, and apparently normally, with freshly killed females. Although these copulations resulted in transfer of a spermatophore, the neck of the spermatophore was never successfully inserted into the common spermathecal duct, and no sperm were introduced into the duct or the spermathecae. Similarly, males of the beetle *Chelymorpha alternans* who copulated with freshly killed females were unable to thread the flagellum up the coiled spermathecal duct of the female (see fig. 1.2; V. Rodriguez, unpub.). Female responses to male genitalic clasping are also apparently necessary for successful copulation in the roach *Nauphoeta cinerea*. When female genitalia were denervated, males often failed to grasp the female properly, and they stayed joined for only a few seconds (Roth 1962). A detailed study of male and female morphology also showed the need for female cooperation in several stages of copulation in two species of planthoppers (Kunze 1957).

Even in vertebrates intromission is sometimes a multistage process. For instance, in the garter snake *Thamnophis sirtalis*, the male first uses small

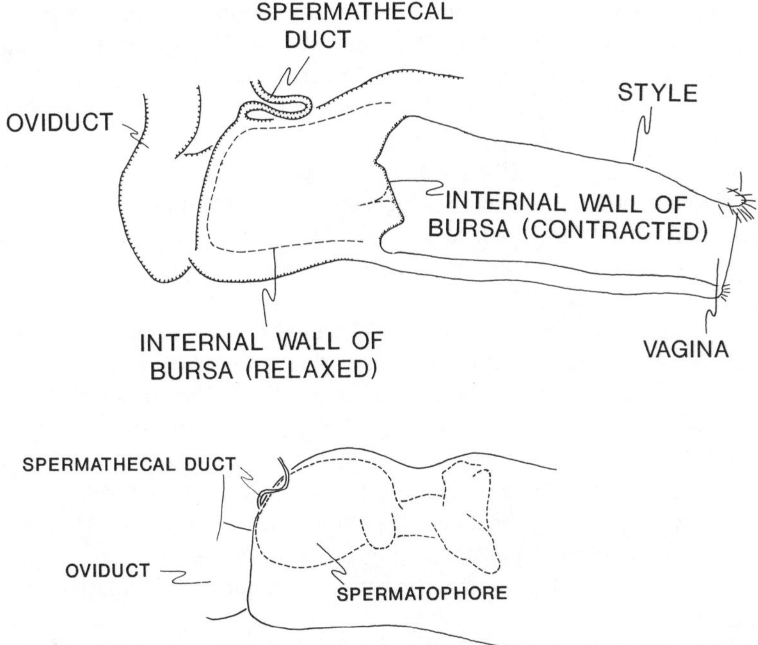

Figure 3.6 The muscular "plug" of a female of the chrysomelid beetle *Macrohaltica jamaicensis*. The female sometimes avoids insemination by a male that has achieved intromission by contracting the powerful muscles in her bursa so tightly that the bursa appears to have an internal plug. This prevents the male from inflating his genitalic sac and forming a spermatophore in her bursa (dotted lines in lower drawing). After thrusting for about a minute with his genitalia, the male aborts the copulation. The female in the upper figure was frozen just after a male withdrew prematurely. The approximate position of the inner wall of the bursa when the male has succeeded in inflating his internal sac is shown with dotted lines in the lower drawing. (After Eberhard and Kariko, in prep.)

hooks at the base of his genitalia to draw the female's cloaca open, and then everts his hemipenis into her cloaca (Crews and Garstka 1982). Possible active or passive female discrimination among males at the stage of opening her cloaca has not been studied. But the existence of the male hooks suggests that at least passive discrimination has been important, especially in view of the many males who simultaneously attempt to copulate with females emerging from hibernation sites (Crews and Garstka 1982).

Even after he has succeeded in inserting his genitalia in the female's body, the male does not necessarily have access to the site he needs to reach in order to give his sperm the best chance of being used. For instance, in the beetle *Macrohaltica jamaicensis*, the male sometimes inserts his genitalia into the lowermost portion of the female's reproductive tract (the vagina) but is not successful in penetrating farther, into her bursa, and pulls back out. Females sometimes close the bursal cavity by contracting the powerful muscles in its walls (fig. 3.6). Because the female's spermathecal duct opens at the inner end of the bursa, and because the male must inflate his internal sac within the bursa to form a sperm-containing spermatophore, copulations in which the male intromits but fails to reach the bursa do not result in sperm transfer (Eberhard and Kariko, in press). Similar "pseudo-copulations," at least some of which are also accompanied by internal contractions of the female, occur in the tsetse fly *Glossina pallidipes* (Jaenson 1979) and the tortoise beetle *Chelymorpha alternans* (V. Rodriguez 1994a).

Nearly half (46%) of the copulations in another leaf beetle, *Labidomera clivicollis*, were "partial" intromissions (Dickinson 1986). In other insects such as the honey bee *Apis mellifera* (Laidlaw 1944; Koeniger 1986, 1991) and the rosechafer beetles *Macrodactylus* spp. (Eberhard 1993a), there are internal valves of various sorts in the female which must be open if the male's sperm and/or his genitalia are to enter. In the honeybee a female movement is apparently necessary to allow a male who has partially inserted his genitalia into her open sting chamber to complete the eversion (Koeniger 1986), and the female may even trigger his ejaculation (Koeniger 1991).

In *Macrodactylus* some copulations terminate before sperm transfer, apparently due to female failure to allow deeper penetration by the male to the mouth of the spermathecal duct (Eberhard 1993a). Elaborate female mechanisms are probably often not necessary. In some groups (e.g., sepsid flies; fig. 2.1) a single egg in the oviduct is enough to exclude them. Very brief "copulations" or "pseudocopulations" occur in sepsids (Parker 1972 on *Sepsis cynipsea*) as well as in other groups (Markow and Ankney on *Drosophila robusta*; R. Rodriguez, in prep., on the lygaeid bug *Ozophora baranowskii*). No sperm is transferred in pseudocopulations of *D. robusta*, nor apparently in *O. baranowskii* since the female subsequently lays only infertile eggs.

Another possible example, in which the cues that may be used by the female to trigger behavior that facilitates male entry deeper into her body have been studied, is the Caribbean fruit fly *Anastrepha suspensa*. Males produced two songs—a calling song that is associated with pheromone production and is attractive at a distance to virgin females, and a different song produced by wing fanning while the male is mounted on the female. The song during mounting often (but not always) continued until the male's long aedeagus was completely threaded through the ovipositor into its final position (Webb et al. 1984; Sivinski and Burk 1989), and was also performed at later stages of copulation, when the female became restless (Webb et al. 1984; Sivinski and Burk 1989).

Female morphology and the apparent need for female cooperation in the process of intromission suggest that the otherwise enigmatic singing by the male serves to induce the female to facilitate the deep penetration necessary for transfer of sperm. The tip of the female's abdomen is long and extensible, and is used to lay eggs in fruits. The female's long vagina is folded into a tight "S" when her ovipositor is not extended (Dodson 1978). There are two reasons to suppose that the female probably must extend her ovipositor for successful intromission to occur: the male first clamps the ovipositor's tip (the aculeus) with his genitalic claspers (see Eberhard and Pereira 1994 on the related medfly, *Ceratitis capitata*), and this tip is hidden and not accessible unless the ovipositor is extended; and then, in the second step, he threads his long, flexible intromittent organ through her long vagina and into her bursa. Threading it through the tight "S" curve of an unextended vagina is probably not possible (see Eberhard and Pereira, subm., for a description of insertion in the medfly, and Solinas and Nuzzaci 1984 on the probable position during insemination in the related melon fly, *Bactrocera* (= *Dacus*) *oleae*).

Approximately 50% of the matings in *A. suspensa* result in little or no sperm being transferred to the female (Mazomenos et al. 1977). There were indirect indications that females caused some early terminations of copulation. The songs of mounted males who were rejected (defined as a male being dislodged from a female's back after at least one second of mounting, usually due to vigorous female movements) were less powerful (had both lower sound pressure levels and less total energy) than those of successful males (Burk and Webb 1983; Webb et al. 1984). The song itself was also different in unsuccessful matings, but this may have been due to female rejection movements.

Rejection by the female increased when the intensity of the song during mounting was reduced by clipping the male's wings (his sound-producing organs; Sivinski et al. 1984). When recordings of the mounting song were played back at normal intensity levels while males with clipped wings were mounted, the rejection frequency went back down; rejection frequencies remained high, however, when the song was played back at low intensity.

Exactly what the male *A. suspensa* was attempting to induce the female to do as he sang after mounting the female is not clear. What is certain (although it was not completely clear in the published descriptions) is that some of the rejections occurred after the male had clamped the female aculeus and inserted his intromittent genitalia into and partway up the female's vagina (J. Sivinski, pers. comm.). The frequencies of pre- and post-intromission rejections were not recorded, however, so further study is needed to determine whether cryptic female choice occurs. In sum, copulatory courtship behavior by male *A. suspensa* apparently serves to induce females to allow the male to achieve the deep penetration necessary for successful insemination. The female uses both the presence of a song and its intensity (which is correlated with the male's size; Webb et al. 1984) as cues to trigger acceptance behavior.

3.5 Sometimes Fail to Transport Sperm to Storage Organs or Fertilization Sites

This mechanism differs from the previous one in that it does not depend on the female preventing the male's genitalia from penetrating to the site where sperm are deposited during successful copulations, or preventing the male from ejaculating. Once he has deposited his semen, however, the female varies the amount of the male's sperm that she transports to the appropriate storage and/or fertilization sites (table 3.4).

The male genitalia of animals with internal fertilization usually do not reach sperm storage or fertilization sites within the female (Eberhard 1985), and many studies have shown that in a variety of groups the female must actively transport sperm if they are to arrive at storage or fertilization sites after copulation (see Hunter 1975, Gomendio and Roldan 1991 on mammals; Lake 1975, Birkhead and Møller 1992 on birds; Davey 1965b, Degrugillier and Leopold 1972, Thibout 1977, Linley and Simmons 1981, Heming-van Battum and Heming 1986, Gillott 1988, Tschudi-Rein and Benz 1990, Sugawara 1993, LaMunyon and Eisner 1993 on insects; Gering 1953, Huber 1993, 1994b, 1995 on spiders; T. Jones 1966, Duggal 1978 on nematodes; see also Eberhard 1985 for references on other groups).

Probably many female transport mechanisms are triggered by stimuli associated with copulation (e.g., see fig. 4.17). For instance, stimulation of the vulva of the rabbit can elicit strong vaginal contractions that can carry fluid introduced into the vagina high up into the uterine horns (Blandau 1973). The number of motile sperm in a rabbit's Fallopian tubes 12 hours after artificial insemination is increased by a factor of four if the insemination is followed by copulation with a vasectomized male (Lambert and Tremblay 1978). Similarly, at least one preejaculatory intromission is necessary to stimulate the female rat to transport sperm from the vagina to the uterus (Adler 1969). Chemical stimulation of the female's genitalia with male seminal products in the bug *Rhodnius prolixus* (Davey 1958), and mechanical stimulation in cows (VanDemark and Hays 1952 and references therein) result in rapid sperm transport (Blandau 1969; see also chaps. 5.2 and 6.4).

There are, of course, exceptions. In some species the male genitalia or his spermatophore penetrate all the way to the storage site (e.g., Blest and Pomeroy 1978 on *Mynoglenes* spiders; Gregory 1965 on the grasshopper

Table 3.4
Possible Cases in Which Females Exercise Cryptic Female Choice by Sometimes Failing to Transport Sperm to Storage or Fertilization Sites

	Criteria						
Species	*I*	*II*	*III*	*IV*	*V*	*VI*	*References*
Melanoplus sanguinipes (Orthoptera)	Y[a]	?	?	Y	?[b]	?	Cheeseman and Gillott 1989
*Scathophaga stercoraria** (Diptera)	Y	Y	Y	Y	Y size	Y	Ward 1993; Simmons and Ward 1991
*Neriene litigiosa**[c] (Araneae)	Y	Y	Y	Y[d]	Y cop. "vigor"	?	Watson 1991
*Rattus norvegicus** (Rodentia)	Y	?	Y?	Y	Y (No.intromiss., male dom., interval betw. ejac. series)	Y	Adler 1969; Chester and Zucker 1970; Matthews and Adler 1979 McClintock et al. 1982a; McClintock 1984

NOTES: Those marked with (*) are discussed in the text.
Each example is classified with respect to criteria for proof that cryptic female choice occurs. See table 3.1 for criteria and key.
[a] Differences were in the movement of contents of different spermatophores from the same male to the spermatheca.
[b] Possible cues for discriminating among males were not checked.
[c] Precopulatory sexual selection by male-male battles also occurs.
[d] Direct paternity analysis was made with genetic markers.

Locusta migratoria). In some he apparently pumps his sperm at least part of the way in under force (Whitman and Loher 1984 on the grasshopper *Taeniopoda eques*). And in a few the sperm swim into storage under their own power (Wilkes 1965 on the wasp *Dahlbominus fuscipennis*). In some species a combination of female transport and sperm movement is involved. In some birds, for instance, females transport both dead and living sperm up the oviduct, but only living sperm are able to enter the sperm storage organs (Lake 1975; see also Thibout 1977 on a moth; Page 1986 on a bee; Degrugillier and Leopold 1972 on a fly).

Appropriate female responses are thus often necessary if a male's gametes are to reach storage and fertilization sites within the female. If a female's sperm transport response is more complete or energetic to some conspecific males than to others, she could exercise cryptic female choice.

Possibly the most complete study documenting cryptic female choice by differential sperm transport is of the Sierra dome spider *Neriene* (= *Linyphia*) *litigiosa* (fig. 3.7; Watson 1990, 1991; Watson and Lighton 1994), although the morphological data are still fragmentary. In this species the first male to mate with a female fathers the majority of her offspring, even

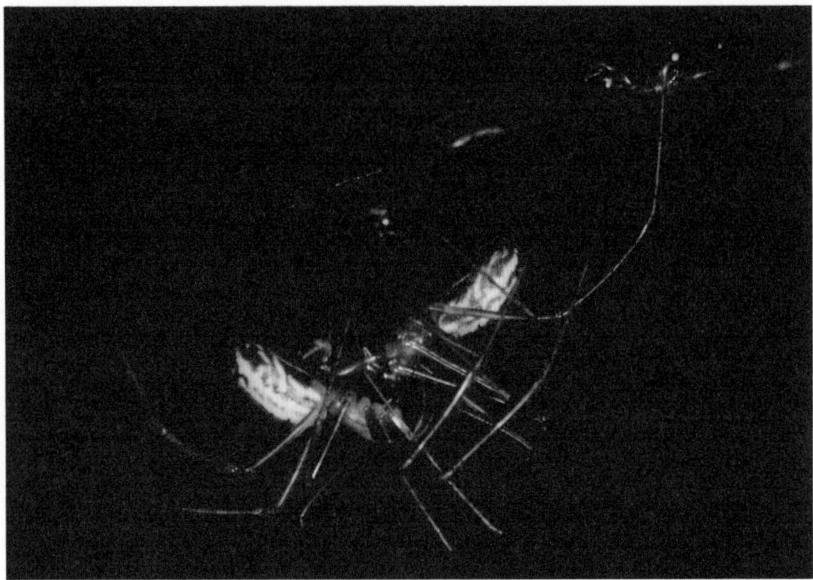

Figure 3.7 A copulating pair of *Nereine* (= *Linyphia*) *litigiosa* spiders. Males of this species compete directly by fighting for access to virgin females, and also indirectly by cryptic female choice. Females grant different degrees of paternity success to second mates on the basis of their courtship vigor (a compound character), perhaps by transporting more or less of their sperm to the spermathecae. (Photo courtesy of Paul Watson)

though the female usually mates with other males subsequently (average number of matings/female in the field = 2.3 ± 0.98, N = 104). As might be expected, males in nature sought out females about to moult to maturity. They often fought with other males for the chance to mate with a newly mature female (on average a male had to fight with at least two males to copulate with a virgin female). In contrast, nonvirgin females were less attractive, and males seldom had to fight to mate with them.

Multiple paternity was common in nature (34–96% of broods, depending on the paternity algorithm used to interpret electrophoretic data). When a female mated twice, the first male usually fathered more of the offspring, but his advantage over the second varied. The fertilization success of secondary males who mated with forty-four marked, free-living, nonvirgin females was correlated positively with both the male's weight, and with his "copulatory vigor" (see below), two uncorrelated traits.

Copulatory vigor is a compound trait associated with the unusual process of copulation that is typical in this family (Linyphiidae) (e.g., van Helsdingen 1965, Austad 1984, and Suter 1990; see fig. 4.6). All spiders have secondary male genitalia on the pedipalps that must be loaded with sperm from the male gonopore in order to inseminate females. In contrast to many other spiders Linyphiid males generally do not load their pedipalps before searching out a female. Instead, after finding and courting a female, the male copulates with her without sperm in his palps (in *N. litigiosa* for 2–6 hr!). After this phase of "preinsemination copulation," the male retires to the edge of the web, charges one or both palps with sperm, and returns to copulate ("insemination copulation"; for another 0.5–1.4 hr in *N. litigiosa*).

Preinsemination copulation in *N. litigiosa* consists of hundreds of separate intromissions, and the insemination phase includes 60–120 more. Each intromission during preinsemination copulation lasts only 1–2 sec, with an average of 6.3 ± 1.8 intromissions/min; and rates varied widely between different males (from roughly 1 to 10/min; Watson and Lighton 1994). Intromission rates were 7–10 times less rapid during insemination copulation. Some intromission attempts failed ("flubs"), perhaps as a result of incorrect orientation, poor mechanical fit, or improper expansion of the palpal membranes.

The male's rate of metabolism rises 1.2–4.5 times over his resting metabolic rate during preinsemination copulation (Watson and Lighton 1994), so copulation is energetically costly. The long series of intromissions during preinsemination copulation may be an endurance test by means of which females judge male vigor. The "copulatory vigor" of the second male to mate with each female was quantified by combining measures of length of preinsemination copulation, rate of intromission, and percentage of "flubs" (the precise contributions of different components were not specified; the rate of intromission gave the highest standard regression coefficients in multiple regressions; P. Watson, pers. comm.).

The mechanism by which males with more copulatory vigor achieved greater paternity was not determined by direct observations and remains to be documented convincingly. The duct through which the sperm enter the female's spermatheca is substantially longer than the male's intromittent organ (G. Hormiga, pers. comm.). Watson proposed that by opening or closing a valvelike structure in the duct near the entrance to the spermatheca, the female controlled access to her spermatheca, and thus entry of the second male's sperm. Spider sperm are generally inactive when transferred to females, with each sperm cell coiled up and enclosed in a membrane (S. Brown 1985; Lopez 1987; Uhl 1993), and this is apparently true in *N. litigiosa* (P. Watson, pers. comm.), so female control of their movement into her spermathecae seems especially likely. Other mechanisms, such as a bias in use of sperm from one of the two spermathecae, are also possible (males mating with nonvirgin females usually insert only one pedipalp on one side of the female's genitalia; P. Watson, pers. comm.). The female genitalia in *N. litigiosa* are also unusual for spiders (see chap. 7.1) in having a relatively long fertilization duct leading from the spermatheca to the oviduct (G. Hormiga, pers. comm.). The significance of this length is unknown. More data will be needed to establish the details of the mechanism of cryptic female choice in this species.

There is an indirect suggestion of a similar role for preinsemination copulations with respect to sperm precedence in another linyphiid spider, *Frontinella pyramitela*. Preinsemination copulations with virgin females lasted only an average of 17.2 ± 6.0 min before the male made a sperm web, whereas it lasted 185.3 ± 100.5 min if she had mated in the previous 24 hr (Austad 1982). Field-collected females do not have recognizable copulatory plugs (G. Hormiga, pers. comm.). The tip of the male's intromittent organ (embolus) may just reach the entrance of the spermatheca, and the "elbow" in the insemination duct, which Watson took to be a possible valve, is absent (G. Hormiga, pers. comm.). Suter (1990) argued that preinsemination copulation serves to inform the male about whether the female has already mated. He did not eliminate the alternative hypotheses, however, that the male was attempting to remove a previous male's sperm or to trigger a female response (such as inducing her to open a valve) that would increase his chances of fertilizing her eggs. The long duration of preinsemination copulations with recently mated females seems paradoxical if this behavior serves only a sensory function for males.

In fact, similar preinsemination copulation behavior occurs in the related spider *Lepthyphantes leprosus* (van Helsdingen 1965), in which the male embolus is apparently too short to reach the female's spermatheca where previous sperm may be stored (G. Hormiga, pers. comm.). Copulations with virgin female *L. leprosus* last an average of 209 min, again seemingly more time than needed for a male to sense if a female is a virgin.

A second group in which transport bias occurs is the lab rat *Rattus norvegicus*. Females often mate with several males in quick succession (McClintock 1984). When a female receives intromissions within the next 10 min after receiving an ejaculation, sperm transport is halted, and the untransported sperm die (McClintock et al. 1982b). In seminatural conditions in which females were free to avoid male sexual advances, the period of inactivity after she received an ejaculation was largely determined by the female (80% of intercopulatory periods ended with the female soliciting another mating, 3% with a male soliciting a female, and 17% were ambiguous; McClintock et al. 1982b). Female rats began to solicit additional intromissions sooner after receiving an ejaculation from a subordinate male (about 55% resumed mating within 10 min after receiving a subordinate's ejaculation; only about 22% after receiving a dominant's ejaculation; McClintock et al. 1982a; fig. 3.8). Thus by remating more quickly after some ejaculations, the female probably decreased the transport of sperm from those ejaculations.

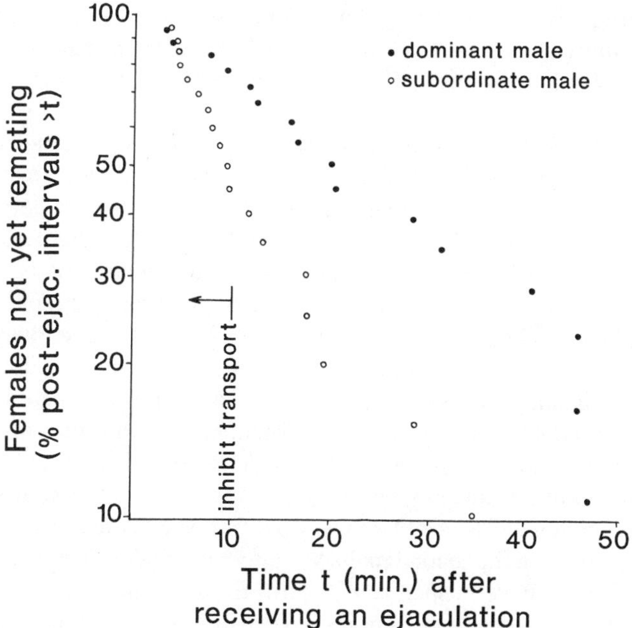

Figure 3.8 Reinitiation of mating by a female rat *Rattus norvegicus* accompanied by several males is less rapid after she receives an ejaculation from a dominant male than after she receives an ejaculation from a subordinate male. Since sperm tranport in rats is inhibited by intromissions that occur in the 10 min following an ejaculation (arrow), more rapid resumption of mating will decrease sperm transport from the first mating. The differences between responses to dominant and subordinate males (indicated by the slopes in the graph) were significant during this critical period of the first 10 min following copulation, but did not differ significantly thereafter. (After McClintock et al. 1982a)

Another probable case of female bias in sperm transport occurs in the dung fly *Scathophaga stercoraria.* This case differs from the others because the female has three different spermathecae, two of which are connected to form a pair, and the bias in transport involves the distribution of sperm among the three spermathecae. Using males with genetic markers and manipulating their sizes by altering larval nutrition, Ward (1993) showed that females who had mated twice showed strong cryptic preferences for larger males (copulations were terminated artificially after 20 min, enough time to guarantee substantial sperm transfer, in order to eliminate the possible effect of smaller males copulating for longer; Ward and Simmons 1991). When the second male was larger than the first, he fathered 100% of the offspring, as contrasted with 48%, 53%, and 64% (for first, second, and third clutches of eggs) when he was smaller. Ward's sample was small (six females), but the same effect of increased P2 values when the second male was larger was demonstrated in larger samples in a previous study using a similar (though not identical) experimental protocol in which copulations were also interrupted after 20 min (Simmons and Parker 1992).

The mechanism responsible for this bias was apparently related to which of the female's spermathecae received sperm. One day after a 20 min copulation with a larger second male, the distribution of sperm between the spermathecae was significantly more uneven, with the paired spermathecae containing disproportionately higher numbers of sperm. The genitalia of male *S. stercoraria* are apparently too large to penetrate the female's spermathecal ducts, and the ducts are so long and thin that it is unlikely that he can pump sperm freely into them (Ward 1993; see Linley and Simmons 1981 for a general discussion of probable female control over sperm movements in thin ducts leading to rigid spermathecae). Ward thus argued that females rather than males must control which spermathecae are filled. The spermathecal ducts are provided with muscles that are presumably controlled by the female. Since there was no difference in the total numbers of sperm stored when the second male was larger or smaller, Ward argued that the female must have flushed some sperm from the first mating out of the singleton spermatheca (other studies have also shown that total numbers of sperm in storage did not increase with additional copulations; Parker et al. 1990). Models in this species that were derived from previous studies of mechanisms of sperm precedence failing to take into consideration possible female roles (e.g., Parker and Simmons 1991; Simmons and Parker 1992) may thus need to be modified.

Ward's study of *S. stercoraria* documents differences in sperm storage patterns that are apparently due to female transport and associates these differences with differences in paternity. Nevertheless, it may give only a partial picture of the processes influencing a male's chances of paternity, due to the experimental protocol. Copulations were all interrupted after 20 min, and the female was unable to oviposit until a day after copulation. While interruptions of copulations are not uncommon in nature, especially at high population densities (Otronen 1993b), delay of oviposition may not be, and many copulations are longer. Copulation in undisturbed pairs averages 35 min in the field (Parker 1970b). On average this length of time is required to fill the female storage organs (with about 700 sperm; Parker et al. 1990), and smaller males normally copulate for longer than larger males (Ward and Simmons 1991; Simmons and Parker 1992). Interruption of copulation may explain the surprisingly high frequencies of unhatched eggs (presumably unfertilized) in the Ward study (1993) (17% when the second male was smaller, 62% when he was larger, which is much higher than the 5.1% infertility rate after a single complete mating in Parker 1970a when oviposition followed much sooner after copulation).

Perhaps in *S. stercoraria* some sperm are kept in the spermathecae for long-term, fail-safe storage, as in the paired spermathecae of *Drosophila melanogaster* (Fowler 1973; Gromko et al. 1984), while other sperm, kept in other portions of the female reproductive tract, are used preferentially for fertilization (as are the sperm in the seminal receptacle in *D. melanogaster*; see chap. 2.2). After some copulations in the mosquitoes *Culex pipiens* or *Aedes taeniorhynchus* (the text was unclear) the female has sperm only in her bursa, not in the spermatheca (Lea and Evans 1972), perhaps due to lack of female transport. Much more detailed knowledge of the fertilization process is needed to determine the significance of bursal vs. spermathecal sperm. In some flies fertilization with bursal sperm seems unlikely. The egg of *Musca domestica* is protected from fertilization by a cap on the micropyle until the last moment, when the cap is removed while the micropyle is very precisely positioned at the mouth of the spermathecal duct, in position to receive only spermathecal sperm (fig. 7.6) (Leopold et al. 1978).

Sperm transport in the pig *Sus scrofa* to the two uterine horns was also asymmetrical (checked using females with cannulae implanted near the Fallopian tubes). When the female remated, the horn that had received less semen following the first mating received a greater amount from the second (Pitkjanen 1959; no data or statistical tests were given, however). This would presumably tend to balance paternity if two males mate once each. Sperm transport in this species is largely due to female contractions and varies with the stage of female estrus. Females could thus cryptically increase or decrease a male's chances of fertilization by altering the asymmetries in transport.

The elaborate cocktail of transport-inducing substances in the semen of some mammals and insects (see chap. 6) is an eloquent indication, however, that males have been selected to induce or reinforce female transport responses. Other species in which male seminal products elicit contractions of muscles in female ducts that probably serve to increase transport of the male's gametes to storage or fertilization sites are discussed in chapter 6.

A second, very poorly studied process of sperm transport in the female involves movement of sperm from storage to fertilization sites. Even such major questions as whether the movement is the result of sperm activity or of transport mechanisms of the female are generally still unanswered. I have not found any cases in which this transport is biased to favor some males over others, although female control of transport has been demonstrated in a few species.

There was a sharp increase in contractions of the spermathecal duct of the cricket *Teleogryllus commodus* when receptors in the genital chamber of the female were stimulated (as they probably are when an egg passes down the oviduct) (see fig. 4.11; Sugawara 1993). Direct observations showed that these contractions resulted in sperm being moved down the duct (Sugawara and Loher 1986). In addition, the female apparently has receptors (possibly at the mouth of the spermathecal duct) which sense when sperm have descended (or emerged) to fertilize the egg (Sugawara and Loher 1986). Female control of transport to fertilization sites was emphasized by the fact that the sperm descended the spermathecal duct tail-first. Still another probable transport mechanism—increase in abdominal pressure due to repeated inflation of the posterior abdominal segments by taking in air—was brought into play while eggs were being laid when no sperm were present in the duct, and when the duct was severed and eggs were being laid (Sugawara and Loher 1986).

A similar female neural control of muscles in the spermatheca and the spermathecal duct (*ductus seminalis*) can cause sperm to be moved from the spermatheca to the duct in the grasshopper *Schistocerca vaga* (Okelo 1979) and was also inferred from a detailed study of female morphology in the semiaquatic bug *Hebrus pusillus* (see fig. 2.9; Heming-van Battum and Heming 1986).

A lack of female control of sperm release, resulting in relatively high sperm wastage when the storage organ is full, has been posited for the seminal receptacle of *Drosophila melanogaster* (Gilbert 1981) and the spermatheca of the blowfly *Lucilia cuprina* (P. Smith et al. 1988). The presence of muscles associated with the storage organ of at least one of these (*Drosophila*, in which Blaney 1970 describes a "well developed layer of visceral muscle") leaves these models open to question. Females could theoretically bias paternity by altering the efficiency of sperm usage (e.g., less efficient use of sperm after a single mating could lead to more rapid remating or proportionally larger numbers of sperm from the second male, and thus loss of paternity for the first male), though to my knowledge, this possibility has never been tested. Female control of sperm use undoubtedly has important consequences for selection on ejaculate sizes.

One further group with demonstrated female control is the male-haploid Hymenoptera, in which females control (in some cases with great precision) the sex of their offspring by fertilizing or not fertilizing their eggs (e.g., Godfray 1994; fig. 1.10). Their obvious ability to manipulate sperm use in response to factors such as host size, prior host-size history, and the presence of other females suggests that they could also manipulate sperm use in response to different males or different mating experiences.

A final possibility is that differences among the sperm themselves affect sperm transport to fertilization sites. This apparently occurs in the sperm from males of the house mouse that are heterozygous for *t* alleles. Sperm carrying the *t* allele arrive in much lower numbers in the Fallopian tubules 4–6 hr after copulation (0–5 *t* sperm vs. 450–700 normal sperm; Braden and Gluecksohn-Waelsch 1958). Using in vitro fluorescent stains in a study of intravaginal artificial insemination in rabbits, Parrish and Foote (1985) showed that differences in the fecundity of semen from pairs of males probably resulted mostly from differences in the abilities of functional sperm to arrive at the fertilization area. Females probably have a role in producing these differences; different females inseminated with portions of the same ejaculate had "vastly different" numbers of sperm attached to the oocytes (Parrish and Foote 1985).

3.6 Sometimes Remate with Another Male

The number of offspring that a male can expect to obtain from a given copulation probably usually decreases if the female mates again. Thus a male will probably often gain reproductively if he can somehow cause females to refrain from remating. In some species the male accomplishes this by direct intervention, using postcopulatory mate guarding, or by physically plugging her reproductive tract (e.g., Parker 1970c; R. Smith 1984; Alcock 1994). Females may have little option in some such cases. In the active male–passive female tradition, there are numerous studies of male behavior that reduces female opportunities to remate (e.g., summary by Alcock 1994 on insects), but fewer on possible female-imposed biases in remating.

In most species, however, males do not physically force the female to refrain from remating. Presumably mate guarding in most species is not feasible, perhaps because it is incompletely effective or is too costly in terms of lost mating opportunities with other females or survivorship, plugging is not physically feasible, and so on. In species in which females have the opportunity to remate and in which sperm from subsequent matings are sometimes used to fertilize eggs that would have otherwise been fertilized by the first male, a male character that induces females to refrain from remating without use of force can give the male a reproductive advantage (table 3.5). There are two different ways in which longer female refractive periods can be advantageous for a male. The longer the time until the next mating, the more offspring can be produced before competition from additional sperm begins. In addition, in many species a longer delay between successive copulations results in lower sperm precedence (lower P2 values). If delays are longer, the competitive advantage of the first male's sperm over that of subsequent males will be greater.

Reduced sexual receptivity following copulation probably often makes adaptive sense for females under natural selection. The female can reduce the dangers of courtship and copulation such as predation risk and exposure to venereally transmitted diseases (Daly 1978), and she can avoid possible reductions in other important activities such as feeding and oviposition. Reduced female receptivity following copulation is clearly very

Table 3.5
Possible Cases in Which Females Exercise Cryptic Female Choice by Sometimes Remating with Other Males

	Criteria						
Species	*I*	*II*	*III*	*IV*	*V*	*VI*	*References*
Decticus verrucivorus (Orthoptera)	Y	Y	Y	Y	Y (sptophyl. size)	?	Wedell and Arak 1989
Kawanaphila nartee (Orthoptera)	Y	Y	?	Y	Y (amt. semen, amt. sptophyl. eaten)	?	Gwynne and Simmons 1990, Simmons and Bailey 1990, Simmons and Gwynne 1991
Oecanthus nigricornis[a] (Orthoptera)	Y	?	Y	Y?[b]	Y (age, fluct. asymmetry)	?	W. Brown, in prep., a,b[b]
Poecilimon veluchianus (Orthoptera)	Y	Y	Y[c]	Y	Y (sptophyl. size)[d]	?	Reinhold and Heller 1993
Stomoxys calcitrans (Diptera)	Y	?	?	?	Y[e] (blood meal)	?	Morrison et al. 1982
Lucilia cuprina (Diptera)	Y	?	?[f]	?	Y (size)	?	Cook 1992, Smith et al. 1989
Musca domestica (Diptera)	Y	Y	?	?	Y (age, size[g])	?	Riemann and Thorson 1969, Riemann et al. 1967
Ceratitis capitata[a] (Diptera)	Y	Y?	Y	Y	Y (fill spthc.)	?	Nakagawa et al. 1971
*Centris pallida** (Hymenoptera)	Y	Y?	?	?	Y (courtship)	?	Alcock and Buchmann 1985
Aphytis melinus (Hymenoptera)	Y	Y?	Y	Y	Y (postcop. ctshp.)	?	Allen et al. 1994
Nasonia vitripennis (Hymenoptera)	?	?	?	?	Y (postcop. ctshp.)	?	van der Assem and Visser 1976
Hylobittacus apicalis (Mecoptera)	Y	Y	?	?	Y (size gift)	?	Thornhill 1976
Panorpa latipennis (Mecoptera)	Y	Y	Y[h]	Y	Y (rapist)	?	Thornhill 1984
*Danaus plexippus** (Lepidoptera)	Y	Y	Y?[i]	Y	Y (sptophyl. size)	?	Oberhauser 1992
Pieris rapae (Lepidoptera)	?	?	?	Y?[j]	Y (size sptophyl.)	?	Sugawara 1979, Obara 1982
*Phidippus johnsoni** (Araneae)	Y	?	Y	Y?[k]	Y (length cop.)	?	Jackson 1980
Anolis carolinensis (Reptilia)	?	?	?	N?	Y (length cop.)	?	Crews 1973, Crews and Silver 1985
*Mesocricetus auratus** (Rodentia)	?	?	Y[l]	Y?[m]	Y (No. ejac. series vag. stim.)	?	Carter 1973, Oglesby et al. 1981, Huck and Lisk 1985

Table 3.5 (*cont.*)

Species	Criteria I	II	III	IV	V	VI	References
Rattus norvegicus (Rodentia)	Y[n]	Y	Y[h]	?	Y (pace intro.[o])	Y	Hardy and DeBold 1972; Erskine and Baum 1982; McClintock 1984
Pavo cristatus	Y	Y	?	Y	Y (male cop. succ.[p])	?	Petrie et al. 1992

Notes: See tables 6.1, 6.3 for numerous other possible examples involving male seminal products. Those marked with (*) are discussed in the text.

Each example is classified with respect to criteria for proof that cryptic female choice occurs. See table 3.1 for criteria and key.

[a] Precopulatory sexual selection by female choice also occurs.

[b] Nonvirgin females are attracted to male songs in nature (W. Brown, pers. comm.). Females remate readily in captivity.

[c] P2 is about 90%, so female remating strongly reduces the first male's reproduction.

[d] Experiments showed a strong effect of complete removal of the spermatophylax from the spermatophore, but did not examine effects of differences in size in the range of natural spermatophylaxes.

[e] Implantations of glands in the ejaculatory duct from males that had more than two blood meals were less effective in inducing oviposition by virgin females. In contrast, extracts of these glands from males that had not fed on blood were less able to switch off female receptivity than those that had had two blood meals.

[f] No data on sperm precedence available.

[g] Remating was more common when the male was very young, and apparently when he was small. The frequency of female remating after mating with a group of 4-day old "smaller" flies was 16%, while it was 5% after matings with a normal-sized group (N = approximately 90 for each group; exact data were not given; Riemann et al. 1967). Riemann et al. (1967) and Riemann and Thorson (1969) suppose that these males transfer less accessory gland material, but they do not give direct data. When males mate with three females during <8 hr, remating is less completely repressed in the third. Inhibition of accessory gland synthesis with cyclohexamide also caused an increase in female remating (Leopold et al.1971a,b) also suggesting a quantitative effect.

[h] Female response strongly alters number of sperm in female; this presumably has an effect on male reproductive success.

[i] Last male sperm precedence occurs in many Lepidoptera (Drummond 1984).

[j] Females remated readily in captivity; remating frequency in nature is unknown.

[k] Remating was relatively common and occurred readily in captivity, suggesting it occurs in nature.

[l] Delay in recovery of receptivity in the female is reproductively significant since first male advantage in fertilization increases with longer delays between copulation of first and second male (Huck et al. 1985). Duration of receptivity can also be reduced from 12–16 hr to 1 hr (Carter 1973). Timing with respect to ejaculations, numbers, and durations of the "long" nonejaculatory intromissions that inhibit female receptivity are variable (Bunnell et al. 1977). Stimulation of the vagina was important in inhibiting receptivity, since (1) injection of a local anesthetic abolished the inhibition, and (2) manual stimulation produced inhibition (Carter 1973).

[m] Females in captivity remated readily, but no field data are available. In the related species *M. neutoni*, the home ranges of several males may overlap >50% that of a given female, so several males may be familiar with the burrow of a given female and available for mating (Huck and Lisk 1986).

[n] Subordinate males ejaculate after fewer intromissions.

[o] Pacing was determined by the female, by moving in and out of the cage containing the male. Differences were found both in overt female behavior and in levels of serotonin and 5HIAA in the brain stem of the female.

[p] Remating was less common after mating with a "preferred" male, which gained at least 70% of the matings at a given lek.

widespread (e.g., references in Huck and Lisk 1986 on mammals; tables 6.1 and 6.3 and chapter 6 on insects and ticks).

Remating can also be advantageous for a female in some circumstances, however. She may be able to replenish sperm supplies, guard against male sterility, obtain better genes for offspring, or obtain resources from the male. If birds are representative examples, the unexpectedly high frequency of species in which apparently monogamous females nevertheless mate with other males argues that the benefits to females of polyandry may in general be greater than has often been supposed. Thus male and female interests may sometimes be in conflict (e.g., Simmons and Gwynne 1991), and females may not respond to all male signals that have evolved to induce resistance to remating.

Any cue from mating that originates in the male and is used by the female to trigger reduced receptivity could be subject to sexual selection. By becoming more choosy and only responding after mating with males who were especially good at delivering such stimuli, females could derive additional advantages from improved offspring quality.

Parker (1984) argued that conflict between male and female over inhibition of female remating is unlikely because males will win out due to the fact that selection is more intense on them than it is on females (see chap. 2.1). I think, on the contrary, that conflict is expected. Furthermore, cases in which it occurs have been documented.

Remating by females is probably often a complex process, involving a number of subsidiary events following copulation such as moving or not moving into an area where additional males may be present, fleeing or not fleeing when a male approaches, struggling or not struggling when he mounts, assuming or not assuming a receptive position, or opening or not opening the genital aperture when he attempts intromission. One such mechanism (maintain genital ducts soft so that second intromissions can occur) is discussed separately in the next chapter (chap. 4.8). The other mechanisms are discussed here as a group. Additional cases in which male seminal products induce refractive behavior in females are discussed in chapter 6. Although the decision by a female to remate is a qualitative, yes-no phenomenon, it probably often has quantitative aspects, such as the duration of her refractory period, the strength of her resistance to courting males, and so forth.

One example, in which male copulatory courtship (see chap. 5.1) apparently induces female refractory behavior, occurs in the solitary bee *Centris pallida* (Alcock et al. 1976; Alcock 1979; Alcock and Buchmann 1985). Male bees search for virgin females by flying low over areas where females are emerging from their underground pupation chambers. When a male locates an emerging female, he digs a shallow pit to free her and immediately grasps, mounts, and mates with her for 5–10 sec as she scrambles to the surface. After withdrawing his genitalia, the male usually flies with her to a nearby tree, where he begins a series of several approximately 60 sec bouts of activity, during which he strokes her vigorously with his middle and hind legs, his antennae, and his abdomen, and in addition produces a rasping, rattling song. After several bouts of stroking and buzzing, the male usually makes one or more additional genitalic contacts (it was not clear whether these involve intromissions or just superficial contact); sometimes he releases the female without any further genitalic contacts.

Dissections of females who had received only the first intromission (while the male dug them out of the ground) revealed many sperm in their spermathecae. Although precise sperm counts were not made, comparisons with females who had received several subsequent genitalic contacts led to the conclusion that "the majority, if not all, of sperm transfer takes place during the short initial intromission" (Alcock and Buchmann 1985, p. 237). They hypothesized that the additional genitalic contacts are tests by which "the male can perhaps determine whether a female will now refuse to copulate with other males," and they investigated this possibility with a simple experiment.

Some females were separated from their mates after the first intromission, before the male began to stroke and buzz. When placed together with new males (in the confines of an insect net in the field), their interactions were very different from those of females whose first mates had been allowed to perform stroking and buzzing. The "interrupted mating" females were often promptly mated by the second male (in 9 of 16 cases on the first genitalic contact). These copulations were usually followed immediately by stroking and buzzing (ten of sixteen cases). Copulation occurred on first genitalic contact in twenty-nine of thirty virgin females (dug up by the researchers and put in a net with a newly captured male), and stroking occurred in thirty of thirty tests. In contrast, "completed mating" females were often briefly contacted repeatedly by the male with his genitalia (presumably the male was attempting to inseminate her but failing), and seldom stroked (one of sixteen) before being released.

There was often intense competition among males for emerging females (fig. 7.5), and some males were displaced by others before the stroking and buzzing phase of mating could occur (Alcock et al. 1976). Alcock and Buchmann (1985) speculate that females may be increasing the chances that, despite the almost instantaneous insemination in this species, the last male they mate is able to hold his own in fights with other males. By not becoming unreceptive to further mating attempts until the male has demonstrated his ability to stay aboard her long enough to stimulate her with postinsemination courtship behavior, she will sometimes avoid mating with smaller, less competitive males. This would explain why males inseminate females so rapidly, but it leaves unexplained the stroking and buzzing performed by mounted males (surely not necessary to inform the female there is a male on her back). In any case, postinsemination courtship does have the effect of reducing the chances that the female will remate.

Studies of the complex mating process in the hamster *Mesocricetus auratus* illustrate that even subtle differences in female receptivity to additional matings can influence male reproductive success. The male hamster performs a series of 12–16 ejaculations, each of which is preceded by a series of nonejaculatory intromissions (see chap. 5.2 for a detailed description; Bunnell et al. 1977). Following the last ejaculatory intromission, the male performs an additional series of 9–27 longer, nonejaculatory intromissions. Only five ejaculations were needed to produce a 100% pregnancy rate. When two males mated with the same female, however, a male's chances of paternity were greater when he performed a higher number of ejaculations (Oglesby et al. 1981; Huck et al. 1985).

Females in one experimental group were allowed to receive five ejaculatory series, those in another received 12–15 ejaculatory series but only a single long postejaculatory intromission, and those in a third group were mated to female satiety (until the female attacked the male after 12–15 ejaculatory series plus 10–20 long intromissions). Then all females were placed with a second male 5 min later and allowed to copulate until the female drove off the male. The number of ejaculations achieved by the second male before being driven off was lower when the female had received higher numbers of long intromissions (Huck and Lisk 1986). Thus a female-determined characteristic of stimulation from the first male (the number of long intromissions permitted before she drives him off) strongly affected the duration of her receptivity to a second male, and thus the second male's chances of paternity. In this species it was not female receptivity per se, but duration of female receptivity that varied and had important reproductive consequences for males.

The salticid spider *Phidippus johnsoni* (fig. 3.9) provides another example in which the level of female acceptance in a first copulation determines the likelihood of rejection of a second male (Jackson 1980, 1992). This species is particularly illuminating due to the discrete variation in mating tactics of males, the extraordinary variability of copulation duration (from 0.3 to 2400 min), and the female's effects on both. When a male encounters an adult female outside her nest, courtship is elaborate and mainly visual, and copulation is relatively brief (av. 14 min but up to a maximum of 1078 min). When a male copulates with an adult female inside her nest, precopulatory courtship is mainly vibratory, and copulation is much longer (av. 110 min, but with a minimum of 1 min). A third type of interaction occurs when the male searches out a nearly mature female and waits to mate with her soon after she moults to the adult stage. These copulations

Figure 3.9 Portrait of a pair of *Phidippus johnsoni* spiders (male is above female). (Photo courtesy of Robert Jackson)

were the longest, averaging 863 min (but with a minimum of 6 min; Jackson 1978, 1980). Copulation duration was determined by the female, as in all of more than two hundred pairs it was the female rather than the male that terminated the copulation, walking or turning away from the male. Insemination is apparently very rapid (it is essentially all-or-nothing), and occurred as early as the first minute of a copulation.

The longer the female's first copulation, the less likely she was to remate when placed outside her nest and exposed to a series of four different males that courted her (fig. 3.10). For instance, the mean duration of the first copulation for females that were receptive on the second day after mating was 94 min, while it was 571 min for those that were not receptive. A sperm precedence study with X-ray sterilized males showed that remating can have a substantial effect on the paternity of a female's offspring.

Male-induced postmating reductions in female receptivity in the monarch butterfly *Danaeus plexippus* are of special interest because they show that a male's material contributions to a female (in this case a spermatophore) can function to inhibit remating, in addition to (or even instead of) as nutrition for the female. Females of this species mate repeatedly (the average lifetime number of matings is probably about 3.0–3.5; Oberhauser

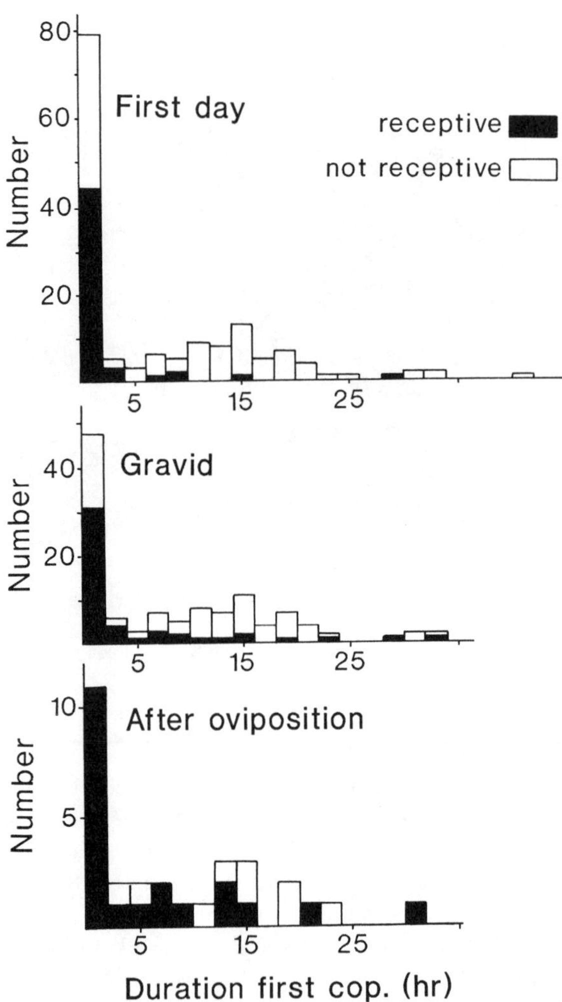

Figure 3.10 The behavior of a female jumping spider *Phidippus johnsoni* during her first copulation is correlated with the likelihood that she will remate. Females rather than males terminate all copulations. After shorter first copulations, there was a greater likelihood that the female would accept a second copulation. This relationship held when the second opportunity to mate occurred on the first day (*top*), on the next day, or, if she refused on the second day, when she was gravid (cumulative numbers of copulations) (*middle*), and also after she had laid a clutch of eggs (*bottom*) (all mating opportunities consisted of presenting the female with a sequence of 4 different courting males, usually separated by more than 4 min). (After Jackson 1980)

1992). By manipulating spermatophore size (through control of the previous mating history of the male) and determining the rate of spermatophore breakdown within the female, Oberhauser (1992) estimated that when a male produced a larger spermatophore he gained more through the delay effect on female receptivity than he did through nutrient contributions (some constant degree of last male sperm precedence, as is typical of lepidopterans, was assumed). Remating tended to occur after only about half the mass of larger spermatophores had been absorbed by the female (it was assumed that the nutritive values of different portions absorbed throughout the period of spermatophore degradation remained constant). In addition, the nutrient effect demonstrated in an earlier study (Oberhauser 1989) was relatively small, and was not even detectable when a single large spermatophore was compared with a small one.

A similar result was found with the zaprochiline tettigoniid *Kawanaphila nartee* (Simmons and Gwynne 1991). The length of a female's refractory period was correlated with the duration of spermatophore ampulla attachment to her genitalia (which in turn was correlated with the amount of sperm and probably other seminal products transferred to her). In females that had been maintained on a poor diet, this effect may have been due to nourishment provided by feeding on another portion of the spermatophore (the spermatophylax; e.g., see fig. 1.5). But in well-fed females the inhibition of receptivity was independent of male nutritional contributions in the spermatophylax.

Finally, it is important to note that a female can bias her possibilities of remating in relatively subtle ways (fig. 3.11) that will not always be easily discovered. For instance, by just moving more or less (*Drosophila melanogaster*; Scott et al. 1988), or by leaving a site where a former mate is trying to defend her (the spider *Nephila clavipes*; Christensen 1990), a female could increase her chances of mating with another male, even if there were no other male in the immediate vicinity at the time.

3.7 Sometimes Reduce Rate or Number of Offspring Produced

It is almost always advantageous for a female in a sexually reproducing species with internal fertilization to refrain from producing young (e.g., ovipositing) until she has been inseminated. As discussed in chapter 2 (and documented in chaps. 5 and 6), male stimuli from copulation often trigger oviposition. These stimuli can also affect the *rate* of oviposition (table 3.6). If a female is likely to remate, then a change in the rate of offspring production following copulation with a given male can alter the proportion of her offspring that is fathered by that male.

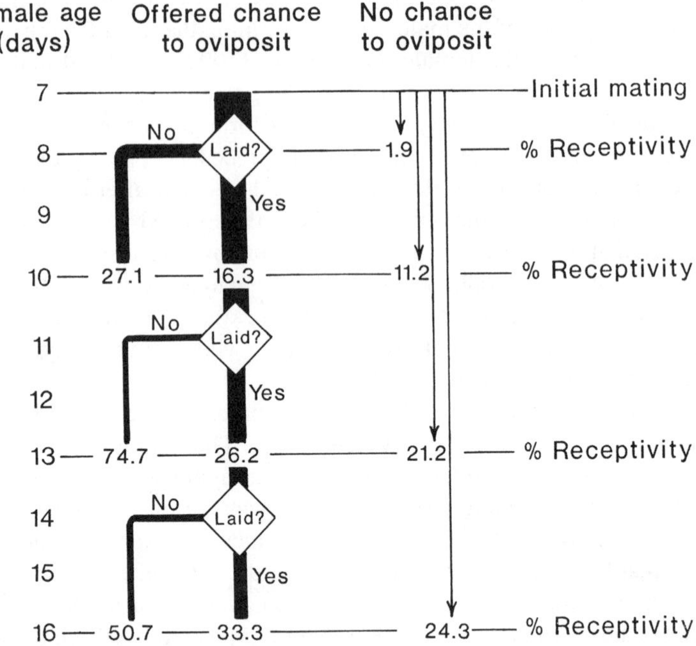

Figure 3.11 Decisions by females of the blowfly *Lucilia cuprina* may produce a compound female effect on male fitness via biases in both oviposition and remating. Those females which failed to oviposit after a given mating were also about twice as likely to mate again, making it especially likely that their former mate would not father their next batch of offspring. Males of this species transfer an accessory gland product or products inhibiting female remating and increasing female tendency to lay eggs that are mature (Barton-Browne et al. 1990). Both effects are apparently dosage dependent, and larger males are better able to render females refractory, at least when they mate with several females in succession (Cook 1992). Thus the illustrated female responses may represent a double premium via cryptic female choice for large male size. The thickness of the lines represents frequency of oviposition and lack of oviposition. (After P. Smith et al. 1989)

Possible biases in the effect on the rate of oviposition are especially interesting, because they are not necessarily predicted by natural selection on females. All other things being equal, one would expect a female to produce as many offspring as soon as she can after she mates. This should be especially true in a species in which female survival is relatively uncertain. Female delays in offspring production (other than those imposed physiologically by limitations in egg synthesis) are probably often due to delays in processes such as locating suitable oviposition sites, the advantages of dispersing eggs, negative correlations between larger reproductive efforts and female survival.

Table 3.6
Possible Examples in Which Females Exercise Cryptic Female Choice by Sometimes Reducing the Rate of Offspring Production or Number of Offspring Following Copulation

	Criteria						
Species	*I*	*II*	*III*	*IV*	*V*	*VI*	*References*
*Erythemis simplicicollis**[a] (Odonata)	Y	Y	Y	Y	Y (territory)	?	McVey 1988
Oecanthus nigricornis (Orthoptera)	Y	Y	?	Y[b]	Y (size)	?	Brown 1993, in prep. a,b
Decticus verrucivorus (Orthoptera)	Y	Y	Y?[c]	Y	Y (sptophyl. size)	?	Wedell and Arak 1989; Wedell 1991
*Harpobittacus nigriceps** (Mecoptera)	Y	Y	?	?	Y (size)	?	Thornhill 1983
Panorpa latipennis (Mecoptera)	Y	Y	Y?[c]	Y	Y (size, rapist)	?	Thornhill 1984
Panorpa vulgaris (Mecoptera)	Y	Y	Y?[c]	Y	Y (size, size of gift[d])	?	Bockwinkel and Sauer 1994
Hylobittacus apicalis (Mecoptera)	Y	Y	Y?[c]	Y	Y (size of gift)	?	Thornhill 1976
Lucilia cuprina (Diptera)	Y	?	?	?	?[e]	?	Smith et al. 1989
*Drosophila melanogaster** (Diptera)	Y	Y	Y	Y	Y (size)	?	Pitnick 1991; Gromko and Markow 1993
Coenorhabditis elegans (Nematoda)	Y	?	Y	Y	?[e]	?	Ward and Carrel 1979
*Ophioblennius atlanticus** (Pisces)	Y?	Y?	Y?	Y?	?[e]	?	Marraro and Nursall 1983
*Afrixalus delicatus** (Amphibia)	Y	Y	Y	Y	?[e]	?	Backwell and Passmore 1990
*Chiromantis xerampelina** (Amphibia)	?	?	Y	Y	?[e]	?	Jennions et al. 1992
Arvicola terrestris (Rodentia)	Y	?	Y	?	Y (dominance)	?	Potapov et al. 1993

NOTES: See tables 6.1 and 6.3 for numerous other possible examples involving male seminal products. Those marked with (*) are discussed in the text.

Each example is classified with respect to criteria for proof that cryptic female choice occurs. See table 3.1 for criteria and key.

[a] See chapter 1.2.

[b] Nonvirgin females in nature are attracted to male songs (W. Brown, pers. comm.). Females remated readily in captivity.

[c] Female response can strongly alter number of sperm in female; this presumably has an effect on male reproductive success.

[d] Females that had been kept without males allowed males with gifts to copulate for long periods, but rapidly terminated copulations with males lacking gifts.

[e] Possible cues for discriminating among males were not checked.

If females at least sometimes remate before laying all their eggs, and if remating results in losses of fertilization opportunities for the first male, then a male-female conflict of interest may arise. Selection on males would favor inducing the female to make especially large reproductive efforts soon after copulation. Selection on females could often favor lower rates of reproductive effort following copulation in order to raise offspring survival and her own life expectancy and total egg production. Male abilities to induce higher rates of offspring production could be favored by sexual selection, and female losses from discrimination against some males could be balanced by indirect benefits (via good viability or attractiveness genes) from superior fathers.

Females of several species apparently adjust the number of eggs they lay soon after mating on the basis of the characteristics of the male. For instance, female *D. melanogaster* that mated once with a small male laid more eggs in the 24 hr after copulation than did females that mated once with a larger male (Pitnick 1991; differences in female sizes do not explain this result, as female size was not correlated with male size, and was not correlated with egg production in two of the three trials). Oviposition on following days was similar, but the cumulative difference remained. Since females often remate, especially as time goes by after the first mating, and since the sperm from the female's last mating usually accounts for most of the fertilizations (Gromko et al. 1984), the increased rate of oviposition immediately following matings with some males probably confers a reproductive advantage. The oviposition-inducing effects of male seminal products in this species are associated with reduced female longevity (Chapman et al. 1995). There thus may be a cost to females of this male effect (Keller 1995), though the balance in terms of eggs laid during a lifetime under natural conditions is not known.

Another study, in which the phrase "cryptic female choice" was coined, also concerned changes in oviposition rates following copulations with different-sized males, in the scorpion fly *Harpobittacus nigriceps* (Thornhill 1983). The effect was the inverse of that in *D. melanogaster*, as females laid significantly more eggs in the first 10 hours immediately following copulation with a larger male than with a smaller male (fig. 3.12). Since *H. nigriceps* females remate frequently (Thornhill 1983), this difference in oviposition rate probably has consequences for male reproductive success.

In some species the male may affect female oviposition rate by nutrient donations, and separating male nutritional effects in which the female may be a passive offspring generator whose output is determined by male donations (the nonpromiscuous mating effort of Gwynne 1984c) from more active female control can be difficult (see chap. 6.6). Changes in oviposition by females of the black horned tree cricket *Oecanthus nigricornis* are

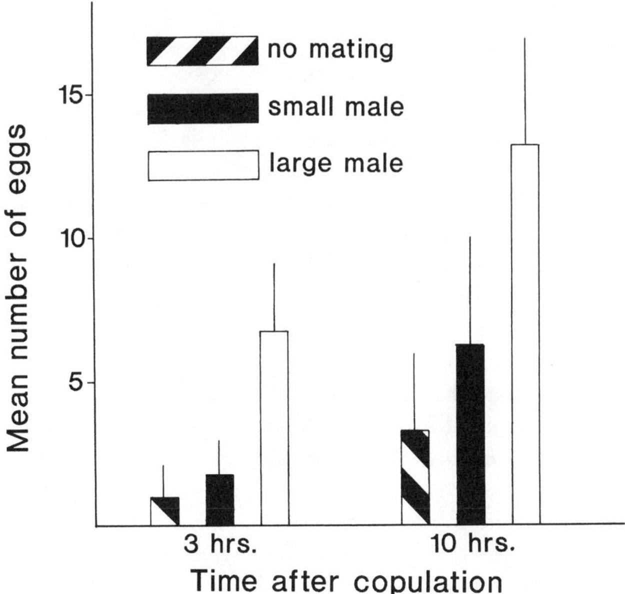

Figure 3.12 A classic case of cryptic female choice by biased oviposition. Female *Harpobittacus nigriceps* scorpionflies exercise cryptic female choice in favor of larger males (av. wing length = 22.1 ± 0.6 mm) over small males (av. wing length = 19.3 ± 0.5 mm) by laying more eggs in the hours immediately after copulation. (After Thornhill 1983)

interesting in this respect, since the female feeds on secretions of the male metanotal gland prior to, during, and following copulation. Females laid more eggs after copulations with larger males, but the difference in female response seems not to be due to nutrition. Male mass did not correlate with the mass of secretion ingested by the female (W. Brown, pers. comm.); there was no effect of the interaction between male mass and courtship feeding on oviposition rate; male mass was not the primary factor correlating with the mass of the courtship gift; and larger males were associated with increased oviposition rates even when gift sizes were held constant (Brown, in prep. a).

Animals with external fertilization have generally been omitted from this book, but cryptic female choice could occur via differences in the numbers of eggs laid with successive males. I have found only cases with suggestive rather than clear documentation, but will briefly discuss them to illustrate how cryptic choice could occur. The leaf-rolling frog *Afrixalus delicatus* is a possible candidate (Backwell and Passmore 1990). Fifteen of approximately one hundred females observed in the field did not lay all of their eggs with the first male that seized them. In seven of these cases the female was seen laying an additional batch of eggs during the next two nights—in one case only 30 min after the first male had apparently released her (in general, a male frog that has achieved amplexus cannot be dislodged by the female; A. S. Rand, pers. comm.). The possibility that the first males in these 15 pairs were somehow different was not checked. Polyandry (and thus a similar opportunity for cryptic female choice) also occurs in another frog, *Chiromantis xerampelina* (see fig. 1.1), in which a female lays 2–4 batches of eggs in the same foam nest in a single night (Jennions et al. 1992).

The frogs just mentioned are apparently unusual in that amplexed males sometimes abandon females before all the female's eggs are laid. Many other anurans show male displacement due to male-male battles, however. If females begin oviposition more quickly when they are in amplexus with some males than when they are with others, they could conceivably exercise choice by manipulating the probability that a given male would be displaced before all her eggs were laid (this would technically not be *cryptic* female choice, as the choice would occur prior to spawning). A possible case of this sort occurs in the toad *Bufo bufo*. Females sometimes wait for up to several days after being seized by a male before spawning, and males in amplexus are often replaced by others (Davies and Halliday 1979). The female delay in this species is especially intriguing because it is probably dangerous. Amplexed pairs are sometimes mobbed by other males, and not infrequently the female is drowned as a result (in four of twenty-one fights the female was dead or dying; Davies and Halliday 1979). Again, no attempts were made to determine whether females delayed longer when amplexed by certain types of males. Other anurans in which female delay in oviposition could influence a male's chances of paternity include *Ololygon rubra* (Bourne 1993) and *Rana sylvatica* (Howard and Kluge 1985). Males of several frog species court the female during amplexus (chap. 5.1), suggesting that female choice via egg production may occur.

Fish with external fertilization may also practice cryptic female choice by withholding eggs. For instance, spawning females of the blenny *Ophioblennius atlanticus* sometimes moved directly from the nest of one male to that of another, while others limited their spawning to a single male's nest (Marraro and Nursall 1983). The male of this species sometimes performs visual displays while a female is in his nest and presumably spawning, perhaps to induce her to lay more eggs. An alternative, nonexclusive interpretation is that he is threatening neighboring males. Natural selection on females to lay more eggs during visits to better oviposition sites and fewer eggs at others could also result in cryptic female choice, which in this case might reinforce the results of sexual selection by direct male-male battles to control superior oviposition sites. In another species of blenny, *Aidablennius sphynx*, it also appears that a female may sometimes lay only a few eggs during visits to some males (Kraak and Van Den Bergh 1992).

Oviposition is induced by products in the male's semen in many other species, and this may also result in sexual selection by cryptic female choice, as discussed in chapter 6.

3.8 Sometimes Forcefully Terminate Copulation before Sperm Are Transferred

In many species females struggle after copulation has begun, and they sometimes appear to cause premature terminations (fig. 3.13; table 3.7). Some forceful terminations occur before males have begun transferring sperm to the female, and in some species they seem precisely timed to prevent it. In the zorapteran insect *Zorotypus barberi*, for example, sperm transfer occurs at a relatively predictable time during copulation. In 74% of ninety-four cases in which females terminated copulations, termination occurred within the 5 sec just before or after transfer would have begun (Choe 1995). About 25% of *Micrathena gracilis* copulations end with the female spider having received sperm in only one of her pair of insemination ducts; one common reason was that the female cut the thread the male used after the first copulation to return to copulate on the other side (Bukowski and Christenson, in prep.). Female spiders in the genus *Argiope* interrupt some copulations more definitively by eating the male after being mated on only one side (Robinson and Robinson 1980; Sasaki and Iwahashi 1995).

Females of the caddis fly *Mystacides azurea* nearly always accepted genitalic coupling by any male who seized her in the mating swarm (only 3 rejections in 215 tandem pairs; the female rejected the male by keeping her abdomen pressed to the substrate). But females often broke off genitalic coupling, and in many cases (28% of 49) this occurred before sperm transfer had occurred (Petersson 1989). Heavier and younger males were apparently more often able to transfer sperm before the female-imposed "deadline." Since pairs in copulation are apparently more susceptible to predation by dragonflies, wasps, and spiders, the female may benefit by avoiding longer copulations (Petersson 1989).

In other species, sperm transfer is gradual, and the duration of a copulation correlates with the amount of sperm transferred. This occurs, for instance, in the scorpion fly *Hylobittacus apicalis*, in which sperm transfer to virgin females does not begin until 5–6 min after copulation starts, and then occurs at a constant rate for the next 15 min (Thornhill 1976; fig. 5.20). Females can apparently terminate copulation whenever they wish. Males allow females to feed on food items they have captured. By copulating longer when receiving larger food items, females exercise cryptic female choice by receiving more sperm from males that offer them larger prey.

Table 3.7
Possible Cases in Which Females Exercise Cryptic Female Choice by Sometimes Terminating Copulation Prematurely

	Criteria						
Species	*I*	*II*	*III*	*IV*	*V*	*VI*	*References*
*Zorotypus barberi** (Zoraptera)	Y	Y	Y	Y	?[a]	?	Choe 1995
Pseudoxychila spp.* (Coleoptera)	Y	Y?	Y[b]	Y[c]	Y (strength[d])	?	R. Rodriguez and Eberhard, in prep.
Macrodactylus spp. (Coleoptera)	Y	Y	Y?	?	?[a]	?	Eberhard 1993
Labidomera clivicollis[e] (Coleoptera)	?	Y	Y	Y	?[a]	?	Dickinson 1986, 1988, in press.
*Hylobittacus apicalis** (Mecoptera)	Y	Y	Y[b]	Y	Y (gift size)	?	Thornhill 1976, 1980b
Panorpa vulgaris (Mecoptera)	Y	Y	Y	Y	Y (saliv. masses)	?	Thornhill and Sauer 1991
*Culicoides melleus** (Diptera)	Y	Y?	Y	Y[c]	Y (virginity)	?	Linley and Adams 1974; Linley and Hinds 1974; Linley and Mook 1975
Drosophila simulans (Diptera)	Y[f]	N	Y	N	? (male genital.)	?	Coyne 1993
Coproica (5 species) (Diptera)	Y[g]	Y	Y	Y	?[a]	?	Lachmann 1994
Mamestra brassicae (Lepidoptera)	?	?	?	?	Y (hair pencils)	?	Birch et al. 1989
*Mystacides azurea** (Trichoptera)	Y	Y	Y	Y	Y? (speed transf.)	?	Petersson 1989
Leucauge sp. (Araneae)	Y	Y	?	Y	?[a]	?	Eberhard, in prep.b
Mesocricetus auratus (Rodentia)	Y	Y	Y	Y?[h]	?[a]	?	Huck and Lisk 1986
Macaca radiata (Primates)	?	?	?	?	?[a]	?	Nadler and Rosenblum 1969

Notes: Those marked with (*) are discussed in the text.

Each example is classified with respect to criteria for proof that cryptic female choice occurs. See table 3.1 for criteria and key.

[a] Possible cues for discriminating among males were not checked.

[b] The relatively large amounts of sperm involved in at least some cases suggest a probable effect on the male's reproductive chances.

[c] Remating was relatively common and occurred readily in captivity, suggesting it occurs in nature.

[d] Freshly captured, more vigorous males were better able to overcome the often violent resistance of freshly collected females.

[e] See chapter 8.1 for a detailed discussion of this species.

[f] Interruptions were observed when females were crossed with *D. melanogaster* males.

[g] Sperm is transferred late in copulations, and some copulations were interrupted early; it was not clear from the descriptions whether males or females were responsible for the interruptions.

Figure 3.13 A copulating pair of the tiger beetle *Pseudoxychila tarsalis*, in which females can affect a male's chances of paternity by brute force. The female sometimes interrupts copulation and forces the male off her back before any sperm are transferred. In other cases she forces the male off her back after he finishes an intromission and before he begins another, and then discards a fresh spermatophore still full of live sperm. Both male copulation behavior, which shows at least three different types of thrusting (the male in the lower photo is thrusting his genitalia deep into the female), and male copulatory courtship, in which the male rubs the female with his hind legs, differ between this species and its close relative *P. bipustulata*. (Photos courtesy of G. Hoebel and R. L. Rodriguez)

[h] Females in captivity remated readily, but no field data are available. In the related species *M. neutoni*, the home ranges of several males may overlap >50% that of a given female, so several males may be familiar with the burrow of a given female and available for mating (Huck and Lisk 1986).

Forceful termination of copulation by the female that results in failure to transfer sperm has also been seen in many other species. Forceful termination may also affect other critical processes in addition to sperm transfer. For instance, females of the lizard *Anolis carolinensis* sometimes break away prematurely from males, and short copulations of this sort may be associated with failure of the copulation to induce the usual female reluctance to remate (table 3.5).

A more subtle case occurs in the biting midge *Culicoides melleus* (fig. 3.14), which has been studied in especially fine detail by Linley and colleagues. At first glance their results appear not to fit the ideas being developed here, so they merit careful discussion.

Males and females of *C. melleus* seem to be *cooperating* in modifying the number of sperm in spermatophores, aborting some copulations before sperm are transferred and limiting the durations of others. Female resistance (at least that observed externally) is mostly expressed through what appears to be stylized communication rather than through physical force. There are several stages in male-female interactions where a male's chances of siring offspring with the female can be reduced: the female sometimes prevents genitalic coupling; the male sometimes releases the female before sperm transfer ("pseudocopulations"); the female sometimes discards the spermatophore before all sperm are transferred; and the male sometimes reduces the number of sperm he ejaculates. Untangling which sex is ultimately responsible for which process is complicated.

Mating occurs on the ground in intertidal, sandy areas in the immediate vicinity of breeding sites. Males are presumed to outnumber females at such sites (Linley and Hinds 1974; J. Linley, pers. comm.). Males leave females soon after copulation ends. A female midge can resist the sexual advances of males by running away, tipping her abdomen ventrally to prevent genitalic coupling, and kicking rearward with her hind legs (Linley and Adams 1974). Females usually kick at courting males prior to genitalic coupling, even those with which they subsequently mate.

A female's tendency to kick at a male after he has already achieved genitalic coupling varies with her age and reproductive status. Both older females and females that had already mated kicked more frequently, more strongly, and for longer (fig. 3.15). Other female resistance behavior, such as restless walking and depressing the tip of the abdomen ventrally, also occurred during copulation (Linley and Hinds 1975a).

Female kicking behavior sometimes apparently caused males to interrupt copulation before spermatophore transfer ("pseudocopulations"), or to reduce the numbers of sperm in the spermatophore. But males were apparently responding to stylized female signals, rather than to brute force. The moment of release was seldom associated with a period of particularly vigorous or well-directed kicking (fig. 3.15; Linley and Adams 1974); and

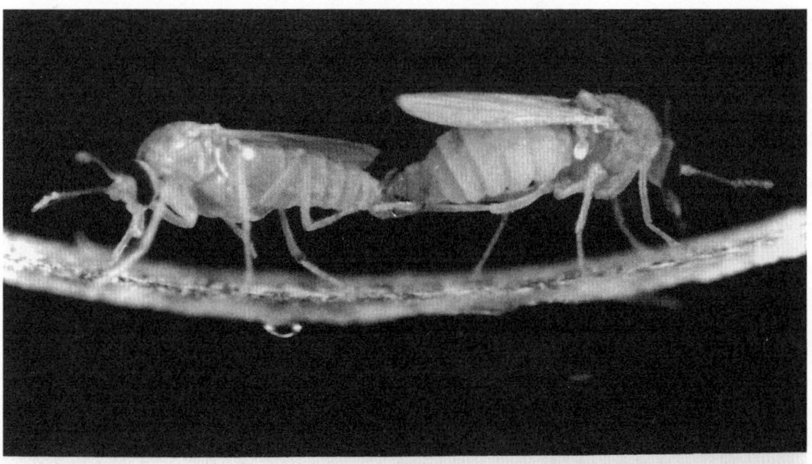

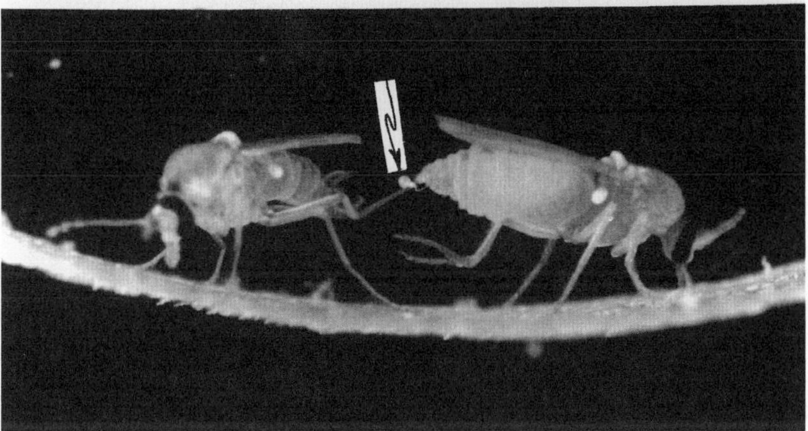

Figure 3.14 Copulation and communication in the biting midge *Culicoides melleus*. *Top*: The female (right) kicks the male's abdomen with her hind legs. Kicking behavior is more energetic when the female has mated previously and induces the male to shorten the copulation and to transfer smaller numbers of sperm in his spermatophore. *Bottom*: Nonvirgin females discard a larger proportion of the male's sperm in the spermatophore (arrow); the pair is in the process of separating. The spermatophore drops free from the female's genitalia soon after copulation ends. (From Linley and Adams 1972; photos courtesy of J. Linley, Florida Institute of Food and Agricultural Sciences)

"Release only rarely appears to be caused by direct force derived from female defensive grooming [kicking]" (Linley and Mook 1975, p. 97). Termination of copulation was probably under male control, since it apparently occurs only when the male releases his hold on the female's abdomen with his genitalic claspers (Linley and Mook 1975).

Pseudocopulations often occurred when a male that had just finished

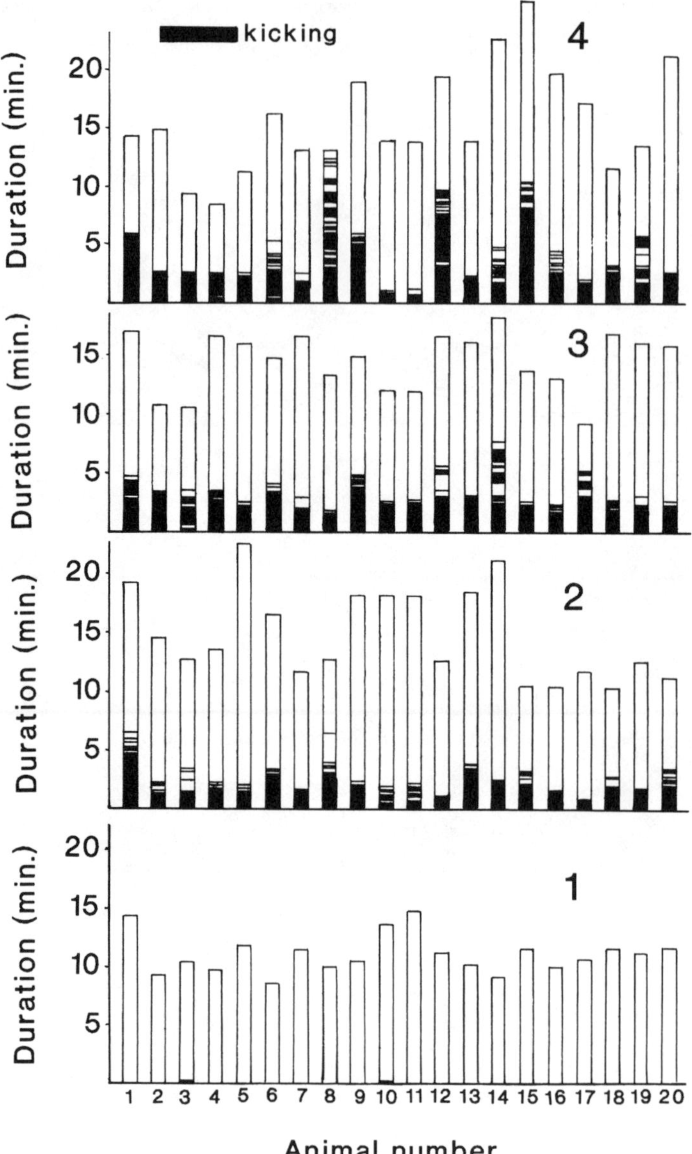

Figure 3.15 Timing and duration of kicking behavior of twenty different female *Culicoides melleus* midges, each of which copulated with four successive virgin males. Kicking (black) induces the male to transfer less sperm, apparently because it informs him of the greater likelihood that the female will discard his sperm instead of transferring it to her spermatheca (see Fig. 3.14). (After Linley and Adams 1974)

mating with a virgin female attempted to mate with her again, or when a different recently copulated female was placed with the male. Recently mated females can nevertheless be inseminated by second matings, as was shown by placing the female with a virgin male; and recently mated males are also physiologically capable of immediately mating again and transferring sperm, as was shown by placing the male with a virgin female.

Experimental removal of a once-mated female's hind legs (thus preventing her from kicking the male) resulted in both longer duration of pseudocopulations (which nevertheless did not result in complete copulations), and a higher frequency of complete copulations resulting from the second male's first mating attempt (Linley and Mook 1975). Female kicking behavior also had the effect of reducing the number of sperm in an ejaculate. When a virgin male mounted a recently mated female (rather than a virgin), he usually transferred a spermatophore despite her stronger kicking, but the number of sperm in the spermatophore was lower than in matings with virgin females (Linley and Hinds 1975a). The role of kicking in producing this reduction was confirmed again by experimentally removing the hind legs of some females.

Why has this communicatory system evolved in *Culicoides melleus*? This question can be broken down into several more specific questions, each of which I will discuss separately.

1. *Why do recently mated males suspend copulation and release kicking females before making a spermatophore, while virgin males and males that have "recovered" from copulation fail to respond to her kicking?*

From the point of view of a male that has just mated, response to female kicking before ejaculating may be a way to avoid investing a major part of his now depleted sperm supply in the same female. If the chances of rapidly encountering a different, recently mated or virgin female were high enough (such encounters may be quite frequent; J. Linley, pers. comm.) when weighed against the likely paternity to be obtained from mating with a female that has already mated (it is about 35%; Linley and Hinds 1975b), then refraining from a complete copulation with a kicking female might be compensated by the advantage of having more sperm for the next female. Since mating occurs at sites where virgin females are emerging, this may be reasonable, though the possibly greater numbers of males than females at these sites argues in the opposite direction.

2. *Why do virgin males reduce their ejaculate size in response to kicking from females?*

Reduction in ejaculate size may be advantageous for the male. When a nonvirgin female mated, she generally discarded the spermatophore with several hundred sperm still in it (despite the fact that there is room in the spermatheca for at least two complete ejaculates of virgin males; Linley and Hinds 1975b; see chap. 8.1). After couplings of similar length with virgin females, the discarded spermatophore was nearly empty. Movement of sperm into the spermatheca is produced by an active transport process of the female (Linley 1981), and second matings after longer delays (96 hr) result in nearly complete emptying of the spermatophore. Thus the residue of sperm in the spermatophore when the animals separate is probably due to female failure to move the sperm. Since nearly 100% of the sperm that leave the spermatophore arrive in the female's spermatheca in *C. melleus* (Linley and Hinds 1975b; Linley 1981), and since sperm precedence is apparently determined by relative numbers of sperm in the spermatheca (Linley 1981), this dumping represents a reduction in the male's chances of paternity (see chap. 8.1 for similar explanations of "undersized" ejaculates in other species).

3. *Why do females attempt to elicit these male responses?*
The benefit to a female of persistent kicking during the first few minutes of copulation, long after her original mate has left, is less obvious (she continues to be likely to kick more than a virgin female up to 70–74 hr after her first copulation; Linley and Adams 1974). This behavior will not function to avoid remating with the same male. Average totals of sperm stored in the spermatheca increase from 803 to 1237 to 1615 for a female that has mated one, two, and three times respectively, so sperm storage problems are also apparently not involved. Female kicking may be at least partially explained by economy in time, since a complete copulation lasts about 10 min while pseudocopulation lasts about 30 sec, or by avoidance of some other risk associated with mating (Daly 1978). It could also be a mechanism to leave room in the spermatheca for sperm from future mates. If the amount of sperm varies systematically between males, this could be a method of favoring some males over others by "testing" them.

In summary, a detailed examination of the adjustments in male copulatory behavior and ejaculate size in response to female communicatory behavior in *C. melleus* reveals that neither male nor female behavior contradicts the arguments being developed here. Of course the fact that such explanations are feasible does not prove that they are correct. More data will be necessary to test the explanations proposed here.

Apparent communicatory behavior by the female during copulation is not limited to these midges. Similar, apparently stimulatory behavior during copulation includes quivering the legs and pedipalps, fidgeting legs, wing buzzing, rhythmic rubbing with the legs, and rhythmic rocking forward and backward in a diplurid spider (Coyle and O'Shields 1990), a pholcid spider (Huber 1994a), tsetse flies (Jaenson 1979), a phorid fly (Wcislo 1990), lygaeid bugs (R. Rodriguez, in prep.), and largid bugs and tridactylid orthopterans (Eberhard 1994).

Female vocalizations also occur during copulation in a variety of other groups of mammals and birds (Diakow 1973; Cox and LeBoeuf 1977; Montgomerie and Thornhill 1989). These are, in contrast with the signals just discussed, "public" rather than "private" signals, and may in some cases function to communicate with other males. For instance, in the baboon *Papio cynocephalus*, context-specific variations in the female's copulation calls suggest that they function to induce other males to copulate with her (O'Connell and Cowlishaw 1994). The female may gain by promoting interejaculate competition, or by creating paternity uncertainty that reduces the chances of subsequent infanticide by males.

3.9 Sometimes Fail to Ovulate

Females of a variety of animals, including many mammals (Jöchle 1975; Milligan 1982), ticks (Galun and Warburg 1967; Khalil and Shanbaky 1975), and insects (Roth and Stay 1961; Chen 1984; Raabe 1986; Gillott 1988) refrain from ovulation (releasing mature eggs from the ovaries so they can be fertilized) until they have mated, and use male stimuli from copulation to trigger ovulation. Differential ovulation responses by the female (table 3.8) could affect male reproductive success in several ways.

In general, if not all copulations are followed by ovulation or if not all mature eggs are ovulated after each copulation, a male trait that increases the likelihood of ovulation would be favored because it would directly increase the number of offspring the male could sire. In addition, greater rapidity of ovulation could improve male reproduction. A male might increase his reproductive success by inducing ovulation that coincided with the peak numbers of his own viable sperm present at fertilization sites in the female. He could also benefit through the reduced probability that additional males would have mated with the female in the meantime.

As in reduction of offspring production (see sec. 3.7), female reduction in ovulation can be particularly dramatic and otherwise paradoxical, since natural selection will generally favor *greater* female fecundity.

An apparent example of modification of ovulation frequency occurs in the lion *Panthera leo*, a species with an unusually low rate of offspring per copulation (a female lion averages about 1500 copulations to produce each litter; Packer and Pusey 1983). Female fertility changes predictably, probably due to changes in ovulation, in ways giving a cryptic advantage to territory-holding males.

When a new male or coalition of males takes over a pride of females, the males kill young cubs as well as those born soon after, and they evict older cubs. This has the effect of causing the females to begin their estrus cycles. Females that come into estrus mate readily with the new males, but the

Table 3.8
Possible Examples in Which Females Exercise Cryptic Female Choice by Sometimes Failing to Ovulate

	Criteria						
Species	*I*	*II*	*III*	*IV*	*V*	*VI*	*References*
Glossina morsitans (Diptera)	Y	Y[a]	?	Y	Y (length of cop.)	?	Saunders and Dodd 1972; Jaenson 1980
*Rattus norvegicus** (Rodentia)	?	?	Y	Y	Y[b] (vag. stim.)	?	Zarrow and Clark 1968
Clethrionomys glareolus (Rodentia)	Y	?	?	?[c]	?[d]	?	Clarke and Clulow 1973; Clulow and Mallory 1974
Microtus agrestris (Rodentia)	Y	?	?	?	?[d]	?	Clarke and Clulow 1973; Milligan 1982
*Microtus ochrogaster** (Rodentia)	?	?	?	?	Y (No. intro., No. thrusts)	?	Gray et al. 1974
Microtus montanus (Rodentia)	Y	?	Y	?	Y (No. ejac. ser.)	?	Davis et al. 1974
*Panthera leo** (Carnivora)	Y	Y	Y	Y	Y (hold pride)	?	Packer and Pusey 1983
Macaca mulatta (Primates)	Y	Y	Y	Y	?[d]	?	Berman et al. 1994
Meriones unguiculatus (Rodentia)	Y	?	Y	Y	Y (No. ejac. ser.)	?	Agren 1990
20 other mammals (5 different orders)	?	?	?	Y/N	?[d]	?	Milligan 1982

NOTES: Those marked with (*) are discussed in the text.

Each example is classified with respect to criteria for proof that cryptic female choice occurs. See table 3.1 for criteria and key.

[a] Experimental shortening of copulation decreased ovulation, but only one of 24 copulations with a conspecific that went to completion resulted in lack of ovulation. Of 7 females inseminated in cross-specific matings with a *G. austeni* male, 1 did not ovulate.

[b] Evidence that ovulation is responsible is indirect. Increased numbers of ejaculatory series (which entail increased vaginal stimulation) result in larger litters, and artificial vaginal stimulation induces ovulation. Females control the number of ejaculatory series both by actively inciting copulations and by leaving a male after one or more copulations.

[c] Multiple sets of corpora lutea in wild females may indicate successive matings (with the same or different males).

[d] Possible cues for discriminating between males were not checked.

[e] Conservative criteria were used for inclusion; an earlier, less conservative list included forty-one species (Jöchle 1975).

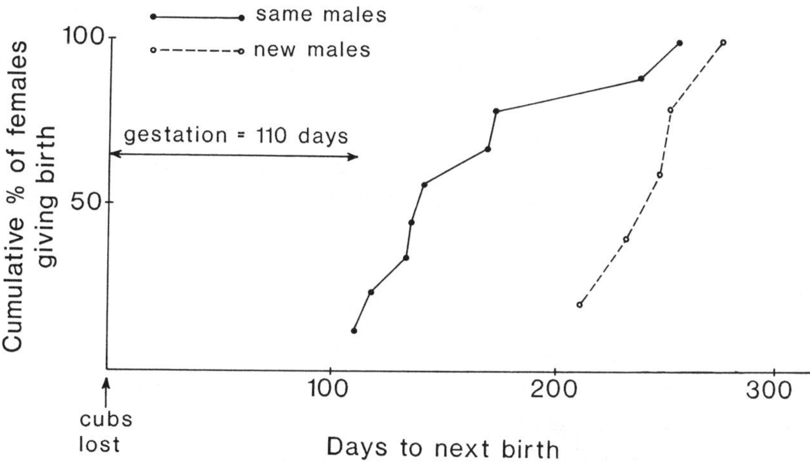

Figure 3.16 An apparent female bias in ovulation. Copulations of female lions which have just lost their cubs (at time 0) are more likely to result in offspring when the male is familiar (solid line) than when the male is new to the female, having just taken over the pride (dotted line). This change in fertility is apparently an adaptation to reduce the female's chances of losing cubs to infanticidal males. It also results in cryptic female choice favoring males better able to maintain long-term control over prides. (After Packer and Pusey 1983)

chances that these matings will result in offspring are unusually low. Females that had lost their cubs when new males took over their pride went through 6–9 estrous cycles before becoming pregnant, whereas most females that lost their cubs under other circumstances (no change of males) conceived in their first or second estrus. The resulting additional average delay was 110 days to the birth of the next litter (Packer and Pusey 1983; Fig. 3.16). A similar increase in delay to conception occurred in females at other reproductive stages (Packer and Pusey 1983).

The female's reduced fertility was associated, paradoxically, with an *increase* in her sexual activity. Copulation frequency (about 2–3/hr) was similar after a takeover, but females initiated significantly more of the copulations and more often consorted with more than one male after a takeover. By a process of elimination, failure to ovulate seems the most likely cause of the reduced infertility. In an extensive sample of female lions, a very high percentage of ovulations (deduced from corpora lutea and scars on the ovary) resulted in implantations (96.5%; Smuts et al. 1978). Pseudopregnancy, abortions, and embryonic deaths are also apparently quite rare (Smuts et al. 1978).

Delayed ovulation is probably related to the fact that a female lion needs to be protected from new, infanticidal males for about 25 consecutive months in order to conceive, gestate, and raise a litter of cubs to independence (Packer and Pusey 1983). Tenure length of males differs dramatically. That of coalitions of 1–2 males is seldom as long as 25 months, that of groups of 3 males is usually more than 25 months, and that of groups of 4–6 males is often longer than the 46.5 months needed to rear two successive litters (Packer and Pusey 1983). Since male takeovers after another group of males ceased defending a given group of females are probably often by relatively small coalitions (92.5% of forty nomadic coalitions seen passing through territories were singleton males or pairs of males), a pregnancy immediately following a change of males would probably be especially unlikely to result in successful raising of young. In addition, the heightened sexual activity of females following a change in males (all females that have lost cubs become receptive) may attract males to the pride's range (no data are available on this point). Packer and Pusey (1983) deduced that the infertility period of females is an adaptation by which females incite competition between different male coalitions and increase their chances of avoiding subsequent loss of their offspring to infanticidal males. They presented a quantitative model indicating that such tactics are feasible.

This pattern of changes in female fertility probably represents cryptic female choice favoring those males best able to maintain long-term control of prides through membership in larger coalitions or because of better fighting ability. Thus in this case sexual selection by cryptic female choice appears to reinforce intrasexual sexual selection, which also favors male aggressive abilities.

In many (perhaps most) mammals, the likelihood of ovulation is increased by copulation, apparently as a result of a surge of luteinizing hormone (LH) associated with coitus (fig. 3.17; Milligan 1982). Milligan listed twenty-two species in seventeen genera in five orders as mating-induced ("reflex") ovulators, and mentioned twenty-four more species (including three additional orders) in which claims of reflex ovulation have been made. Jöchle (1975), using less restrictive criteria, listed forty-one species of mammals (seventeen were not included in either of Milligan's lists) in thirty-one genera.

Even in many species that are usually classified as spontaneous ovulators (e.g., cow, mouse, rat, sheep, swine, rhesus, perhaps also humans), copulation and/or stimuli mimicking copulation (e.g., stimulation of the clitoris of the cow) can cause ovulation to occur earlier than it would have otherwise (e.g., Zarrow and Clark 1968; Jöchle 1975), or cause ovulation when the normal cycle has been blocked (e.g., in rats: Milligan 1982). In some species some females ovulate spontaneously while others do not and copulation can increase ovulation rates. For instance, in the llama *Lama*

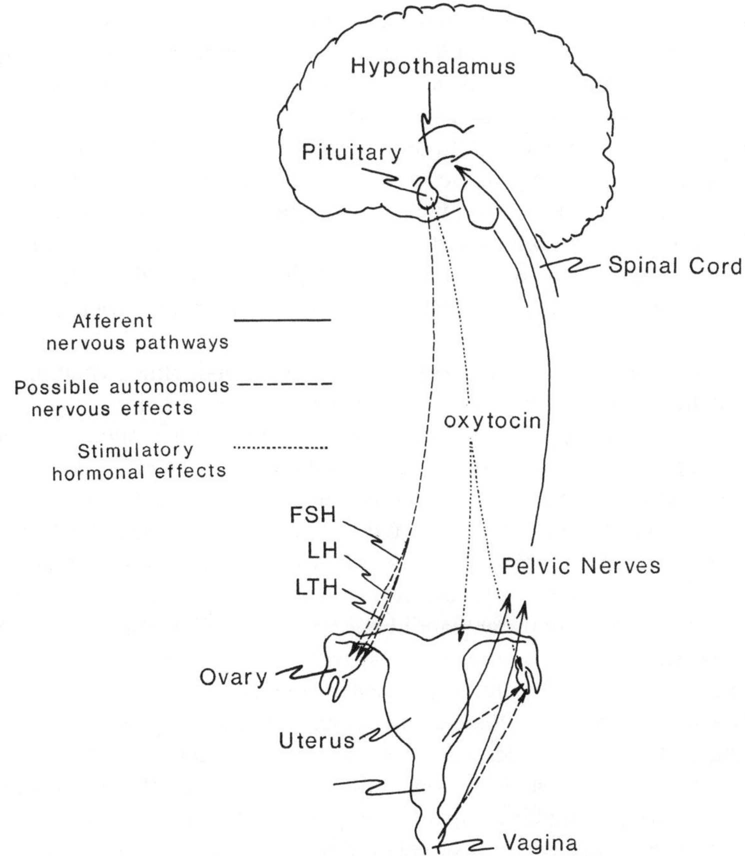

Figure 3.17 The multiple pathways by which stimuli from copulation induce ovulation in mammals. Coital stimuli are commonly either necessary to induce ovulation or to induce it to occur earlier than it would have otherwise. (After Jöchle 1975)

glama, 15.0% of twenty unmated females ovulated, while 82% of thirty-three mated females ovulated (England et al. 1969).

Induction of ovulation can be reproductively important for males of various species. For instance, short and long-term ovulatory failure is apparently extremely widespread in humans (Wasser and Isenberg 1986), and failure to respond to normal ovulatory stimuli is "not uncommon" in rabbits, cats, mink, and alpacas (Milligan 1982).

In addition to possible effects of copulation on female hormone titers, the exact moment of ovulation may also be influenced more directly by inducing contractions of the ovary. Such contractions have been observed in both reflex ovulators (cats) and spontaneous ovulators (humans) (Palti and Freund 1972). When the tension produced by an ovarian contraction exceeds the tensile strength of the stigma area of the mature follicle (which is presumably gradually weakened in a spontaneous ovulator by the breakdown of collagen), the stigma opens, the follicular contents are extruded, and ovulation takes place (Palti and Freund 1972). A male's ability to induce more or stronger contractions of the ovary while his sperm are in position for fertilization could thus result in increased chances of paternity. Milligan (1982) summarized studies of stimuli that induce ovulation in mammals, saying that "in general it appears that any stimulus not involving intromissions tends to be very much less effective than unrestricted coitus" (p. 18).

The importance of copulatory stimulation in inducing ovulation has been demonstrated experimentally by interrupting copulations at different stages in the prairie vole *Microtus ochrogaster*. A larger number of preejaculatory intromissions and thrusts resulted in a greater percentage of females ovulating (fig. 3.18). The importance of thrusting per se was indicated by differences among the females in a given experimental group that received different numbers of thrusts. Thus, of the females permitted eight intromissions, those that ovulated received significantly higher numbers of thrusts (av. 28.8) than those that did not (av. 17.2) (Gray et al. 1974).

As is predicted if sexual selection is involved (see chaps. 5.1, 5.2, 6.2 for detailed discussions), there is apparently considerable variation between closely related species in the type and range of stimuli capable of inducing ovulation. For instance, females of some rodents respond to artificial mechanical stimuli, but others do not (Milligan 1982). The effects are sometimes relatively complex. When female voles *Microtus ochrogaster* were limited to receiving twenty-five pelvic thrusts without an ejaculation from a male, conspecific males proved to be much more effective in inducing ovulation than were males of *M. pennsylvanicus* or *M. montanus* (Kenney et al. 1978).

Milligan (1982) notes that in these and other groups, such as cats, rabbits, and mink, relatively little stimulation is necessary to induce ovulation in most experimental females, and argues that species-specificity in stimuli may not be biologically important. This conclusion is weakened, however, by the fact (noted by Milligan himself) that the physiological state of a female can alter the likelihood that she will ovulate, and that induction in the females used in most tests was probably relatively "easy," since the females were in good health and well fed (see chap. 1.8, "Morphology"). Even among such females in the tests a few did not respond, and advantages could accrue to a male able to induce ovulation in such "difficult" females. For example, female lab rats responded more often with ovulation when coital stimulation was increased (Aron et al. 1966). There were also

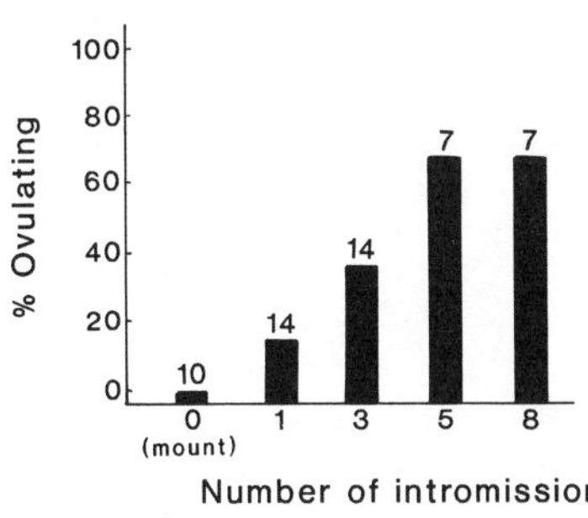

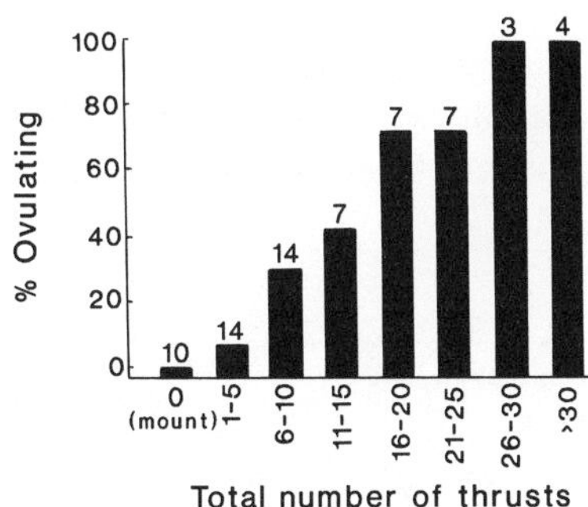

Figure 3.18 The likelihood of ovulation in the prairie vole *Microtus ochrogaster* increases with increased genitalic stimulation by the male. Control copulations averaged 9.5 intromissions and a total of 42.7 thrusts. When copulations were interrupted at different stages, the greater the stimulation the male had delivered, the greater the proportion of females which ovulated (sample sizes above each bar). (after Gray et al. 1974)

differences in responses to coital stimulation between females of different genetic strains (Aron et al. 1966).

The question of how many eggs are ovulated when ovulation occurs has been studied less often. In at least one species the number of eggs varies with male stimulation. When female rats were artificially stimulated by rapidly inserting and withdrawing a glass rod from the vagina in several bursts lasting 15 sec, a highly significant increase in the number of eggs ovulated resulted from five such bursts as compared with one or two bursts (Zarrow and Clark 1968).

3.10 Sometimes Fail to Mature Eggs (Vitellogenesis)

Another reproductive process that females may advantageously delay until after copulation has occurred is the development and maturation of the eggs in her ovary. Improvements in several different aspects of a male's ability to induce vitellogenesis, including increasing the proportion of his mates in which he induces vitellogenesis, increasing the rapidity with which they mature their eggs, and increasing the number of eggs they mature, could have important reproductive payoffs for males. Vitellogenesis often consists of several subprocesses, such as feeding, mobilization of reserves such as fat bodies, and synthesis of egg materials. These mechanisms will be lumped together in this discussion (table 3.9).

The red-sided garter snake *Thamnophis sirtalis* provides an example in which vitellogenesis is triggered by stimuli associated with copulation,

Table 3.9
Possible Cases in Which Females Exercise Cryptic Female Choice by Sometimes Failing to Perform Vitellogenesis (Mature Eggs) or Performing It More Slowly

	Criteria						
Species	*I*	*II*	*III*	*IV*	*V*	*VI*	*References*
Coenorhabditis elegans (Nematoda)	Y	Y	Y	?	Y (mutant male)	Y	Ward and Carrel 1979
*Thamnophis sirtalis** (Reptilia)	?	?	?	Y	?	?	Schwartz et al. 1989; Mendonça and Crews 1990

NOTES: The species marked with (*) is discussed in the text. Each example is classified with respect to criteria for proof that cryptic female choice occurs. See table 3.1 for criteria and key.

when the snakes emerge from hibernation in the spring. A two-stage response occurs in the female. First, about 15 min after copulation, there is a relatively quick and acute surge in the blood of prostaglandin PG F2 alpha; later, starting about 4 hr after copulation and peaking at about 24 hr, there is a surge of estradiol-17 beta (Whittier et al. 1987; Whittier and Crews 1989). In addition to other possible effects, the estrogen may cause oviductal contractions that transport sperm (Halpert et al. 1982). The hormonal responses trigger long-term ovary development and maturation of eggs, which occurs over the next several weeks.

By transecting the female's nerve cord midway between the ovary and the cloaca, thus denervating the cloaca, and also by applying anesthetics topically to the cloaca, Mendonça and Crews (1990) showed that stimuli from copulation (as opposed to male courtship or hormonal effects of his seminal products) were responsible for triggering both the surge of estrogen and ovarian development. This example is incomplete as a demonstration of cryptic female choice, as no attempts were made to determine whether some males triggered these processes more effectively than others.

Several additional cases in which male seminal products induce vitellogenesis in insects and ticks are discussed in chapter 6.

Notes

1. The question of whether or not there must be genetic variation in a character if selection is to act on it in the present is controversial (Endler 1985). My interest here includes the past, as it is also of interest to know whether cryptic female choice influenced the present states of male and female characters. Since cryptic female choice could have exerted influence in the past even though there is no genetic

variation in the present (in fact, it could have been *responsible* for the present lack of genetic variation), this criterion is conservative. As noted in chapter 2, natural selection may favor female discrimination on male traits that affect cryptic female choice even during periods when lack of genetic variation in males makes changes in gene frequency due to sexual selection impossible.

4 Other Mechanisms of Cryptic Female Choice

4.1 Sometimes Fail to Prepare Uterus for Embryo Implantation

In mammals, initiation of the luteal hormone cycle causes changes in the cells lining the uterus which are necessary for embryo implantation to occur. In some species, such as the guinea pig, the luteal cycle is spontaneous, occurring as part of estrus (Chester and Zucker 1970). In several species of rodents, however, stimulation of the female's vagina during copulation is necessary to trigger the luteal cycle (table 4.1). In both the rat *Rattus norvegicus* and the mouse *Mus musculus*, two different types of experiments have demonstrated that variation in both the number and pacing of intromissions affects the likelihood that the luteal cycle will be triggered.

One type of experiment manipulated the amount of copulatory stimulation received by the female prior to ejaculation. A male rat was allowed to perform several preejaculatory intromissions with one female, and then the female was removed and replaced with another. The second female usually received an unusually low number of intromissions before the male ejaculated. And the first female, when placed with a second male, usually received an unusually high total number of intromissions before ejaculation. The luteal cycle was triggered in 67% of females receiving 6–9 intromissions (including the ejaculatory intromission), and in 86% of those receiving 10–16 (Chester and Zucker 1970). Edmonds et al. (1972) confirmed this effect by finding that the number of intromissions (none ejaculatory) correlated with the probability of triggering the luteal cycle. By allowing some females to receive especially high numbers of preejaculatory intromissions (≥14) but not an ejaculation, and then injecting semen into the female's uterus, Adler (1969) showed that neither the ejaculatory behavior of the male nor the copulatory plug deposited during ejaculation were necessary to induce the luteal cycle. Since subordinate males performed an average of only 4.4 intromissions (compared with 6.1 for dominant males; McClintock et al. 1982a; fig. 4.1), such differences may have important reproductive consequences for males. Interpretation is complicated, however, by the fact that females are often accompanied by "rushes" of multiple males in nature (McClintock 1984); one male might induce the luteal cycle but fail to father any offspring. This factor would certainly dilute but

Table 4.1
Possible Cases in Which Females Exercise Cryptic Female Choice by Sometimes Failing to Prepare Uterus for Implantation.

	Criteria						
Species	*I*	*II*	*III*	*IV*	*V*	*VI*	*References*
*Mus musculus** (Rodentia)	?	?	Y	?	Y (ejac. resp., cop. thrusting)	?	McGill 1962; Land and McGill 1967; McGill et al. 1968; Diamond 1970
*Rattus norvegicus** (Rodentia)	Y	Y	Y	Y	Y (pattern introm.)	Y	McClintock 1984
*Mesocricetus auratus** (Rodentia)	?	?	Y	Y?[a]	Y (vag. stim.)	?	Huck and Lisk 1985; Huck et al. 1986a,b
Peromyscus maniculatus (Rodentia)	Y	Y	Y	Y	Y (No. intro.)	Y	Dewsbury and Baumgardner 1981; Lovecky et al. 1979; Birdsall and Nash 1973
*Peromyscus eremicus** (Rodentia)	?	?	Y	?	Y (No. postejac. introm.)	?	Dewsbury and Estep 1975
Microtus agrestis (Rodentia)	?	?	?	?[b]	Y (vag. stim.)	?	Milligan 1982

NOTES: Those marked with (*) are discussed in the text.

Each example is classified with respect to criteria for proof that cryptic female choice occurs. The criteria are as follows I: Do females treat some males differently from others? II: Are differences in female treatment of males likely to occur in nature? III: Do differences in female treatment of males result in differences in male reproduction? IV: Do females mate with more than one male in nature? V: Do females use male characteristics to determine how to treat males? VI: Are variations in the male character used by females determined genetically? (Y = yes; Y? = probably yes but no direct data; N = no; ? = no data available)

[a] Females in captivity remated readily, but no field data are available. In the related species *M. neutoni*, the home ranges of several males may overlap >50% that of a given female, so several males may be familiar with the burrow of a given female and available for mating (Huck and Lisk 1986).

[b] Multiple sets of corpora lutea in wild females suggest successive matings (with the same or different males).

not necessarily eliminate the selective advantage to a male of inducing the female's luteal cycle.

The timing between intromissions in rats is also important. One method of altering the pacing of the intromissions was to allow the female to escape periodically from the male during mating. In such cases the female often retired for a short time after each preejaculatory intromission, then returned. This resulted in longer delays between intromissions. Longer delays resulted in higher percentages of females initiating the luteal cycle after lower numbers of intromissions (Gilman et al. 1979).

One particularly striking effect was that female rats have relatively long-term "memories" of copulatory stimulation. Ten preejaculatory intromis-

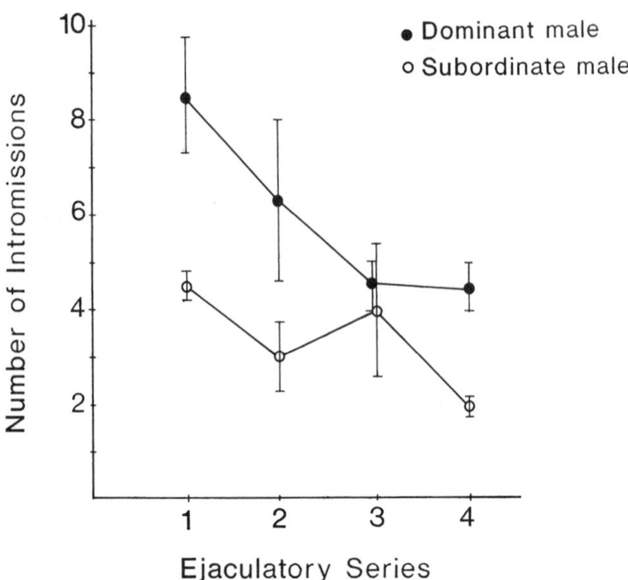

Figure 4.1 Social status of a male lab rat influences the number of intromissions preceding ejaculation and thus the probability that he will trigger the female's luteal cycle. Higher numbers of intromissions (6–9) preceding ejaculation are needed to trigger the female's luteal cycle, which prepares her uterine lining for implantation of the embryo. Subordinate males ejaculate more prematurely. Rapid ejaculation may be a holdover from the wild, to avoid being displaced from the female before having a chance to inseminate her (tame lab rat males seldom fight over females). When a subordinate male is placed alone with a female, he switches to the higher numbers of intromissions before ejaculation typical of dominant males. (After McClintock et al. 1982a)

sions resulted in induction of the luteal cycle more often than five similar intromissions, even when the first five were separated by 4 hr from the second five (Edmonds et al. 1972). The importance of vaginal stimulation was confirmed by the finding that cutting the nerves connecting the vagina with the CNS abolished the female response (Carlson and DeFeo 1965 in Zarrow and Clark 1968).

Similar experiments with mice (McGill 1962; Land and McGill 1967; McGill et al. 1968) showed that, in contrast, the most effective trigger for the female luteal cycle was the dramatic male ejaculatory response, which can last for up to 25 sec, and includes an increase in the rate of thrusting, shuddering, grabbing the female with all four limbs, and swelling of the penis, especially at the tip. Thrusting during preejaculatory intromissions did not trigger the cycle. Matings with males from which the coagulating glands and seminal vesicles had been surgically removed showed that the sperm plug (deposited during the ejaculatory reflex) was also not involved in inducing the luteal cycle. Although female mice may mate with more than one male during an estrus (see sec. 4.5), a female is generally accompanied by a single male at a time, so the male advantage of inducing the luteal cycle would not be as diluted as in rats.

In a second type of experiment, the vagina of an estrous female was stimulated artificially. DeFeo (1966) showed that stimulating the vagina of an estrous rat with a rod reliably induced the female luteal cycle, and he mentioned that other techniques such as electrical stimulation of the cervix have similar effects. When an estrous female mouse was stimulated in different patterns with a brass rod attached to a vibrator, the combination of number, interval, and duration of insertions that most closely resembled normal male copulatory behavior was associated with most frequent triggering of the luteal cycle (Diamond 1970). Use of an artificial penis and then a cotton swab (to simulate the copulatory plug) resulted in increased pregnancy rates from artificial insemination (other female processes, such as sperm transport and ovulation may also have been involved, however, in affecting pregnancy rates; Leckie et al. 1973).

Different races of both rats and mice vary in both the numbers of intromissions preceding ejaculation and the time between intromissions, as well as in other aspects of mating behavior (McGill and Ransom 1968; McGill 1962, 1977; McClintock 1984). Since these strains were derived from wild populations, it is likely that there are genetic differences between males in nature for these characters (McGill 1977).

The golden hamster *Mesocricetus auratus* provides an even more striking example of male induction of the luteal cycle. The male performs several series of 2–3 sec intromissions, and each series terminates with an ejaculatory intromission that is slightly longer and is accompanied at first by especially rapid thrusting (Bunnell et al. 1977). The male usually makes 12–14 ejaculatory series of this type, and then performs 10–24 longer, postejaculatory intromissions; during each of the latter, he thrusts rhythmically for 10–30 sec (Bunnell et al. 1977; Huck and Lisk 1985). These longer intromissions are not associated with sperm transfer, though seminal fluid may be emitted (Bunnell et al. 1977). The female finally ends the copulation by turning on the male and attacking him (Huck and Lisk 1985).

When the types of intromissions a female received were experimentally manipulated, it emerged that the postejaculatory intromissions (or possible emission of seminal products during such intromissions) were particularly effective in producing pregnancy initiation, apparently by triggering the female's luteal cycle (Huck and Lisk 1985; Huck et al. 1986b). The number of shorter, preejaculatory intromissions also showed a positive correlation with the probability of triggering the female's luteal cycle. The effect of postejaculatory intromissions was particularly strong if the male was limited to few preejaculatory intromissions and a single ejaculation, as occurred if the male had mated recently with another female (Huck and Lisk 1985). The effect was also strong when the male was allowed only five ejaculations as opposed to thirteen, when the female was mated later in her receptivity period, or when she was older and multiparous (Huck et al. 1986b). It is not clear how often matings in rapid succession by males occur under natural conditions; females associate with only one male at a time and terminate interactions by attacking males (Oglesby et al. 1981).

Vaginal stimulation may also be involved in triggering additional processes that affect the likelihood that hamster pregnancies will be carried full term. Although the effects of vaginal stimulation in inducing pregnancy initiation in artificially inseminated females were mimicked by injection of progesterone, only 18.8% of the injected females delivered live offspring versus 79.2% in the stimulated females (Diamond 1972).

Postejaculatory intromissions also occur in other rodent species (Dewsbury 1975). A pregnancy-inducing effect of such intromissions was demonstrated in the cactus mouse *Peromyscus eremicus*, as most females that did not receive them failed to conceive (Dewsbury and Estep 1975). Similar effects from intromissions may occur in other rodents. The number of preejaculatory intromissions correlated positively with the likelihood of female deermice (*Peromyscus maniculatus*) becoming pregnant or pseudopregnant (female's luteal cycle triggered, but without fertilized eggs; Dewsbury and Baumgardner 1981). Prolonged mating (several ejaculatory series) is necessary to trigger the development and maintenance of functioning corpora lutea in the vole *Microtus agrestis* (Milligan 1982).

Crosses between species of *Peromyscus* mice revealed apparent genetic effects on the number of intromissions preceding ejaculation (Lovecky et al. 1979), and different species groups of *Peromyscus* show different patterns of mounting and thrusting during copulation (Langtimm and Dewsbury 1991). These data suggest that, just as in lab rats and mice (above), variation in male behavior patterns that affects the probability of successful implantations may be under genetic influence.

4.2 Sometimes Impede Plugging of Reproductive Tract

Males of a variety of animal groups deposit substances of one sort or another in the external genitalic openings of females following copulation (e.g., Voss 1979; Dewsbury and Baumgardner 1981 on rodents; Hope 1974 and Hass 1990 on nematodes; Abele and Gilchrist 1977 on an acanthocephalan worm; Jackson 1980, Austad 1984, and Masumoto 1993 on spiders; Aiken 1992 on dytiscid beetles; Ehrlich and Ehrlich 1978 and Drummond 1984 on butterflies; Markow and Ankney 1988 on *Drosophila*; Hartman and Loher 1974 on a grasshopper; Hartnoll 1969 and Diesel 1991

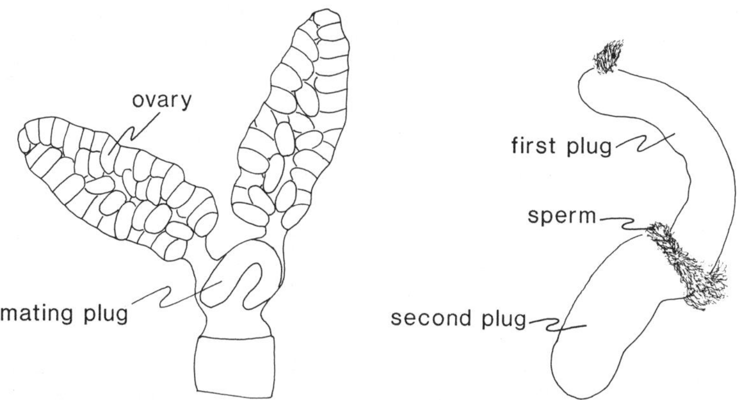

Figure 4.2 A mating plug, made by a male *Anopheles gambiae* mosquito, which appears to temporarily exclude sperm from other males. The plug lodges in the female's common oviduct (*left*). It remains intact in 80% of the females 12 hr later, but is gone in nearly all females 36 hr later. Intact plugs apparently block the entry of the second male's sperm; at least some sperm (presumably of the second male) are trapped between the first and second plugs from a doubly mated female (*right*). (After Gillies 1956)

on brachyuran crabs; Devine 1975, 1977 on snakes). Several nonexclusive functions have been proposed for these "mating plugs" in the context of improving male reproduction. Plugs may reduce sperm loss (or ejection) immediately following mating in mice *Mus musculus* (McGill et al. 1968) and the honeybee *Apis mellifera* (Laidlaw 1944). They may stimulate the female to perform various reproductive processes such as sperm transport or ovulation, as in the mouse (Leckie et al. 1973). They may bind up and thus immobilize sperm from previous males, as in the beetle *Anthonomus grandis* (Nilakhe and Villavaso 1979). And they may make subsequent intromission and/or insemination by other males more difficult, as in several rodents and worms (Voss 1979; Abele and Gilchrist 1977).

In some species the plug is large, hard, and adheres tightly to the female, and probably physically blocks access by subsequent males (Figs. 4.2, 4.4, 4.6). In others it is small relative to the female opening or duct and does not prevent subsequent intromissions (fig. 4.3), or it is soft and adheres poorly and is thus easily displaced by subsequent males (e.g., Busse and Estep 1984 on the monkey *Macaca nemestrina*). In plugs that function to block subsequent males, only partial effectiveness is expected to be the general rule. If plugs become completely effective, males will be selected to completely avoid nonvirgin females, and the selection that favors plugs will end and plugs will gradually return to a partially effective state (Parker 1984). In many groups the plugs are indeed not 100% effective in preventing insemination by subsequent males (e.g., Jackson 1980 on a salti-

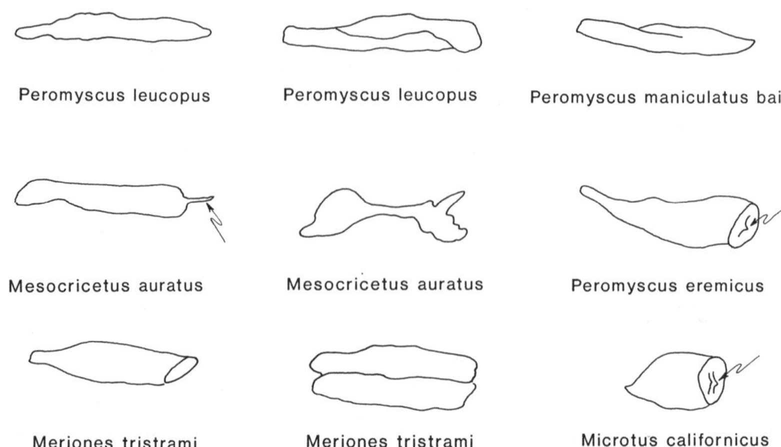

Figure 4.3 Vaginal copulatory plugs of several species of rodents. Multiple plugs found in the female after two copulations in three species are in the *middle*. If plugs make remating more difficult, the effect is not dramatic, though it is possible that some did not have time to harden after the first copulation. In fact, when two male *M. auratus* mated with the same female for equal amounts of time, the second male sired more offspring than the first (Oglesby et al. 1981). Even the hardened plugs of *P. maniculatus* are "dislodged rather easily and frequently" (Dewsbury and Baumgardner 1981). Whatever their function(s) (chastity enforcement, sperm retention, and induction of transport or the luteal cycle are commonly mentioned), female effects on the time before the plugs are discarded (see text) could affect male reproductive success and thus result in cryptic female choice. Arrows mark projections into the cervix. (After Dewsbury and Baumgardner 1981).

cid spider; Dewsbury and Baumgardner 1981 on five species of rodents) (fig. 4.3). In some animals they appear to have no effect whatsoever (Ehrlich and Ehrlich 1978 on many lepidopterans; Ikeda 1974 on *Drosophila mercatorum*). In some species of *Drosophila* their blocking ability, if it exists, may be selectively irrelevant since a female at least usually refuses to remate until after the plug has become degraded within her reproductive tract (T. Markow, pers. comm.).

Females are known to affect plug formation in two different ways—by making it more difficult for the male to deposit a plug, and by producing plugs of their own (table 4.2). For instance, in several *Drosophila* species the epithelial cells of the female's vagina apparently swell and occlude her genital opening by themselves or with their secretions for up to 8 hr after mating (Grant 1983; Markow and Ankney 1988 and references). Possibly similar "insemination reactions" with active female participation may occur in brachyuran crabs (Hartnoll 1969; Diesel 1991). In the mosquitoes *Aedes aegypti* (Spielman et al. 1969) and *Anopheles* spp. (Giglioli and Mason 1966), the mated female releases an abundant secretion, but at least

Table 4.2
Possible Cases in Which Females Exercise Cryptic Female Choice by Sometimes Making It Difficult or Impossible for the Male to Plug Her Reproductive Opening

	Criteria						
Species	*I*	*II*	*III*	*IV*	*V*	*VI*	*References*
Agelena limbata (Araneae)	Y	Y	Y	Y	Y (plug size)	?	Masumoto 1991, 1993
Cyrtodiopsis whitei (Diptera)	Y	Y	Y	Y	?[a]	?	Lorch et al. 1993, Burkhardt et al. 1994
Sciurus spp. (Rodentia)	Y	Y	?	Y	?[a]	?	Koprowski 1992

NOTES: All are discussed in the text.
Each example is classified with respect to criteria for proof that cryptic female choice occurs. See table 4.1 for criteria and key.
[a] Possible cues for distinguishing among males were not checked.

in *Anopheles* it seems to have no role in plug formation or digestion (Giglioli and Mason 1966). Females produce part or all of the plug in some vespertilionid bats, and in *Didelphis* opossums (Voss 1979). A less well documented case occurs in some scorpions, in which females apparently make a plug some time after having mated (Polis and Sissom 1990).

Plugs thus apparently benefit females in at least a few species. Several authors (e.g., R. Smith 1984; Diesel 1991) have argued that plugs might benefit females because they would reduce disturbance resulting from competition among males to mate with them. Although such an effect might eventually result from very effective plugs, this benefit seems unlikely to explain the origin and gradual improvement of plugs, since reduction in male attempts to mate would only be expected after highly effective plugs evolved. Another possible female benefit is that she could prevent loss of sperm, especially in animals in which the female is unable to close her external genitalic opening or does not move sperm deeper into her body.

A further, less direct possible benefit for females is that of a filtering device for males. By selectively favoring males that were better able to form effective plugs or induce her to form a plug herself, a female could increase the chances that her sons could form such plugs. Selectively facilitating plug formation for some males but not others could also be a mechanism by which females favor some other male character.

An example of cryptic female choice via selective plugging occurs in the spider *Agelena limbata*. Masumoto (1991, 1993) demonstrated that the large size of the area where the plug is deposited around the entrances to the insemination ducts of females can result in greater effectiveness of the plugs of larger males. He also showed that differences in plugging success can influence male reproductive success in nature.

As is typical of many spiders, the female has two sperm storage organs (spermathecae). Each spermatheca has a separate entrance from the outside (via the insemination duct) and an exit leading to the oviduct (via the fertilization duct) (see chap. 7.1; fig. 7.3). A male can thus plug the female's entrances after inseminating her, without affecting her ability to fertilize eggs and oviposit. During copulation, a male *A. limbata* first inseminates the female, and then deposits a brown liquid from his secondary genitalia (pedipalps), in the cavity where both insemination ducts open (the "atrium"). The liquid soon hardens into a plug. In some cases, the plug filled the atrium and covered the openings of both insemination ducts completely; in others it covered only a portion of the atrium. Incomplete plugs were common (38% of fifty lab matings with virgin males and females; the use of only virgin males, which may have greater reserves of material, may overestimate the sizes of plugs in nature). Smaller males more often made incomplete plugs. The ratio of male/female size was larger when complete plugs were made, and females with complete plugs tended to be slightly though not significantly smaller. After depositing a plug, males in the field usually left the female within a day, presumably to search for another mate (male remating was documented in the field; Masumoto 1991).

If a second male subsequently attempted to mate with the female, he first used his palp to attempt to hook the plug and remove it. Complete plugs were not removable, and all offspring were sired by the first male ($N = 4$; sperm precedence was determined by sterilizing males with gamma radiation). If the plug was incomplete ($N = 6$), however, the second male hooked it with his pedipalp and removed it, then mated with the female and deposited a plug of his own. In these pairs, the second male fertilized an average of 62.9% ± 11.4% of the female's eggs (there was no increase in the hatching rate of eggs of doubly inseminated females). In a second set of observations, five of nine males paired with females with incomplete plugs succeeded in removing the plugs, while zero of ten paired with females with complete plugs were able to remove the plug. Masumoto concluded that the design of the female's genital openings is such that it enables the female to bias the fertilization of her eggs in favor of larger males. The size of the female atrium is apparently correlated with female body size (the coefficient of correlation of cephalothorax width [a measure of body size] with maximum width of the epigynum was 0.678, $N = 40$). Larger females (which would be particularly important for male reproduction since they probably tend to produce larger numbers of eggs; Eberhard 1979) may discriminate more strongly in favor of larger males.

Cryptic female choice in *A. limbata* is selectively important in the field. Of thirty-nine individual females whose sexual history was followed, 31%

remated (Masumoto 1991; females in the field may also exercise overt, precopulatory choice, as they failed to copulate with some additional males that courted them). Incomplete plugs were common in the field. Of twenty females collected at the end of the mating period (guarding egg sacs), 35% had incomplete plugs. This may underestimate the frequency of incomplete plug formation, since some incomplete plugs in the field were presumably removed from some of these females and replaced by complete plugs.

A species that may combine female plug removal with still another cryptic female choice mechanism (removal of spermatophores before all sperm are transferred; see sec. 4.4) is the stalk-eyed diopsid fly *Cyrtodiopsis whitei*. Copulations frequently ended before the 40 sec necessary for sperm transfer to begin (75 of 180 copulations in the lab were ≤40 sec; median duration for copulations resulting in sperm transfer was 54 sec; Lorch et al. 1993). Females sometimes remated in the space of as little as 30 min in the field. Lorch et al. suggest that short copulations occurred when the female had not yet discarded the spermatophore from a previous copulation, since the spermatophore blocks the entrance to the paired spermathecae (Kotrba 1990). Females sometimes discarded spermatophores almost immediately following copulation (20% were discarded within 2 min after copulation ended, >50% within 30 min after copulation; Kotrba 1991 in Lorch et al. 1993), and Lorch et al. (1993) speculate that females may exercise choice between males by altering the delay in discarding the spermatophore. Some discarded spermatophores still contained sperm (M. Kotrba, pers. comm.), so sperm dumping is an additional possible cryptic choice mechanism in this species. There were no direct confirmations that spermatophores acted as plugs and caused short copulations, nor were there any attempts to check possible correlations between the length of delay in discarding the spermatophore and characteristics of the male. Larger males sire more offspring than smaller males when they are kept together, despite not mating more frequently or transferring more sperm (Burkhardt et al. 1994), so female choice occurs in this species.

There are indications that female rodents may also sometimes actively influence the length of time copulatory plugs are present in their genital openings. For instance, the plug in a female paca *Agouti paca* is usually expelled slowly over about a day, and the female eats the exposed portion as it emerges (N. Smythe, pers. comm.). Occasionally, however, it is expelled soon after copulation, still intact (fig. 4.4). Female *Rhinolophus ferrumequinum* bats also sometimes expel plugs (Fenton 1984).

Female *Sciurus* squirrels may discriminate among males by removing the plugs of some but not of others. Plugs were removed by the female (pulled from the vagina with the teeth) soon after copulation in 73% of twenty-two copulations in *S. niger*, and 50% of twenty-six copulations in *S. carolinensis* (Koprowski 1992). Apparently the female either removes a

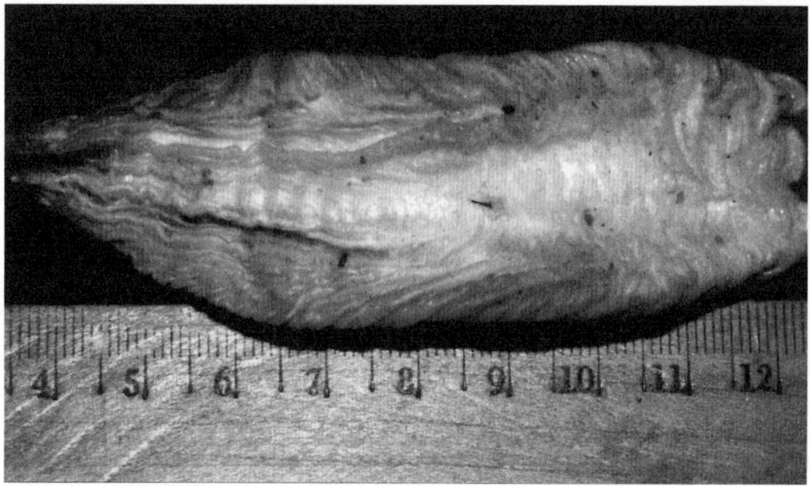

Figure 4.4 A copulatory plug that emerged from a female paca (*Agouti paca*) almost immediately after copulation ended. Plugs from other matings stayed in place in the female's vagina for more than a day. The possibility that females were responsible for the occasional expulsion of the plug and that cryptic female choice resulted has not been explored. (Scale in cm; photo courtesy of N. Smythe)

plug immediately, less than 30 sec after copulation ends, or leaves it in place for several hours. The plugs are large, and hard enough to present obstacles to additional matings (Koprowski 1992); they could also prevent sperm emission. In either case, female removal of some plugs but not others may influence male reproductive success.

There is another, less obvious type of removal that is widespread in rodents. Copulatory plugs adhere tenaciously to the vaginal mucosa and are covered with a sheet of epithelial cells when they emerge from the female. The vagina is often invaded by large numbers of leucocytic cells shortly after copulation, which produce a shedding of the vaginal lining (Voss 1979). Variations in leucocyte numbers could affect the length of time a plug remained in place, even when the female did not physically remove it.

In some species the male may use his own body as a plug. Thus, for instance, the common finding that nematode males remain coiled around the female's vulval region for up to several minutes after sperm transfer is finished (fig. 4.5; e.g., Jones 1966 on *Pelodera teres*; Duggal 1978 on *Panagrellus redivivus*; Somers et al. 1977 on *Rhabditis pellio*; and Rehfeld and Sudhaus 1985 on *R. marianane* and *R. synpapillata*) may be related to the frequent observation of sperm "leaking" out of recently mated females (chap. 3.2) (in the last two species of *Rhabditis*, the male also leaves a mass of material in the vagina, which in some cases but not others deters further inseminations; Rehfeld and Sudhaus 1985). The female can influence the

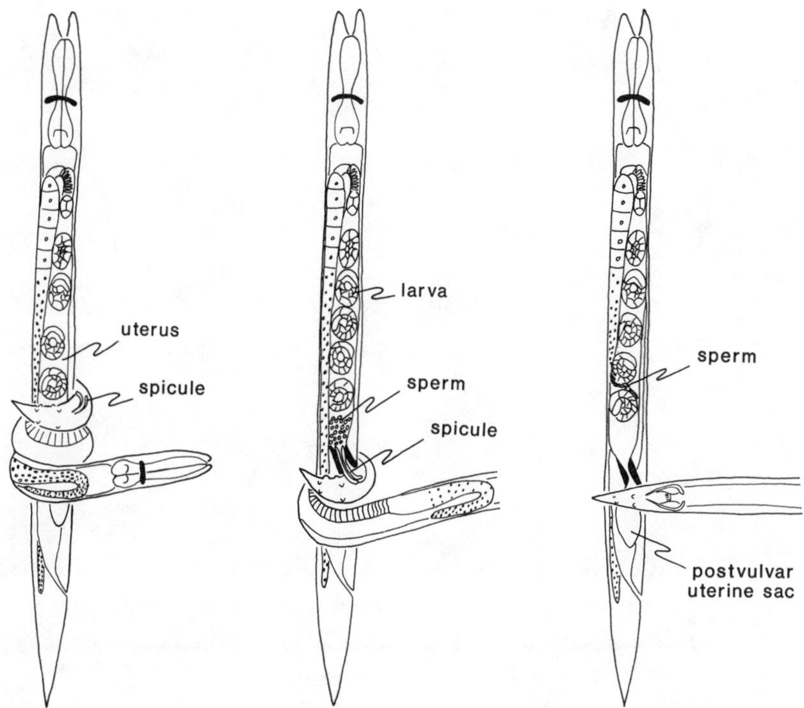

Figure 4.5 Possible use of the male's own body as a sperm plug. Positions of male and female *Panagrellus redivivus* nematodes during courtship (*left*), sperm transfer (*middle*), and after sperm transfer (*right*). Females generally initiated separation, and if the female was quiet the male sometimes stayed attached for up to 15 min after sperm transfer. Since sperm ejection by the female occurs in several other nematodes (e.g., fig. 3.3), the male's prolonged postinsemination contact with the female in this and other nematodes (e.g., Somers et al. 1977 on *Rhabditis pellio*) may serve to reduce the chances she will discard his sperm. (After Duggal 1978)

length of postcopulatory contact by the male, as she sometimes wriggles free of the male's clasp soon after sperm transfer, and sometimes she does not (e.g., Duggal 1978; Rehfeld and Sudhaus 1985), possibly giving rise to cryptic female choice.

4.3 Sometimes Impede or Fail to Carry Out Plug Removal

Since males of some species deposit mating plugs, which make intromission more difficult for subsequent males, the ability to remove such plugs could also be expected to evolve in males and also females. Such adaptations have been observed. For example, males of a linyphiid spider species

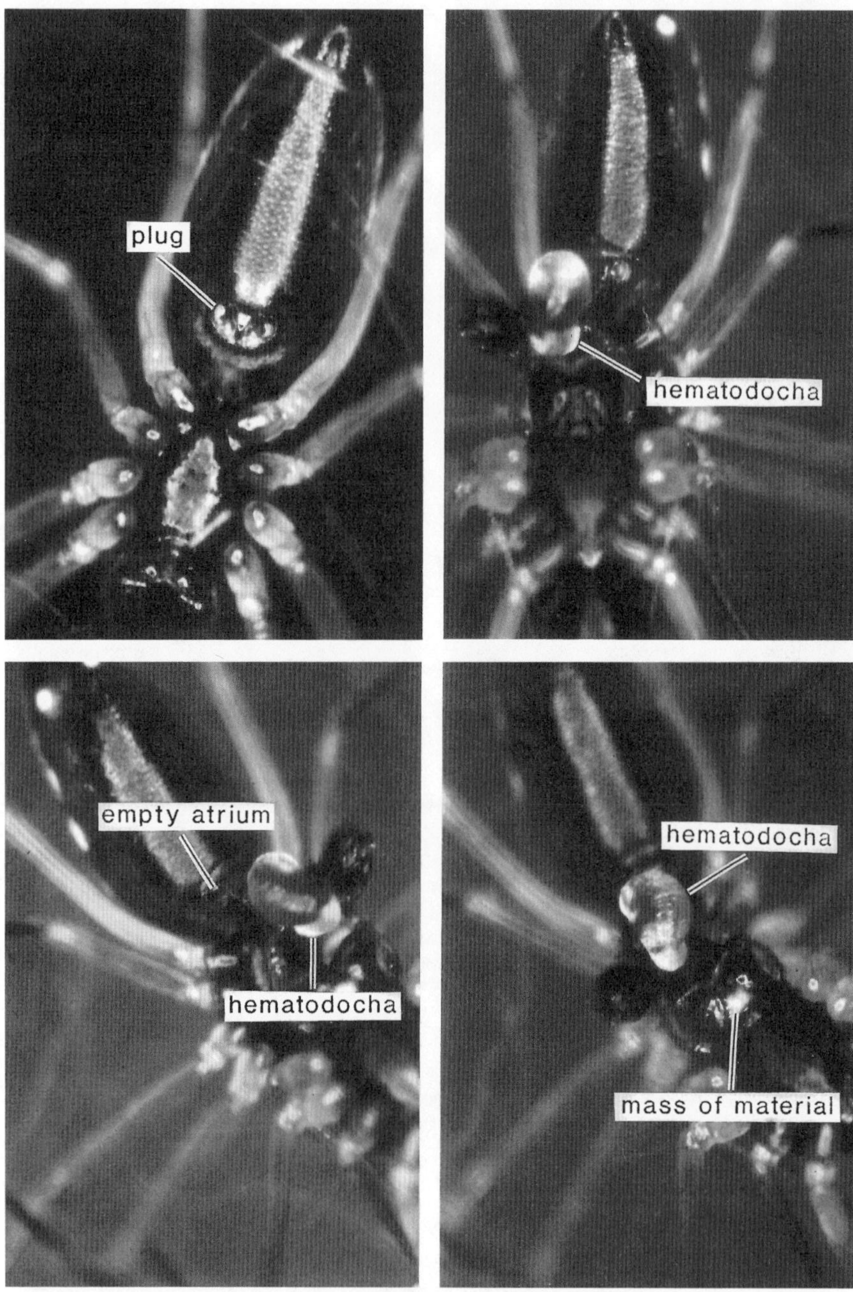

Figure 4.6 Copulatory plugs are often incapable of excluding intromissions by subsequent males. Virgin females of an unidentified linyphiid spider (near *Diplothyron*) have no plugs. At the end of copulation, a mass of material fills each of the female's genital openings (atrium), and after a day this liquid hardens into a solid, light-colored mass (*upper left*). If the female mates again, the second male uses his genitalia to remove the female's

near *Diplothyron* apparently actively dissolve and remove plugs (fig. 4.6; A. Berghammer, unpub.). Since plug removal by a male can require that he maintains sustained access to the female's genitalia, females could influence the success or failure of removal attempts by simply altering the length of copulation. A spider species in which this occurs is the salticid *Phidippus johnsoni* (Jackson 1980). The plugs in this species apparently adhere only loosely to the female genitalia, and they can be dislodged with an insect pin. Females determined the length of all copulations, and longer copulations more often resulted in removal of plugs than did shorter ones. Some second copulations (ten of twenty-two using sterile males) resulted in offspring from the second male (Jackson 1980), so the female-influenced opportunity to remove a plug and inseminate the female may be selectively important for the male.

In other species, such as the grasshopper *Chorthippus curtipennis*, the female herself dissolves the male plugging structure (Hartmann and Loher 1974). Variation in the timing of such removal by the female could affect the male's chances of paternity (no variation was noted in *C. curtipennis*, but sample sizes were small).

4.4 Sometimes Remove Spermatophore before Sperm Transfer Is Complete

In some crickets and tettigoniid orthopterans, males attach a spermatophore to the female's abdomen, and most of the spermatophore, including the sperm reservoir (the ampulla), remains outside the female. Sperm are gradually injected into the female due to an increase in the pressure within the ampulla (Khalifa 1949). In a number of species the female turns and pulls the spermatophore from her genitalia and begins to consume it before all of the sperm in the ampulla have entered her body (table 4.3). By altering the delay before she begins to consume the spermatophore, a female can exercise cryptic female choice.

Females of the cricket *Gryllus bimaculatus* mate repeatedly, both with the same male and with different males (Simmons 1986). Sperm from dif-

---

plugs (*upper right*) (the hematodocha is an inflatable membrane in the male genitalia). First the plugs are softened and liquefied and then removed in small pieces. The pieces adhere to the male's pedipalp, and after each insertion, the male cleans the material off when he passes his pedipalp through his mouthparts. In the pair at the *upper right* (still early in copulation), one intact plug is still visible. Later, as insemination begins, the plugs are apparently gone (*lower left*) and a drop of light-colored plug material has accumulated in his mouthparts (*lower right*); it is later discarded. Presumably the male is the source of both the plug and the material that liquefies it, but this has not been confirmed. (Photos by W. Eberhard, from an unpublished study by A. Berghammer)

Table 4.3
Possible Cases in Which Females Exercise Cryptic Female Choice by Sometimes Removing the Spermatophore before Sperm Transfer Is Finished

	Criteria						
Species	*I*	*II*	*III*	*IV*	*V*	*VI*	*References*
*Gryllus bimaculatus** (Orthoptera)	Y	Y	Y	Y	Y (size)	?	Simmons 1986
*Gryllodes supplicans** (Orthoptera)	Y	?	Y	Y	?[a]	?	Sakaluk 1984
*Decticus verrucivorus** (Orthoptera)	Y	Y	Y	Y	Y (size, sptophyl.)	?	Wedell and Arak 1989; Wedell 1991
Poecilimon veluchianus (Orthoptera??)	Y[b]	Y	Y?[c]	?	Y (sptophore size)	?	Reinhold and Heller 1993
Oecanthus nigricornis (Orthoptera)	Y	Y	?[d]	Y?[e]	Y (length fed by male)	?	W. Brown in prep. a,b
Hemideina femorata (Orthoptera)	Y	Y	?	Y	Y? (defend female[f])	?	Field and Sandlant 1983

NOTES: Those marked with (*) are discussed in the text. Each example is classified with respect to criteria for proof that cryptic female choice occurs. See table 4.1 for criteria and key.

[a] Possible cues for discriminating among males were not checked.

[b] Authors defined successful sperm transfer as >20% of that in the reservoir of the spermatophore, and found that females eat about 20% of the spermatophores before successful transfer occurs. If one were to define "successful" as, say, >75%, the rate of successful transfer would be substantially less.

[c] Assuming that sperm dilution determines paternity, as in the related *Decticus* (Wedell 1991), reduction in the percentage of sperm transferred from the spermatophore to the spermatheca, which was shown to occur, is likely to reduce the male's chances of paternity.

[d] Length of time the spermatophore was attached to the female (before she began to eat it) was correlated with the length of time the female fed on the male metanotal gland secretions; neither of these variables correlated with male size. Transfer of sperm from the spermatophore to the female was gradual, but time needed for complete transfer was not determined. Sperm precedence patterns from different attachment times were not determined.

[e] Remating occurred readily in captivity, and inseminated females are attracted to male calls in nature (W. Brown, pers. comm.).

[f] Only two of thirty-five spermatophores were consumed, one just after the male mated and failed to defend the gallery; this female immediately mated with a new male.

ferent copulations apparently mix within the female's spermatheca, and sperm are apparently used randomly with respect to the male's identity except for small effects when small amounts are transferred (Simmons 1987a). Thus a male's chances of paternity are correlated with the proportion of the sperm in the female's spermatheca that are his. The female's spermatheca is membranous and expands with successive matings, allowing her to store all the sperm received during her lifetime (Simmons 1986).

Females of this species favor larger males via both overt female choice and cryptic female choice. Males are territorial, and females exert overt choice by spending more time in the territories and copulating more often with larger males (Simmons 1986). Cryptic choice also occurred, because when a female mated with a smaller male, she was more likely to turn and eat the spermatophore before it had emptied its sperm into her (Simmons 1986). Females sometimes began to eat a spermatophore less than 2 min after it was attached. Since transfer of sperm is more or less linear over the approximately 60 min required for total transfer (Simmons 1986), early spermatophore consumption by the female must sometimes substantially reduce the number of sperm transferred.

An additional effect may result from transfer of an oviposition-inducing substance in the male's semen, as in other crickets (see table 6.1; chap. 6.2). Premature removal of spermatophores (after 10 min) in *G. bimaculatus* resulted in a long-term reduction in the number of eggs laid (Orshan and Pener 1991). This may be an all-or-nothing female switch, however, which does not result in selection on males in natural contexts. Simmons (1988) showed that moderate increases in the amount of spermatophore whose contents a female received did not affect egg production.

The female's tendency to be subsequently attracted to male calling songs is also thought to be more strongly reduced when greater amounts of semen are stored in her spermatheca (Loher et al. 1993). Egg size is also larger when more seminal fluid and sperm are received by the female (Simmons 1988; chap. 4.10). Thus early female consumption of the spermatophores of smaller males may reduce these males' chances of reproduction in four different ways: reduce paternity of subsequent offspring, reduce oviposition rates, reduce egg size, and raise remating rates. Since body size in male *G. bimaculatus* was heritable in a natural population (Simmons 1987b), female choosiness may affect offspring quality.

This is an instructive case in which to consider the effects of female-imposed "rules of the game," because the reproductive morphology and physiology of this species are relatively well studied. Obviously there must be strong selection on male *G. bimaculatus*, especially on smaller males, to produce spermatophores that empty as rapidly as possible. Simmons (1986, 1987a) discusses the time needed for complete transfer as if it were a "given," but it is also informative to consider the role of females in influencing the rate of transfer. The spermathecal duct of *G. bimaculatus* has a muscular envelope (Yasuyama et al. 1988). Although the tip of the spermatophore tube reaches the spermatheca, active female transport of sperm into the spermatheca may occur as in *Acheta* (= *Gryllus*) *domesticus*, in which seminal fluid also arrives in the spermatheca after mating (Khalifa 1949). A striking bit of evidence favoring the possibility of female influence in sperm transport is that the spermatophore of *G. bimaculatus* has a substance that dramatically increases the rate of contraction of the muscles in the region of the spermathecal duct nearest the genital chamber (Kimura et al. 1988). Selection would have presumably favored such a male trait only if induction of the female response were important to his reproductive success (that is, if females sometimes failed to respond appropriately). Incomplete female responses of this sort may occur in the cricket *Teleogryllus commodus*, where incomplete transfer of spermatophore contents to virgin females sometimes occurs (Loher et al. 1981).

Such simple female traits as a weaker wall of the spermatheca, relaxation of the spermathecal wall (which is sheathed in muscles; Loher et al. 1993), or perhaps filling the spermatheca less full of the clear liquid observed in the spermathecae of virgin *Teleogryllus commodus* crickets (Ai et al. 1986), could also speed sperm transfer. Variations of this sort affecting sperm transport may have caused the substantial variation among copulations in the rates of sperm transfer in *G. bimaculatus* (Simmons 1986).

A second type of female-imposed morphological "rule" in *G. bimaculatus* is that the ampulla must stay outside the female's body. Presumably if the female vagina or bursa were larger in diameter or more easily expanded, as in many other orthopteroid insects (Davey 1960b), the male would be able to place the ampulla of the spermatophore inside rather than outside the female. In combination with female consumption of the ampulla, this constitutes an important female-imposed rule of the game, as noted by Simmons (1987a). The female's extensible spermatheca (which can hold the sperm from up to thirty or more spermatophores) would favor larger sperm numbers and ampulla volumes; but since sperm transfer is slow, female ampulla consumption places an upper limit for males on effective ampulla size.

A possible alternative explanation for reduced ampulla size is that males need to conserve sperm and accessory gland resources for subsequent copulations with other females. This is not likely in *G. bimaculatus*, however, because spermatophores are relatively small (about 0.18% of the male's body weight; Simmons 1988), and males often mate repeatedly in rapid succession with the same female (pairs kept together for 7 hr mated an average of 5.3 ± 0.2 times; Simmons 1988; see also Simmons 1986). Obviously the male has more material available and is willing to invest it in the same female. The chance that a female might walk away before the male has transferred all the spermatophores he is prepared to give her would, in the absence of female spermatophore consumption behavior, result in selection favoring larger spermatophores (especially for smaller males, which are more likely to be quickly abandoned).

Rapidity in sperm uptake is also selectively important to males of other species of spermatophore-transferring gryllids and tettigoniids, which produce large masses of sperm-free material that covers or "protects" the ampulla. Such a mass (a spermatophylax) can increase the time a female takes to reach the sperm reservoir in the ampulla when she starts to eat the spermatophore (Gwynne 1983). Again, however, the "protection" depends on a behaviorally modified sexual environment (the female-determined order of consumption of spermatophylax and ampulla; fig. 1.5; see also below).

The tettigoniid *Decticus verrucivorus* illustrates the possibility of complex interactions between spermatophore consumption and other mechanisms of cryptic female choice. Males produce a large spermatophylax which is so positioned that when the spermatophore (which averages 9.5% of the male's weight) is attached to the female, the spermatophylax blocks the access of the female's mouth to the sperm reservoir. When the male and female separate after mating, the female first removes the spermatophylax and consumes it, then bends over a second time to remove the sperm-containing ampulla and eats it. Artificial manipulations of the weights of spermatophylax material consumed between 0 and 300 mg (normal mean is 104.4 ± 22 mg) revealed a positive relationship between the size of the spermatophylax and the duration of the attachment of the ampulla to the female (Wedell and Arak 1989). The spermatophylax is apparently not nutritious, since its consumption had no effect on female longevity, fertility, or egg size (Wedell and Arak 1989).

Sperm transfer is thought to be gradual (no data were presented for this species, but it is gradual in other species in this family: *Anabrus simplex*, *Requena verticalis*, and *Kawanaphila nartee*; Gwynne 1983, 1986; Simmons and Gwynne 1991). Since a larger spermatophylax takes longer to consume, spermatophylax size is probably correlated with the amount of sperm transferred to the female's spermatheca (Wedell and Arak 1989). A sperm precedence study with irradiated males showed a positive correlation between the percentage of a female's offspring sired by the second male and the relative size of the second male's spermatophylax compared with that of the first male (Wedell 1991). Sperm from the two males apparently mixed freely in the female's spermatheca, and the likelihood of paternity is probably a simple function of the relative numbers of sperm of the two males. In addition, longer durations of ampulla attachment (before the female seized it) were also correlated with longer female refractory periods and more rapid oviposition during this period (Wedell and Arak 1989). Thus, just as in *G. bimaculatus*, a male's chances of paternity are influenced in multiple ways by the length of time the ampulla is attached to the female.

The variations in the length of time the ampulla of *D. verrucivorus* is attached seem to differ from the case of *G. bimaculatus* in that the delay is partially imposed by the male. This is an illusion, however. The female could easily discard the spermatophylax as soon as she removed it from the ampulla (e.g., drop it or wipe it off her mouthparts), and then pull off the ampulla. Instead, females consistently refrained from seizing the ampulla until after having consumed the spermatophylax—a clear "rule of the game" by which male reproductive success is determined in this species

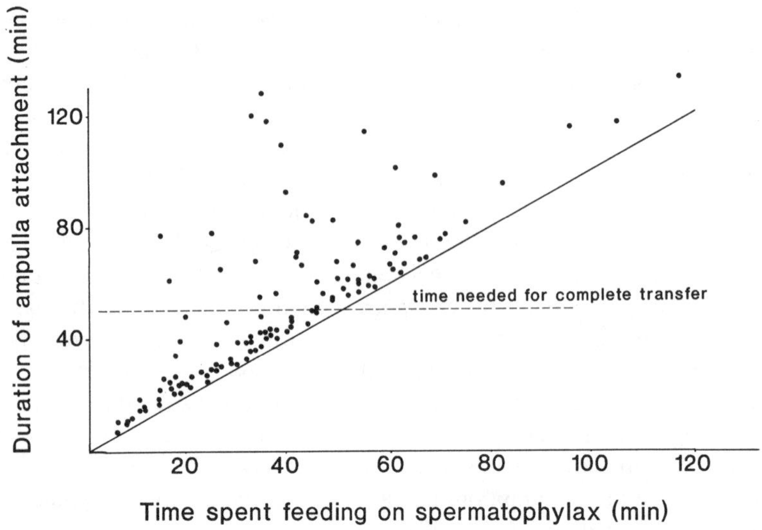

Figure 4.7 The interplay between female decisions and male inducements determines a male's insemination success in the cricket *Gryllodes supplicans*. The dotted line marks the approximate ampulla attachment time (about 50 min) needed for all the sperm in the ampulla of the spermatophore to be transferred to the female. The male's inducement to leave the ampulla attached was the spermatophylax, on which females feed before pulling off and eating the ampulla. In some pairs, the female removed the ampulla within a minute after finishing the spermatophylax, but in others she waited for more than an hour (in the graph this delay is the vertical distance of each point above the slanting line). Another female decision was to drop the spermatophylax before finishing it (after an average of only 10.5 min), then eat the ampulla. Note that there was almost no delay in the cases of very short feeding times (see also fig. 1.5). (After Sakaluk 1984)

(e.g., see fig. 1.5). That females need not necessarily adhere to this rule is illustrated by the fact that females of the cricket *Gryllodes supplicans* often "broke" the rule. They dropped the spermatophylax before fully consuming it in 36% of fifty pairings; spermatophores that were only partially consumed weighed on average more than those that were eaten completely (Sakaluk 1985; fig. 4.7). Female failure to "abide by the rules" may have thus been biased with respect to the male (heavier males tended to produce heavier spermatophores, though there was a substantial amount of variation; Sakaluk 1985).

There was also a great deal of variation in *G. supplicans* in the effects of spermatophylax size that may have stemmed from differences in female efforts to eat the spermatophylax (fig. 4.7). Some variation was apparently due to the substantial variation in the time for consumption of the spermatophylax, which varied from 85 to 250 min ($N = 43$), and the time between finishing the spermatophylax and seizing the ampulla (which ranged between 1 and 163 min). Thus both the speed of consumption of the sperma-

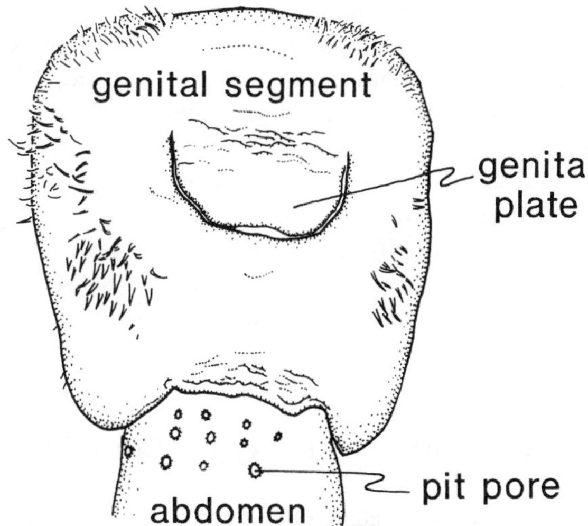

Figure 4.8 The "pit pores" of a female copepod *Labidocera aestiva* are part of a possible female spermatophore removal mechanism. The male's spermatophore is glued to the female's abdomen near her gonopore, over the pit pore area. At the base of each pore is a reservoir that is associated with glandular tissue. The glands produce a substance that is released onto the female's surface after the spermatophore is deposited, apparently modifying and weakening the cement that binds the spermatophore to the female. (From Blades and Youngbluth 1979)

tophylax and the delay in seizing the ampulla varied widely, suggesting two further possible mechanisms of female choice.

It would not be surprising if males of crickets and katydids are found to include substances in the spermatophylax that are designed to manipulate female behavior to increase the delay before the ampullae are eaten. They might use phagostimulants that increase the female's interest in feeding on it (e.g., mimics of natural feeding stimulants), components that cause her to consume the spermatophylax more slowly (e.g., the sticky froth of *Anubrus simplex*; Gwynne 1983), and perhaps even arrestants of further feeding (e.g., Dethier 1988; Lee and Bernays 1990).

Female mechanisms of spermatophore removal occur in other groups. In the calanoid copepod *Labidocera aestiva*, the female probably determines the duration of adhesion of a spermatophore (Blades and Youngbluth 1979; Blades-Eckelbarger 1991). The male cements his spermatophore to an area near the female's gonopore that is covered with unusual "pit-pores" (fig. 4.8). The pores are connected with secretory cells which produce a substance that is thought to modify the spermatophore's cement and loosen its attachment to the female. Transfer of the immobile, aflagellate sperm from the spermatophore occurs gradually for about 30 min and is brought about by hydrostatic pressure developed by specialized sperm that swell by taking up water, forcing their companions into the female. After copulation the female removes the spermatophore by inserting the long exopods of her fifth legs under the attachment structure of the spermatophore. Modification of the timing of either the rejecting movements of the fifth legs or the extrusion of material from the pit pores could interrupt sperm transfer. A striking aspect of preliminaries to spermatophore deposition is the male's assiduous and prolonged stroking of the female's pore area with his highly modified, scrub-brush-like fifth-leg endopod. A similar field of cuticular pores with secretory tissue below the site of spermatophore deposition occurs in females of another copepod, *Undinula vulgaris* (Blades-Eckelbarger 1991).

4.5 Sometimes Abort Zygotes (Bruce Effect)

In some animals in which females brood their offspring, a female that suspends care for one brood by aborting them often produces another brood sooner. Conversely, a female in the process of raising offspring is usually not likely to conceive new offspring. Thus, for example, female mammals do not enter into estrus while they are pregnant, and in many species they do not enter estrus until they finish lactating.

Abortions are quite common in some mammals (Wasser and Barash 1983). For example, approximately 78% of all human pregnancies abort spontaneously (Wasser and Isenberg 1986). Some abortions are not related to the male and probably result from adaptations evolved under natural selection on females. For instance, female coypu (*Myocaster coypus*) selectively abort small litters with predominantly female offspring when the female is in especially good condition and is able to raise the presumably more costly successful male offspring (Gosling 1986). Other female mammals abort when environmental conditions are so poor that adequate care for the offspring is impossible (Wasser and Barash 1983).

As noted by Andersson (1994), male abilities to induce or prevent abortions could result in reproductive benefits. For instance, when a male encounters a female while she is caring for a previous brood, her offspring can represent impediments to the male's reproduction. There may thus be an advantage to a new male of eliminating such offspring via abortion or direct infanticide, as occurs in birds (Lifjeld et al. 1993) and mammals, including primates (Hrdy 1977), rodents (Huck et al. 1982), and carnivores (Bertram 1975).

Females of species with infanticidal males have evolved an array of countermeasures, including false estrus, hiding their offspring, inhibition of ovulation (see chap. 3.9 on lions), and premature elimination of doomed litters. Females use stimuli from males to cue these processes, so a male's

Table 4.4
Possible Cases in Which Females Exercise Cryptic Female Choice by Sometimes Aborting Zygotes (Bruce Effect)

	Criteria						
Species	*I*	*II*	*III*	*IV*	*V*	*VI*	*References*
*Mus musculus** (Rodentia)	Y	Y?	Y	Y?	Y (dominance, high testost.)	?	Hoppe 1975; Bronson 1979; Huck 1982
Microtus pennsylvanicus (Rodentia)	Y	Y?	Y	Y	?[a]	?	Clulow and Langford 1971; Mallory and Clulow 1977; Webster et al. 1981; Boonstra et al. 1993
*Microtus agrestis** (Rodentia)	Y?	?	Y	Y?	?[a]	?	Clarke and Clulow 1973
Dicrostonyx groenlandicus (Rodentia)	Y	Y?	Y	?	Y (dominance)	?	Mallory and Brooks 1978
*Mustela vison** (Carnivora)	Y	?	Y	Y	?[a]	?	Hansson 1947

NOTES: Those marked with (*) are discussed in the text.
Each example is classified with respect to criteria for proof that cryptic female choice occurs. See table 4.1 for criteria and key.
[a] Possible cues for discriminating among males were not checked.

abilities to trigger (and inhibit triggering of) these processes could have important consequences for his reproduction (table 4.4).

Thus, just as in the other processes already discussed, differences between conspecific males in their abilities to induce females to respond, or to reduce their responses to the signals of other males, could result in sexual selection on male signaling abilities. In the case of inducing abortions, an increase in the ability of a previous male to induce females to ignore abortion-inducing stimuli of new males would result in cryptic female choice, while an increase in the ability of a new male to trigger the female response would result in classic, precopulatory female choice (more matings). Both abilities will be discussed together here, as the possibility that they are under sexual selection has generally not been considered before for either one (see, however, Huck 1982).

When impregnated females of various genera of rodents, including house mice (*Mus*), field mice (*Peromyscus*), voles (*Microtus*), gerbils (*Meriones*), and lemmings (*Dicrostonyx*), are exposed to an unfamiliar male (or to his urine), they abort (Clulow and Langford 1971; Stehn and Richmond 1975; Mallory and Brooks 1978; Webster et al. 1981; Schadler 1981; Labov 1981; Marchlewska-Koj 1983; Albone 1984). This "Bruce effect" is best understood in the laboratory mouse *Mus musculus*, in which the pregnancy block results from a failure of the young embryo to implant in the uterus. The female's accessory olfactory system, involving her vomeral nasal organ, is used to sense what appears to be a nonvolatile pheromone, perhaps produced in the sexually dimorphic portions of the Bowman's capsule of the male kidney (Crabtree 1941; Hoppe 1975) via direct contact (Keverne and Rosser 1985). The male odor may be influenced by the major histocompatibility complex (Yamazaki et al. 1983). As little as 15 min of exposure to strange males for 3 days is enough to produce a pregnancy block (Milligan 1980). Blocks in voles can occur much later, up to 17 days into pregnancy, when the embryos are already implanted and well developed (Stehn and Richmond 1975; Mallory and Clulow 1977; Schadler 1981).

There are striking differences among the male cues that elicit female responses as well as the responses themselves in related species. In *Microtus agrestis* voles, abortions can be induced by direct male-female contact, and the cue is not in the male urine (Milligan 1980). In *M. pinetorum*, in contrast, the soiled bedding of a strange male was enough to block pregnancy (Schadler 1981). In *M. pennsylvanicus*, an additional copulation with a different male has the effect of blocking pregnancy from a previous copulation. When a female was allowed to mate with a series of males, each male interrupted the pregnancy that resulted from the preceding male and induced a further ovulation (with a new generation of corpora lutea in her ovaries; Clulow and Mallory 1974). Since male signals are expected to evolve rapidly under sexual selection (e.g., chaps. 5.1; 5.2; 6.2, "Interspecific Diversity of Male Products"), such divergence is not unexpected.

Females are not passive receivers of these male stimuli. Female lab mice of different ages show differing degrees of avoidance of male chemical signals. Prepubertal females avoided male odors, immediately prepubertal females showed a weak attraction, and adult females were strongly attracted (Drickhamer 1985). In contrast, adult females that were in their first 4–8 days of pregnancy (when the Bruce effect can occur) strongly avoided the odor of bedding soiled by strange males (Drickhamer 1989). In addition, female responses to males varied, and some females failed to abort when others did (e.g., 12% of thirty-three females carried their pregnancies to full term after sustained exposure to a strange male; Schadler 1981).

When one takes into account the propensity of males of these species to kill young pups when they take over a territory, it seems very likely that both the strange male and the female stand to gain reproductively from the Bruce effect. The male gains because the female will be available to raise his offspring more rapidly, both because she will come into estrus again and because implantation and gestation will be shorter if she has no other offspring (Mallory and Brooks 1978; Webster et al. 1981; Labov 1981).

Several characteristics could increase a male's reproduction in the context of a new male's ability to block pregnancies, or a previous male's ability to inhibit the blocking response: effectiveness at lower concentrations; greater ability to attract females to contact his urine or inhibit attraction to that of other males; shorter exposure times needed to elicit female responses; longer duration of activity; or blockage effective later in a female's pregnancy.

The female stands to gain under natural selection by not wasting time and energy on offspring that are likely to be killed by a new male. In addition, by selecting a sire who is especially good at inducing pregnancy blocks, the female could increase her chances of having sons that are superior in this respect. Conversely, any male trait that decreased the susceptibility of a female with which he had conceived offspring to pregnancy blocks by other males would be favored, and females favoring such males could be favored indirectly via greater reproduction of their sons.

Biased female responses to cues from different conspecific males occur in the house mouse. Urine from dominant males had a stronger blocking effect than that from subordinate males (Huck 1982), and males receiving testosterone injections were also more effective blockers (Hoppe 1975). Such biases would result in selection favoring dominant males.

There are also genetic (strain) differences among males in their abilities to block pregnancies and in female susceptibilities to being blocked (Hoppe 1975; Furudate and Nakano 1980; Marchlewska-Koj 1983). Interestingly, males of strains with the least responsive females were less able to induce pregnancy blocks with at least one other strain of females (Furudate and Nakano 1980; Marchlewska-Koj 1983).

Differences in male abilities to protect against blocks have been tested less often. Dominant males in one study were marginally better in their ability to protect against blocks (Huck 1982; $P < 0.038$ with one-tailed chi-squared test).

Doubt has been expressed regarding the significance of these findings. Some authors (e.g., Bronson 1979; Keverne and Rosser 1985) have raised the possibility that pregnancy blocks in *Mus* and other groups are artifacts of captivity, with no selective consequences in nature. I agree with Milligan (1980), Labov (1981), and Huck (1982) that these objections are unlikely to be important. The objection that changes of males, and thus male infanticide, are unlikely to occur in the wild is, on the face of it, unconvincing, since males are not immortal, and there must be a certain rate of turnover of territorial males. In fact, Bronson (1979) summarizes feral *Mus musculus* reproductive ecology as "characterized by temporal, spatial, and social instability" (p. 265).

In addition, demonstrations that male mice in captivity tend to kill the young of unfamiliar females but not those of familiar females (Labov 1981;

Huck et al. 1982), and that they mate more with unfamiliar females whose pups they have just killed (Huck et al. 1982), are in accord with this hypothesis. Similar selectivity in male infanticide also occurs in the collared lemming *Dicrostonyx groenlandicus* and the meadow vole *M. pennsylvanicus*, both of which also show the Bruce effect (Mallory and Brooks 1978; Webster et al. 1981). Abortions are apparently common in the field in some species. Some wild-trapped females *Microtus* had 2–3 generations of corpora lutea in their ovaries, as is typical in captivity after matings with 3–4 males in succession and the attendant abortions in the lab (Mallory and Clulow 1977; Milligan 1982; Marchlewska-Koj 1983).

Labov (1981) argues that the fact that urine of subordinate male *M. musculus* was just as effective as that of dominant males in inducing pregnancy blocks rules out the possibility that sexual selection by female choice has played a role in the evolution of the Bruce effect. Later data (Huck 1982) differ, suggesting that urine from dominant males is more effective in blocking pregnancies. Even if the urine of subordinates is equally effective, however, Labov's argument misses the point that by responding to a male (whatever his former history) that has been able to displace the previous resident male and deposit his urine at a site where the female will encounter it, the female may favor male genotypes that are better able to effect such displacements.

Another, less complicated case of male-induced abortion that is under female control comes from Hansson's classic study of the mink *Mustela vison* (Hansson 1947). In nature the male is apparently usually associated with the female only long enough to copulate. If offered a new male every other day during her approximately 3-week mating season, a female sometimes remates (from one to nine times, av. 2.9 matings per female). Females are in complete control of whether a copulation will occur, and even when confined in captivity they often reject the advances of a male. Copulation, or just the male mounting and riding the female, induces ovulation about 36 hr later. If a second mating occurs more than 6 days after the first, the zygotes from the first mating are nearly always aborted (twenty-five of twenty-seven females), and the female ovulates again. By accepting the advances of a second male, the female thus nearly always causes the abortion of the first male's offspring.

The case for cryptic female choice is incomplete in this species, however, as female mating frequencies in nature are unknown, and Hansson did not make any attempt to associate differences in female responses with differences among males. One possibly interesting variation he observed was the length of copulation, which ranged from <30 min to 14.5 hr (av. 64 min); average durations more than doubled as the season progressed, going from 49 min before 10 March to 114 min after 26 March.

4.6 Biased Use of Stored Sperm

Several female traits might produce biases in sperm usage. As noted in chapters 1 and 2, conditions within the female storage sites or reproductive ducts could bias arrival at fertilization sites due to differences in transport, maintenance, or mortality of sperm; the ability of sperm to penetrate the eggs once they have arrived could also be affected. Complex events, including clumping of sperm and cyclic alternation of usage of different males' sperm or biased use of sperm at different storage sites for fertilization, may occur within the female (fig. 4.9; Page 1986). Thus the mechanism of selective sperm usage, as with several others discussed earlier, is really a composite of several possible mechanisms. Males producing sperm (and seminal plasma) resulting in better survival and arrival at fertilization sites, and better abilities to penetrate eggs, would tend to outreproduce those whose sperm lack these abilities if females mate with more than one male (table 4.5).

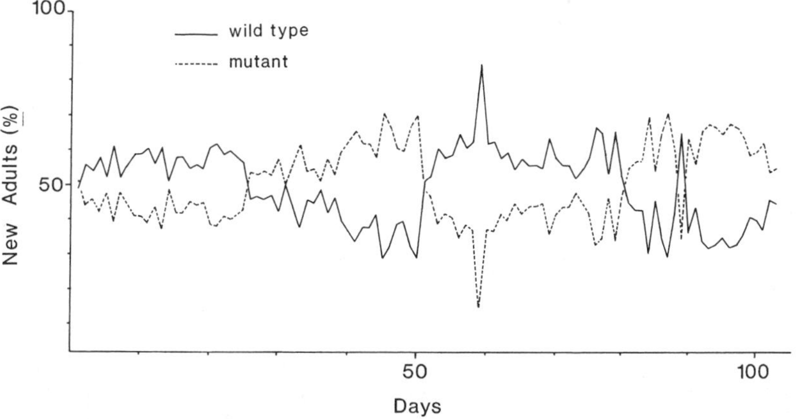

Figure 4.9 Complex patterns of sperm usage from the spermatheca of a queen honeybee *Apis mellifera*. The queen was artificially inseminated with equal parts of semen from a male homozygous for wild-type and a male homozygous for a body color mutation (1 microliter each). The body colors of the adult offspring born each day (starting on the fifth day after insemination) were noted. The overall frequencies were nearly equal (49% wild-type, 51% mutant), but the frequencies on 70 of 110 days were significantly different from this. The average number of bees counted each day was 258 ± 112 (range 54–870). Long-term sperm clumping was suggested by the fact that offspring of the same male tended to prevail on consecutive days. Similar short- and long-term clumping patterns occurred in all 11 other test colonies, and other studies of this species have also shown incomplete sperm mixing (reviewed in Page 1986). (After R. B. Raphael de Almeida, Master's thesis, University of São Paulo)

Table 4.5
Possible Cases in Which Females Exercise Cryptic Female Choice by Biasing the Use of Stored Sperm in Favor of Certain Males

	Criteria						
Species	*I*	*II*	*III*	*IV*	*V*	*VI*	*References*
*Chorthippus parallelus** (Orthoptera)	Y	Y	Y	Y?[a]	Y (geogr. race)	Y	Bella et al. 1992
*Podisma pedestris** (Orthoptera)	Y	Y	Y	Y	Y (geogr. race)	Y	Hewitt et al. 1989
Allonemobius spp.* (Orthoptera)	Y	?	Y	Y?[a]	Y (species)	Y	Gregory and Howard 1994
*Drosophila melanogaster** (Diptera)	Y	?	Y	Y	Y (seg. dist. gene transloc., geog. race)	Y	DeVries 1964; Zimmering and Fowler 1968; Denell and Judd 1969; Childress and Hartl 1972; Gromko and Markow 1993
Macrohaltica spp. (Coleoptera)	Y	?	Y	Y	Y (species)	Y	Eberhard et al. 1993; Eberhard and Ugalde, in prep.
*Mus musculus** (Rodentia)	Y				Y (*t* allele)	Y	Bateman 1960 in Denell and Judd 1969; Bronson 1979

NOTES: Those marked with (*) are discussed in the text.
Each example is classified with respect to criteria for proof that cryptic female choice occurs. See table 4.1 for criteria and key.
[a] Remating was relatively common, and occurred readily in captivity, suggesting it occurs in nature.

Winnowing processes may also work among the sperm of a single ejaculate, causing those with some types of phenotypic malformations to be left behind (Cohen 1975; Cummins 1990; see chap. 7.1). If the genotypes of sperm influence their own phenotypes, selection could also favor some sperm genotypes over others from the same male. Genotype expression is common in the pollen tubes of plants and may be a basis on which discriminations are made by females. But expression of the genes in the sperm cells of animals is at best very rare, as their DNA is strongly compacted and thus unavailable for transcription (Mulcahy 1975; Baccetti 1991). In fact, the one portion of the sperm genome that is clearly active, the mitochondrial DNA, is generally not included in the zygote's genome. Intraejaculate phenotypic winnowing could occur both immediately after sperm are deposited by the male and (perhaps to a lesser extent—see below) after sperm have been stored in a female. Genotypic winnowing (if it occurs) would seem most likely after the sperm has arrived at the egg and its DNA has been decondensed (see sec. 4.11).

An alternative possibility is that differences in usage of stored sperm are due to direct competitive male-male interactions among sperm themselves, or perhaps to the effects of other components of the ejaculate that accompany them (e.g., Silberglied et al. 1984). For instance, some sperm might have a greater ability to position themselves near the exit of the storage organ or the ability to inactivate stored sperm from other males. This could involve "sperm competition" *sensu strictu*.

Although most previous publications on "sperm competition" have emphasized the male side, in fact female and male control of sperm interactions are not easily disentangled. In practice, discrimination between female and male determination of biased sperm usage seems to me to be impossible based on much of the data available. Many previous discussions of sperm competition are in fact open to the alternative interpretation of cryptic female choice (see chap. 1.9).

For instance, the female's morphology, the chemical milieu inside her storage organs, or other behavioral characteristics such as contractions of muscles in her reproductive ducts can and probably often do influence the outcome of male-male competition. Females determine the playing field and the rules of the game for sperm competition. For a simple example, consider the reduction of sperm mobility, which has arisen repeatedly in different animal groups (Baccetti and Afzelius 1976). This is only a feasible option when female morphology and behavior are such that sperm are transported to storage and fertilization sites by the male himself, or (more often) by the female. Only when transport is relatively certain (or motility has no effect on migration in the female) could selection against motility occur. In females with high rates of leakage of sperm from storage sites (due, for example, to lack of valves in the female ducts as in *D. melanogaster*; Gilbert 1981), selection might act against especially motile sperm (there is evidence in *D. melanogaster* that sperm "leak" from different strains of females at different rates; Childress and Hartl 1972).[1]

Unfortunately, there are few experimental studies of preferential use of sperm to consult for illumination on female versus male roles. It is technically difficult to study the behavior and ultimate fate of sperm within a female (see, for example, the thoughtful discussion by Overstreet and Katz 1977 of the difficulty of distinguishing sperm storage sites or reservoirs in female mammals from accumulations of sperm that are destined to be eliminated). The most powerful techniques depend on genetic markers, which are of course available for relatively few species. Thus only a limited sample of species is available to check for biased sperm usage. Cases of possible interejaculate biases will be discussed first, followed by possible biases among sperm from a single male.

Interejaculate Biases

Female involvement in biased sperm usage can be established experimentally (if individual females differ) by showing that a bias exists, and that different females have different biases. Several possible cases of this sort occur in *Drosophila melanogaster*, a species in which sperm do not clump appreciably within the female's storage organs (Prout and Bundgaard 1977; Gromko et al. 1984; fig. 1.12).

By counting the number of sperm stored 3 hr after a single mating, and comparing them with the lifetime adult offspring production from similar crosses, Zimmering and Fowler (1968) found that sperm of males of one strain (Oregon-R) were used more efficiently by females of their own strain than by females of a different strain (*yellow*). Females of both strains stored equal numbers of sperm (av. = 494 in Oregon-R, 515 in *yellow*). But the ratios of number of adult offspring/number of sperm (presumably a measure of sperm use) averaged around 0.7 when an Oregon-R male was crossed with an Oregon-R female but only about 0.4 when an Oregon-R male was crossed with a *yellow* female. Subsequent tests with females of several more strains (Zimmering et al. 1970) showed further variation, with ratios from 0.31 to 0.88. The mechanism or mechanisms responsible for these differences were not determined. Possible differences in offspring viability were unfortunately not checked.

A second case of bias in *D. melanogaster* was interpreted as involving a selective deactivation of sperm from particular males, somewhat similar to a mammalian immune response. DeVries (1964) found that when she counted sperm in the female's major storage organ (the ventral receptacle) after males and females of different geographic strains and their hybrids were crossed, the amounts varied with different combinations of males and females. The amount of sperm from males of a given strain was correlated with the strain of the female; and the amount of sperm stored in females of a given strain varied when they mated with males of different strains. In addition, patterns of sperm remaining in the ventral receptacle up to 29 days after copulation gradually changed according to the combination of strains.

These data are not completely convincing. Some differences were not documented with statistical tests, and alternative interpretations were not ruled out. For instance, differences in sperm numbers received by females could be due to different numbers of copulations during the 24 hr period males and females were kept together (it was assumed they would mate only once, as is usually the case in *D. melanogaster*; Gromko et al. 1984). Or ejaculate sizes may have differed when males were paired with different females (this possibility must be taken seriously in the light of facultative adjustments of ejaculate size in the medfly *Ceratitis capitata* and the beetle *Tenebrio molitor*; Gage and Baker 1991; Gage 1992). Differences in sperm retention times might be due to different rates of oviposition (females were kept on minimal medium, where they laid "few" eggs), or to expulsion of sperm some time after copulation (Wheeler 1947 in DeVries 1964).

It is clear, however, that different combinations of strains gave different results with respect to the quantity of sperm in the female's ventral receptacle, and differential viability or activation of sperm is a strong possible explanation. Killing of the sperm of some males but not others due to apparent "immune" responses in the female reproductive tract is well known in humans (Mann and Lutwak-Mann 1981). Another possible factor is the behavior of the sperm themselves (which can be influenced by the female; Katz and Drobnis 1990). For instance, the sperm from different strains of mice differed in its hyperactivated behavior (swimming behavior in the near vicinity of the egg; Olds-Clarke 1990).

There are some less direct observations in other groups suggesting that interactions between sperm from different males are influenced by the female. When females of different geographic subspecies of the grasshopper *Chorthippus parallelus* were each mated to one male of their own subspecies and to one of the other (the order was varied), homogametic matings resulted in more offspring in both types of female (paternity was determined by examining the chromosomes of 9-day-old embryos; Bella et al. 1992). Females of one subspecies showed a stronger bias toward homogamy than those of the other. Both hybrid mortality and differences in copulation duration were tentatively eliminated as possible explanations of the biases. Similar homogametic bias occurred in two parapatric races of another grasshopper, *Podisma pedestris*, and evidence again pointed to preferential fertilization as the mechanism responsible (Hewitt et al. 1989). In this species there was no difference between the two types of females with respect to the bias, but there were significant differences among individual females in the strength of the bias.

A second way to determine fertilization biases is to compare relative volumes of sperm from different males that are clumped in coherent masses versus fertilization frequencies. A possible example of biased use comes from the damselfly *Mnais pruinosus* (Siva-Jothy and Tsubaki 1989a). The percentage of paternity for the last male a week after copulation sank to about 50%, even though the relative volume of his sperm in the spermatheca was substantially higher (estimated at 89.8% from the area occupied in fig. 8 of Siva-Jothy and Tsubaki 1989a; estimated volumes, which were calculated by multiplying the area by the square root of the area, were 428.1 vs. 48.5). Perhaps sperm emission by the female (E. Gonzalez, pers. comm.; see fig. 1.4), position effects, or inefficient sperm usage in fertilizing eggs were responsible for this discrepancy.

Similar cases of biased sperm use also occur in pairs of closely related species. These are relevant to possible cryptic female choice inasmuch as differences among species may be exaggerations of differences between members of a single species. In ground crickets of the genus *Allonemobius* (Howard and Gregory 1993; Gregory and Howard 1994), the species *A. fasciatus* and *A. socius* are sympatric in an extensive mosaic contact zone in the eastern United States. Relatively few hybrid individuals occur in areas of sympatry, even though their survivorship under natural conditions is not consistently lower than that of purebred individuals at the same sites (Howard et al. 1993). The lack of hybrids is paradoxical, because females from at least some mixed populations do not move preferentially toward conspecific songs (which differ slightly), and fertile hybrids are produced in abundance when the species are crossed in captivity.

The lack of hybrids is probably largely due to preferential use of conspecific sperm by females that have been inseminated by males of both species, or increased abilities of sperm to directly dislodge or otherwise inactivate sperm of the other species ("sperm competition" *sensu strictu*) when within a conspecific female. When females of a species are crossed successively with a male of each species, nearly all of the offspring are fathered by the conspecific male regardless of the order of mating (97–99% for *A. fasciatus* females and 90–100% for *A. socius* females; Gregory and Howard 1994). Direct observations of sperm when the female's spermatheca was broken open in Ringer's solution suggested that sperm may be less active in females of the other species (Gregory and Howard 1994; small sample sizes precluded statistical tests).

Similar preferential success of homospecific rather than heterospecific sperm in an otherwise freely hybridizing pair of species also occurs in the chrysomelid beetle genus *Macrohaltica* (Eberhard and Ugalde, in prep.). The many possible mechanisms for these differences are yet to be explored; it is clear, however, that the type of setting in which the competition occurs (i.e., the identity of the female) has a strong effect on which sperm win out.

A possible (though not necessary) advantage to females of biased sperm usage is increased offspring viability. This appears to occur in the adder *Vipera berus* (Madsen et al. 1992), though the relative importance of male and female roles in sperm usage in this species are unclear (Parker 1992).

Intraejaculate Biases

The different portions of the phenotype of a sperm cell (the acrosome, the tail, etc.) are thought to be produced by genes expressed in the testes of the male, rather than by the haploid genome of the sperm itself. The sperm genome is generally thought to remain completely repressed until some time after it fuses with that of the egg (e.g., Mulcahy 1975; Sivinski 1984; Cohen 1991). Thus, while sperm from different males could well bear phenotypic markers that allow them to be discriminated within the female, discrimination among sperm from a single male would seem to be much less likely. Nevertheless, there is evidence that such intraejaculate discrimination does occur.

Perhaps the strongest indication of female selectivity in this sort of bias comes from males bearing a pair of translocations in *D. melanogaster*. Childress and Hartl (1972) found that once-mated females of one strain produced progeny that were successively more and more biased toward one of two types of sperm produced by the male. They ruled out the possibility that differential postzygotic survival was responsible, since the bias in offspring was affected by a previous mating from which the sperm had been exhausted. They reasoned that active female choice rather than direct sperm competition was involved. The analogy to an immune reaction (DeVries 1964) seems especially appropriate here. This type of data, combined with the nearly total absence of studies designed to check for intraspecific discrimination, led Gromko et al. (1984) to speculate that in *Drosophila* gamete selection during sperm storage may represent an important but largely unexplored mechanism of natural or sexual selection.

In an additional experiment with *D. melanogaster*, females of different strains were crossed with males that were heterozygous for segregation distorter genes. A slight but significant degree of difference in distortion was obtained in the offspring (Denell and Judd 1969). Differences due to differential zygote mortality were ruled out. It was not clear whether differences were due to differences in sperm transport, storage, viability, ability to effect syngamy, or some other characteristic. A similar example of possible intraejaculate discrimination in sperm usage was found in the use of sperm bearing X and Y chromosomes in this same species. In sperm from younger males, Y-bearing sperm seemed to be preferred, while X-bearing sperm were more often used from older males (Mange 1970).

One further example of intraejaculate bias comes from the house mouse *Mus musculus*, in which female genotype has an effect on *t* allele recovery in her offspring (Bateman 1960 in Denell and Judd 1969).

It is worth reemphasizing the low number of species in which enough is known of their genetics and in which experiments appropriate to sense biases in sperm usage have been performed. An additional bias toward underestimating the frequency of this mechanism is that differences in sperm utilization have been discovered accidentally; the genetic markers used in these studies (body color, etc.) were only incidentally linked with differences in sperm usage.

4.7 Sometimes Move Previous Male's Sperm to a Site Where the Current Male Can Manipulate Them

Even if she does not discard the sperm of a previous male, a female can modify their fate during subsequent copulations simply by moving them to a site in her reproductive tract where the current male can manipulate them (table 4.6). Thus in a species in which the male's genitalia do not reach the female's sperm storage organs, male removal or inactivation of a previous male's sperm is feasible only if the female cooperates. I will include here examples in which a female participates actively by moving (or failing to move) a previous male's gametes. As discussed in chapter 1 with reference to sperm removal in odonates and crickets, "passive" female participation, resulting in cryptic female choice, can also occur when female morphology (without any overt behavior) causes differences in the abilities of males of her species to deal with these features and thus results in differences in male reproduction. Since the selective contexts in which such design features evolved are generally unknown, it is prudent to omit them here, even though this may result in an underestimate of the frequency of cryptic female choice via this mechanism.

Male ability to induce the female to move sperm to a site where the male can cause them to be discarded occurs in the katydid *Metaplastes ornatus* (von Helversen and von Helversen 1991). Copulation in this species usually proceeds as follows. The male first attracts the female from a distance

Table 4.6
Possible Cases in Which Females Exercise Cryptic Female Choice by Sometimes Moving a Previous Male's Sperm to a Site Where the Current Male Can Manipulate Them

	Criteria						
Species	*I*	*II*	*III*	*IV*	*V*	*VI*	*References*
Metaplastes ornatus (Orthoptera)	?	?	Y	Y	?	?	von Helverson and von Helverson 1991
Drosophila melanogaster (Diptera)	Y[a]	Y?	Y	Y	Y (esterase-6)	?	Gilbert 1981; Scott and Richmond 1990; Gromko and Markow 1993; Scott and Williams 1993

NOTES: Both are discussed in the text.
Each example is classified with respect to criteria for proof that cryptic female choice occurs. see table 4.1 for criteria and key
[a] Effect stronger when female remates soon after first mating. At least in captivity, females do begin remating as soon as 2 hr after the first mating (Gromko et al. 1984).

by singing. When she approaches, he moves under her, grasps the base of her ovipositor with his cerci, and inserts his subgenital plate into her genital chamber (Phase I coupling). After moving his subgenital plate back and forth rhythmically in the genital chamber, he pulls away. As his genitalia are withdrawn, the barbs and spurs on his subgenital plate cause the lower portion of the female's reproductive tract to be pulled from her body inside out (fig. 4.10). The female then cleans the exposed portion of her genital

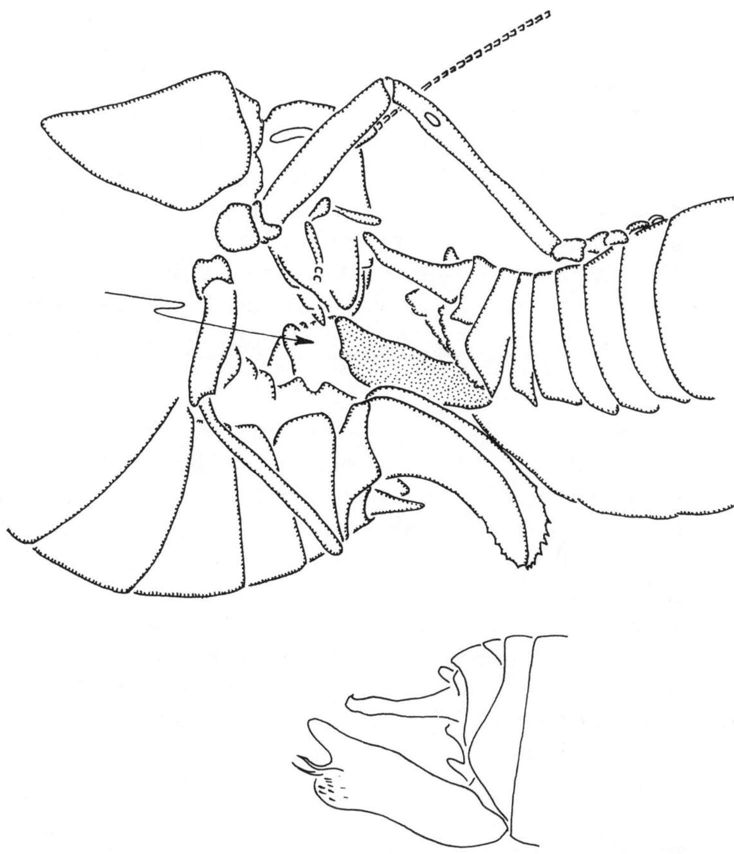

Figure 4.10 A female *Metaplastes ornatus* katydid (*top*) cleans a previous male's sperm from an everted portion of her reproductive tract (arrow), which the current male has pulled from her body with his subgenital plate (stippled). The subgenital plate has several spines and hooks (*below*). It is rubbed back and forth in the female's oviduct, apparently mimicking the passage of an egg and causing movement of sperm from the spermatheca to the oviduct; it also pulls the oviduct out so the female will clean off the sperm. After rubbing and then pulling out the oviduct up to nineteen times, the male introduces his own sperm into the female. (After von Helversen and von Helversen 1991)

chamber with her mouthparts and pulls it back inside her body. The male resumes singing and couples again with the female. There are an average of 5.8 ± 4.3 (range 2–19) Phase I couplings per pair. Finally, the male's subgenital plate is flexed ventrally, and he brings his protruding copulatory organs into contact with the female's genital opening in a single Phase II coupling. The male makes a large spermatophore (av. 22% of his weight), attaches it to the female's genitalia, and leaves. The female takes several hours to consume the spermatophylax, during which time the male's sperm move (or are moved) into her spermatheca.

When copulation of field-collected females was interrupted just after Phase I ended (as the male was preparing to transfer his spermatophore), the female's spermatheca contained an average of only about $0.18 \pm 0.24 \times 10^6$ sperm ($N = 9$), or about one-seventh of the average contents of a full spermatheca ($1.23 \pm 0.43 \times 10^6$). Since the male's genitalia do not reach the spermatheca at any stage, this removal was not produced directly by the male. The duration of Phase I, which varied widely (av. 31 ± 15.7 min), may correlate inversely with the amount of sperm left in the female when it ended (the sample was too small to be certain). Males may sometimes perform copulations with shorter Phase I durations in order to reduce the danger of predation, since copulation occurs during the day, and pairs are easy to detect (O. von Helversen, pers. comm.).

Von Helversen and von Helversen (1991) argue that sperm removal from the female in this species results from the male having in a sense "fooled" the female. The male's subgenital plate is about the size of an egg (fig. 4.10), and they propose that stimuli from its movement in the female's genital chamber mimic the stimuli produced by an egg as it descends the oviduct. The female's "normal" response to such stimulation—to cause sperm to be moved down the spermathecal duct to the genital chamber where they can fertilize the passing egg—is used by the male to induce her to discard the sperm of previous males. This hypothesis was tested by moving an egg back and forth in the genital chamber in the same way the male moves his subgenital plate. When the egg was then removed and rinsed in Ringer's solution, hundreds of sperm were found. In contrast, when an egg was introduced into the chamber but not moved, only a few sperm were found.

The structures in the female that probably produce the responses just described have been discovered in the cricket *Teleogryllus commodus* (Sugawara 1993) and the grasshopper *Schistocerca vaga* (Okelo 1979). Sensory neurons in the genital chamber at the mouth of the spermathecal duct apparently sense the presence of an egg and cause stimulation of nerves leading to muscles in the walls of the spermatheca and the spermathecal duct (fig. 4.11). The resulting rhythmic peristaltic contractions (in the grasshopper) or sharp twitches (in the cricket) cause sperm to be moved

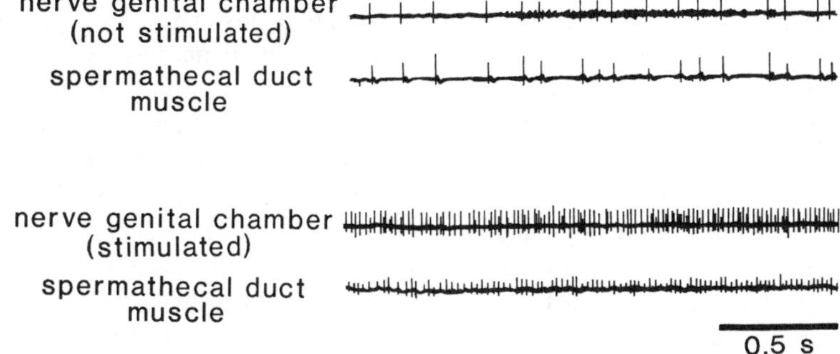

Figure 4.11 Female control of sperm usage in the cricket *Teleogryllus commodus*. Stimulation of the oviduct increases muscle activity in her spermathecal duct, causing sperm to be moved from storage to the fertilization site. When the genital chamber (the portion of the oviduct where the spermathecal duct empties and fertilization is presumed to occur) is not stimulated, a muscle contraction follows after every action potential in the nerve from the receptors in the wall of the genital cavity (*top*). When the receptors in the genital cavity are stimulated, as they would be by an egg descending the oviduct, the rate of firing of nerves from the receptors increases, resulting in an increase in the rate of muscle activity in the spermathecal duct (*bottom*). (After Sugawara 1993)

out of the spermatheca, down the long spermathecal duct (in the cricket they descend tail-first, emphasizing the probable female control of this process), and into the genital chamber.

Further information suggests that female *M. ornatus* are probably not entirely "fooled." The form of the male subgenital plate varies in different species in this genus (O. von Helversen, pers. comm.). This relatively rapid divergence suggests that the form of the subgenital plate may be under sexual selection, since presumably the form of the eggs of different species is relatively similar. It seems likely that female responses to stimuli from the subgenital plate originated from mimicry of the stimuli from eggs but have gone on to include additional stimuli.

This example is still incomplete as a demonstration of cryptic female choice, which would require documentation of differences in female responses to different male phenotypes (i.e., possible differences among males in rubbing behavior, singing, genitalic morphology, etc.; variations per se did occur, as the duration of Phase I copulation and the number of sperm remaining after Phase I ended were quite variable: av. ± SE = $.180 \pm .240 \times 10^6$).

A second species in which females may displace stored sperm to a site where they are manipulated by a second male is the fruit fly *Drosophila melanogaster*. The male genitalia of *D. melanogaster* reach only to the openings of the ducts leading to the sperm storage sites (two spermathecae, plus the ventral receptacle; Gromko et al. 1984). Nevertheless, some second matings cause sperm from the first mating to be lost. By using a genetic strain of males unable to transfer sperm in second matings, Scott and Richmond (1990) showed that when remating occurred 6 hr after the first mating, probable female sperm supplies from the first male were reduced by about 30%. Further experiments using males with inactive accessory glands suggested that seminal fluid is involved in sperm reduction (Harshman and Prout 1994; Chapman et al. 1995), perhaps via the sperm activator esterase-6 (Gilbert 1981; Richmond and Senior 1981; Gromko et al. 1984). Lost or inactivated sperm from the first male are thought to be normally replaced in the female storage organs by sperm from the second male (Scott and Williams 1993).

The male "manipulation" could be as simple as dilution. Male *D. melanogaster* normally transfer at least 2–4 times more sperm than are stored (Gromko et al. 1984; precise numbers vary in different reports; Gilbert 1981 found that 4000–6000 were transferred, and that about 4–7 hr later approximately 700 were stored in the seminal receptacle and 400 more in the two spermathecae). Any sperm from the first mating that were displaced into the uterus would thus be greatly outnumbered by sperm from the second male. When a portion of this mixture then moved (or was moved) into storage, the first male's sperm would have a reduced chance of being included. This interpretation is as yet uncertain (e.g., Harshman and Prout 1994). Unfortunately, most of the data on numbers of stored sperm in these experiments were only indirect estimates, based on number of adult progeny. Since the number of offspring could be affected by several other factors, such as oogenesis, oviposition, and sperm use efficiency in addition to sperm number transferred, conclusions can only be tentative.

There is ample opportunity for females to also participate more actively in this process, most obviously by contraction (or lack of contraction) of the muscular sheath surrounding the ventral receptacle. Participation of this sort could explain the relatively high variance in sperm precedence values often seen in *Drosophila* (Gromko et al. 1984; see fig. 1.12). Whether or not female participation of this sort occurs seems not to have been asked. Sperm number in the ventral receptacle has been modeled on the assumption that sperm leak out randomly (Gilbert et al. 1981a), but it is likely that females use the receptacle muscle to move sperm (this would seem to be its most logical function).

Other insects, and probably many other animals, also have sense organs whose positions within the female reproductive tract suggest that they are stimulated by eggs that are in position to be fertilized (e.g., Siva-Jothy 1987a on dragonflies; Okelo 1979 and Sugawara 1993 on orthopterans). Male stimulation of such sensillae during copulation could cause females to move stored sperm to sites where, for instance in the case of dragonflies, they could be more easily removed. It is common for insects (and presumably many other groups) to have anatomical arrangements that suggest that sperm are moved to fertilization sites by muscular contractions of the spermathecae and/or their ducts (Davey 1965b).

4.8 Sometimes Make Subsequent Sperm Transfer More Difficult Morphologically

In some species of arthropods, mating is easier when the female is newly mature and still soft. By altering the rate at which the cuticle of her genitalia hardens after copulation a female of such a species could alter the chances of paternity for subsequent males. A possible example of this sort occurs in the giant golden silk spider *Nephila clavipes* (fig. 4.12; data summarized in Christenson 1990).

A male *N. clavipes* often searches out an immature female that is about to moult to maturity and waits on her web. He fights with other males there for the opportunity to mate with the female as soon as she moults. Such matings are veritable orgies of repeated copulation. During the first 48 hr, the pair spends up to 28 min of each hour in repeated copulations, each of which involves many rhythmic inflations of membranes (hematodochae) on the male's palps (genitalia) that probably drive an intromittent sclerite deeper into the female (average 43.1 ± 13.6/min; Christenson and Cohn 1988). Apparently two different portions of the male genitalia are inserted into grooves and ducts in the female genital plate (Schult and Sellenschlo 1983; see, however, Huber 1993 for criticism of the technique used in this study).

By interrupting matings at different stages, it was determined that all of the male's sperm (which are immobile) were usually transferred during the first 3 hr of this mating marathon (in fact, they may be transferred within the first coupling of each of the male's palps; T. Christenson, pers. comm.). Large males had more sperm in their palps and transferred larger amounts to the female's spermathecae (the av. was about 320,000—to fertilize the female's total of about four hundred eggs; Cohn 1990).

The cuticular lining of a virgin female's insemination ducts (especially in the deeper portions; T. Christenson, pers. comm.) as well as that of the spermathecae hardened only gradually, over the space of several days. Male copulation behavior was different with older virgin females. Insertions of his palps were shorter, and the inflations and contractions of his genital hematodochae that accompanied intromission were both less rapid (24.8/min vs. 43.1/min) and less regular (Christenson and Cohn 1988).

Females did not give any overt signs of greater resistance during second matings (Christenson and Cohn 1988), but these copulations nevertheless

Figure 4.12 A tiny male *Nephila clavipes* (arrow) clings to the abdomen of a female. After mating soon after she moults to maturity, the female's genital ducts sometimes harden quickly, making insemination by subsequent males more unlikely to produce offspring. In other cases, females harden less rapidly, making subsequent inseminations more likely to result in offspring. (Photo courtesy of Michael Robinson)

resulted in the transfer and uptake of a much smaller proportion of the male's sperm. An average of 62% of the male's available sperm were transferred, and only 21% of these, or 14% of the male's total, reached the spermathecae, as compared with 100% and 62%, respectively, for first copulations (subsequent work has indicated that there were technical problems in counting sperm because sperm sometimes adhere to vials, and 62% is probably an underestimate; T. Christenson, pers. comm.).

Females often remate several days after mating. Experiments using sterilized males showed that the second male normally fathered an average of 18% of the female's offspring in such cases. This percentage doubled (36%), however, when the first male was allowed to copulate for only 3 hr (more than enough time to transfer all his sperm) and then removed. Premature interruptions of first matings undoubtedly sometimes occur in nature, since 15% of all females in the field left their original website (and thus the males there) within 24 hr of moulting to maturity (Cohn et al. 1988). Displacements of males (presumably due to aggression among males) also sometimes occur in the field (at least 5%, and up to 22% of 211 pairs observed in the field; Cohn et al. 1988).

The mechanism responsible for the lower sperm precedence values after longer first matings is thought to be related to maturation of the female and attendant increases in the hardness of her genital ducts. Preliminary data suggest that stimuli of some sort associated with long copulations just after the final moult (presumably copulation itself or seminal products transferred by the male) favored the sclerotization process in the female. The spermathecae and their ducts in unmated females are less sclerotized than those of females of the same age that mated soon after their final moult (Higgins 1989). Exactly how the hardness of female ducts might affect sperm transfer and uptake (smaller numbers of sperm released?) was not determined, and the possibility that other changes in the female caused changes in male fertilization success via mechanisms other than transfer success (e.g., sperm transport by the female, survivorship of sperm in her ducts) was not ruled out (Christenson 1990). An experimental manipulation, in which one of the male's palps was removed so that he could inseminate only one of the female's two spermathecae, suggested direct local involvement of male sexual products in producing sclerotization rather than general bodywide changes in the female. The mated spermatheca was more sclerotized (darker) than the virgin spermatheca 1, 7, and 17 days after mating ($N =$ 2, 1, and 1; Higgins 1989). Over longer periods of time even unmated spermathecae became sclerotized (Higgins 1989).

A hint that female cooperation (transport?) is involved along with the number of sperm the male transfers comes from cases in which sperm depletion in the second male was checked. In only two of the eight males that obtained high paternity rates (>66%) was the male even partially depleted of sperm; there was no correlation between a second male's depletion and his paternity success (Christenson and Cohn 1988).

Progressive hardening of female reproductive structures that reduces male access is not restricted to *Nephila* spiders. In several groups of crabs the opercula of the female must be decalcified each reproductive cycle before insemination is possible (Salmon 1984; Adiyodi 1988). The possibility that males influence this process is apparently untested.

4.9 Sometimes Resist Male Manipulations That Result in Discharge of His Spermatophore

Males of some animals make sperm packets (spermatophores) which they attach to the substrate, and then attempt to induce the female to take the sperm up into her reproductive tract. In some groups, including scorpions (Polis and Sissom 1990), some pseudoscorpions (Weygoldt 1969), and cephalopod molluscs (Wells and Wells 1977), the spermatophore is "explosive," in the sense that when triggered (normally by the female), it injects or inserts its sperm into the female. The male thus does not depend completely on the female to take up his sperm (she sometimes fails to do this in other groups which lack explosive spermatophores, such as salamanders; see Arnold 1972 on *Amblystoma* and *Plethodon*). In some cases the male manipulates the way the spermatophore contacts the female so as to increase the chances that the sperm will be introduced into her (in the pseudoscorpion *Dactylochelifer* the male has a specialized first tarsal claw, with which he "helps" the female take up the spermatophore; Weygoldt 1969).

In the scorpion *Bothriurus flavidus* the male manipulates the female to cause his spermatophore to discharge, and the female sometimes resists (Peretti, in press). As is typical of scorpions, the male seizes the female's pincers with his, and then "promenades" with her until they find a suitable site for him to deposit a spermatophore. In *B. flavidus*, one of the characteristics of such a site is an object against which the male can push with his telson once he has deposited the spermatophore and it has coupled with the female's genitalia. This push causes the female's body (still held with his pincers) to move posteriorly, and this movement in turn causes the spermatophore to pivot so that the sperm in their reservoir are inserted into the female (see drawings of a congeneric species in fig. 4.13). In some matings (six of twelve observed), the female allowed the male to execute the promenade and deposit his spermatophore but then resisted the final steps. In two, the female did not allow the male to position her with her genital opening over the spermatophore; and in four others she allowed the spermatophore to enter her gonopore but then resisted the push by the male that would have caused his sperm to be injected into her.

Female resistance at this final stage, causing failure in sperm transfer, may not be uncommon. A. Peretti (pers. comm.) has seen it in other bothriurids, including two other species of *Bothriurus*, *Timogenes elegans*, and *Urophonius iheringi*, although it was uncommon in the buthid *Zabius fuscus*. Even females that have not resisted the male during the promenade sometimes do not allow themselves to be "seated" on the male's spermato-

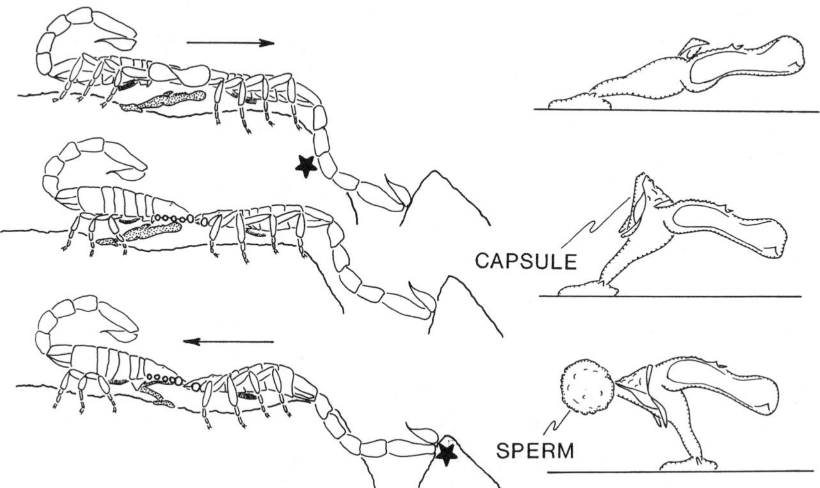

Figure 4.13 Sperm transfer in the scorpion *Bothriurus bonariensis* is not always a smooth process (animals at *left*, with spermatophore stippled; detail of spermatophore at *right*). First the male (on the right) pushes down on the substrate with his telson (star in *top* drawing) to pull the female forward (arrow) so her genital opening is over the spermatophore. The female lowers her body (*middle*) to engage the spermatophore and cause eversion of the internal capsule (some legs and palps are omitted for clarity). Then the male again uses his telson (star, *bottom*) to push the female backward (arrow), bending the spermatophore and causing the ejection of a drop of sperm into her body. In some pairs of *Bothriurus*, the female shows little or no resistance until the last step, when she opposes the male's attempts to push her backward, thus preventing the "firing" of his spermatophore. (After Peretti 1992)

phore, or they resist the male's push that causes it to fire. Males sometimes make repeated attempts to seat the female, and this can lead to the female stinging the male, finally causing him to release her. Since it can take a male a week or more to produce the next spermatophore (Polis and Sissom 1990), rejections of this sort are likely of selective importance to males.

A related phenomenon occurs in the pseudoscorpion *Cordylochernes scorpioides* (J. and D. Zeh, pers. comm.). The male holds the female during the approximately 4 min while he is depositing a spermatophore on the substrate. He then pulls her over it, and she depresses her abdomen so that the sperm packet adheres to her gonopore. A long, hooked tube everts from the sperm packet into the female's reproductive tract, and the sperm are transferred into her body. If the female is "restless" during spermatophore production, the spermatophore is sometimes apparently normal in form but contains few or no sperm. Nonreceptive females generally terminate matings by refusing to be pulled over the spermatophore, which they then sometimes eat. The possibility that these female behavior patterns were responses to differences between males was not investigated. Spermatophores also occasionally "misfire" in other groups (e.g., Robinson 1942 on the tick *Ornithodoros moubata*), but the possible role of the female is unknown.

4.10 Sometimes Invest Less in Each Offspring

After copulation, females of many animals invest resources such as time, energy, and materials in their offspring. Variation in such investments could affect male reproductive success in several ways. If the male does not care for the offspring, reduction in the female's investment (say a reduction in the sizes of the eggs she lays, or in the amount of care she gives the offspring after they are born) could reduce the number of his offspring that survive to reproduce (table 4.7). In species with paternal care, the analysis is somewhat more complex, because a decrease in female care of the brood could also raise the parental demands on the male. This could lower his chances of caring for another brood at the same time or in the future.

There is probably often a conflict between male and female interests in female investment in offspring. With the possible exception of species with lifelong pair bonds or strictly monogamous females, a male would generally benefit from greater immediate investment by the female, as this would maximize the number of her offspring sired by him. The female's interests would, in contrast, include longer-term considerations, because selection would act more directly on her total, lifetime production of offspring. If she remates, some later offspring would be sired by other males. Such conflicts may be especially important in species in which male parental care occurs by feeding the female (see Simmons and Gwynne 1991; Oberhauser 1992).

One obvious female investment in offspring which might be altered is egg size. In the katydid *Requena verticalis* the size but not the number of eggs the female produced increased with greater donations of spermatophylax material from the male (Gwynne 1984b, 1988a,b). The larger eggs produced offspring that were more likely to survive the winter (Gwynne 1988a), so males producing a larger spermatophylax probably gained a reproductive payoff via nutrient donation.

This correlation persisted unaltered, however, in the standard test experiment for nutrition effects, as well-nourished females showed the same response as poorly fed females (Gwynne 1988a). This argues that female manipulation of egg size may not depend simply on nutrition from the male (see discussion in chap. 6.6), although it may be that the spermatophore contains specialized nutrients which are critical for larger eggs and which are available only from the male (Gwynne 1988a). Unfortunately, there

Table 4.7
Possible Cases in Which Females Exercise Cryptic Female Choice by Sometimes Investing Less in Each Offspring

	Criteria						
Species	*I*	*II*	*III*	*IV*	*V*	*VI*	*References*
*Gryllus bimaculatus** (Orthoptera)	?	?	Y?[a]	Y	? (sem. prod.)	?	Simmons 1988
Kawanaphila nartee (Orthoptera)	Y	?	?	Y	Y (size sptop.)	?	Simmons 1990; Simmons and Bailey 1990
*Requena verticalis** (Orthoptera)	Y	Y?	Y	Y	Y (size sptop.)	Y	Gwynne 1988a,b
Frontinella pyramitela (Araneae)	Y	Y?	Y?[a]	Y	Y (length cop.)	?	Austad 1982; Suter and Parkhill 1990
Poephila guttata[b] (Aves)	Y	Y?	Y?	?	Y (attractiveness)	?	Burley 1988

NOTES: Those marked with (*) are discussed in the text.
Each example is classified with respect to criteria for proof that cryptic female choice occurs. See table 4.1 for criteria and key.
[a] Larger offspring probably survive better, but this was not demonstrated.
[b] Both male and female alter their parental investment on the basis of their own attractiveness and the behavior of their partner. The potential interactions are too complex to analyze confidently, and this may well not be a case of cryptic female choice.

seems to be little known regarding the physiological mechanisms in female insects that cause changes in the sizes of mature eggs, except that older females often produce smaller eggs (Kimura and Tsubaki 1985 and references therein).

Another species in which egg size varies is the cricket *Gryllus bimaculatus*. By manipulating both the number of inseminations and the female's opportunity to consume the empty spermatophore, Simmons (1988) showed that egg size is increased both by sperm or other products transferred to the female's reproductive tract, and by the spermatophores she eats. As discussed in section 4.4, females were more likely to allow complete transfer of spermatophore contents when mating with larger males (Simmons 1986), so larger males may be more likely to obtain the benefits of larger eggs.

The question whether the male effect on egg size is due to stimulation or to nutrition (or to both) is not entirely (nor easily) resolved in this species, but more than nutrition is probably involved. This can be illustrated quantitatively in terms of the weights of spermatophores and eggs produced. Spermatophores are relatively small (about 1–2 mg, or about 0.18% of the male's body weight). Increases in egg size were, in contrast, relatively dramatic (from about 0.78 mg/egg after one insemination to about 0.86 mg/egg after 5–6 inseminations, or an increase of about 0.08 mg/egg). Using tables and graphs in Simmons (1986, 1988), one can calculate that this amounts to a total increase of about 215 mg of eggs as a result of receiving about 2–4 mg of semen (assuming, conservatively, that half the weight of a spermatophore is semen; see figure in Simmons 1986).

Clearly, unless males are supplying some special, otherwise limiting nutrient rather than simply additional protein as hypothesized by Simmons (1986), nutrition is unlikely to be an important factor in this species. Simmons (1986) also reached the conclusion that the spermatophore and its contents should be considered as mating effort rather than parental effort (nutrition), because the reproductive benefits a female derives from a given spermatophore may not occur until after the male's chances of fertilizing a large proportion of her eggs are gone, due to further matings.

The data are still inconclusive, however, because egg fertility was lower for females that received only one spermatophore; perhaps some eggs were lighter because they were infertile (L. Simmons, pers. comm.).

A further hint of male involvement in induction of female brood care comes from the rat *Rattus norvegicus*. When the female received higher numbers of intromissions, her mammary gland development was higher 1–5 days later (females received 1–3, 4–20, and >40 intromissions; all received sperm and a vaginal plug; Dilley and Adler 1968). The possible effect of this response on male reproductive success is uncertain, since it is not clear whether greater development of mammary glands so early in pregnancy results in greater female investment in the nursing young. There were also methodological problems. It was unclear whether different delays after copulation were equally distributed among the treatment groups, and if any females received more than one ejaculation.

There is a theoretical complication in all these cases, which may be especially important where male nutritional donations are not involved. Females that invested more heavily in the offspring of males with greater abilities to induce such investment could presumably benefit from having sons with greater reproductive success. But they could also suffer the disadvantage of having their own future reproductive success lowered because of the greater investment in the current offspring. One would expect, for instance, that, in the absence of male effects, the female would already be laying eggs of the size best adjusted to her own reproductive interests. A female would thus stand both to gain and to lose by altering her parental investment with different males, and it is not obvious which of these factors would weigh more. The arguments in chapter 2 regarding the lack of cost of cryptic female choice among males may not apply.

This complication, plus the relatively small number of possible cases in table 4.7, might suggest that the differential investment in offspring is a relatively unimportant mechanism of cryptic female choice. It is difficult to judge the importance of the lack of concrete examples, since relatively few studies check egg size (for example, in nearly all of the studies in chap. 6 in which male effects on female egg production were checked, eggs were counted but not weighed or measured).

Empirical data make it clear, however, that this type of presumed cost of cryptic female choice is not infrequently outweighed by other factors. This is because similar costs are also associated with two other mechanisms already discussed: changes in the rates of ovulation, and of oviposition after copulation with different males (chaps. 3.7, 3.9; see also chap. 6). There is abundant evidence, from a variety of groups, that females differentially modify rates of both ovulation and oviposition in response to differences between males, in spite of the probable costs.

4.11 Choose among Sperm That Have Reached the Egg

An egg is often encountered by many sperm, and as noted in chapter 1.9, the processes that lead to the fusion of egg and sperm pronuclei are largely performed by the egg. Active egg processes in mammals include the following: the egg may produce pseudopodia to actively ingest the sperm; the sperm head is drawn into the egg surface and clothed in a vesicle derived from the vitelline membrane, then drawn deeper and enclosed in vesicles produced by the egg (Deuchar 1975); and the sperm's highly condensed chromatin is decondensed, and thus restored to an active state, by oocyte factors that reduce protamine disulfide bonds and promote removal of protamine and its replacement with egg histones. The new nuclear envelope of the sperm is probably also made by the egg (Perreault 1989). The maternal chromosomes must be present for this decondensation process to proceed (Wright et al. 1989). Other active processes of the egg, such as addition of material to the rapidly swelling sperm nucleus, streaming of egg cytoplasm, sperm nucleus migration utilizing egg-synthesized microtubules, and complex folding of the sperm's tail inside the egg, occur in insects and echinoderms (Carré and Sardet 1984; Retnakaran and Percy 1985; Longo 1987; Karr 1991). Clearly there are several female mechanisms by which some sperm could be favored over others. Just as copulation does not always lead to insemination and fertilization, so sperm contact with the egg may not always result in fusion of their nuclei unless subsequent female processes are performed. Differences in the abilities of sperm from different males to trigger these processes in the egg could result in sexual selection.

A possible example of this type of selection occurs in the ctenophore (comb jelly) *Beroe ovata* (Carré and Sardet 1984). The large (1 mm) trans-

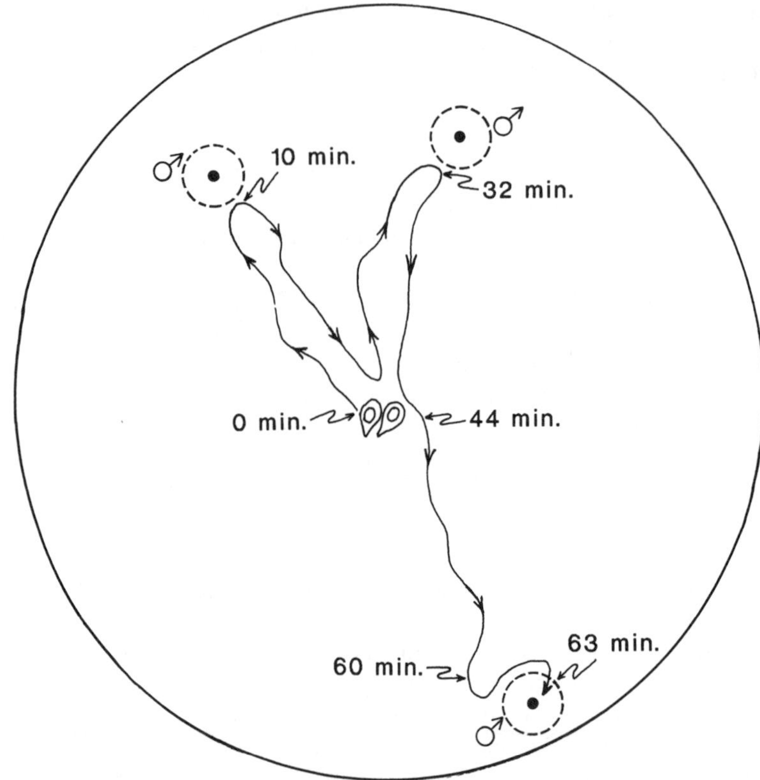

Figure 4.14 The path traced by the choosy egg pronucleus of the ctenophore *Beloe ovata* egg, which "investigated" the pronuclei (large dots) of three different sperm that had entered the egg, fusing with the third. Data are from a time-lapse video recording; dotted lines show areas free of cortical granules surrounding each sperm pronucleus. (After Carré and Sardet 1984)

parent eggs are shed into the sea, where they are fertilized. When a sperm cell reaches the egg, the sperm head is incorporated rapidly into the egg (in less than 1 min) but remains near the site of entry. If only a single sperm enters the egg, the egg pronucleus then migrates from its maturation site (marked by the polar bodies) to the sperm nucleus, and the two fuse.

Often, more than one sperm enters a given egg, however. When several sperm have entered the egg, fusion of egg and sperm pronuclei is often delayed while an extraordinary drama is played out (fig. 4.14). The egg pronucleus migrates, at about 18 μm/min, toward a sperm entry site (not necessarily the closest one). Often it stops just short of the clear zone surrounding the sperm pronucleus, pauses, and then returns to the maturation site; it then moves out to approach a different sperm pronucleus. Eventually the egg pronucleus enters the clear zone around a sperm pronucleus, and fusion occurs.

A number of questions remain to be answered before this can confidently be accepted as a case of cryptic female choice. What determines whether a given sperm pronucleus will be accepted for fusion or rejected? Do sperm from different males have different chances of being accepted? Do sperm acceptance patterns differ among eggs? Among females? Are eggs in nature often polyspermic? But the behavioral observations certainly suggest that egg pronuclei are somehow testing and choosing different sperm pronuclei (Carré and Sardet 1984).

A more difficult question is whether or not such selective behavior by eggs is a bizarre biological freak. The "standard" account of syngamy emphasizes active male roles ("sperm penetration of the egg") but in fact there are many more active processes performed by the egg. In addition, the fusion of egg and sperm is an extremely ancient process, and there has been a great deal of time to evolve abilities to induce discrimination, and abilities to discriminate. The egg-sperm asymmetry probably results from an ancient dimorphism, in which sperm cells discarded much of their cytoplasm and condensed (and thus deactivated) their DNA, presumably to make them more mobile and cheaper so they could be broadcast more widely (Parker et al. 1972).

These considerations suggest that discrimination at the level of the egg may also occur in other animals. Its discovery may have been impeded by the technical difficulties of checking for selective egg behavior. Arrival of multiple sperm at the same egg is probably not unusual. For instance, in birds multiple sperm are often found on the membranes of eggs (Birkhead 1994), and several sperm often enter the micropyle of an insect egg (Retnakaran and Percy 1985). But deciding whether selective processes by the egg determine which sperm cell is pulled in and eventually fuses with the egg pronucleus appears to be a technically daunting task. Another factor that could also lead to underestimates of egg selectivity is the "fertilization myopia" of biologists (chap. 1.8). Events such as the failure of an egg pronucleus to fuse with a sperm may be neglected.

4.12 Cryptic *Male* Choice

In a small number of animals the usual male-female roles are reversed. Females court more or less coy males, and males care for the offspring. In one group, the seahorses and pipefish, females place their eggs in special cavities in the male, where he then fertilizes them. There is indirect evidence in the sex-reversed pipefish *Syngnathus typhle* that males may sometimes limit the number of eggs transferred by females during copulation

(Berglund 1993). Females initiate courtship and lay their eggs in the male's pouch, where he provides the offspring with nutrients and oxygen. In the presence of a predator, more eggs were laid during copulations than when no predator was present (Berglund 1993). The fish altered other aspects of their reproductive behavior when predators were present in ways apparently designed to reduce the likelihood of being preyed upon. The larger number of eggs transferred may result in reduced predation cost for both male and female due to a reduced need for subsequent courtships (Berglund 1993).

The fact that lower numbers of eggs were transferred when no predator was present suggests some sort of limitation, presumably due either to female restraint or to male limitation of oviposition. If males somehow limit female oviposition, then the possibility arises (but of course is not proven) that limitation by males may vary in response to differences among females and result in selection on such differences.

4.13 Undetermined Mechanisms

There are relatively strong indications that cryptic female choice occurs in several additional species, although the choice mechanisms have not been established (table 4.8). Two of these cases, chimpanzees and tree swallows, will be discussed here.

Chimpanzee (*Pan troglodytes schweinfurthii*) mating systems are flexible, and all adults of both sexes appear to have the potential to engage in any of three different patterns (Tutin 1979; Goodall 1986). All three types of mating occur when the female is at the peak of her estrus (maximum tumescence of her external genitalia). Ovulation is believed to occur within one day of the last day of maximum tumescence.

The first type of copulation is "opportunistic." A receptive female is mated repeatedly (up to fifty times per day; Goodall 1986) by all of the up to fourteen males that accompany her. Males show relatively little aggressive interference with copulations by other males (during only one of 830 copulations in the Tutin study). In the second, "possessive" type of mating, a dominant male establishes a short-term relationship with a female (1 hr—5 days), during which he prevents or attempts to prevent lower-ranking males from copulating or from finishing copulations with her. In the third, "consortship" type of mating, a single male (often not the highest ranking male) induces a female to temporarily leave the rest of the group and maintains exclusive sexual access to her for from 3 hours to 3 months (Goodall 1986).

Establishing and maintaining a consortship involves a combination of male coercion and female cooperation (Goodall 1986). Consortships are usually initiated when other males are nearby, and if the female does not cooperate by accompanying the male away from the group when the other males are not watching and by remaining silent, the rest of the males will follow along. Threats or attacks on the female are sometimes included in a male's attempts to initiate or maintain a consortship (Goodall 1986). It is common, however, for a male to fail repeatedly in attempts to induce a female to leave the group (Tutin 1979; Goodall 1986). Some males fail more consistently than others.

Thus consortships require active female cooperation, and appear to involve female preferences for certain males. Consorting with a male in this way is probably somewhat dangerous for both the female and any offspring accompanying her. Individuals isolated in this way from the rest of the group are apparently at increased risk of injury and death from males of adjoining communities (Tutin 1979; Goodall 1986).

Tutin (1979) carefully followed the sexual behavior of receptive females at maximum tumescence in the wild, and observed 1137 copulations. Opportunistic mating was the most common, accounting for 73% of the copulations (in 68% of the 1200 observation hours); 25% of the copulations occurred during possessive mating (in 18% of the observation hours); and only 2% were during consortships (in 14% of the observation hours). A female chimpanzee with a large party of opportunistically mating males mated about 30 times/day, while one in consortship typically mated only about 5–6 times/day (Goodall 1986).

By counting back in time from observed births during the study, Tutin estimated that at least 50% of the pregnancies during her study period were initiated during consortships. Thus it appears that about 2% of the copulations led to about 50% of the pregnancies. It seems that a female chimpanzee is more likely to become pregnant from copulating with a male with which she has chosen to consort. Additional data (Goodall 1986) tend in the same direction. Consortships more often led to conceptions than did group mating situations, and longer consortships were also more likely than shorter ones to result in pregnancies. The mechanism responsible (ovulation? sperm transport? abortions?) is unknown.

Several male behavioral characteristics correlate with a male's chances of inducing a female to form a consort pair or be defended: the amount of time the male spent in the same group of animals as the tumescent female during opportunistic mating; the proportion of that time which the male spent grooming maximally tumescent females during opportunistic mating; and frequency with which the male shared food with females (Tutin 1979).

The sample in this study was small, and interpretation of the data is complicated by the fact that it is not clear how Tutin chose females for observation (consortships are more difficult to detect; Goodall 1986), and

Table 4.8
Possible Cases of Cryptic Female Choice
in Which the Mechanism Used to Exercise Choice Was Not Determined

	Criteria						
Species	*I*	*II*	*III*	*IV*	*V*	*VI*	*References*
Culex pipiens[a] (Diptera)	Y	?	Y[b]	?	?[c]	?	Lea and Evans 1972
Aedes taeniorhynchus[a] (Diptera)	Y	?	Y[b]	?	?[c]	?	Lea and Evans 1972
Utethesia ornatrix[d,e] (Lepidoptera)	Y	Y	Y[f]	Y?	Y (size sptop.[g])	?	LaMunyon and Eisner 1993, 1994; LaMunyon 1994
Pseudaletia unipuncta (Lepidoptera)	Y	Y	Y	Y	?[h]	?	Svard and McNeil 1994
Panorpa latipennis (Mecoptera)	Y	Y	Y	?	Y (gifts[i])	?	Thornhill 1980a
Aphytis melinus (Hymenoptera)	Y	Y?	Y[f]	Y	Y (postcop. court.[j])	?	Allen et al. 1994
*Tachycineta bicolor** (Aves)	Y	Y	Y	Y	Y? (extrapair)	?	Venier et al. 1993; Lifjeld et al. 1993
Malurus cyaneus[k] (Aves)	Y	Y	Y	Y	Y?		Mulder et al. 1994; Mulder and Magrath 1994
Peromyscus eremicus[l] (Rodentia)	Y	Y?	Y	?	Y (No. postejac. introm.)	?	Dewsbury and Estep 1975
Mesocricetus auratus[m] (Rodentia)	Y	Y?	Y?	?[n]	Y (No. ejac. ser. male odor)	?	Oglesby et al. 1981; Huck and Lisk 1985; Tang-Martinez et al. 1993
Dicrostonyx groenlandicus (Rodentia)	Y	?	Y	?	Y (strange male[o])	?	Mallory and Brooks 1978
Meriones unguiculatus (Rodentia)	Y	?	Y	Y	Y (intrauter. pos.[p])	?	Clark et al. 1992
Homo sapiens (Primates)	Y	Y	Y	Y	? (psych. conflicts in female[q])	?	Mastroianni 1958; Kroger and Freed 1950
*Pan troglodytes** (Primates)	Y	Y	Y	Y	Y (consortship)	?	Tutin 1979
Macaca fuscata (Primates)	Y	Y	?[r]	Y	Y (cop. stim., domin. rank)		Troisi and Carosi 1994

NOTES: Those marked with (*) are discussed in the text.
Each example is classified with respect to criteria for proof that cryptic female choice occurs. See table 4.1 for criteria and key.

[a] Mating with tethered virgin females in captivity. It was not clear whether sperm from the male was ejected by the female, or whether males were unable to deposit sperm in the female. Both female tendency to allow genitalic coupling and for insemination (judged by the presence of sperm in the spermatheca) to occur once coupling was achieved increased in older females (fig. 1.11). Occasional females were found with sperm in the cloaca or the bursa, but not in the spermathecae, suggesting that sperm ejection occurred.

[b] So much sperm was dumped or so few sperm were found in the female afterward that reduction in male paternity probabilities seems inevitable.

[c] Male success changed with female age.

[d] The authors suggest that female transport of sperm is involved in the demonstrated favoritism for the male with the larger spermatophore, but several alternative mechanisms are also possible.

[e] Precopulatory sexual selection by female choice also occurs.

[f] Direct paternity analysis was made with genetic markers.

[g] Size of spermatophore is correlated with size of male. Frequent remating of females in the field and long delay in male to recover spermatophore size suggest males in the field often transfer less than the maximum spermatophore size for their body size, so ability to synthesize spermatophores may also be favored.

[h] Reduced sperm usage for the second of two males (a lower P2 value) was correlated with larger numbers of eggs laid in the four days between the first and second mating. The mechanism responsible for this trend, which is the opposite of what would be expected on the basis of sperm depletion in the female, is not known. Males apparently increase female egg production with nutritionally or hormonally active seminal products, so lower P2 values may favor males that are better able to induce female oviposition.

[i] Females that were "raped" instead of being enticed to mate with a gift of food less often had sperm in their spermathecae 24 hr later.

[j] The percentage of offspring of doubly mated females that were sired by the first male increased if the first male performed postcopulatory courtship.

[k] Authors suggest that females control fertilization because extra-pair paternity did not correlate with extra-pair courtship by males, and because one male lineage was especially successful in fathering extra-pair offspring. Fertilization biases could have resulted from biases in copulations (no data given), and thus may not be due to cryptic female choice.

[l] Authors speculate that lack of sperm transport may have been responsible for failed copulations.

[m] Multiparous females had smaller litters than virgin females when both were allowed to receive only a single ejaculation, and they had smaller litters than multiparous females allowed multiple ejaculations. Possibly, differences in sperm transport or survivorship within the female were responsible. Reduced sperm numbers (from males that had ejaculated repeatedly just previously) were associated in other observations with smaller litters. Females had larger litters when mated with males with whose odor they had been familiarized (via soiled bedding). In nature, dominant males successful in defending territories and visiting resident females would presumably be better at familiarizing females with their odors. Possible female mechanisms include larger numbers of eggs ovulated, better fertilization success, and higher percentage of implantation.

[n] Females in captivity remated readily, but no field data are available. In the related species *M. neutoni*, the home ranges of several males may overlap >50% that of a given female, so several males may be familiar with the burrow of a given female and available for mating (Huck and Lisk 1986).

[o] Females take longer to bring to term offspring of strange males (perhaps due to delayed implantation or delayed ovulation).

[p] Litter sizes after first pairing by male were significantly larger for males whose intrauterine positions had been between two males (2m) than for those whose intrauterine positions had been between two females (2f). 2m males had higher levels of circulating testosterone, so even though intrauterine position is not heritable, the differences in litter sizes may exercise selection on genetic differences between males.

[q] Evidence from correlations (rather than experimental demonstrations of cause-effect) suggest that female psychological conflicts (which are probably to some extent affected by male traits) are sometimes responsible for human infertility. For instance, conception not uncommonly ensues for a childless couple after nothing more than an examination with the premise of a diagnostic program. Female tubal spasms are suspected to cause some infertility.

[r] Male traits correlated with the likelihood of female orgasm; the effect of female orgasm on male reproductive success, if any, is not known.

Figure 4.15 Copulation in the tree swallow *Tachycineta bicolor*, a species in which a male's frequency of copulation is not necessarily correlated with his frequency of paternity. Copulation often occurs in exposed sites, and is accompanied by vocalizations, so at least part of a female's mating history can be determined relatively accurately. As suggested by the precarious copulation position, females can easily avoid male copulation attempts, either by not raising their tails or by flying away. Females thus actively determine whether or not extra-pair copulations occur. The disproportionately high percentage of fertilizations resulting from extra-pair copulations (as compared with within-pair copulations) suggests that the females may bias sperm use in favor of extrapair males. (From Dunn et al. 1994b; photograph courtesy of R. Robertson)

the balance of observation hours may not accurately reflect the frequencies of the different mating contexts. Nevertheless, the more than an order of magnitude of difference between number of conceptions and number of copulations suggests there was a real effect.

Turning to birds, the tree swallow *Tachycineta bicolor* offers another striking contrast between the frequencies of observed copulations and paternity, in which the paternity bias agrees with apparent female preferences for mates. A female tree swallow must raise her tail to permit copulation, and females can easily reject most copulation attempts (fig. 4.15; Venier et al. 1993). Female rejections consistently caused males to desist at least temporarily, and frequently males gave up and departed after a single rejection. So male harassment seems of little importance, and neither population density nor breeding synchrony affected the frequency of extra-pair paternity (Dunn et al. 1994). Only one of 651 copulations observed may have been forced.

Most female matings near nest boxes were with the male with which the female was paired (93.1% of 651 copulations during *ad libitum* observations, 96% of 69 copulations during 158 hr of systematic observations of forty-seven focal pairs; Venier et al. 1993). Timing of within-pair and extra-pair copulation did not differ with respect to the time of day or the day of the female's egg-laying cycle (Venier et al. 1993). The female played an active role in initiating twelve of the observed forty-five extra-pair copulations as she moved into the male's territory (usually onto his nest box; Venier et al. 1993).

Both the copulation behavior and the paternity of the nestlings were studied in twelve monogamously paired females (Lifjeld et al. 1993). Copulation observations (during sample periods in the morning and afternoon) started 8 days before the female's first egg and ended on the morning when the next to last egg was laid (when the last fertilization probably occurred). Paternity was determined using multilocus DNA fingerprints. The two types of data showed a dramatic discrepancy. Of 70 observed copulations of these females, not a single one was with an extra-pair male, yet 38% of 86 nestlings were fathered by extra-pair males. All nestlings were offspring of the female of the pair, and none of the extra-pair fathers had been associated with the female early in the season, so neither intraspecific brood parasitism nor early mate switching was responsible for extra-pair offspring. Thus the extra-pair offspring resulted from truly "illicit" matings. Even if one uses the 4% extra-pair copulation frequency from other studies (Venier et al. 1993), extra-pair copulations are apparently much more likely to produce offspring than within-pair copulations.

An alternative explanation is that extra-pair copulations also occurred away from the vicinity of the nest boxes where they were not observed. It is likely that such copulations occur, since many extra-pair offspring were not sired by neighboring males (of 63 extra-pair offspring in a grid of nest boxes with twenty-three broods, only 21% were sired by a male resident on the grid; Dunn et al. 1994a,b).

There are several indications, nevertheless, that some copulations were much more effective than others. Some females raised substantial numbers of extra-pair offspring but nevertheless copulated at high rates with their mates near the nest box. For instance, one female mated an average rate of 1.6 times/hr with her mate during the observation periods, yet produced 100% offspring from other males; another mated 1.2 times/hr with her own male and produced >60% extra-pair offspring. Very high numbers of extra-pair copulations away from the nesting area would be necessary if one were to explain these rates of extra-pair fertilizations by copulation frequency.

Similarly, when mate switching was induced experimentally in a sample of ten females, replacement males gained very little paternity despite the fact that some of them copulated at a normal rate with the resident female during a portion of the female's fertile period (Lifjeld and Robertson 1992).

Finally, even if one takes into account only the proportion of extra-pair offspring sired by the nearest neighboring male (33% in 1990, 19% in

1991; Dunn et al. 1994a), they represent 10.7% of the total of 205 offspring sampled in the two years, still substantially higher than the observed frequencies of extra-pair copulations with neighboring males (4%, 0%; Vernier et al. 1993; Lifjeld et al. 1993).

One possible way in which females could bias fertilization probabilities in favor of extra-pair males would be to time extra-pair copulations more favorably with respect to the chances of fertilization (copulation involves female cooperation). The conclusion of Lifjeld et al. (1993) that "females control the sperm transfer success of copulations" (that is, whether or not copulation results in sperm transfer; J. Lifjeld, pers. comm.) is open to further interpretation since several other mechanisms could be involved, such as differences in sperm degradation, ejection of current or previous sperm, sperm entry into storage sites, sperm movement from storage sites to fertilization sites, or timing of ovulation. These results from tree swallows fit the perhaps general trend for relative numbers of extra-pair copulations to correlate poorly with relative numbers of extra-pair offspring (fig. 4.16; Dunn and Lifjeld 1994).

A possible alternative explanation for different rates of fertilization in both chimpanzees and tree swallows from different copulations is that sperm numbers in different ejaculates vary (Gomendio and Roldan 1993). For instance, numbers are higher in some species when other males are in the vicinity of the copulating pair (Gage 1991; Gage and Baker 1991), the male has been out of contact with the female (Baker and Bellis 1989b), or the male is not likely to have other copulation opportunities (Birkhead and Fletcher 1992). Such changes in ejaculate size seem unlikely to explain at least the chimpanzee and the tree swallow data because (a) the trends seem too strong, (b) in the chimpanzee greater conception rates (lack of sperm from competing males) occur in conditions in which *smaller* ejaculates would be expected, and (c) extra-pair fertilizations in tree swallows are no less common when the extra-pair male is presumably somewhat sperm-depleted due to copulations with his own female (i.e., when breeding is more synchronous; Dunn et al. 1994a).

A final note of caution is needed. In both the chimps and the swallows there was an apparent correlation between paternity bias and overt female preferences for males. Sexual selection by female choice may often involve a series of male characters being used one after another. For instance, the male *Oecanthus nigricornis* cricket attracts the female from a distance using both pheromones and a call, then feeds her during copulation, and sings and vibrates to her following copulation. If females are using male signals as indicators of superior male viability genes, then at least a general concordance in female preferences is expected (a male preferred on the basis of Stimulus A will also tend to be preferred on the basis of Stimulus B). If, however, the payoff to females is good attractiveness genes, I see no theoretical reason that would insure that the female criteria at different

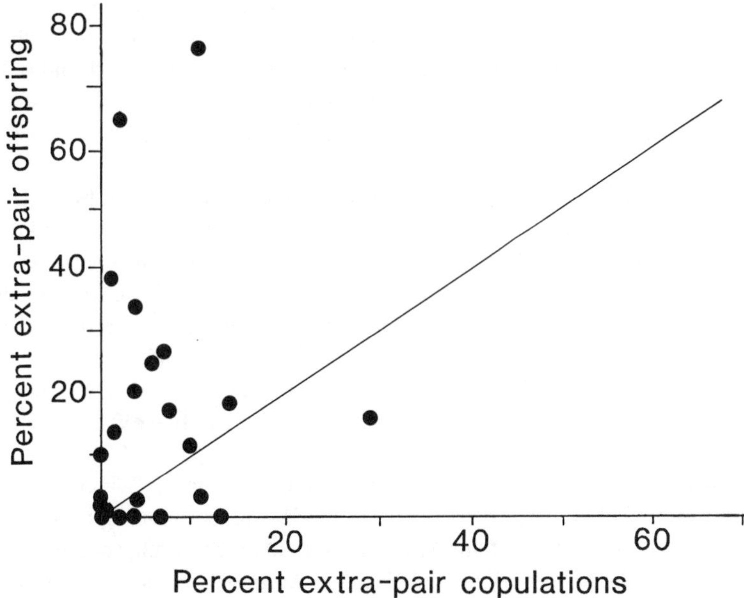

Figure 4.16 A possible trend in fertilization biases in birds is in accord with predictions from cryptic female choice. In birds with pair bonds, the proportion of extra-pair copulations by the female shows no significant relation with the proportion of extra-pair offspring she produces (line shows expected relationship if the two proportions were equal and all copulations were equally likely to result in offspring; an additional analysis correcting for effects of phylogeny showed a weak but significant correlation; Birkhead and Møller 1995). Cryptic female choice is one possible explanation for this lack of correlation, but there are several possible biases in the data (e.g., extra-pair copulations may be more cryptic), so interpretation is uncertain. It is nevertheless striking that the major deviations from expectations are toward more rather than less extra-pair offspring than expected (points above the line in the graph); since female birds probably usually determine whether or not extra-pair copulations occur (e.g., Lifjeld et al. 1993), this trend is in the direction expected if cryptic female choice is occurring. (After Dunn and Lifjeld 1994)

stages should be in accord with each other. In other words, cryptic female choice (at least on the basis of good attractiveness genes) is not ruled out by a lack of correlation between female criteria in overt female choice and paternity biases.

4.14 Discussion of Chapters 3 and 4

Concrete examples of mechanisms by which cryptic female choice may occur were presented in this and the preceding chapter. A total of more than one hundred possible cases are included in the tables, illustrating twenty different mechanisms. The number of species in the tables is an underestimate. Male effects on female oogenesis, oviposition, and remating resulting from products in their semen have been documented in more than fifty additional species of insects and ticks (chap. 6; see tables 6.1 and 6.3). My search of the literature was also incomplete, and there are probably other cases, especially in the older literature, which I generally did not check.

In addition, several of the female mechanisms are composites. For example, failure to remate can be due to female movement patterns that keep them away from sites where males are, lack of production or release of attractant pheromones, production or release of antiaphrodisiac substances, flight from approaching males, or physical resistance to copulation attempts. These processes can be under separate control in the female; for example, depletion of pheromone, cessation of calling, and resistance to mating attempts are under separate control in the moth *Helicoverpa zea* (Kingan et al. 1993a,b).

I also expect that the list of mechanisms is incomplete, since several mechanisms only occurred to me after having read other reports (e.g., W. Brown, in prep., a), and I surely have not made a complete search of the literature on animal reproductive behavior. As more species are studied, further mechanisms will probably be uncovered. Several possible candidates are worth checking: failure to seize and break open a male spermatophore, hold it in the appropriate orientation, or squeeze out its contents with contractions of the bursa, as occurs in females of the bruchid beetle *Acanthoscelides obtectus* (Huignard 1971), the staphylinid beetle *Aleochara curtula* (Gack and Peschke 1994), and several lepidopterans (Drummond 1984; Sugawara 1979; Tschudi-Rein and Benz 1990); changes in mortality rates of sperm inside female storage organs (see Yamagishi et al. 1992 on sperm mortality in the melon fly *Bactrocera cucurbitae*; Lake 1975 on stimulation of gland cells in storage sites by sperm and on different rates of sperm mortality in different strains of chickens; and Kuster and Davey 1983 on the control of muscles that express gland secretions into the lumen of the spermatheca of the bug *Rhodnius prolixus*); restraint from oviposition while sperm are in the path of descending eggs and subject to being flushed out of the female (Ward and Carrel 1979 on nematodes; Uhl 1994a on spiders; Birkhead and Møller 1992 on birds); inhibition of coagulation or adhesion of mating plugs; inhibition of female aggression toward males thus giving the male an opportunity to copulate more times (Carter and Getz 1993 on *Microtus* voles); activation of sperm in some storage sites but not others just prior to fertilization opportunities (Lake 1975 on birds); and changes in efficiency of sperm usage (number of sperm/egg fertilized; e.g., P. Smith et al. 1988 on *Lucilia cuprina* blowflies; Okelo 1979 on a grasshopper; Leopold and Degrugillier 1973 and Leopold et al. 1978 on the housefly; Harbo 1979 on the honeybee; and Drummond 1984 on lepidopterans for evidence of female control of sperm release). If one counts different subprocesses separately and includes these other likely candidates, there are surely at least 30–40 different mechanisms by which females could affect the reproduction of males after allowing them to achieve genitalic coupling.

The critical reader will be struck by how few of the examples listed in the tables and discussed in the text meet all of the criteria necessary to constitute convincing proofs of cryptic female choice. I think, however, that the same reader will notice that the dates of nearly all the studies are quite recent and would admit that there are very few species for which the many kinds of detailed, sometimes difficult-to-acquire data needed for a proof are presently available. I suspect that this lack of critical data is partly due to the fact that very few research plans have been specifically designed to test for cryptic female choice. I hope this book will help change this situation. In any case, the examples serve to illustrate that the different mechanisms are at least feasible for particular real animals.

One type of commonly missing data in the tables relates to possible genetic differences between conspecific males in their abilities to induce female responses. Genetic variation has usually been discovered whenever it has been searched for (Lewontin 1974), even in traits under sexual selection (see chap. 2.3). It is thus probable that many of the male traits listed in the tables as cues for possible cryptic choice mechanisms show genetic variation in nature. In fact, it is perhaps overly conservative even to list the criterion of genetic variation among males because it limits attention to selection occurring in the present. Even if no variation currently exists, cryptic female choice could have occurred in the past and may even have been the cause of reduced variation in the present.

A second type of commonly missing data concerns variation in female responses based on differences among males. There are two possible reasons for this deficiency. The first is a simple dearth of studies. For instance, in almost none of the more than one hundred species in which no differences were established between males that correlated with differences in female responses did the author report an attempt to check for such a correlation. In addition, female discrimination may be based on male characters that are difficult to observe, or at least seldom checked. Examples include possible differences in the quantity and quality of seminal products that induce female reproductive processes such as sperm transport, oogenesis, or oviposition; and differences in the fertilization ability of sperm, genitalic form or behavior, and copulatory courtship.

The second possible reason is that cryptic female choice mechanisms may sometimes be applied in nonselective ways (e.g., female remates simply to increase genetic variance among her offspring or to increase the probability that the sperm she has received is viable, irrespective of any male traits). Nonselectivity is a null case and will be very difficult to prove (current data from birds does not favor the genetic variance idea; e.g., Birkhead 1993).

One apparent pattern seen in the tables is that criteria for female preferences are more often related to male size than to other possible criteria. I suspect that this is an artifact, caused by a bias in researchers who check male size or weight for correlation with insemination and fertilization success more readily than they check other possibly important variables (see Andersson 1994 for a similar trend in classic, precopulatory female criteria). Obviously, male size is an important possible factor, though the frequent supposition that larger is better is not necessarily true (see, for instance, the demonstration by Witter et al. 1994 of *reduced* escape ability of heavier *Sturnus vulgaris* birds). Male size is also easier to quantify and check than many other possible variables. Very few studies seem to have included searches for correlations with traits *other* than male size (or, in a few cases, male age). Among the rare exceptions, in which the male's copulatory courtship and/or his genitalic morphology was checked rather than only his size, the spider *Neriene litigiosa* and the beetle *Chelymorpha alternans* gave dramatic correlations with cryptic female choice (Watson 1991; V. Rodriguez 1994a; V. Rodriguez et al., subm.; chap. 7.2).

Species differences can serve as a guide to the particular aspects of males and their displays that have been under selection. It appears that many characters other than male size have been under selection. For instance, intrageneric differences in courtship performed during copulation include different sites where the female is touched, different rhythms of touches, and different appendages that are used to tap, stroke, or vibrate her (Eberhard 1994; chap. 5.1). In fact, male size is probably relatively *un*important in some female choice contexts. Species-specific genitalia, which are probably also used in copulatory courtship (Eberhard 1985; chap. 7.2), tend to be more weakly correlated with body size than are other parts of the male's body (see chap. 8.3).

Another pattern in the tables is that some types of female mechanisms such as remating and oviposition are more common than others. This may also be an artifact of the questions asked by researchers. It is usually technically easier to determine how readily a female will remate or how many eggs she will lay than it is to determine, say, the rates of sperm transport or the proportions of ejaculates that reach storage organs. For instance, despite the relatively small number of species in table 3.4, the fact that there are substances in the semen of both insects and mammals that stimulate smooth muscle contractions (see chap. 6), apparently in order to induce females to transport gametes, argues that lack of gamete transport by females has indeed been an important and widespread selective force (fig. 4.17). It is only reasonable that in the early stages of research in a field

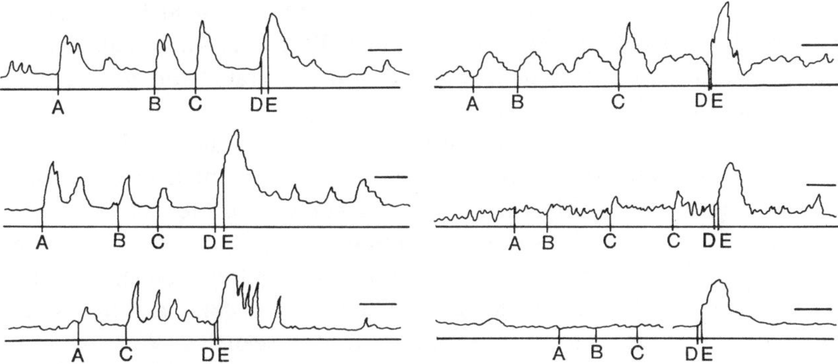

Figure 4.17 The poker face of a cow can hide dramatic responses inside. Recordings of the pressure (vertical displacements in the graphs) within a female's reproductive tract (using a balloon inserted in her uterus) showed that powerful contractions occur in response to stimuli from males (three estrous cows on *left*, three postestrous cows on the *right*). A: bull brought into sight of the cow; B: bull nuzzles her vulva and hindquarters; C: bull mounts briefly but is prevented from intromitting by leading him away; D: bull mounts, intromits, and ejaculates (E) and is then immediately led away. Presumably such contractions of the uterus result in sperm transport, as occurs in other mammals (VanDemark and Hays 1952; see also chapters 3.4, 5.2, 7.1). Internal female responses of this sort to male stimuli are seldom studied and may be much more common than present data would suggest. Timescale markers = 1 min. (After VanDemark and Hays 1952)

such as cryptic female choice, easier questions will tend to be asked more frequently; but I hope that the more difficult questions will receive increasing attention.

An additional point to keep in mind is the overwhelming importance of fertilization success for determining male fitness and the relatively small sample sizes of many studies, and their resulting insensitivity. Even an increase of 0.1% in the number of eggs the female lays soon after copulation, or in female resistance to remating, could result in significant selection on males (e.g., fig. 1.11). For a number of reasons, including the current academic environment that usually provides only short-term financial support and rewards rapid publication, most observations of phenomena related to cryptic female choice involve only relatively small samples (often 20–30/treatment) that are not capable of documenting weak male effects on females. In general, data capable of testing for subtle effects are simply not available (Hinton 1974 makes this same point).

The possible selective importance of subtle effects is especially important since it is quite clear that female choice often involves combinations of different criteria and mechanisms. For instance, females of the scorpion fly *Panorpa latipennis* discriminate cryptically against males that attempt to copulate with them without offering food by reducing sperm transfer, by reducing the rate of oviposition following copulation, and by remating more readily (Thornhill 1984). The use of multiple cryptic female choice mechanisms has also been demonstrated in the lab rat *Rattus norvegicus*, and the hamster *Mesocricetus auratus*.

Precopulatory female choice probably also occurs along with cryptic choice in many of the species in the tables. For instance, male black-horned tree crickets *Oecanthus nigricornis* attract and then court females for up to about an hour with calling songs, a variety of substrate vibrations, and pheromones; both pheromones and calling songs attract females from a distance (Bell 1980). Precopulatory female choice occurs, as females are preferentially attracted to songs produced by larger males (W. Brown, in prep. b). In addition, the male feeds the female with metanotal gland secretions prior to, during, and following copulation, and repeatedly vibrates and stridulates when she pauses during postcopulatory feeding (Bell 1980). After this multimedia barrage of stimuli, the female exercises cryptic female choice by ovipositing more rapidly following copulations with larger males by sustaining greater delays before remating after copulations with younger males that have greater fluctuating asymmetry (W. Brown, in prep. a). Similarly, sexual selection by male-male battles, precopulatory female choice, and cryptic female choice all appear to occur in the fly *Cyrtodiopsis whitei* (Burkhardt et al. 1994).

4.15 Summary

A conservative count gives twenty different female mechanisms which could result in cryptic female choice, and which have been shown to occur in approximately one hundred different species: sometimes discard sperm from the current male; sometimes discard sperm from previous males; sometimes prevent complete intromission or ejaculation; sometimes prevent arrival of ejaculated sperm in storage organs; sometimes remate with another male; sometimes reduce the rate of oviposition or number of offspring produced following mating; sometimes terminate copulation prematurely; sometimes fail to ovulate; sometimes inhibit egg maturation (vitellogenesis); sometimes fail to prepare the uterus for implantation following copulation; sometimes impede plugging of the female genitalia, which prevents other males from mating with her; sometimes impede removal of plugs; sometimes remove the spermatophore (with its sperm) from the genitalia before sperm transfer is complete; sometimes abort developing zygotes; sometimes fail to use some stored sperm; sometimes move sperm from previous males to a site where it can be manipulated by the present male; sometimes harden the genital ducts following copulations so that sperm transfer by subsequent males will be more difficult; sometimes resist male manipulations that cause his spermatophore to discharge its contents in her reproductive ducts; sometimes fail to invest equally in all offspring; and sometimes fail to fuse with the pronucleus of the sperm cells that have entered the egg.

Other likely mechanisms remain to be studied, and a more likely number of possible cryptic female choice mechanisms is thirty or more. Apparent female bias, in which female responses are triggered more effectively or completely by some conspecific males than by others, has been demonstrated for sixteen different mechanisms.

Taking into account the limits of our present knowledge, I think the examples presented in these chapters are sufficient to make the case that females of at least some species can and do perform a variety of processes that probably result in sexual selection by cryptic female choice on males. The scope of taxonomic groups represented is wide, ranging from ctenophores and nematodes to humans, and including snakes, lions, scorpions, spiders, and mosquitoes in between. These data are in accord with the theoretical arguments made in chapter 2 that female control of reproductive events affecting male fitness is likely to evolve.

The question remains, however, whether cases of cryptic female choice are rare, widely scattered biological curiosities, or whether they are examples of a common, pervasive phenomenon in animals with internal fertilization. The next three chapters present evidence that cryptic female choice is common and widespread.

Notes

1. One possible advantage of cleistogamic, or encapsulated, immobile sperm—that of reduced energy demands on the male and female for sperm maintenance—is ruled out by the timing of sperm activation in some groups. Thus in the spider *Nephila clavipes* the sperm are activated on the first day after copulation, even though the female generally does not oviposit until about 6 weeks later (Christenson 1990). Another possible advantage of cleistogamic sperm, which is to my knowledge untested, is that they are able to survive better in especially hostile conditions within the female.

5 Evidence That Cryptic Female Choice Is Widespread, I: Copulatory Courtship and Related Phenomena

The previous chapters suggest that cryptic female choice is likely to occur in a number of scattered species, but the question remains whether cryptic female choice is common. Is it a biological curiosity, limited to a few unusual species? Or is it a pervasive trend in animal evolution? This chapter and the next two chapters approach this question from the direction of behavior (chap. 5), physiology (chap. 6), and morphology (chap. 7). All three suggest that cryptic female choice may be surprisingly widespread.

5.1 Copulatory Courtship

One way to estimate the frequency of cryptic female choice is to examine the courtship behavior of males during and immediately following copulation ("copulatory courtship"). All common interpretations of male courtship behavior assume that its basic function is to induce the female to respond in a way that favors the male's reproduction. If the need to elicit female cooperation ends when intromission begins, then one would predict that courtship would cease with the achievement of intromission. If cryptic female choice occurs, with females sometimes cooperating with male efforts to fertilize their eggs and sometimes not, one would predict that the male would often court after beginning copulation. Thus further male attempts to elicit female responses with additional courtship after copulation has begun can be considered evidence that cryptic female choice occurs (Eberhard 1991a, 1994).

Observations of copulatory courtship should give a *conservative* indication of the frequency of cryptic female choice because (1) cryptic female choice could be based entirely on courtship prior to intromission, which could also serve to induce postintromission female processes. For instance, it has long been known in some birds, mammals, and lizards that male precopulatory courtship induces critical female processes such as ovulation or early estrus (Lofts and Murton 1973; Follett 1984; Jöchle 1975; P. Licht 1984; McComb 1987). (2) Special male behavior capable of inducing favorable changes in female responses may not have arisen, even though cryptic female choice occurs. For example, females might favor heavier males by cryptic female choice, and variations in male behavior might not be capable of accentuating her perception of his weight. Also not taken into account are the strong possibilities that (3) males court females internally, utilizing their genitalia (Eberhard 1985), and that (4) male seminal products act to induce favorable changes in female responses (Eberhard and Cordero 1995; chap. 6). In addition, (5) some male clasping organs that are immobile as they hold the female nevertheless serve as courtship devices (Robertson and Paterson 1982; Belk 1984; Eberhard 1985). Because external movements are relatively easily observed, a survey checking for externally visible copulatory courtship in the mating behavior of an array of different taxonomic groups should provide a preliminary—albeit probably conservative—estimate of how common cryptic female choice may be.

I have performed two such surveys, one using a blindly chosen sample of previously published firsthand accounts of the mating behavior of 302 species of insects in 231 genera and 102 families (Eberhard 1991a), and the other using direct observations of copulation in another blindly chosen sample of 131 species in 102 genera and 49 families of insects and spiders (Eberhard 1994).

Strict, conservative criteria were used in both surveys to distinguish courtship behavior from other kinds of male behavior:

1. The behavior was repeated, in general outline if not in minute detail, both during a given copulation and in different copulations. In most cases behavior that was not repeated rhythmically was not counted as courtship.
2. The behavior was appropriate to stimulate the female. For instance, waving legs or antennae which did not consistently touch the female and which were out of her sight was not counted unless the movements were so energetic and jerky that they caused her body to vibrate. Movements that caused sexually dimorphic structures of the male to rub or tap against the female were counted because such structures tend to be involved in courtship (West-Eberhard 1983, 1984; Eberhard 1985).
3. The behavior was mechanically "irrelevant" to the male's problems of staying physically coupled to the female. For instance, movements of the male's legs associated with mechanical problems of alignment with the female as she moved, with pushing his genitalia deeper into her body, or with fending off another approaching individual were not counted.
4. The behavior was not performed in contexts suggesting other functions. Cleaning movements and apparent threats elicited by the approach of other individuals were not counted.
5. Rhythmic thrusting or twisting movements of the male's genitalia were not counted. In none of the species studied were the mechanics of copulation understood well enough to distinguish movements involved in

sperm transfer per se (or possible sperm removal). Not counting this very common type of movement probably results in an underestimate of the frequency of copulatory courtship, since genitalic movements clearly trigger female reproductive responses in some species (see sec. 5.2). Rhythmic rubbing or tapping movements of male genitalia on the surface of the female, which clearly were not involved in sperm transfer, were counted, as were rapid vibrations of the entire genital capsule that caused the female's body to vibrate.

6. Male behavior that, judging by the context, may have induced the female to refrain from capturing him after copulation ended (two species of spiders) was not counted. Such behavior may not influence female usage of the male's sperm.

Several additional factors probably led to underestimates of the frequency of occurrence of copulatory courtship. Repeated intromissions involving the same pair were not counted as courtship, although they may function this way (see Jöchle 1975 and Dewsbury 1988 on nonejaculatory intromissions that can trigger ovulation; see also next section). The possibilities of substrate vibrations and odors were also generally not taken into account.

It should be kept in mind that definitive proof that a given behavior pattern functions as courtship would require demonstration (ideally experimental) that the behavior induces females to respond in ways that increase the male's reproductive success. Such data are lacking for nearly all the copulatory courtship behavior patterns seen in both surveys; indeed, they are absent in the overwhelming majority of studies in which *pre*copulatory behavior by males has been interpreted (usually on the basis of criteria 1–4 above) as courtship.

Both surveys showed that copulatory courtship behavior is surprisingly common: it occurred in at least 36% of the 302 species in the literature survey, and in 81% of the 131 species observed directly (there was no overlap in the species in the two studies). The numbers change very little when analyzed at the level of genus (34% and 79%) or family (43% and 76%), respectively. This, along with the evolutionarily labile nature of the male behavior (see below), indicate that phylogenetic bias in male behavior was probably not responsible for overinflated estimates (see below for females).

The male behavior patterns and the structures they employed varied widely. Male behavior included singing, tapping, rubbing, hitting, kicking, waving, licking, wetting with secretions, biting, feeding, rocking, and shaking (e.g., figs. 2.11, 5.1, 5.18, 5.19). The male's mouthparts, his antennae, his first, second, and third legs, his wings, parts of his genitalia that stayed outside the female, his abdomen, and his entire body were all used.

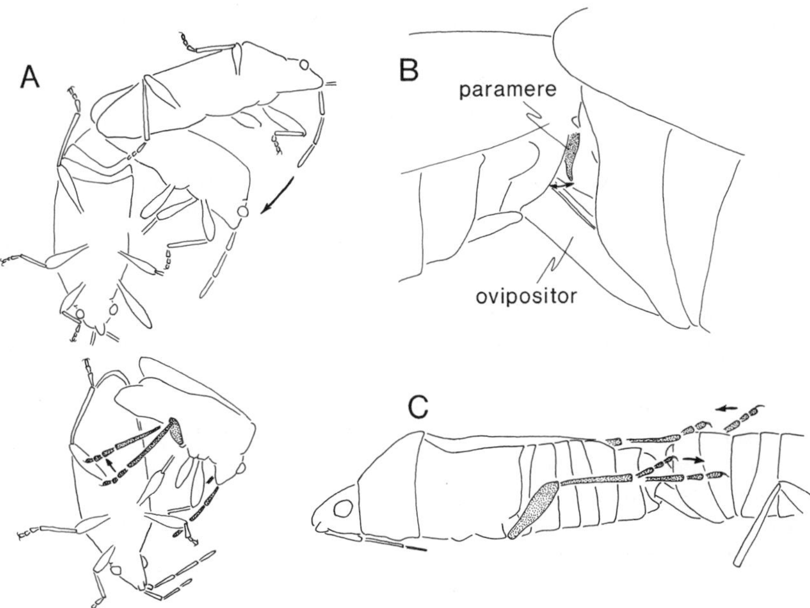

Figure 5.1 The minor but persistent movements that constitute copulatory courtship in the lygaeid bug *Xyonysius* sp. near *basalis* have no obvious relationship with the overall vigor of the male; nor do male movements involve contact with areas of the female that are likely to be contacted by the chemical sense organs of subsequent males. The male periodically turns (swinging on his genital capsule) to briefly tap the female with his antennae and rub her with his hind leg (stippled in A) on the ventral surface of her abdomen. He also squeezes her ovipositor rhythmically with his parameres (stippled in B) and taps and rubs her abdomen with his hind legs (stippled in C). Antennal contact by subsequent males is generally limited to the female's dorsum, so none of the male copulatory courtship movements seem appropriate for marking the female chemically. The male movements are also apparently undemanding in terms of energy or agility, and they seem ill-chosen to give the female information on his overall vigor (see chap. 2.8). (After R. Rodriguez and Eberhard 1994)

The estimates from direct observations are probably more accurate, because there are several reasons to suppose that previous studies were biased against observing and reporting copulatory courtship (Eberhard 1991a, 1994). Most previous published observations were made in theoretical contexts in which courtship after copulation has already begun is paradoxical and unintelligible and thus may have been considered trivial and not worth reporting (male courtship was generally thought to function either to reproductively isolate related species [e.g., Dobzhansky 1970], or in classic precopulatory sexual selection by female choice). A particularly dramatic illustration of underreporting comes from the intensively studied fly *Dro-*

sophila melanogaster. Many studies of its sexual behavior were published before its quite energetic, prominent copulatory courtship was first reported (Robertson 1982). In my own observations of both *Macrodactylus* beetles and *Leucauge* spiders, many aspects of copulatory courtship behavior went ignored and unperceived until they became theoretically intelligible in the context of cryptic female choice (for a similar discussion of failure to report observations because of theoretical biases, see McKinney et al. 1984 on extra-pair copulations in birds; in this case the theoretical bias was species-level selection).

In addition, the direct observations were made with the specific objective of checking for male courtship behavior, and revealed some relatively inconspicuous behavior patterns (e.g., repeated, infrequent bouts of tapping or stroking, relatively small but sustained vibrations and rocking movements) that would have been easily overlooked in the previous observations. Concrete manifestations of this bias were the larger numbers of different types of copulatory courtship behavior patterns/species in the direct observations (56% vs. 36% had more than one; $P < 0.001$ with chi-squared test) and the slightly greater frequency of reporting behavior patterns not seen prior to genitalic coupling (65% vs. 56%; $0.2 > P > 0.1$; Eberhard 1994).

One might think that some of this male "courtship" behavior was simply "twitches," or other inadvertent, selectively insignificant behavior, perhaps resulting from the male's presumably often excited state during copulation. There are several reasons to reject this idea. In the first place, 44% of 147 different presumed copulatory courtship behavior patterns in the literature, and 35% of 125 others in the direct observations, also occurred prior to genitalic coupling; they were thus apparently continuations of classic, precopulatory courtship (though the possibility that they were also selectively insignificant prior to copulation was not tested). That these same behavior patterns should so often be performed during copulation is not predicted by the "inadvertent twitches" hypothesis but is in accord with the hypothesis of a courtship function.

In some cases the contexts in which the male behavior was performed also argued for a courtship function. In two cases from the literature (Alcock and Buchmann 1985 on *Centris* bees; Humphries 1967 on *Ceratophyllus* fleas) experiments demonstrated a female response to the male behavior. In several species the male began to perform the behavior whenever the female began to move, and she often responded by stopping (Eberhard 1991a, 1994). The context of copulatory courtship suggests similar interpretations in others (e.g., postcopulatory bouts of courtship vibrations occur in *Limnoporus* waterstriders while the male is attempting to maneuver the female to an oviposition site; Wilcox and Spence 1986). The copulatory courtship behavior was in many cases so stereotyped and persistent that it seems unlikely that it was simply incidental, insignificant "twitching." The uniformity in the behavior of different males of the same species that so often occurred is also not expected for incidental twitches.

Finally, copulatory courtship behavior shows a tendency to diverge relatively rapidly, as would be expected if it is under sexual selection (West-Eberhard 1983), but not if it constitutes selectively insignificant twitches. In all twenty-one genera of insects and spiders in which more than a single species was observed directly (fifty-six species in Eberhard 1994; plus five species in the sphaerocerid fly genus *Coproica*, Lachmann 1994; and two species in the lygaeid bug genus *Ozophora*, R. Rodriguez, in prep.), there were intrageneric species differences. In fact, qualitative *intra*specific differences between different populations are now known in four species: the vervet monkey *Cercopithecus aethiops* (Gartlan 1969), the bee *Nomia triangulifera* (Wcislo et al. 1992), and the stinkbug *Mormidea notulata* and the weevil *Nicentrus lineicollis* (Eberhard 1994). Quantitative differences in the duration of copulatory courtship also occur in different strains of the tsetse fly *Glossina pallidipes* (Jaenson 1979).

Another possible alternative interpretation for some of the male behavior is that it does not constitute courtship, but rather the means by which the male applies his own odor (e.g., his own sex-specific cuticular hydrocarbons) to the female, thus making her less attractive to other males. Such transfer occurs in *D. melanogaster* (though without overt male "application behavior"; Scott et al. 1988). Similar transfers of male surface substances whose behavioral significance (if any) is unknown may also occur in tsetse flies *Glossina* where male-produced compounds vary between species, and stable flies *Stomoxys* (Carlson and Langley 1986).

Although this alternative interpretation to courtship is not reasonable for several types of male copulatory courtship behavior, such as vibration, rocking, singing, lifting, waving, and shaking, it could be feasible for some of the relatively common patterns, such as rubbing, tapping, and licking. There are, however, several reasons to doubt that this is a general objection to the courtship hypothesis.

In the first place, the conditions under which antiaphrodisiacs are likely to evolve may be somewhat restrictive, as a subsequent male is only expected to be "deterred" when it is to his own advantage to avoid the female. Second, if antiaphrodisiac odors are being applied (especially long-lasting odors), one would expect males to concentrate their rubbing, tapping, etc., on those portions of the female most likely to be contacted by the olfactory sense organs of subsequent males. This seems not to be the case in many of the species in the survey of direct observations (e.g., fig. 5.1; Eberhard 1994).

Of a total of 118 behavior patterns in eighty-nine species, 32% resulted in contact with areas on the female that were unlikely to be touched by chemical sense organs of subsequent males (antennae, mouthparts, front legs of Diptera, legs in spiders), and 37% with areas that were likely to be touched; in 32% the behavior of males prior to copulation was not known in sufficient detail to determine whether subsequent male contact was likely. Most of the cases in which contact was likely were "defaults," in the sense that the male used the same portions of his body with sense organs (mouthparts, antennae) to perform courtship. In only a few cases (<8%) did the male rub or tap with a different part of his body an area of the female that was especially likely to be touched by the sensory structures of subsequent males.

A further expectation of the odor-marking hypothesis is that tapping and kicking with the legs would be much less common than rubbing, as they would presumably be less effective in spreading a male substance. Again the data are not in accord. Of seventy-one behavior patterns in sixty-five species in which tapping, rubbing, kicking, squeezing, pressing and pushing occurred, 38% did not involve rubbing (the five species described as both rubbing and tapping were all counted as rubbing). In addition, in close to half of the cases of rubbing (43%), the male rubbed the female at a site that chemical sense organs of subsequent males were unlikely to touch.

These data refer only to general trends, and much further research will be necessary to examine the possibility of chemical marking. It may well turn out that some tapping or rubbing represents application of male marking compounds (of course the stimuli that accompany application of an anti-aphrodisiac odor could still be under sexual selection by cryptic female choice). It seems likely, however, that most male copulatory courtship movements are not simply incidental aspects of male marking behavior.

A final alternative is that male copulatory courtship functions to induce female reproductive responses, but that different types of behavior are equally effective, so that rapid divergence is the result of genetic drift rather than sexual selection (D. Zeh, pers. comm.). Experimental manipulations will be needed to test this idea. Nevertheless, the relatively large direct reproductive payoffs to males from affecting female responses suggest that such a precise neutrality may be unusual.

In almost none of the species in either survey has the function of male copulatory courtship been studied. There is, however, a possible association with male genitalic design. Given the difficulty that males of many species have in reaching sites in the female where sperm are to be deposited (see chap. 7.1), male copulatory courtship may often represent attempts to induce the female to allow deeper penetration. In many groups of Coleoptera and also in Apoidea (bees), males often have largely membranous extensions of their genitalia that must be inflated within the female (e.g., see fig 7.10); and females in at least some beetles and bees have muscles in their reproductive tracts that may be able to prevent this inflation (e.g., fig. 3.6; Eberhard 1993a, and Eberhard and Kariko, in press, on beetles; Roig-Alsina 1993 on bees). In two related groups (meloid beetles, sphecid wasps) the male genitalia do not have such inflatable sacs, and females presumably cannot exclude males as easily once they have achieved intromission. In both meloids and sphecids the frequency of copulatory courtship in literature accounts was unusually low: 7% of fourteen species of meloids versus 62% of forty-two species of other beetles; 0% of five sphecid species vs. 48% of twenty-eight apids (neither meloid beetles nor bees happened to be included in the survey of direct observations, in which species were chosen mostly on the basis of the ease of observation and large population sizes). The timing of male behavior in at least some species is in agreement with this function. Thus male *Macrohaltica jamaicensis* beetles vibrate their antennae only during the first part of copulation, when they have still not successfully inflated the internal sac in the female bursa (Eberhard and Kariko, in press).

In sum, these conservative surveys suggest that cryptic female choice may be extremely widespread. This conclusion depends, however, on the assumption that the species in the samples are truly representative of other animal groups with respect to copulatory courtship. Evaluating this assumption is complex. There is reason to suspect that both surveys (and for that matter, the entire published literature on copulation behavior) probably do have a bias, favoring those species in which females mate repeatedly. In species in which females remate less often, an observer would be less likely to be watching when a female acceded to a male's precopulatory courtship. Such a bias toward multiple mating in females probably favors inclusion of species in which cryptic female choice is more likely to occur: cryptic female choice among males is possible only if females sometimes mate more than once. The effects of cryptic female choice on male fitness are also likely to be larger when females mate with more males.

The question thus becomes how widespread multiple mating by females is in nature. This is still an open question for most species because of the difficulty of obtaining convincing data (Eberhard 1985; see also chap. 9.4), so it is difficult to judge the extent of the bias in the surveys. In those few groups in which minimum estimates of female remating frequencies in the field have been extensively documented, either from vestiges left by males that have transferred material to the female (e.g., spermatophores in butterflies and moths; Ehrlich and Ehrlich 1978; Drummond 1984), or from genetic analyses of the offspring of single females (Westneat et al. 1990 and Birkhead and Møller 1992 in birds; Mock and Fujioka 1990 in vertebrates in general), species with at least occasional multiple mating by females seem to be much more common than those in which females mate with only a single male (see Eberhard 1985 and chap. 9.4 for a more complete discussion).

In sum, in two large samples of insects and spiders that may have been biased toward species in which cryptic female choice is an evolutionarily feasible possibility, observations of external male behavior suggest that it is relatively common. Similar surveys of mating behavior have not been attempted in other taxonomic groups, so it is not certain whether these high frequencies of copulatory courtship are typical of other kinds of animals (the discussion in the next section suggests, however, that genitalic courtship during copulation may be common in mammals). I know of no reason that would suggest a priori that insects and spiders should be unusual in this respect.

I have found scattered examples of apparent copulatory courtship in a wide array of other animals: the insectivore *Tenrec ecaudatus* (ritualized biting and clapping of mouth while head moves from side to side during the 5 min intromission; Eisenberg and Gould 1970); the tree shrew *Tupaia longipes* (the male "quivers" the female with his front legs; Conway and Sorenson 1966); the marsupial mouse *Antechinus flavipes* (the male performs periodic "remarkable sinuous lateral wriggling" of his tail and biting of female's scruff and thrusts with his genitalia about once every 4 min during the approximately 5 hr of copulation; Marlow 1961); the dasyurid marsupial *Smithopsis* sp. (the male rubs his chin on the back of the female's head, apparently calming her when she begins to run about; Ewer 1968); the Uganda kob *Adenota kob* (fig. 5.2; Buechner and Schloeth 1965); the stump-tailed macaque *Macaca arctoides* (the male chatters and barks, then bites the female slowly and strongly—the bite "look[s] rather like the restraining . . . bite which mothers do to babies," not "like . . . bites seen in fighting;" Blurton-Jones and Trollope 1968; see also fig. 2.10 and table 5.1); the little brown bat *Myotis lucifugus* (the male gives a unique copulation call during at least the first 5–10 min of copulation [Barclay and Thomas 1979], despite the fact that such calls may attract the attention of other bats and cause the copulation to be interrupted [Thomas et al. 1979]); the rat *Rattus norvegicus* (the male vocalizes for 3–4 min after ejaculation; Sachs and Barfield 1976); the ring dove *Streptopelia risoria* (the male makes a "kah" vocalization immediately following cloacal contact; Cheng et al. 1981); the boat-billed heron *Cochlearius cochlearius* (the male vocalizes at the moment of cloacal contact and bill-claps immediately following copulation; Alvarado 1992); the oystercatcher *Haematopus ostralegus* (the male makes soft wee-wee noises during cloacal contact; Heg et al. 1993); the goldeneye duck *Bucephala clangula* (the male performs four different postcopulatory displays: rotation, steaming, bathing, and wing-stretch; Dane and van der Kloot 1962); the African black duck *Anas sparsa* (the

Figure 5.2 Postcopulatory display by a male Uganda kob, *Adenota kob thomasi*. The male of this lek-mating species performs a complex postcopulatory display lasting up to 5 min. His behavior includes waving his erect phallus up and down in jerky movements, whistling loudly, licking and nuzzling the female's vulva and inguineal region, and (usually following nuzzling) raising his front leg under her inguineal region with his phallus still erect (drawing). Sometimes the male pushes his neck against the female, and then "with obvious gentleness" clamps her between his chin and his stiff front leg. Females are promiscuous (up to seventeen copulations observed in 7 hr), and move from territory to territory on the lek; after having a postcopulatory display directed toward her, the female may not interact again with a particular male. Buechner and Schloeth (1965) speculate that the male's postcopulatory courtship may serve to induce the female to transport his sperm in her reproductive tract. (After Buechner and Schloeth 1965)

male performs two postcopulatory displays: mallardlike "bridling," and the lateral posture; McKinney et al. 1978); the vasa parrot *Coracopsis vasa* (male feeds female; Wilkinson and Birkhead 1995); the tortoises *Gopherus* sp. and *Geochelone elegans* (the male vocalizes in both; Weaver 1970 in Crews 1980; Rao and Rao 1990); the sidewinder snake *Crotalus cerastes* (male nudges the female with his head and waves his tail; Klauber 1972); possibly the Mexican milk snake *Lampropeltis triangulum* (the entwined tails twitch back and forth—it was not certain whether the male or the female was responsible for the movement; Gillingham et al. 1977); the internally fertilizing fish *Poecilia reticulata* (male makes up to 20–30 short, sharp jerky forward and upward movements of the whole body immediately following withdrawal of his intromittent organ (Liley 1966; Kodric-Brown 1992; see also E. Clark and Aronson 1951 for a discussion

	Before Copulation	During Copulation	After Copulation
Stroking	x	-	x
Rubbing	x	-	x
Poking	x	x	-
Body Lifting	x	-	x
Bouncing	x	-	x
Beating	x	-	x
Kicking	x	x	-
Carapace caressing	x	-	x
Assisting in moult	x	-	-
Leg encircling	x	-	x
Leg interposing	x	x	-
Palpating	x	-	x
Maxilliped tap	-	x	-

Figure 5.3 Possible courtship movements made by male tanner crabs *Chionoecetes bairdi* before, during, and following copulation. Female crabs are apparently barraged by stimuli during and following copulation. Body lifting, beating, and kicking occur when the female resists, and induce at least temporary quiescence. (After Donaldson and Adams 1989)

of apparent and real copulations in this species—it is possible some displays followed unsuccessful intromissions, which are common); the scorpion *Buthriurus flavidus* (the male uses his chelicerae to massage those of the female during sperm transfer from the spermatophore; Peretti, in press; see also Polis and Sissom 1990); the millipede *Glomeridella minima* (the male secretes a substance on which the female feeds, and which apparently functions as an arrestant; Haacker 1974); the tanner crab *Chionoecetes bairdi* (the male makes a variety of apparent courtship movements both during and following copulation; fig. 5.3; Donaldson and Adams 1989); and the nematode *Cruznema lambdiensis* (the male sweeps his body in wide arcs and oscillates the anterior region vigorously; Ahmad and Jairajpuri 1981; fig. 5.4).

The mammals in which the male bites the female's neck during copulation (e.g., Ewer 1968; Eisenberg and Gould 1970) are particularly interesting because this male behavior may constitute a sensory trap (see fig. 2.10). Ewer (1968) notes that such grips probably serve both to forcefully hold the female and to help the male align his body correctly. She emphasizes, however, that in the Carnivora the young are carried by the nape of the neck by the female, and that they respond to such grips by becoming limp and

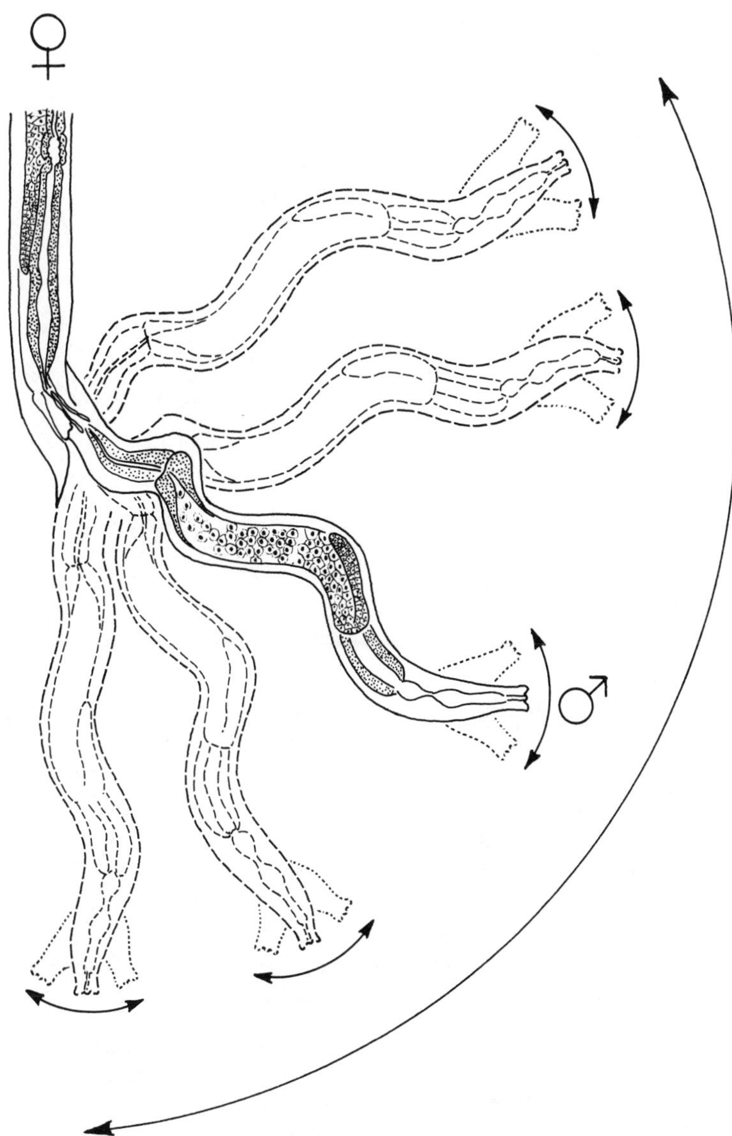

Figure 5.4 Apparent copulatory courtship occurs even in animals as morphologically simple as the nematode *Cruznema lambdiensis*. After introducing his intromittent organs into the female, the male energetically swings his body widely from side to side, at the same time oscillating the anterior portion of his body vigorously. The male movements cease just before sperm transfer begins (individual sperm cells are shown in the middle drawing of the male). (After Ahmad and Jairajpuri 1981)

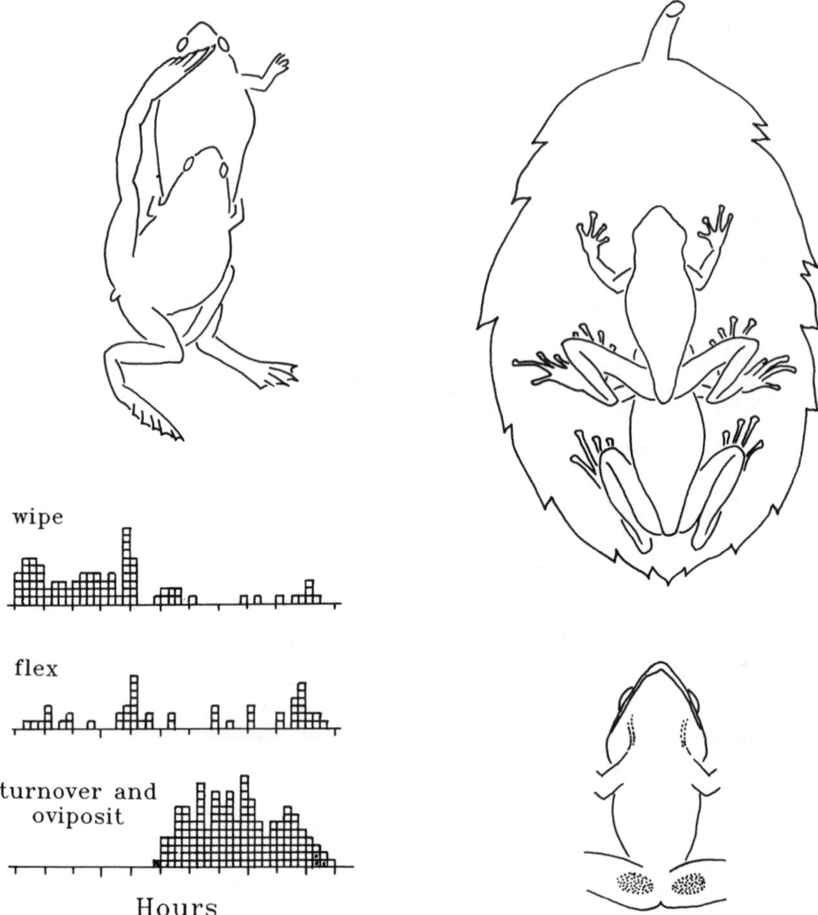

Figure 5.5 Copulatory courtship probably also occurs in some animals with external fertilization such as frogs. *Left*: in *Hymenochirus boettgeri*, the male first seizes the female in amplexus, then wipes his hind feet repeatedly across her head (top) during the long period while she is laying eggs. He also repeatedly flexes his front legs, simultaneously lifting his head up and down on her back in quick bobs. The number of times these actions were executed during 15 min periods in one mating is shown at the lower left (boxes with dots indicate times the female turned over as if to lay eggs, but failed to deposit any). Both wiping and flexing may function to elicit oviposition. (After Rabb and Rabb 1963) *Right*: In the frog *Mantidactylus liber*, the male does not grasp the female, but instead sits on her head and shoulders while she is laying eggs near the lower edge of a vertical leaf. This unusual posture, which is poorly designed to bring the male's sperm into contact with her eggs (his sperm are thought to dribble down her back), is associated with another unusual trait—the presence of sexually dimorphic glandular areas on the male's thighs (stippled in drawing at the lower right of a different species of the same genus). The gland products are presumed to stimulate the female to ovulate or oviposit. (Top after Blommers-Schlosser 1975; bottom after Duellmann and Trueb 1986)

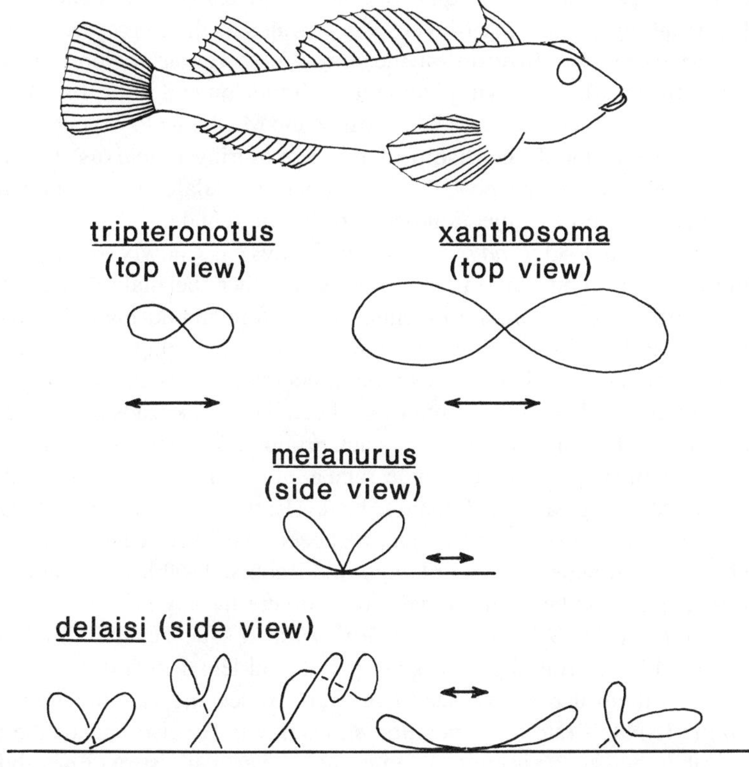

Figure 5.6 Copulatory courtship apparently occurs in some fish that have external fertilization. Male *Tripterygion* blennies perform "figure-8 swimming" in front of the female during spawning (which is a prolonged process, as the female lays her eggs one by one). In accord with the hypothesis that this male behavior is under sexual selection, the swimming patterns differ between congeneric species. Males of *T. tripteronotus* and *T. xanthosoma* make figure-8 patterns along the bottom, with those of the first species being tighter with respect to the male's length (double-headed arrows). Males of *T. melanurus* make slanting figure-8 patterns, and those of *T. delaisi* swim in slanting figure-8s as well as in many other patterns. (After Wirtz 1978; fish after Thresher 1984)

passive. Males may be using such programmed responses in females to elicit female cooperation during copulation.

Even in frogs and toads (fig. 5.5) and fish (fig. 5.6), which practice external fertilization and in which many of the possible mechanisms of cryptic female choice mentioned in chapters 3 and 4 are not feasible, some males apparently perform "copulatory" courtship after they have already succeeded in seizing females in amplexus: males call during amplexus in *Pachymedusa dacnicor* (Bagnara et al. 1986), *Rana aurora* (Licht 1969), *Boophis goudoti* (Blommers-Schlosser and Blommers 1984), and *Bufo americanus* (accompanied in this species by squeezing movements of the male's front legs; Price and Meyer 1979). Males of the pygmy toad *Mertensophryne micranotis* drum on the female's sacroiliac region with the long fourth toe after achieving amplexus (Grandison and Ashe 1983), and the male also calls during amplexus. Salthe and Mecham (1974) cite examples of apparent tactile courtship of this sort during amplexus in seven additional genera, and propose that it serves to stimulate ovulation or oviposition, or to suppress rejection behavior by the female.

In the Surinam toad, *Pipa pipa*, the female lays eggs slowly over a period of more than an hour, and after he has seized her the male deluges the female with apparent courtship stimuli both before and during oviposition (Rabb and Rabb 1960). He occasionally makes a pumping movement by repeatedly tightening his forearms around her, and at the same time makes a staccato up and down movement of his head. He also kneads the female's abdomen with his fingers just before each group of 3–5 eggs is laid. Late in the process of oviposition, the male repeatedly swings his extended hind leg forward alongside the female's back (Rabb and Rabb 1960). Leg swings usually occurred after the pair had been to the surface to breathe, or just before the female laid a set of eggs, and Rabb and Rabb speculated that this behavior stimulated the female to complete the laying. The male of another pipid frog, *Hymenochirus boettgeri*, also squeezes the female repeatedly with his front legs and swings his hind legs forward, but in this case wipes them across her head (fig. 5.5). Squeezing is more frequent when the female's rate of oviposition slows down. He also kneads the female's abdomen as she is emitting eggs, and occasionally strokes her abdomen with the fingers of his hand. The male may also produce stimulatory chemical products from his postaxillary subdermal glands (Rabb and Rabb 1963).

Secretions of the male's femoral glands may stimulate female frogs in *Mantidactylus liber* and an unidentified species of *Gephyromantis* (Blommers-Schlosser 1975; Blommers-Schlosser and Blommers 1984; fig. 5.5). Duellmann and Treub (1986) mention sex-specific male glands in a variety of species on areas of the male that contact the female during amplexus and are thought to stimulate the female to ovulate or oviposit (Duellmann and Treub 1986). The possibility remains that some may also function as threats or weapons when males embrace each other in male-male wrestling combat. In some cases, however, such as in femoral glands (fig. 5.5), an aggressive function seems very improbable because of their locations.

It is interesting to note that the male's calling during amplexus may be costly for the male in both *P. dacnicolor* and *M. micranotis*, as calls attract other nearby males that sometimes displace the first male from amplexus and sometimes apparently fertilize some of the female's eggs (Bagnara et al. 1986; Grandison and Ashe 1983). Thus it seems likely that the male calls must confer some sizable advantage to the calling male which more than compensates these losses.

Finally, comparative data in one group are in accord with the hypothesis that the male copulatory courtship functions to influence cryptic female choice. The cryptic female choice hypothesis predicts that copulatory courtship should tend to occur in species with polyandrous females, but not in those with monandrous females. This pattern obtains in the genus *Equus*. Males of the polyandrous Grevy's zebra (*E. grevyi*) and wild asses (*E. asinus*) emit loud postcopulatory calls, while the males of horses (*E. caballus*), mountain zebra (*E. zebra*), and plains zebra (*E. burchelli*), whose females generally mate with only a single male each estrus period, are silent (Ginsberg and Rubenstein 1990). Intraspecific comparisons of matings involving polyandrous versus monandrous females of *E. grevyi* show the same trend: seven times more postcopulatory calling and nonintromittent mounting activity followed copulations with polyandrous females (males were able to observe directly whether females had mated with other males; Ginsberg and Rubenstein 1990).

5.2 Genitalic Movements during Copulation as Courtship

Evidence of Widespread Occurrence

Genitalic thrusting was not counted as possible courtship behavior in the surveys of copulatory courtship in insects and spiders. Genitalic movements were, however, quite common. Rhythmic movements of the male genitalia within the female occurred in at least 63% of the 107 species of insects and spiders observed directly (Eberhard 1994). Previous descriptions of insect behavior have clearly been strongly biased against reporting male thrusting (0% in 302 species in the literature survey; Eberhard 1991a). They also occur in other insects (e.g., Wcislo and Buchmann, in press, on the bee *Dieunomia heteropoda*; Clements 1963 on the mosquito *Culiseta inornata*).

Similar intravaginal thrusting movements are widespread and common in mammals (e.g., Tavolga and Essapian 1957; Eisenberg and Gould 1970; Mitchell 1979; Dixson 1987; Dewsbury 1972, 1975, 1988; Dewsbury and Pierce 1989). For instance, in 84.6% of the 104 species surveyed by Dewsbury (1972) for which there were sufficiently detailed data, the male performs either multiple intromissions or multiple thrusts before ejaculation. Thrusting also sometimes occurs after ejaculation has occurred. For instance, in the thick-tailed bushbaby *Galago crassicaudatus* intromission is maintained for up to 260 min after ejaculation, with the male performing intermittent bursts of thrusting (Eaton et al. 1973).

There are also scattered reports of rhythmic thrusting or pushing in other groups, including rattlesnakes (Klauber 1972), gopher turtles (Crews 1980), parrots (Wilkinson and Birkhead 1995; J. Eberhard, unpub.), millipedes (Haacker 1969a,b, 1970), acaroid mites (for up to 45 min; Griffiths and Boczek 1977), and nematodes (Ahmad and Jairajpuri 1981; Rehfeld and Sudhaus 1985). Other genitalic movements also occur. The helical penis of the boar (*Sus scrofa*) rotates rhythmically to the right and to the left during copulation (Larsen 1986). In the zoraptera *Zorotypus barberi* the male intromittent organ "throbbed constantly" while inside the female (Choe 1995), while in the earwig *Skalistes inopinata* the male periodically pulled and/or twisted on the genitalic union (Briceño and Eberhard 1995). The penis of the bull often coils during ejaculation (Chenoweth 1986). In some whipscorpions (Uropygi) the male thrusts rhythmically for up to 2 hr with his spermatophore, which he holds with his chelae (Thomas and Zeh 1984).

Repeated insertions and withdrawals of the male genitalia prior to definitive intromission also occur in many mammals (e.g., Dewsbury 1972, 1988; Dixson 1987; Dewsbury and Pierce 1989), and some insects (Lewis and Wang 1991; Eberhard 1993a; Eberhard and Kariko, in prep). Although published descriptions of mating in a few groups clearly show that genitalic thrusting does not occur (e.g., Crews 1973 on the lizard *Anolis carolinensis*; Eberhard 1994 on some insects), thrusting is probably substantially more common than indicated by the available literature. The case of *A. carolinensis* is instructive in this context, because despite the lack of thrusting, genitalic stimulation from intromission is known to reduce the female's tendency to remate (Crews and Silver 1985).

In fact, rhythmic genitalic thrusting is undoubtedly even more common than careful direct observations would suggest. "Cryptic thrusting" has been seen by closely examining species in which males have semitransparent genitalia, and/or the female's body wall is partially transparent. It occurs, for instance, in seven species of melolonthine and chrysomelid beetles (Eberhard 1993a,b, 1994; fig. 5.7), the cercopid leafhopper *Zulia pubsescens* (Eberhard 1994), and the lygaeid bugs *Ozophora baranowskii* and *O*. sp. near *parva* (R. Rodriguez, pers. comm.). Internal portions of the male's genitalia move back and forth rhythmically inside the female's body, or even in some cases into and back out of her body, while the external portions of his genitalia remain motionless.

Evidence of Courtship Effects

Experimental evidence that genitalic movements during copulation function to induce female reproductive responses that favor the male's reproduction (i.e., that thrusting functions as courtship) was presented in chapters 3 and 4 for certain species of mammals and spiders: genitalic

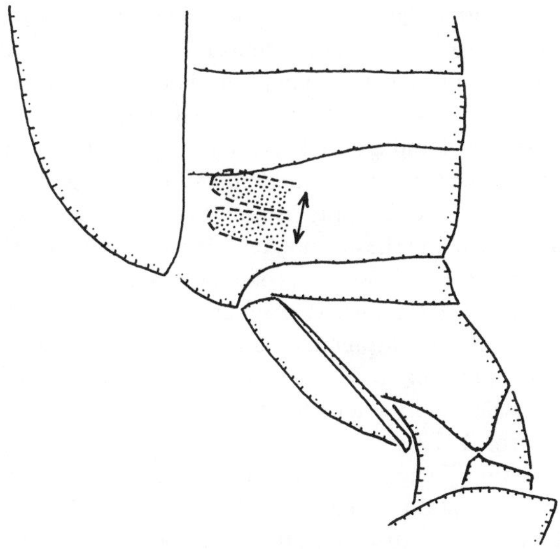

Figure 5.7 Cryptic genitalic thrusting by a male a *Phyllophaga vicina* beetle. Internal male genitalic structures can be seen moving rhythmically back and forth through the semitransparent outer wall of the male's genitalia. The female's abdominal sternites are also semitransparent, and a dark structure within her abdomen (stippled in drawing; probably her bursa) moved every time there was a movement within the male's genitalia. Parts of the male's genitalia extend into the bursa during copulation, and are apparently involved in transfer of seminal products other than sperm. (After Eberhard 1993b)

movements increase sperm transport (chap. 3.5), inhibit female receptivity to further mating (chap. 3.6), increase ovulation (chap. 3.9), and induce the female luteal cycle (chap. 4.1; see also chap. 6). In some species (e.g., the hamster *Mesocricetus auratus*, the rat *Rattus norvegicus*, the mouse *Mus musculus*), studies employing application of local anesthetics, severance of nerves, males mounting without intromission, and manual stimulation showed that mechanical stimulation of the female's genitalia with those of the male (rather than the act of mounting) provided the important cues (R. Carlson and DeFeo 1965; Carter 1973; Leckie et al. 1973).

Many other rodents perform both ejaculatory and nonejaculatory intromissions, and nonejaculatory intromissions appear to "synchronize, persuade, and orient at the same time as they manipulate the partner" (Dewsbury 1988, p. 218). For instance, postejaculatory intromissions with intravaginal thrusting are the rule in the cactus mouse *Peromyscus eremicus*. When the intromissions were prevented, nearly all (90%) females failed to initiate the functional luteal phase; and when the male performed them in lower numbers, the probability that the luteal phase would be triggered but that the female would nevertheless fail to give birth (due to lack of sperm transport?) was increased (Dewsbury and Estep 1975).

Dewsbury (1988) speculates that the functional significance of the multiple-intromission patterns typical of rodents is that they generate long mating sequences that could aid the female in choosing particularly strong, dominant males, since they increase the likelihood of male-male interactions before they are completed. This type of explanation, in the tradition of "good viability genes" arguments, attempts to link copulatory behavior with natural selection. It could apply in some social species, such as guinea pigs in the genus *Cavia* and rhesus monkeys *Macaca mulatta*, where subordinate males ejaculate more rapidly, probably to avoid interruption before sperm transfer (Rood 1972; Curie-Cohen et al. 1983; Manson 1994; see, however, the discussion of intrageneric differences below). It is not particularly convincing, however, in the many species such as *Mus musculus* and other rodents in which more than one male is seldom with the female at a given time (Bronson 1979).

Rhythmic, twisting, or prying genitalic movements (fig. 5.8) may also influence male fertilization success in the cellar spider, *Pholcus phalangioides*, but the chain of cause and effect may be more complicated (Uhl 1994c). Males that twisted their genitalia more rapidly early during copulation copulated longer ($r_s = 0.63$, $P = 0.006$, range 11–141 min). The mechanical fit between male and female genitalia (the width of the female opening vs. the width of the two male intromittent structures—the procursi) was also correlated with copulation duration ($r_s = 0.50$, $P = 0.03$). Analyses of partial and multiple correlations showed that mechanical fit did not influence copulation duration directly, but instead via the frequency of palpal movements.

The length of copulation was at least partially determined by the female, as 41% of seventeen copulations with virgin females ended when the female twisted her body and pushed the male away with her legs (further, "internal" rejections could also occur, and termination of copulations with nonvirgin females was even more often initiated by the female). Sperm transfer is probably slow and gradual, since encapsulated sperm are ejaculated by being displaced from the male genitalia by gland products, and histological studies revealed no reservoirs or accumulations of gland products that would allow rapid release (Uhl and Rose, unpub.). The hypothesis of gradual transfer is supported by the number of young produced by females allowed to mate only once: females that mated for longer times produced more offspring ($r_s = 0.72$, $P = 0.02$). These data suggest that copulations with more rapid genitalic twisting may result in the transfer of larger numbers of sperm (as well as longer periods of stimulation that might lead to greater female investment in offspring), and thus presumably increase the male's chances of paternity.

This interpretation is as yet tentative. The sample sizes in this study were only modest. Sperm precedence patterns have been worked out only quali-

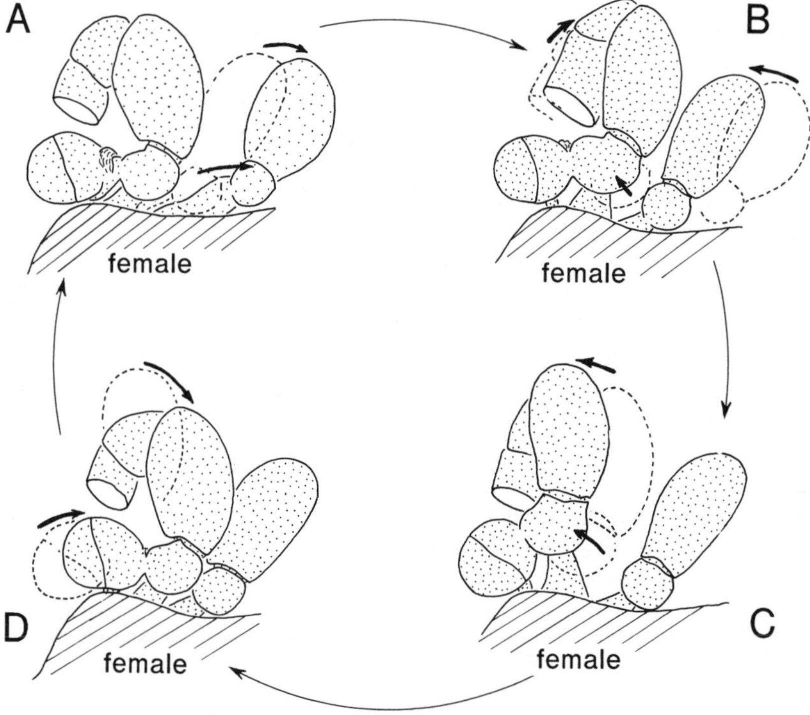

Figure 5.8 Stylized representation of the complex rhythmic twisting and prying movements of the powerful male genitalia (stippled; note large diameter of basal segments of palp) of the cellar spider *Pholcus phalangioides* during copulation. Greater rates of palpal movements during the first portion of copulation may induce the female to allow the male to copulate longer and thus transfer more sperm. (After Uhl 1994c)

tatively (marked sperm from the second male to mate with a female were found in her storage area). Factors that affect sperm emission from the female during copulation (see fig. 3.1) are unclear. Direct counts of the number of sperm to document gradual transfer are lacking.

There are several other, less direct indications that genitalic movements during copulation constitute courtship. Female orgasm in the monkey *Macaca fuscata* was more likely to occur when several measures of stimulation during copulation (copulation duration, number of mounts per copulation, number of pelvic thrusts per copulation, and number of male ejaculations) were greater (Troisi and Carosi 1994). The possible influence of female orgasms on male reproductive success has not been established. In humans, female orgasm is associated with an increase in circulating oxytocin, which both promotes contractions of smooth muscle (as in the uterus and vagina) and affiliative behavior in a number of mammals (Thornhill et al. 1995). In some species details of the male's behavior eliminate some other possible explanations. The male of the nematode *Cruznema lambdiensis* first uses his spicules (intromittent organs) to probe the female rhythmically. Once the spicules have penetrated the female's vulval opening, they make repeated thrusts deep into the vagina. They do not reach deep enough to contact sperm stored in the female, and so do not directly remove sperm (Ahmad and Jairajpuri 1981; see fig. 5.4 for another species). The male's spicules were *retracted*, however, just as sperm transfer began. Similar thrusting, which is also part of neither sperm removal nor sperm transfer per se, also apparently occurs in *Rhabditis*, *Pelodera*, and *Cylindrocorpus* nematodes (Duggal 1978; Ahmad and Jairajpuri 1981; Rehfeld and Sudhaus 1985), and "rapid spiracular action" also occurs during copulation in *Cylindrocorpus longistoma* (Chin and Taylor 1961).

If stimulation of female genitalia serves as courtship, it might be predicted that in some groups the males would utilize other portions of their bodies to stimulate female genitalia. This indeed occurs. For instance, in all ticks studied to date, spermatophore transfer is always preceded by the male inserting some or all of his mouthparts into the female's vagina. In at least some species he moves them back and forth there, sometimes for hours at a time (Oliver 1982; Feldman-Muhsam 1986). Both the movements (which are not necessary for ingestion in ticks, whose mouthparts are generally embedded in a hard, cementlike secretion during feeding (Moorhouse 1969; Kemp et al. 1982), and the fact that the mouthparts do not reach deep into the female where sperm are stored (e.g., Kemp et al. 1982 on *Boophilus*) indicate that removal of previously deposited sperm is not occurring (deposition of a spermicide cannot be ruled out, but this would still not explain why the mouthparts are moved back and forth). Only following this phase, which is generally thought to be stimulatory in function (e.g., Feldman-Muhsam 1986), does the male produce a spermatophore and insert it into the female's vagina.

Another group in which males move their chelicerae in and out of the female vagina before sperm is transferred are rhodacarid mites (Lee 1974). In these species sperm are deposited elsewhere rather than in the vagina, and thus the possibility of sperm removal can be discarded.

Similarly, male solifugids of the genus *Eremobates* perform preinsemination thrusting or chewing in the female genital opening with the needlelike fingers of their chelicerae (Muma 1966). These movements are similar, but not identical, to movements used later to introduce the sperm into the female. In *Metasolpuga picta* the male continuously probes the female genital opening with his chelicerae after introducing the spermatophore (Wharton 1987), a movement perhaps analogous to the genitalic thrusting of other groups. Preinsemination chewing and thrusting might serve to remove sperm from previous males (one male *E. palpisetosulosus* withdrew

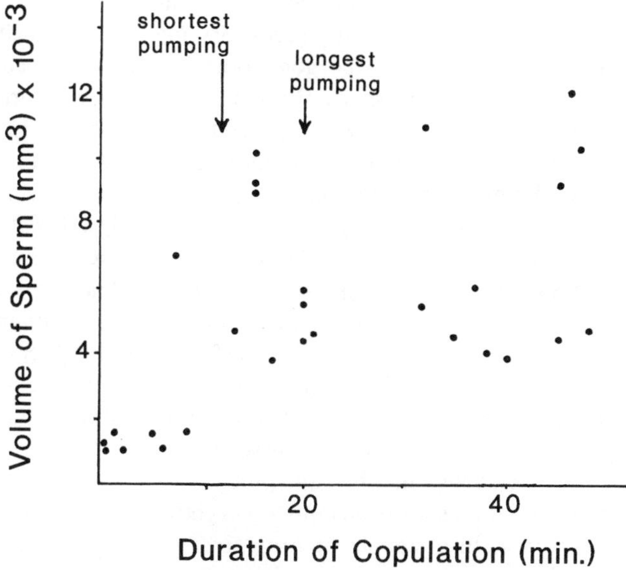

Figure 5.9 Sperm removal or genitalic courtship? Genitalic "pumping" behavior in many odonates is part of the process by which males remove or reposition sperm from previous males before transferring their own. The possibility that such movements also have a courtship function is suggested by the fact that when *Ischnura verticalis* males mate with virgin females (most matings in this species are with virgins, since females apparently seldom remate), they sometimes perform pumping movements well into the period during which they are transferring sperm (the longest pumping period was about 20 min, while sperm transfer begins at about 10 min). (After Fincke 1987)

from the female with a droplet of fluid on his chelicerae); but in another species males also inserted and chewed at the female's anus before beginning on her genital opening (Muma 1966).

These observations provide only indirect indications of a courtship function and are in many cases open to alternative interpretations (e.g., thrusting provides the male with sensory information on female readiness to receive and accept his ejaculate, or is part of a sperm removal or sperm-killing process). It is worth keeping in mind, however, that even male genitalic movements (and morphology) that result in such processes may also have courtship effects on females. A possible example is the male genitalic "pumping" movements that have been associated with sperm removal in a number of odonate species (e.g., Waage 1979, 1984, 1986). Stimulation of the female with pumping movements could explain the otherwise puzzling 12–20 min of pumping while the male of the dragonfly *Ischnura verticalis* is transferring his own sperm (fig. 5.9); it is unlikely that he is removing sperm from a previous male because females are usually monandrous (Fincke 1987).

A second type of evidence of probable stimulatory functions are "preinsemination" copulations involving secondary male genitalia (including pedipalps in linyphiid spiders, and gonopodia in some millipedes), that have not first been filled with sperm. In one species that was already discussed (the linyphiid spider *Nereine litigiosa*) these intromissions function as courtship (chap. 3.5). An alternative interpretation is that the behavior serves to sense (Suter 1990) or to remove previous sperm. In the millipede *Cylindroiulus punctatus* the male genitalia are moved rhythmically during the preinsemination copulation (stimulating the female? removing sperm?) and are immobile during insemination copulation (Haacker and Fuchs 1970). The highly stylized, rhythmic patterns of preinsemination insertions in linyphiids (van Helsdingen 1965; Suter and Renkes 1984; Suter 1990; A. Berghammer, unpub., on *Diplothyron* sp.) are more in accord with a courtship function than sensory or removal functions, but further study will be necessary to obtain more definitive answers.

A third type of evidence involves the predicted association between female monogamy and a lack of selection by cryptic female choice on male copulation behavior. In a sample of forty-six species of primates, more elaborate copulations (multiple intromissions preceding ejaculation, prolonged intromissions that lasted more than 3 min or were maintained after ejaculation, contrasted with short intromissions in which the male withdrew promptly after ejaculation) tended to occur in species in which females are likely to mate with more than one male during a given estrus (when females live in a social group with more than one sexually active male or are dispersed; Dixson 1987; see also Dixson 1991b). Simpler copulations tended to occur in those species in which the female's social situation precludes her mating with other males (member of a monogamous pair, or member of a single male's harem). This trend remained after correcting (at least partially) for possible phylogenetic bias (Verrell 1992).

The predicted trends only occurred, however, in ten of an additional twenty-one species in eleven additional genera that were later added to the analysis (Dewsbury and Pierce 1989). Nevertheless, combining the two sets of data,[1] the trend remained for more elaborate copulations to occur in species in which females are more likely to mate with more than one male per estrus (51% of forty-six species vs. 18% of twenty-two species) ($P < .01$ with chi-squared test). This analysis must be regarded with caution, since it does not take into consideration possible phylogenetic inertia (species may have particular combinations of characters because they inherited them rather than because those combinations are advantageous; Ridley 1983; Felsenstein 1985). The importance of inertia is uncertain, however, since copulation behavior is evolutionarily labile in at least some groups; differences exist between several pairs of closely related (con-

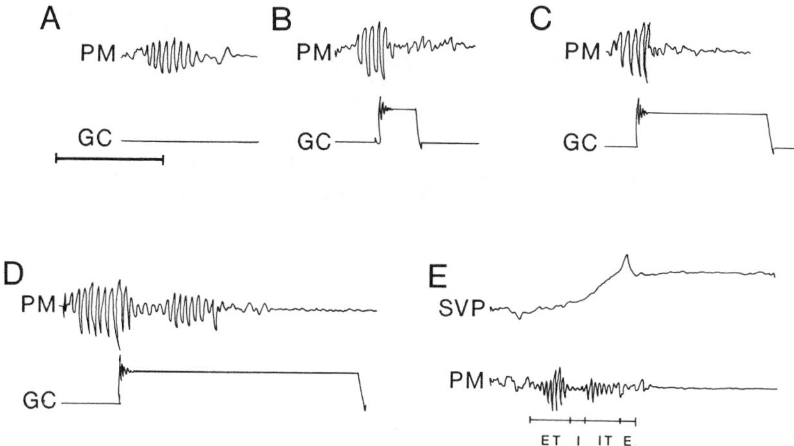

Figure 5.10 Complex stereotyped thrusting behavior during copulation in the rat *Rattus norvegicus*. Pelvic movements (PM) of a male were recorded with an accelerometer and were associated with electrically recorded genitalic contact (GC) and pressure recorded within the seminal vesicle (SVP) to show ejaculation. A: Mount and typical spindle-shaped series of thrusting movements without genital contact; B: series of thrusts which ended soon after a short, nonejaculatory intromission began (upward displacement in GC record); C: thrusting stopped soon after an ejaculatory intromission began; D: thrusting continued, but with lower and more constant amplitude, during a long ejaculatory intromission; E: seminal vesicle pressure rose gradually during extravaginal thrusting (ET), intromission (I), intravaginal thrusting (IT), to peak at ejaculation (E). Scale marker = 1 sec. (After Moralí and Beyer 1992)

generic) species (Dixson 1987; Dewsbury and Pierce 1989). In any case, if there is a trend, it is in the direction predicted by sexual selection.

Still another type of evidence comes from the relatively stereotyped nature of the copulatory behavior of some species. If copulatory thrusting constitutes courtship, it should often be relatively organized and stereotyped in at least some respects, and also often relatively complex with respect to patterns of timing and amplitude. An especially detailed demonstration of both complexity and organization comes from the studies of Morali and Beyer (1992) on the rat *Rattus norvegicus* (fig. 5.10). Some aspects of male behavior were strikingly invariable. For instance, the coefficient of variation of the duration and frequency of individual pelvic thrusts was <10%, and a burst of thrusts usually began and ended with low amplitudes (fig. 5.10A); the durations of penile insertions were much more variable (coefficient of variation 37%).

Patterns of male behavior recorded in similar detail for rabbits were also stereotyped but quite different. Up to seventy rhythmic, constant-amplitude, high-frequency extravaginal thrusts occurred, and were apparently

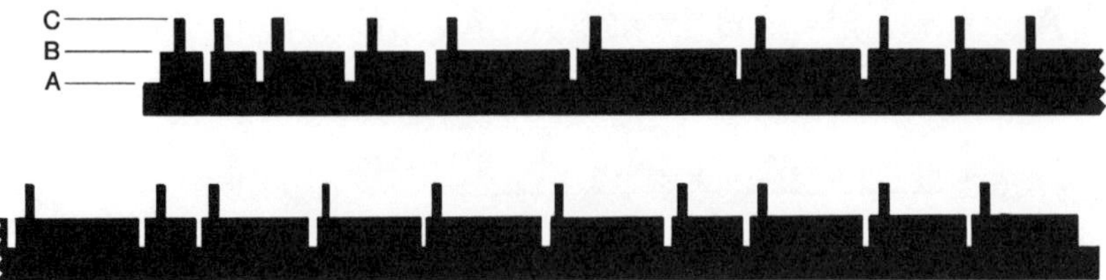

Figure 5.11 Highly ordered pattern of movements producing different degrees of insertion of the male genitalia (pedipalp) of the spider *Mynoglenes diloris* after it has locked onto the female's genitalia. A, B, and C represent different degrees of inflation of pedipalp membranes, which are associated with different degrees of coupling and progressively deeper penetration of the female. Judging by the results of artificial modification of the male palp, ejaculation probably occurs (up to 27 times in a single intromission) during stage C. (After Blest and Pomeroy 1978)

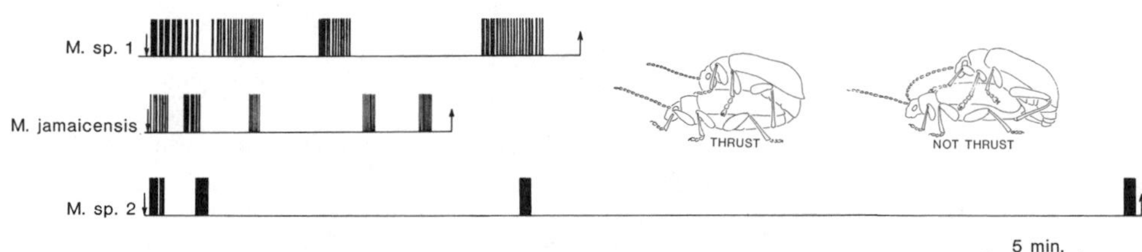

Figure 5.12 Species-specificity in the patterns of genitalic thrusting in three species of *Macrohaltica* chrysomelid beetles. Each vertical bar represents a thrust, in which the male pushes his genitalia deep into the female; the width of the bar is the duration of the thrust. Intromission begins with the downward-pointing arrows, and ends with the upward arrows. (After Eberhard and Kariko, in prep.)

necessary to induce lordosis and thus permit intromission (strictly speaking, this thrusting may constitute classic, preintromission courtship). Low frequency and nonrhythmic thrusting rarely or never elicited lordosis, even in highly estrous does (Morali and Beyer 1992). Spiders are another group with highly regular patterns of intromittent "thrusting" (fig. 5.11; Gering 1953; van Helsdingen 1965; Blest and Pomeroy 1978).

If genitalic movements during copulation function as courtship, then a further expectation is that the genitalic movements of closely related species will often differ. This is because sexual selection often produces a relatively rapid divergence in related species (West-Eberhard 1983). This prediction is confirmed by intrageneric variation in genitalic movements in a scattered array of insects and spiders, primates, and rodents. Three species in the chrysomelid beetle genus *Macrohaltica* differ in both rhythms and durations of thrusts (fig. 5.12; Eberhard 1994; Eberhard and Kariko, in press). Durations of intromissions differ in a pair of species of the tiger beetle genus *Pseudoxychila* (R. Rodriguez and Eberhard, in prep.), as do

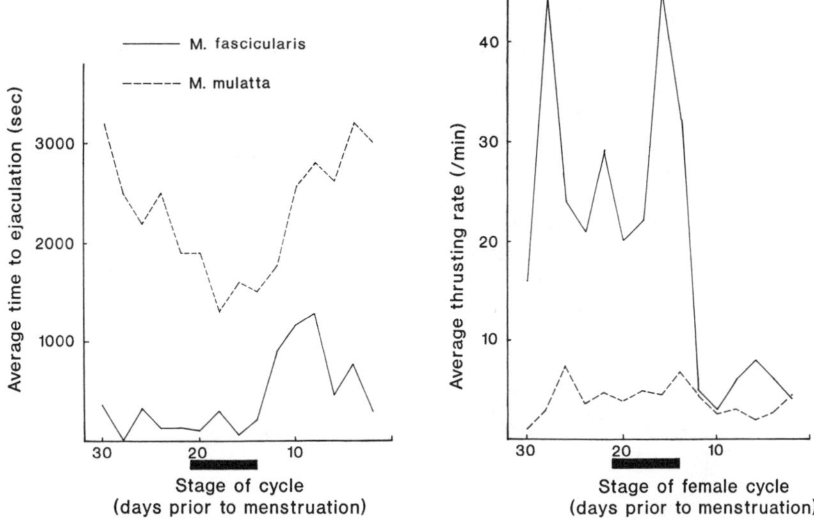

Figure 5.13 Male copulation behavior in *Macaca* monkeys varies among species, as expected to often occur if this behavior is under sexual selection (see also table 5.1). Male behavior also varies with the estrus condition of the female (black bar indicates presumed period of ovulation). Perhaps different male behavior when females are not fertile means that copulations at such times have different functions. (After Zumpe and Michael 1983)

the durations and patterns of "preliminary genitalic contacts" that precede true copulation in two species of the fly genus *Sepsis* (Parker 1972). The strength, frequency, and pattern of changes in timing and strength during copulation of rhythmic contractions of the male postabdomen that compress the female abdomen varied in five species of *Coproica* flies (Lachmann 1994). Two species of the spider *Ixeuticus* differ dramatically in the number of palpal insertions per mating bout (average 121 and 3, respectively; in this case a possible species isolation function was shown to be very unlikely because none of seventy-five mixed species pairs copulated; Willey 1992). The frequency with which the embolus (intromittent organ) of male *Mynoglenes* spiders was thrust deep into the female was lower in *M. titan* than in *M. diloris* (see fig. 5.11), but total intromission was probably longer (Blest and Pomeroy 1978). Intra*specific* differences in the number and duration of insertions occur in different populations of the spider *Nesticus cellulanus* (Huber 1993).

In nine species of macaque *Macaca* monkeys, differences occur among at least some pairs of species with respect to number of intromissions per ejaculation, numbers of thrusts per intromission, and times between intromissions (fig. 5.13; table 5.1). One *Macaca* species (*arctoides*) also differs from the others by having an approximately 1–3 min "postejaculatory pair-

Table 5.1
Two Illustrations of the Intrageneric Diversity Predicted by the Hypothesis That Male Behavior Is under Sexual Selection by Cryptic Female Choice

A. Average values for copulation behavior in nine species of *Macaca* monkeys (multiple values are from different studies).

Species	*I*	*II*	*III*	*IV*	*V*	*VI*	*VII*	*VIII*	*IX*
radiata	1,1	10	Clonic jaw movements	3.2/sec	7, 5, 17			Y	Y[a]
arctoides	1	15–175	Lipsmack, vocalize, bite female neck at ejac., clatter teeth		15–170[a] 39.3	1–15[b]		Y	Y[c]
mulatta	6–8,[d] 5–25, 1–20, 7.4, 10.7	8.5 8.5	Chew, teeth gnashing	2/sec 4.4/sec	less than *nemestrina*	26	less than *nemestrina*	Y	Y[a]
fuscata	8–30 17, 9.8, 5–45				1–8			?	Y[a]
fascicularis	1–3[f], 2.9–6.6[e]	8.5–6[c]	Vocal squeaks, open mouth, screech	2.3, 3–5[f]	8.2, 4–5, 17–13[f]			Y	?
silenus	3.5	9.3	Bared teeth (occasionally)	1	3–16	205	Y		Y
nemestrina	8.8, 10	9.5	None/ vocalize at very end	2.5	13, 12	179, 96	>*mulatta*	N[h]	Y
nigra	1		Hoarse barks		12.2			Y	Y[i]
sylvanus	1	8.7		1.4	9.1	—		?	?

Notes: Note also the possible correlation between special male facial expressions during copulation and females looking back at the male (fig. 1.9). Some variability among studies may stem from differences among races, and differences in the estrous conditions of different females (Zumpe and Michael 1983). References for the different species are as follows: *arctoides*—Blurton-Jones and Trollope 1968; Michael and Zumpe 1971; Kanagawa et al. 1972; Mitchell 1979; Estep et al. 1986; Bruce and Estep 1992; also Tokuda 1961; Hanby et al. 1971; and Enotomo 1974 in Kumar and Kurup 1985; *fascicularis* (= *irus*)—de Benedictis 1973; Kanagawa et al. 1972; Kurland 1973; Jewett and Dukelow 1978; Deputte and Goustard 1980; Shively et al. 1982; Zumpe and Michael 1983; *fuscata*—Michael and Zumpe 1971; *mulatta*—Michael and Zumpe 1971; Nadler and Rosenblum 1973; Mitchell 1979; Shively et al. 1982; Zumpe and Michael 1983; Missakian et al. 1969; Michael et al. 1967 in Kumar and Kurup 1985; *nemestrina*—Tokuda et al. 1968; Nadler and Rosenblum 1973; Mitchell 1979; *nigra*—Dixson 1977 (mostly from a single male); *radiata*—Michael and Zumpe 1971; Nadler and Rosenblum 1973; Glick 1980; Shively et al. 1982; *silenus*—Kumar and Kurup 1985 (ejaculation was cryptic so data are tentative); and *sylvanus*—Taub 1982. There were strong intraspecific differences among individual males in *nemestrina* and *mulatta*; see Nadler and Rosenblum 1973; Michael and Saayman 1967. Probable relationships, based on electrophoretic data, are as follows:

Table 5.1 (*cont.*)

(((*radiata*(*arctoides, mulatta*)) *fuscata, fascicularis* (*silenus* (*nemestrina, nigra*))) *sylvanus*) (Cronin et al. 1980); the species are arranged in this order in the table.
I = average number of mounts/ejaculation; II = average duration of a mount (sec); III = male vocal or facial signals?; IV = rate of thrusting/sec; V = number of thrusts/mount; VI = interval between mounts (sec); VII = post-ejaculatory interval (min); VIII = female looks back at male during copulation; IX = female reaches back to grasp male (usually on the leg) during copulation.

[a] Occurs at moment of ejaculation.
[b] Male typically remains quiet, but still intromitted for approximately 1.5–2 min after ejaculation.
[c] Shorter duration if another male mates with the female afterwards.
[d] Begins early in mount.
[e] Modal number, not mean, for first and second ejaculation with the same female.
[f] Values for single- and multiple-mount copulations, respectively.
[g] Values for first and second ejaculation with the same female.
[h] Photo shows female looking straight forward (Nadler and Rosenblum 1973).
[i] Sometimes.

B. Qualitative differences in copulation behavior in eleven species of *Peromyscus* mice

Species group	*Species*	*Preejac. Intromiss.*	*Intravag. Thrusting*	*Stereotyped Dismount*	*Postejac. Intromiss.*
mexicanus					
	mexicanus	N	N	N	N
	melanocarpus	N	N	N	N
	yucatanicus	N	N	N	N
melanophrys					
	melanophrys	N	N	N	N
crinitus					
	critinus	Y	N	Y	N
leucopus					
	leucopus	Y	N	Y	Y
	gossypinus	Y	N	Y	Y
maniculatus					
	maniculatus	Y	N	Y	Y
	polionotus	Y	N	Y	Y
	eremicus	Y	Y	Y	Y
	californicus	Y	Y	Y	Y

Notes: All but the last two species are in the subgenus *Peromyscus*. All data are from Langtimm and Dewsbury 1991. Y = yes; N = no.

sit" in which vaginal intromission is maintained (Estep et al. 1986). There are also dramatic differences in the morphology of the penis between two of these species (Dixson 1987). In at least four of the macaque species, sexual selection by cryptic female choice has ample opportunity to work, since almost all conceptions occurred during estrus cycles during which the female mated with more than one male (Kuester et al. 1992; Menard et al. 1992; Altmann et al. 1988).

In *M. mulatta* subordinate males both in captivity and in the field showed a highly significant decrease in the number of preejaculatory intromissions when compared with dominant males (probably an adjustment to reduce the chances of being interrupted by a dominant male before ejaculation; Curie-Cohen et al. 1983; Manson 1994). Postejaculatory pair-sits of subordinate *M. arctoides* were also more likely to be interrupted by other dominant males (Estep et al. 1986).

A similar pattern of intrageneric diversity occurs in the copulation behavior of *Peromyscus* mice, even when only qualitative rather than quantitative data are compared (table 5.1). Similar rapid divergence in male mating behavior occurs in two subspecies of another rodent, the bank vole *Clethrionomys glarcolus* (McGill 1977). One (*C. g. britannicus*) occurs throughout the mainland of Great Britain. The other (*C. g. skomerensis*) occurs only on the island of Skomer, 1.5 miles off the Welsh coast. The two populations have been isolated for only about 7000–9000 years. Males of the mainland *britannicus* form exhibited more intromissions and mounts before ejaculation and shorter intervals between intromissions, and they resumed copulation sooner after ejaculation than did *skomerensis*.

A final example, in which both female responses and male stimulation may be species-specific, comes from voles in the genus *Microtus*. The copulation behavior of male montane voles (*M. montanus*) differs from that of male prairie voles (*M. ochrogaster*) with respect to average intromissions per ejaculation (28.6 vs. 13.8), average number of thrusts per intromission (2.2 vs. 3.8), and average interval between intromissions (14.9 sec vs. 24.0 sec) (Kenney et al. 1978). When female prairie voles were allowed to receive twenty-five intromissions from either a conspecific prairie vole male or a cross-specific montane vole male, those paired with conspecifics were more likely to ovulate (Kenney et al. 1978). Similar cross-specific pairings with males of a third species, *M. pennsylvanicus*, resulted in even lower rates of ovulation. Possible effects of other stimuli such as pheromones or ultrasonic vocalizations on female responses were not considered, however, so it is not certain that differences in male copulation behavior produced the differences in female response (Kenney et al. 1978).

When these differences were first noted, Dewsbury proposed that interspecific differences in mammals may be adaptations to different habitats (Dewsbury 1975) or species isolating mechanisms (Dewsbury 1972). But the reasons why larger or smaller numbers of intromissions or of thrusts would be more or less advantageous in one environment or another were not apparent, and there seem to be no simple correlations between copulation and habitat or feeding habits in rodents (Dewsbury 1975). The likelihood also seems remote that the species-discrimination cues acting prior to copulation fail often enough to make details of male copulation behavior selectively important in this context. In addition, some effects of species differences on female responses in *Microtus* are too subtle to be of likely importance in preventing gamete wastage by females (Kenney et al. 1978). Dewsbury has more recently emphasized possible roles in intraspecific communication (of which male attempts to influence cryptic female choice would be one type; Dewsbury 1988).

I know of only one comparative study of copulation behavior that failed to find divergence among congeneric species (Gering 1953 on three species of *Agelenopsis* spiders). Even here one of the species may have differed (there were no statistical tests) in the frequency of pulsation of genitalic membranes (hematodochae) during the period of maximum penetration, and spiders in another genus in the same family (*Hololena*) may differ in at least the duration of hematodochal expansion (Fraser 1987). The significance of the pulsations is uncertain; they may be associated with flow of semen from the tip of the male organ (see, however, Lopez 1987), and/or they may have resulted in further extensions the female's already stretched bursa.

A further aspect of copulation that could alter the stimuli received by females is the overall duration of copulation. Sizable differences in this variable also occur in closely related species (e.g., Schaller et al. 1985 on giant pandas and red pandas; Dewsbury and Pierce 1989 on several primates; Grant 1983 on *Drosophila*; Cowan 1986 on *Ancistrocerus* wasps; Eberhard and Kariko, in press, on *Macrohaltica* beetles; fig. 5.12).

Finally, it is worth noting that stimulation from other aspects of copulation in addition to genital movements per se can have courtship effects. For instance, ovulation is induced in the mink *Mustela vison* by the male mounting and riding the female; intromission is not necessary (Hansson 1947).

Multiple Ejaculations

If copulation per se constitutes courtship, it may help explain why in many species males mate and ejaculate repeatedly instead of mating only once and passing a larger ejaculate (Lifjeld and Robertson 1992 on tree swallows; Davies 1992 on passerines; Hunter et al. 1993 on birds in general; Michael and Zumpe 1971 on primates; Dewsbury 1988 on rodents; Sharman et al. 1966 on marsupials; Ewer 1968 on other mammals; Blest and Pomeroy 1978; M. Robinson 1982 and Austad 1984 on spiders; Gregory 1965, Pickford and Gillott 1971, and Cheeseman and Gillott 1989 on grasshoppers; Dickinson, in press, on plant beetles; Loher 1981 on a cricket; Parker 1984 on other groups of insects; Haacker 1969a on a millipede; A. Cohen and Morin 1990 on ostracod crustaceans; fig. 5.14). Several hypotheses have been proposed, including spreading out the times at which the male introduces sperm into the female throughout her fertile period (thus compensating for possible sperm mortality; Parker 1984), increasing

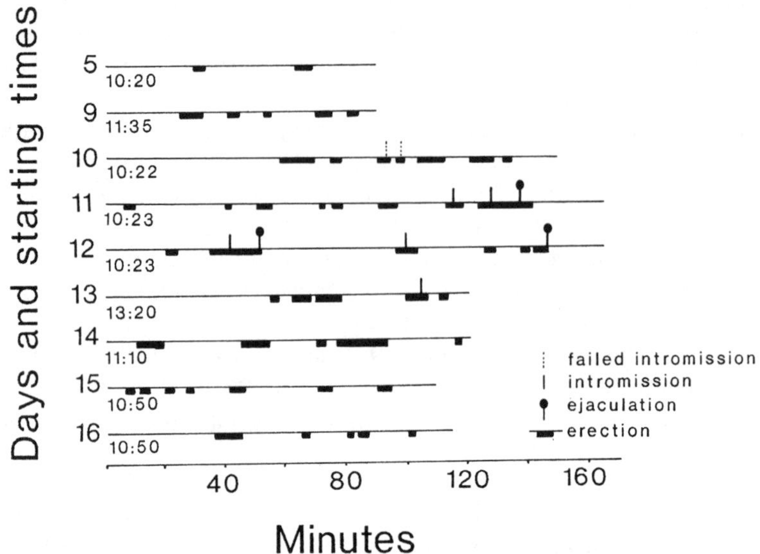

Figure 5.14 Record of sexual activities of a male elephant (*Elephas maximus*) during the estrus period of a female, showing both intromissions without ejaculations (days 11–13), and multiple ejaculations (day 12). (After Eisenberg et al. 1971)

the amount of sperm from previous males that is removed, or increasing the amount of sperm transferred (Thomas and Zeh 1984; Otronen 1994) because either limited male supplies or limited female receiving structures have a bottleneck effect on the sizes of single ejaculates.

Compensation for sperm mortality may be important in some animals such as gelada baboons (Dunbar 1978), monogamous birds (Birkhead and Møller 1992; Hunter et al. 1993), and termites (Imms 1964), in which females mate repeatedly with only a single male. It seems especially likely in the termites, where the female's fertile period lasts up to several years (Imms 1964).

In some species females may use copulation to achieve other social or communicatory ends to influence males with which they will have to interact in the future, for example to form, strengthen, or defend a pair bond (Petrie 1992), to induce a non-pair male to aid in brood care (Davies 1985, 1992), to obtain access to resources controlled by a male (Hunter et al. 1993), or to induce a male to refrain from killing her offspring (Hrdy 1977). In such cases, repeat copulations may have the effect of increasing the effectiveness of the communication.

However, repeated copulations occur in many groups that have effective long-term sperm storage, no postcopulatory male-female bonds, and no clearly limited female fertile period. Multiple ejaculations in many species are relatively closely spaced, the female gains no resources from the male, and then the male and female separate (references above; also Bunnell et al. 1977 and Dewsbury 1982 on rodents; Thornhill and Alcock 1983 and Eberhard 1991a, 1994 on insects; Ikeda 1974 on *Drosophila mercatorum*; Friedel and Gillott 1977 and Whitman and Loher 1984 on grasshoppers; Thomas and Zeh 1984 on a pseudoscorpion; Diesel 1991 on brachyuran crabs; Wells and Wells 1977 on octopod mollusks). These data argue against both the sperm mortality and communication hypotheses as general explanations.

With regard to an increase in number of sperm transferred, it is likely that larger numbers of sperm often confer a competitive advantage. The question remains, however, why several small ejaculates instead of one larger one are transferred. In a few cases small female openings or a relatively small volume of the cavity into which the male ejaculates could oblige males to transfer smaller volumes of ejaculate. Thus in brachyuran crabs, small spermatophore size may be necessary for successful transfer through small female openings and/or ducts (Diesel 1991). In *Macaca* monkeys the female's vagina appears to be filled to overflowing with semen after each ejaculation (J. and S. Manson, pers. comm.). Even in these cases it could be argued that females are probably capable of transporting ejaculates up their reproductive tracts, so that "overflows" are not really inevitable (see, for example, Baker and Bellis 1993a on humans). The possibility of limited female acceptance of sperm can explain the occurrence of undersized ejaculates (see chap. 8.1).

But there are many other species in which the male transfers rapid multiple ejaculates, where sperm removal seems physically impossible, and where physical constraints on ejaculate volume do not occur. For instance, the reproductive tract of the female hamster *Mesocricetus auratus* does not have a small, rigidly fixed volume, and larger ejaculates would be advantageous since dilution effects of competing sperm are probably important in determining paternity (Oglesby et al. 1981; Huck et al. 1985). A larger, single ejaculate would surely be a more efficient, less dangerous, and more certain tactic. It would both reduce additional effort by the male and avoid the danger of female rejection of subsequent mating attempts, which sometimes occurs in this species (Huck and Lisk 1986). Nevertheless, a male hamster typically performs about 10–15 "ejaculatory series," each of which includes several discrete intromissions (Bunnell et al. 1977). Similar examples from distantly related groups that apparently lack physical constraints in the female and in which sperm removal seems improbable, but multiple ejaculations nevertheless occur, include *Cylindrocorpus* and *Rhabditis* nematodes (Chin and Taylor 1961; Rehfeld and Sudhaus 1985), the plant beetle *Labidomera clivicollis* (Dickinson 1986), *Drosophila mercatorum* (Ikeda 1974; the second copulation begins before sperm begin to enter the female storage organs), the elephant *Elephas maximus* (fig. 5.14; Eisenberg et al. 1971), and several rodents (Dewsbury 1972). A particularly dramatic example is the barnacle *Balanus*, where an "enormous excess of spermatozoa over the number required to fertilize the ova" is transferred in the first copulation (Barnes et al. 1977), and copulation is a dangerous process because the penis is exposed and defenseless against predators (outside the breeding season the penis is often lost or regresses in size; Barnes 1992). The male nevertheless usually mates 6–8 times in succession with the same female over the course of 30–60 min, eventually transferring about 50% of his body weight to her (Barnes et al. 1977).

If copulatory behavior and/or seminal products (see chap. 6) function as courtship to induce the female to use the male's gametes, then repeated copulations may represent male courtship to influence cryptic female choice. Such an effect has been demonstrated in a few species. Stimulation of ovulation by repeated copulation may be common in mammals in which cyclic corpora lutea produce no or only insignificant amounts of progesterone for maintenance of gestation (Jöchle 1975; see also Milligan 1982; chap. 4.1). The ability of females to "sum up" copulatory stimuli with respect to triggering ovulation or the luteal cycle has been demonstrated both in rats (Edmonds et al. 1972) and tsetse flies (Saunders and Dodd 1972; Chaudhury and Dhadiala 1976).

Copulation Duration

A second mystery that may be explained in some species by copulation functioning as courtship is the "overly long" duration of copulation in some species. Suter and Parkhill (1990) list seven possible explanations for lengthy copulations,[2] two of which involve cryptic female choice mechanisms. The simple idea that long copulations are somehow necessary to transfer larger amounts of sperm and seminal products does not fit many observations. For instance, comparisons between species of *Drosophila* show no correlation between the amount of material transferred and the duration of copulation (Pitnick et al. 1991). Thus, *D. pachea* has a relatively long copulation (41 min) but transfers the smallest ejaculate of any *Drosophila* species known; *D. mojavensis* transfers about fourteen times as much ejaculate in only 2.5 min (Pitnick et al. 1991).

Many factors, such as increased danger from predators, time lost from other activities such as feeding, oviposition, territorial defense, and search for other mates, and the risk of interruption before ejaculate transfer was completed would probably often favor brief mating, both for males and for females. Despite these multiple, and probably frequently important disadvantages, however, copulations in many species are relatively long. One widely cited possible advantage is guarding the female from the attentions of other males (e.g., Alcock 1994). In some species behavioral details seem to favor this interpretation. For instance, copulations in the bug *Nezara viridula* and the millipede *Alloporus uncinatus* are longer when other males are nearby; McLain 1980; Barnett and Telford 1994). As noted in chapter 1.9, however, the same changes in copulation times would also be predicted if longer copulations serve to influence cryptic female choice mechanisms (for instance, to reduce the delay to oviposition, or increase the female's resistance to further mating).

A specific test of the significance of copulation duration by taking into account seven different hypotheses was performed by Suter and Parkhill (1990) on the spider *Frontinella pyramitela*. They were able to eliminate six hypotheses, including transfer of oviposition stimulants or receptivity inhibitors, and defense of the female from other males or from predators, and support a seventh—increased transfer of a male substance that caused the female's offspring to be larger. This trait might or might not be under sexual selection by cryptic female choice (see chap. 6.7). A further cryptic female choice hypothesis, which was not considered, is that longer copulations resulted in greater stimulation of the female to receive or use the sperm.

A related question involves the duration of sperm transfer. In some species it is long and drawn out, and longer copulations result in larger numbers of sperm being transferred. Transfer is relatively slow and gradual, for instance, in *Rhabditis* nematodes (Rehfeld and Sudhaus 1985), the dog *Canis familiaris* (Ewer 1973), the beetle *Diabrotica virgifera* (Lew and Ball 1980), the zorapteran *Zorotypus barberi* (Choe 1995), the cricket *Gryllus bimaculatus* (Simmons 1986), the melon fly *Bactrocera cucurbitae* (Yamagishi and Tsubaki 1990), and the scorpion fly *Hylobittacus apicalis* (Thornhill 1984; fig. 5.20). In the weevil *Hypera postica*, sperm transfer has not even begun after 30 min and occurs only gradually over the rest of the approximately 80 min copulation (Le Cato and Pienkowski 1972).

In many other species, in contrast, transfer is quite rapid, occurring over the span of a few seconds even though the copulation itself may last many minutes or even hours, e.g., the last 30 sec of a >60 min copulation in *Glossina* flies (Wall and Langley 1993), and the first minutes of copulations that last up to many hours in the spider *Phidippus johnsoni* and the millipede *Alloporus uncinatus* (Jackson 1980; Barnett and Telford 1994). A slow, drawn-out transfer would appear to be disadvantageous. The more rapidly the male can make the transfer, the better. This is particularly true for some of the species just mentioned, in which females sometimes interrupt transfer before it is complete (e.g., *Hylobittacus apicalis*, *Gryllus bimaculatus*, *Zorotypus barberi*, possibly *Coproica* spp.; Thornhill 1984; Simmons 1986; Choe 1995; Lachmann 1994; see also chaps. 3.8, 4.4).

Some cases of slow sperm transfer may constitute examples of female-imposed "rules of the game" that have evolved to select for other male traits. For instance, in the scorpion fly *Hylobittacus apicalis* the number of sperm found in the spermatheca is very tightly correlated with the duration of copulation (fig. 5.20; Thornhill 1984). One possible reason for the tightness of the correlation is that the variance in male ability to transfer sperm is very small. In other words, all males may have more than enough sperm ready, and the reason for the slow transfer may be a female-imposed barrier that males are unable to overcome, allowing the female to favor males on the basis of the length of time males are able to induce them to copulate (in this case, by giving them larger nuptial meals). Other, perhaps more typical cases in which there is more variation in sperm transfer rates (e.g., in melon flies; fig. 5.20) may represent cases in which some males but not others are able at least partially to overcome female-imposed barriers, and/or that females are judging males on traits other than copulation duration and raising or lowering barriers accordingly, or males differ widely in sperm supply.

A note of caution is needed to end this section. Male copulation behavior is variable, and both learning and activities just prior to copulation can influence this variation (Diakow 1973). Females as well as males are involved in copulation, and some patterns of male behavior are actually due to the behavior of the female (e.g., Diakow 1973 on cats). A clear illustration comes from data on the rat *Rattus norvegicus*. When domestic strains of rats were compared with wild rats, the domestic rats performed fewer intromissions prior to ejaculation (av. 5–6 vs. 13–14), and the interval between intromissions was shorter (about 3 min vs. 4 min). As shown in fig. 5.15, when domestic males were paired with wild females, they behaved more like wild males. In the inverse cross, wild males behaved like domestic males. Differences in female solicitation behavior may at least partly explain these results (McClintock 1984). Solicitation, in turn, is affected by the size of the enclosure in which mating takes place and the female's ability to temporarily escape from the male (Gilman et al. 1979; McClintock 1984). At a finer level, a female rat apparently often moves her perineal region in the direction of the thrusting penis (Morali and Beyer 1992). By doing so, the female may thus modify the amount of stimulation she receives from any given thrust. Females also modify some aspects of male copulation behavior in different races of the bank vole *Clethrionomys glareolus* (McGill 1977).

On the other hand, some differences in male copulation behavior are probably independent of the female. Some aspects of male bank vole copulatory behavior did not vary with different races of female (McGill 1977). Reciprocal crosses among inbred lines of *Drosophila melanogaster* with different mating durations showed that the male was responsible for most of the variation (MacBean and Parsons 1966). Many studies of changes or differences in male copulatory behavior have not taken into account, however, the possibility that differences are due to differences in female behavior.

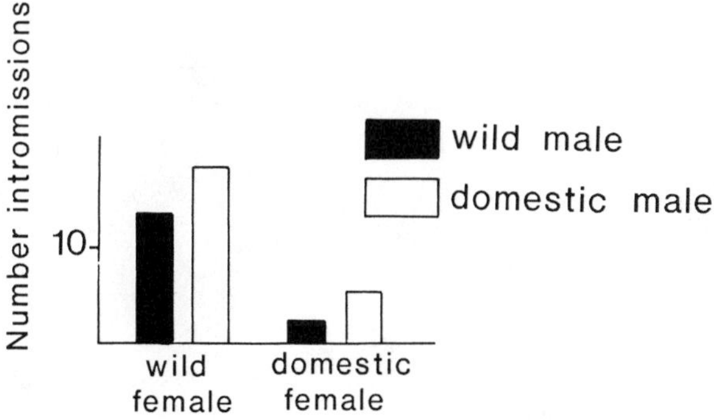

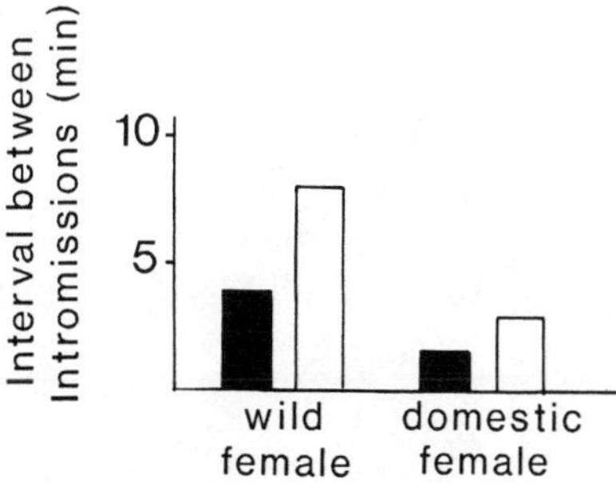

Figure 5.15 A male's copulation behavior can be influenced by his mate's behavior. Males of both domesticated and wild strains of the rat *Rattus norvegicus* made more than twice as many intromissions before ejaculating and waited longer between intromissions when mating with wild females than with domestic females. In some rodents the male adjusts his copulation behavior according to the female's condition in ways that increase his chances of siring offspring. (After McClintock 1984)

5.3 Other Evidence of Copulation as Courtship

Clasping (Genitalic and Nongenitalic) as Courtship

In many animals the male clasps the female just before or during the first stages of copulation. Often the male uses structures that are specialized for this function, and often these are species-specific in form (e.g., the bursae of nematodes, the antennae of fairy shrimp, calanoid copepods (fig. 5.16A) and sminthurid collembolans, the wings of scorpion flies, the legs of diplurid spiders, sepsid flies, and nomiine bees (Blades 1977; Blades and Youngbluth 1979; Coyle 1985b, 1986; Eberhard 1985; Coyle and O'Shields 1990; Wcislo et al. 1992; Wcislo and Buchmann, in press; Wcislo and Engels, unpub.). Clasping often continues throughout copulation.

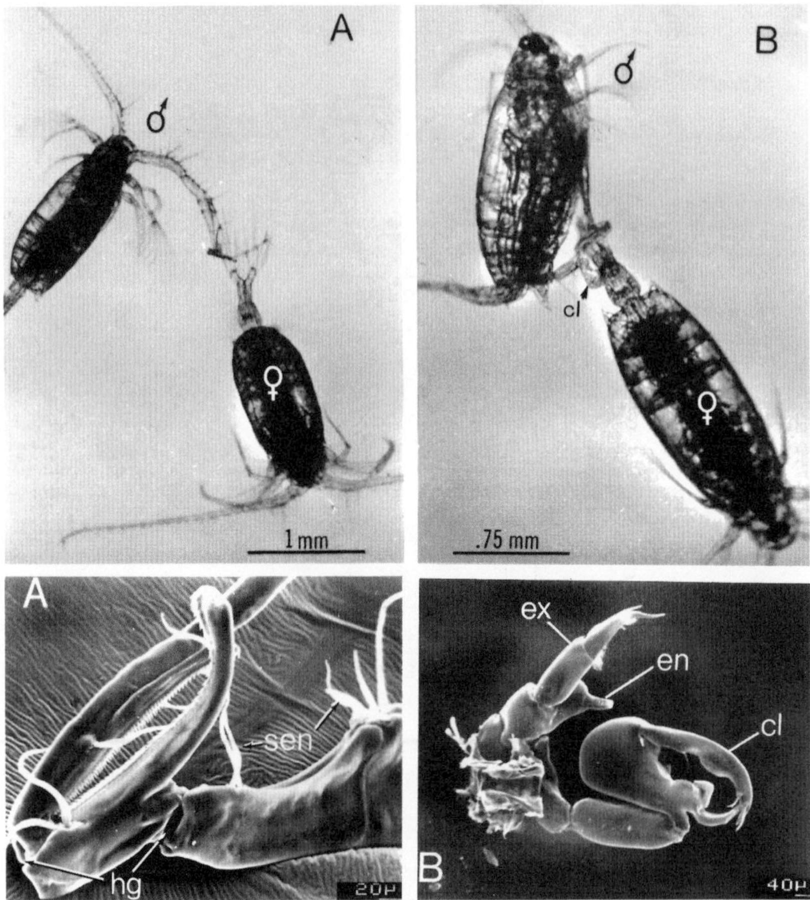

Figure 5.16 Two different elaborate clasping organs occur on males of the copepod *Labidocera aestiva*. Both the male's right antennule (A) and the distal portion of his right fifth leg (B) are species-specific in form (the right leg, which clamps the female, is shown along with the left leg at *lower right*). The male first grasps the female on her caudal setae with his right antennule, then attempts to grasp her caudal furcae with his right fifth leg. If successful (*upper right*), he releases his hold with his antennule, rhythmically strokes her abdomen and probes around the edges of her gonopore with his left fifth leg, and then finally attempts to attach a spermatophore to her. (From Blades and Youngbluth 1979; courtesy of P. Blades-Eckelbarger)

Several types of data indicate that male clasping of the female sometimes functions as courtship. Although most of these studies involved precopulatory discrimination (e.g., clasping induced the female to mate), they are nevertheless probably also pertinent to discussions of cryptic female choice. If a given type of stimulus sometimes has the effect of inducing

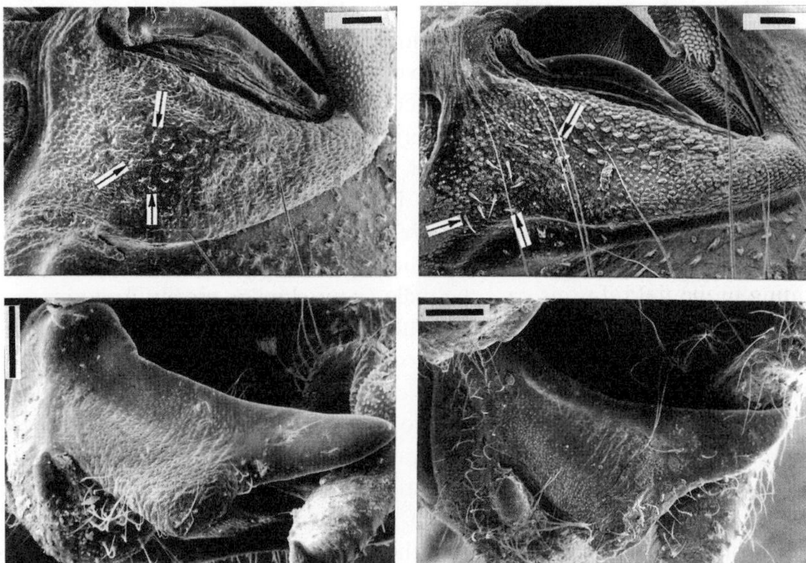

Figure 5.17 Male clasping organs (*bottom*) and the portions of the female which they grasp (*top*) in two species of damselflies in the genus *Coenagrion*, showing that the females are well endowed with peglike sensory structures (arrows) in the area contacted by the species-specific male structures. Thus females can probably sense the male structures, increasing the plausibility of the idea that the clasping organs are under selection by female choice. (From Battin 1993b)

favorable responses in females *prior* to copulation, it is reasonable to suppose that it may also sometimes trigger responses when received during or after copulation (recall that 40% of the 272 copulatory courtship behavior patterns in the two surveys of copulatory courtship also occurred prior to genitalic coupling; see sec. 5.1). The data linking clasping with courtship are relatively strong in some groups, and species-specificity in clasping structures is extremely widespread. Therefore it is important to take clasping into consideration, even though as yet the connection with cryptic female choice is indirect.

Experiments with dragonflies and damselflies (Krieger and Krieger-Loibl 1958; Loibl 1958; Robertson and Paterson 1982; Tennessen in Tennessen 1982), and fairy shrimp (Belk 1984) in which the stimulatory properties of both genitalic and nongenitalic male clasping organs were altered (by direct modification) resulted in reduction of the female's tendency to allow intromission as in normal pairing. In other odonates, special mechanoreceptors occur in just the area of the female's body that is clasped by the male, giving additional, indirect evidence favoring the same conclusion (fig. 5.17; Robertson and Paterson 1982; Battin 1993a).

Males of the water strider *Gerris odontogaster* clasp the female prior to and during copulation using ventral processes on the male abdomen (fig. 2.8). The length of the male's ventral process influenced how much female resistance the mounted male could endure without being dislodged (Arnqvist 1989). Males with longer processes were more likely to be found riding females in the field (Arnqvist 1989; since copulation occurs early after the male mounts, these males had presumably mated with the females). There was no significant correlation between the length of the processes and male size, and male size did not correlate with the frequency of riding in the field. In this case the male "courtship" is his ability to physically overcome or endure the forceful resistance that females usually put up soon after the male mounts. This resistance is part of "the rules of the game" imposed by females (in this case, a precopulatory rule), because females are more subject to predation when they are being ridden by a male (Arnqvist 1989, in press). Natural selection on females to resist being mounted apparently results in selection favoring longer abdominal processes in the male. Sexual selection favors larger processes, but natural selection acts against them because they make moulting to the adult form more likely to fail (Arnqvist 1994).

Thornhill (1984) and Thornhill and Sauer (1991) discussed the function of species-specific clasping organs in another group, the *Panorpa* scorpion flies. After presenting reasons to reject the idea that the claspers function to prevent takeovers of the female by other males, as chemical courtship organs, or as species-recognition devices, they concluded that the claspers function to force copulations on unwilling females. This idea does not easily account for the combination of the relatively uniform morphology of the female structures that are seized, and the complexity and species-specific differences in the male claspers. A more likely modification is that the male clasping organs combine both force *and* persuasion to increase the male's chances of inseminating recalcitrant females via mechanical stimulation while clamping the female (the female inspection of male scorpion-flies prior to clamping is an additional possible source of stimuli). Thornhill and Sauer (1991) rejected this hypothesis, but on the basis of a sharp underestimate of the diversity of possible mechanisms of cryptic female choice.

In general, when male structures (such as legs, wings, antennae) are specialized (morphologically modified) for a clamping function, they tend to evolve "unnecessarily" elaborate forms, which are frequently species-specific in form (e.g., fig. 5.16). Similar elaboration sometimes also occurs in male organs that contact the female during copulation but do not clasp her, e.g., the setae of the male apical terga in the zorapteran *Zorotypus barberi* (Choe, 1995); the chelicerae of pholcid spiders that press on the

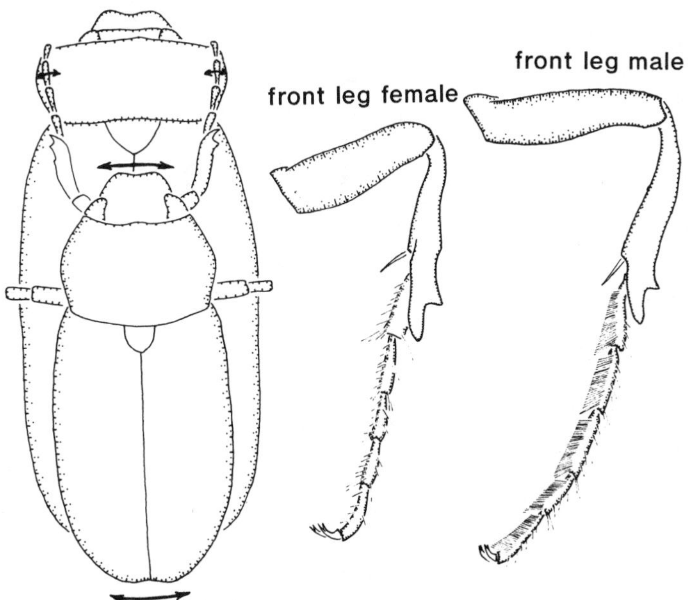

Figure 5.18 Secondary sexual characters which have diverged relatively rapidly are used to stimulate a female during copulation. The male of the beetle *Phyllophaga aequata* waggles from side to side during copulation (arrows), thus rubbing the female with his long front legs and the dense pads of setae on their ventral surfaces. These secondary sexual characters are present only in the subgenus to which this species belongs, and waggling during copulation has also only been observed in this subgenus. (After Eberhard 1993b)

female's abdomen (Huber 1994a; Uhl et al. 1994; see Eberhard 1985 for an extensive though very incomplete list). The important point is that the species-specific forms of nongenitalic male organs specialized to contact females in sexual contexts often show the pattern of rapid evolutionary divergence that is common in characters under sexual selection. This suggests that they probably often function as courtship devices.

Since contact of this sort usually continues throughout copulation, it is reasonable to suppose that in at least some cases such structures continue to have a courtship function during copulation. In fact, in some species the male structures move actively against the female *during* copulation in ways that suggest just such a function (Eberhard 1985; also the modified abdominal sternites of male *Phyllophaga* beetles, and the modified front tarsi of the *aequata* group of *Phyllophaga* beetles: fig. 5.18; Eberhard 1993b); the setose ventral surfaces of male *Macrodactylus* rose chafer beetles (fig. 5.19; Eberhard 1993a); the setose ventral surfaces of the legs of male

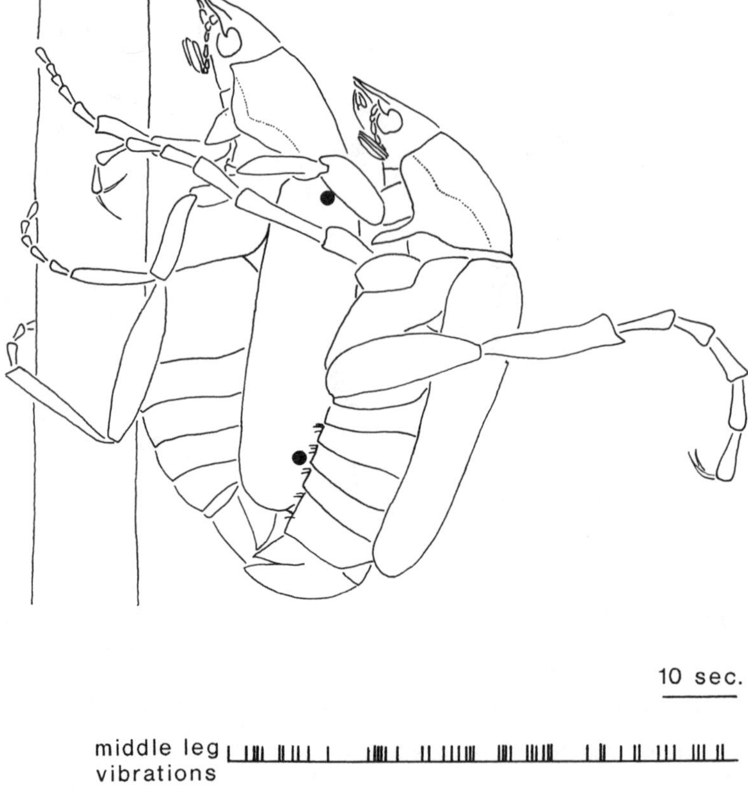

Figure 5.19 A barrage of stimuli, in "three-part harmony," occurs during copulation in the beetle *Macrodactylus costulatus*. Each vertical bar in the graphs represents a male courtship movement. In addition to the three types of stimuli in the graphs, species-specific structures marked with dots in the drawing (the male's front legs, spines on his ventral surface) also hold and/or rub against the female at the same time, possibly providing additional stimuli. Even further simultaneous stimulation may come from movements of other parts of the male genitalia within the female. (After Eberhard 1993a)

Metamasius sp. beetles (Eberhard 1994); and some modifications of the male cerci of earwigs (Briceño and Eberhard 1995). Nongenitalic male contact structures may also function during post-copulatory contact (e.g., the elongate front legs and bristles on the tarsus of the cerambycid beetle *Monochamus scutellatus* in which females in contact with males laid eggs about 1.5 times more rapidly than those that were alone; Hughes 1981).

Males also clamp females with their genitalia. Not uncommonly, physical coupling between male and female is a multistage process, with early stages involving clasping and clamping and only the last stages involving penetration to the site in the female where the male must deposit his semen (chap. 3.4). Male genitalic claspers are often elaborate and species-specific in form, and it is probable that they are also under selection by cryptic female choice (Eberhard 1985; see also chap. 7.2). In some species careful observations show that conspecific males sometimes must repeatedly attempt to achieve coupling. For instance, reports of repeated failures to lock or engage the female's genitalia ("flubs") are reported for several spiders: *Lepthyphantes leprosus* (van Helsdingen 1965); *Mynoglenes* spp. (Blest and Pomeroy 1978); *Hololena adnexa* (Fraser 1987); *Nephila clavipes* (Christenson and Cohn 1988); *Neriene litigiosa* (Watson 1991); *Nesticus cellulanus* (Huber 1993); *Anyphaena accentuata* (Huber, in press); *Diplothyron* (?) sp. (A. Berghammer, unpub.); see also Huber 1995 on other species. In one of these species (*N. litigiosa*), higher frequencies of "flubs" are part of the lower male "mating vigor" that correlates with lower sperm precedence values (Watson 1991). A stimulatory function for male genitalic clamping structures is also suggested by the diverse, sustained, and rhythmic squeezing movements of the male surstyli of some sepsid flies (Eberhard and Pereira, in press).

Copulation outside Female Fertile Periods

If copulation behavior functions as courtship, it might be expected to occur in some species in contexts in which the original function (sperm transfer to fertilize eggs) is not feasible but male reproductive success is nevertheless affected. Females might also use copulations to court males. Such nonreproductive copulations occur in a variety of bird species, and in at least some there is circumstantial evidence that they have an effect on forming or reinforcing pair bonds. Although these nonreproductive copulations do not necessarily involve cryptic female choice, they are pertinent to the discussion here because they demonstrate that, in still another set of species, copulation can serve a communicatory, courtship function.

In several birds, copulation occurs when the female is infertile. For instance, in kestrels (*Falco naumanni*) copulation frequency has two peaks: about 65 days and about 5 days before the female oviposits (Negro et al. 1992). Since sperm are thought to survive for only about 7 days in the female's reproductive tract, copulations in the first peak almost certainly do not lead to fertilizations. Similar copulations outside the females' probable fertile periods, or when the birds' gonads are not producing gametes, occur in at least six other raptors, six species of ducks, seven seabirds, moorhens, little owls, house sparrows, white storks, razorbills, spotted sandpipers, oystercatchers, and starlings (Birkhead et al. 1987; Wagner 1991; Negro et al. 1992; Birkhead and Møller 1992; Tortosa and Redondo 1992; Heg et al. 1993). Copulations also occur in apparently nonreproductive contexts in the Arabian babbler *Turdoides squamiceps* (Zahavi 1988).

An extreme case occurs in the hoatzin *Opisthocomus hoazin*, which has two types of copulation (Strahl 1988). "Reproductive" copulation occurred during nest building and within several days of oviposition. "Display" copulations, which were shorter and included a modified male copulation call and usually no cloacal contact, were associated with cooperative territory defense.

One indication that early copulations in birds function as courtship is that copulations were more frequent in kittiwake (*Rissa tridactyla*) pairs in which birds had changed partners from the year before than in pairs in which the same birds had been paired the previous year (Chardine 1987). Frequencies of copulation attempts are especially high immediately after initial pair formation in ospreys *Pandion haliaetus* (Birkhead and Lessells 1988), and very early in the breeding cycle of goshawks *Accipiter gentilis* (Møller 1987a).

An especially well-studied example is the oystercatcher *Haematopus ostralegus*. The timing of extra-pair copulations, the sex which initiated copulation attempts, and paternity patterns furnished convincing evidence that some copulations serve to establish and reinforce pair bonds rather than to reproduce, and that females used copulations to court males (Heg et al. 1993). There was no apparent female bias in paternity assignment (1 of 65 chicks were fathered by an extra-pair male, as compared to a predicted 2 of 65 on the basis of the overall frequency of extra-pair copulations [3.3%] during the 30 days prior to oviposition). Individuals that engage in nonreproductive copulations tend to form reproductive pairs in the following breeding season in spotted sandpipers *Actitus macularia* (Colwell and Oring 1989) and perhaps the razorbill *Alca torda* (Wagner 1991).

Particularly clear cases of probable female manipulation of males via copulations occur in the dunnock *Prunella modularis* and the alpine accentor *P. collaris*. Females that are associated with more than one male actively solicit copulations, especially with the subordinate male, and the dunnock females continue to solicit after the final egg of the clutch has been laid (Davies 1985; Davies et al. 1995). It appears that such copulations serve to increase the chances that her offspring will be cared for by all the males (Davies 1992; Hartley et al. 1995). An almost identical situation occurs in the barbary macaque *Macaca sylvanus*. Female behavior ensures that nearly all the mature males in her troop copulate with her during each estrus, and males provide extensive and possibly very important paternal care (Taub 1980). The Galapagos hawk *Buteo galapagensis* is still another species with the same combination of multiple male care and polyandry that results in equal chances for males to sire offspring (Faaborg et al. 1995).

In mammals, copulation outside female fertile periods also occurs in several primates, including humans, chimpanzees *Pan troglodytus*, pygmy chimpanzees *P. paniscus*, the lowland gorilla *Gorilla gorilla gorilla*, *Macaca* monkeys, patas monkeys *Erythrocebus patas*, and vervet monkeys *Cercopithecus aethiops* (J. Kaufmann 1965; Dixson 1977; Mitchell 1979; Goodall 1986; Chism and Rowell 1986). Indirect evidence in the chimpanzee suggests that copulation may have a pair-reinforcing function. At least 16 of 20 copulations with females completely lacking sexual swellings occurred during consortships (Goodall 1986), while only 2% of 1475 copulations (in a separate study) were during consortships (Tutin 1979). Oxytocin, which is often released in mammals as a result of genitalic stimulation, has the effect of promoting social bonds in some rodents (Carter and Getz 1993).

Female lions have also been observed to copulate when pregnant and when lactating (Schaller 1972; Smuts et al. 1978). Smuts and colleagues speculated that such copulations function to reinforce social bonds. Pairs of the porcupine *Hystrix indica* copulate repeatedly every night during estrus, pregnancy, and lactation. This high frequency of copulation may be associated with pair-bonding, as this species is also unusual in being highly monogamous (Sever and Mendelssohn 1988). Copulation also occurs outside the breeding season in the llama *Lama glama* (England et al. 1969). Mice *Mus musculus* and prairie voles *Microtus ochrogaster* also sometimes copulate while pregnant (Nalbandov 1964 in Land and McGill 1967; Carter and Getz 1993). A pair-reinforcing function in the vole was suggested by the fact that females that were already paired with males copulated for relatively shorter times (sometimes only a few minutes), compared with the excessively long period of mating (30–40 hr) by a female in her first estrus (the female typically ovulates at about 12 hr, and successful pregnancy can occur shortly thereafter). This species has clear male-female pairs, while in the nonpairing species of *Microtus* the mating period lasts for a few hours only (Carter and Getz 1993). Another, very different species which both copulates outside the breeding season and lives in male-female pairs is the cockroach *Cryptocercus punctulatus* (Nalepa 1988).

There are other possible explanations of copulations outside a female's fertile period. Thornhill (1984) suggested that in some social species copulation outside the female's fertile period may function to disguise the female's fertility from other males (to the advantage of the male paired with her), or from the male himself so the female can mate with other males (to the advantage of the female). But most of the species mentioned are not social, so an explanation of this sort seems unlikely in most cases.

Birkhead and Møller (1992) argued that in birds many cases of copulations outside the female's fertile period could be explained by the fact that both the male and the female are unable to predict the start of the female's fertile period. This "mistake" hypothesis seems unlikely to provide a general explanation. The relatively precise timing of copulations to coincide with fertile periods of females in many birds (Birkhead et al. 1988, 1989; Birkhead and Møller 1992), even to the hour of the day (Møller 1987c), the precision of timing of forced extra-pair "rapes" which may be strictly insemination attempts rather than courtship (Sorenson 1994), bimodal distributions of copulation times with widely separated peaks (e.g., the kestrels), active female solicitation of copulations outside fertile periods in some species (above; also Hunter et al. 1993), and clear external manifestations of estrus in some mammals (e.g., sexual swellings in females) argue against the general applicability of the mistake hypothesis (though exceptions may well occur).

One other possible explanation of mating outside periods of female fertility also does not necessarily involve communication. The female may manipulate the male to her own advantage, reducing the probability that he will mate with other females and thereby infect her venereally or reduce his investment in her offspring, or fail to transfer sufficient sperm or other seminal products to her (Petrie 1992). The question of whether male copulations with a female outside her fertile period have the presumed effect of reducing the probability that he will copulate with other females needs to be answered before this explanation can be evaluated.

5.4 Stimulation Necessary to Trigger Ejaculation

It is reasonable to expect that males will be selected to refrain from ejaculating until their genitalia are positioned appropriately within the female so that their sperm will have the best chance of arriving at storage or fertilization sites. Stimulation from copulation probably often serves to inform the male of the appropriate time to ejaculate. But in many species, males delay ejaculation until long after their genitalia have reached sperm deposition sites. Stimuli from thrusting behavior after complete penetration are probably important in triggering ejaculation in many male mammals, as is illustrated by the stimulatory treatments widely used to cause ejaculations in order to obtain semen for artificial insemination (Morrow 1986), and by the inhibition of ejaculation when the male genitalia are deenervated (Diakow 1973; Crews 1973). The question is, why do males so often require much more genitalic stimulation to trigger ejaculation than that associated with appropriate positioning of their genitalia within the female?[3]

I will argue (see also Adler 1969 on rats; Diamond 1970 on mice) that delay in ejaculation is often the male side of the coin of genitalic copulatory courtship: by waiting until the female has been sufficiently stimulated, the male refrains from depositing his sperm until he has induced (or at least attempted to induce) the female to perform the responses necessary to increase the chances that his sperm will be used to fertilize her eggs. Just as ejaculation occurring before intromission is disadvantageous, so it is also disadvantageous to ejaculate before the female is ready to accept and use the male's sperm (transport, store and nourish sperm, ovulate, oviposit, etc.).

Natural Selection on Ejaculation Time

Delays in ejaculation after the male's genitalia reach the sperm deposition site are probably very common. In many mammals, for instance, the male thrusts and/or intromits several times before ejaculating (e.g., Michael and Zumpe 1971 on primates; Dewsbury 1982 on rodents; Ewer 1973 on carnivores). There is often no indication that the male has been able to penetrate any deeper on the thrust or intromission when ejaculation occurs.

Delay in ejaculation is not limited to mammals. A dramatic example from insects is the tsetse fly *Glossina pallidipes*, in which copulation lasts on average 69 min (Jaenson 1979), but sperm transfer occurs only during the last 30 sec (Wall and Langley 1993; the exact location of the male's genitalia within the female at different stages of copulation seems to be unknown). Five species of flies in the genus *Coproica* also transfer sperm only near the end of their 9–33 min copulations, and copulations are sometimes interrupted (presumably by the female) before transfer (Lachmann 1994). Some *Rhabditis* nematodes delay several minutes after achieving intromission before ejaculating (Rehfeld and Sudhaus 1985). Males of some spiders and millipedes execute complete "dry run" copulations with up to hundreds of intromissions before seminal products are transferred (e.g., Austad 1984; Haacker and Fuchs 1970; see chap. 3.5).

At first glance, delays in ejaculation are puzzling, especially in species with long copulation times. Natural selection on both males and females would seem to favor a shorter time to ejaculation, because it would reduce the risk that the pair would be killed or separated by predators. There would often be additional selection on the male to shorten copulation in order to reduce the chances that a series of thrusts, intromissions, or copulations would be interrupted prior to ejaculation by other males or by female resistance (e.g., Tsubaki and Sokei 1988 on frequent female resistance during long copulations in the melon fly *Bactrocera cucurbitae*, which can break pairs apart; see chap. 3.8; Hong 1984). Nevertheless, substantial delays in ejaculation, as well as only gradual transfer of sperm (fig. 5.20) are common.

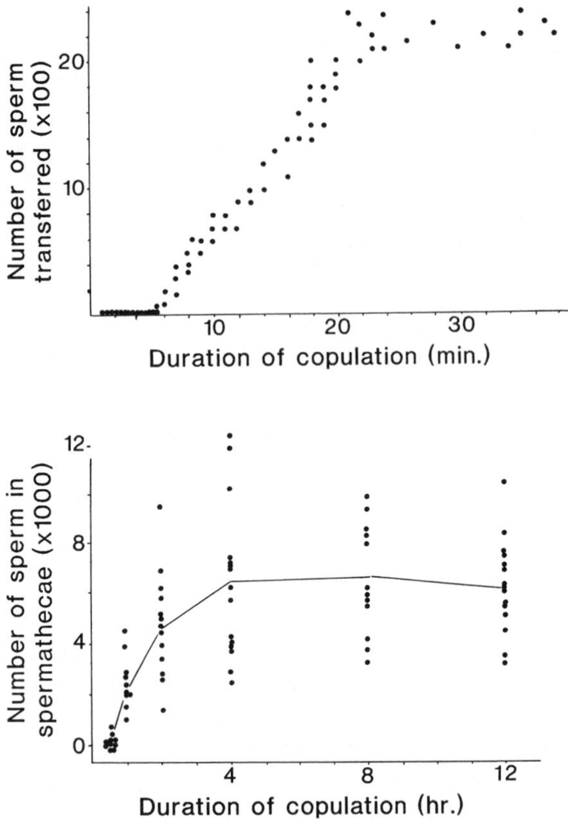

Figure 5.20 The suspiciously strict correlation (*top*) between the number of sperm found in the female spermatheca after matings that lasted different lengths of time in the scorpionfly *Hylobittacus apicalis* (after Thornhill and Alcock 1983), compared with more typical dispersion in the number of sperm transferred to the female's spermathecae in the melon fly *Bactrocera cucurbitae* (after Yamagishi and Tsubaki 1990). The almost complete lack of scatter in the ascending portion of the upper graph suggests a female-imposed barrier preventing more rapid insemination. This gives a premium to longer copulation times, which, in turn, result when the male provides the female with a larger nuptial meal. All scorpion fly males appear to have more than enough sperm ready to be transferred (thus reducing variance stemming from males). The variance in the lower graph is presumably due to either variation in female uptake or male transfer (or both).

Evidence of Sexual Selection

The most convincing evidence to support the idea that delays in ejaculation function to increase the chances that the female will accept and use the male's sperm comes from intraspecific flexibility in delay to ejaculation. In some species the male's ejaculatory delay changes in accord with social circumstances in a way that appears to increase the likelihood that the female will respond in a way that increases the chances that he will fertilize her eggs. For instance, males of the deermouse *Peromyscus maniculatus* adjust the number of intromissions prior to ejaculation according to the female's recent mating history. When a female mates with different males in close succession, she is less likely to initiate the luteal cycle than if she receives the same number of matings from the same male (Dewsbury and Baumgardner 1981). Female initiation of the luteal cycle is also less likely if the male makes fewer intromissions prior to ejaculation. When a male mates with a female that has recently mated with another male, he increases the number of intromissions prior to ejaculation (Dewsbury and Baumgardner 1981). A similar effect also occurs in a different subspecies of *P. maniculatus* (Clemens 1969). Thus males apparently modulate ejaculation delays to compensate for the change in the female's response thresholds.

The delay in ejaculation in male rats *Rattus norvegicus*, guinea pigs *Cavia*, and the rhesus monkey *Macaca mulatta* varies with the male's social status. Copulation in all three groups consists of a series of short, nonejaculatory intromissions followed by an ejaculatory intromission. Subordinate males of all three have lower ejaculatory thresholds (lower number of intromissions precede ejaculation; Rood 1972; McClintock et al. 1982a; Curie-Cohen et al. 1983; Manson 1994). A subordinate male runs a greater risk of having his copulation interrupted by a dominant male in the monkey (Manson 1994), and the same is probably true for the others. In the rat higher numbers of intromissions increase the male's chances of triggering the female's luteal cycle, which prepares the lining of her uterus for implantation of the embryos (Chester and Zucker 1970; Edmonds et al. 1972; see also chap. 4.1; no data on possible female responses are available for the guinea pig or the monkey). Ejaculation thresholds were raised (as predicted) in both rats and monkeys when subordinate males were no longer dominated by others (Thor and Carr 1979 in McClintock 1984; Curie-Cohen et al. 1983; Manson 1994).

Another aspect of delays in ejaculation in rats is especially interesting because at first glance it appears to contradict the idea that male ejaculatory thresholds are adjusted to produce appropriate stimulation of the female prior to sperm transfer (data summarized in McClintock 1984). When a male's mounting opportunities are experimentally manipulated by removing and introducting females to his cage, the male ejaculates with the fewest intromissions when the time between intromissions is approximately 3.5 min. In contrast, the most effective time interval between intromissions for triggering the female luteal cycle is 10–15 min. McClintock (1984) argues that this apparent lack of agreement between female and male criteria is, paradoxically, evidence of agreement. *Rattus norvegicus* has been bred in captivity for many generations in groups, and mating in these groups is often a strange affair. Even though there are subtle effects of male dominance status on copulation (above), males often "share" females, "taking turns" performing intromissions with a single estrous female. When more than one estrous female is present, the females also tend to take turns soliciting the successive mounts and intromissions of a given male's series. The pattern of intromissions performed by a given male can be (and usually is) different from the pattern of intromissions experienced by any individual female. Thus a male sometimes ejaculates in a female that has received few or none of his preejaculatory intromissions.

McClintock (1984) concluded that "copulation is timed in a pattern that does match the optimal patterns for triggering ejaculation and the progestational state" (p. 7). This seems still uncertain, due to lack of several important types of data (the numbers of males and females in breeding colonies of domesticated rats; the degree of variation in female copulation experiences in different social contexts, which can reach the extremes of twenty successive intromissions before an ejaculation, and two successive ejaculations without any other intromissions; the effect of long retention of excitation in the female, up to at least four hours: Edmonds et al. 1972; and the degree of coordination of estrous cycles of females in breeding colonies). Nevertheless, it is clear that the lack of agreement between male ejaculatory criteria and female criteria for triggering the luteal cycle does not necessarily contradict the argument that male ejaculation criteria should result in stimulation of the female that tends to maximize her reproductive responses.

It is interesting to note that under natural conditions female rats have an active role in determining male copulation behavior. When females were allowed to move away from the male between mounts, the rate of intromissions was reduced. The female's luteal cycles were triggered with fewer intromissions when she could control the pace (Gilman et al. 1979), and males ejaculated after fewer intromissions (McGill 1977). Females that controlled the rate also ended their estrus more quickly (Erskine and Baum 1982), thus perhaps reducing their chances of mating with other males. The male behavior (mount the female as soon as she appears) is apparently an appropriate response to trigger female reproductive processes.

The complex chain of triggering stimuli leading to ejaculation in the cricket *Gryllus bimaculatus* raises other important points related to intraspecific variation. The first step is for the male epiphallus to hook onto the female's subgenital plate. Ablation and stimulation experiments by Sakai et al. (1991) showed that when hairs on the male's epiphallus were stimulated as a result of this contact, the male pulled the female plate down. This movement brought the male genitalia into position to stimulate sensillae on the female's subgenital plate, inducing her to evert her copulatory papilla, so that it was inserted into the cavity formed by the male's endophallus. This in turn stimulated male sensilla in this cavity, triggering several processes. The male gripped the papilla with his ectoparamere claspers; moved his spermatophore into position, to be attached to the female; and inserted his grooved rod into the female's papilla to guide his spermatophore. He then pushed the spermatophore into the female, inserting its attachment plate into her copulatory chamber.

The stimulus-response chain in these crickets was not inflexible, however, and variations illustrated how copulatory behavior can evolve. When the afferent input from the female sensillae on the subgenital plate was eliminated, the female failed to extend her copulatory papilla. Most males (75%) spent extra time (up to 100 sec) pulling down the female's subgenital plate, then abandoned her; but 25% were able to proceed successfully with coupling and spermatophore insertion even with the female's papilla in an abnormal position. Such a behavioral (or morphological) ability to compensate for lack of female cooperation could be favored by selection. Accounts of copulation that fail to note variations of this sort can obscure potentially important aspects of behavior (see chap. 1.8).

Alternative Explanations for Preejaculatory Delays

One possible reason for long preejaculatory delays is that males are attempting to use their genitalia for other functions, such as directly removing or repositioning sperm inside the female (Waage 1979, 1984, 1986; Ono et al. 1989; Yokoi 1990; Gage 1992) or removing a mating plug (Jackson 1980). In many animal groups, however, the male's genitalia do not reach the female's storage areas (Eberhard 1985), and direct manipulation of other males' gametes in the female is not possible (see, however, chap. 2.2). For instance, the male genitalia of mammals generally do not reach any possible sperm storage sites (e.g., Ginsberg and Huck 1989), yet delayed ejaculation is widespread. Plugs that must be removed for remating are also relatively uncommon and are not known in several of the groups mentioned above as having long preejaculatory delays.

In addition, many species have long copulations with virgin females. Suter (1990) hypothesized that males of the spider *Frontinella pyramitela* may evaluate female sperm stores during preinsemination intromissions. Other interpretations of his data are possible (see chap. 3.5; also sec. 5.2), and the sensory hypothesis seems unlikely in groups like mammals in which the male's genitalia do not reach sites in the female where previously deposited sperm are stored. Thus neither direct plug or sperm removal nor sperm-sensing attempts explain why males of many species do not ejaculate when their genitalia first reach the appropriate site within the female.

5.5 Summary

Males of a variety of animal groups move actively during copulation. In many cases the male's behavior almost certainly stimulates the female and appears designed to do so. It is very likely that "copulatory courtship" of this sort has been underreported in previous studies of sexual behavior. A focused survey, using conservative criteria in 131 species of insects and spiders, found that copulatory courtship occurred in just over 80% of the species. Copulatory courtship even occurs in animals with external fertilization such as frogs and fish, where it may function to induce the female to increase the number of eggs she lays.

In addition, rhythmic movements of male genitalia within the female are apparently common in many groups. This type of behavior has also undoubtedly been underreported. In some groups experimental manipulations have demonstrated that genitalic movements typical of those of the males of the species induce physiological processes in the female that increase the male's chances of fertilizing her eggs. Indirect, less conclusive evidence from other species, of relative stereotypy and rapid divergent evolution, is also in accord with the idea that genitalic thrusting functions to induce females to perform critical reproductive processes. The idea that copulation can function as courtship is also supported by the apparently regular occurrence of copulation outside of female fertile periods in some species. Several lines of evidence, including direct experiments, indicate that in some species early stages of copulation that involve clasping the female have the effect of inducing females to cooperate in sperm transfer.

Ejaculation thresholds, the male side of stimulation during copulation, have presumably evolved to increase the likelihood that the male's sperm will be accepted and used to fertilize the female's eggs. Males of many species delay ejaculation longer than is necessary for their genitalia to penetrate to where sperm are to be deposited. This delay is probably disadvantageous with respect to possible predation, and also because it increases the likelihood of interruption by the female or other males. But it is in accord with the idea that males often need to stimulate females after intromission but before ejaculation in order to increase the likelihood that their gametes will be used to fertilize her eggs. In some species male ejaculation thresholds vary facultatively in ways that support this idea.

Notes

1. As noted in the text, Dixson's sample size, including species with short intromissions, was forty-six, not twenty as stated by Dewsbury and Pierce.

2. They are the following: reduce female receptivity to remating; engage in copulatory guarding against intrusion by other males; increase sperm precedence by manipulating stored sperm; reduce predation; enhance female foraging; facilitate oviposition or oogenesis; and engage in nuptial feeding.

3. For examples of exceptions, in which ejaculation is very rapid, see Schenkel 1965 and Ewer 1968 on mammals; Wilkes 1965, Eberhard 1974, 1975, and Alcock and Buchmann 1985 on wasps and bees.

6 Evidence That Cryptic Female Choice Is Widespread, II: Effects of Male Sexual Products on Females

The previous chapter discussed male behavior during copulation and its possible influence on cryptic female choice mechanisms. This chapter discusses similar effects of the substances the male transfers to the female along with his sperm. Just as the behavior that a male performs during copulation can serve to induce female responses that increase his chances of fertilizing her eggs, so the chemical products in his semen often induce favorable responses in female physiology and behavior.

Females often use stimuli from copulation and insemination to trigger egg maturation, ovulation, oviposition, and reduced receptivity to additional matings. A fifth process, sperm transport, is less often studied but is probably also usually induced by copulation. It makes adaptive sense for the female to use mechanical or chemical stimuli associated with copulation to cue these processes. Failure to perform any of them will render mating ineffectual, while initiation of processes such as ovulation and oviposition before being inseminated is also usually disadvantageous. Since transport of sperm to storage and/or fertilization sites may also be costly in terms of energy and risk of venereal infections, it is probably also best repressed until needed. Efficiencies in investment of time and effort could also make costly investment in egg maturation prior to insemination disadvantageous. And avoiding unnecessary additional matings will generally help a female reduce expenditure of time and effort as well as minimize the multiple dangers of courtship and mating (Daly 1978; Scott and Richmond 1985).

Given these obvious naturally selected advantages for females, it is not surprising that substances in male semen often influence several female reproductive processes. What is less expected—and constitutes striking evidence favoring the general importance of cryptic female choice—is that the details of how and where these substances act suggest that males have evolved to attempt to usurp female control and manipulate female responses to their own advantage. Such attempts at manipulation are predicted consequences of cryptic female choice. If females had been consistently "cooperative" (giving complete or maximum responses to all males, or at least not varying their responses in accord with the abilities of the male to elicit them), there would not have been selection favoring such male traits.

Before beginning, a word about levels of analysis is in order. Some female "traits" that can be labeled with a single term are probably the result of several different, at least partially independently controlled subprocesses. For example, the number of eggs laid by a female is a commonly measured variable, but egg number is in fact determined by the rate of feeding, mobilization of reserves such as fat deposits, vitellogenesis (egg synthesis), ovulation (movement of the egg out of the ovary to the oviduct), and oviposition *sensu strictu*. In turn, each of these processes can be complex. For instance, oviposition *sensu strictu* can include seeking out appropriate oviposition substrates, extrusion or insertion of the ovipositor, and movement of the egg down the oviduct and the ovipositor in response to stimuli from the substrate.

Other processes are also complex. Rejection of mating attempts by females in *Drosophila melanogaster* involves kicking, warding off the male with raised legs, extruding the ovipositor so as to make genitalic coupling impossible (Burnet et al. 1973), and production of antiaphrodisiac substances (Mane et al. 1983; Scott 1986). As noted by Gillott (1988), most studies have not distinguished among these types of process. Thus unless otherwise specified, terms like "oviposition" and "remating" are used *sensu latu* in table 6.1 (see end of chapter) and in the discussions below; they probably hide a diversity of more complex interactions.

6.1 Two Hypotheses

As noted above, it is reasonable to suppose that the influence of male seminal products on female reproductive functions might have evolved under natural selection on females. It is clearly advantageous for females to cue the initiation of some reproductive processes on successful mating. Females could evolve to use any chemical transferred during copulation as a signal that insemination had occurred. This is the context in which most previous authors have considered the evolution of the effects of male accessory gland products on female behavior and physiology (e.g., review of Gillott 1988).

It is also possible, however, that such male substances have evolved under sexual selection by cryptic female choice. Males whose substances were more effective at triggering female responses could have a reproductive advantage over others mating with the same female. And females that screened their mates more rigorously for such abilities (for example, by raising their threshold of response to a male product) could also derive several possible reproductive advantages. One possible female benefit, nutrition from male seminal products, could evolve under either natural or sexual selection, and will be discussed in section 6.7. More rigorous screening could also benefit the female by improving her sons' abilities to stimulate other other females. If a male's production of such triggering substances were linked with his overall vigor, more rigorous screening could also benefit the female from improved viability genes in her offspring. If the effects of male substances were sometimes disadvantageous (say, oviposition induction were so powerful that some eggs were laid at inappropriate sites), females might also benefit from raising response thresholds and thus retaining more control over their own bodily processes (e.g., being able to hold off ovipositing until finding the best possible site).

Ways to distinguish between the natural selection and the sexual selection hypotheses are described in the following section on insects. Some discussions of data are relatively detailed, because the interpretations of the data are new; the original studies were not focused on the patterns that are discussed here.

Keep in mind that sexual selection is only expected in species in which females mate more than once. The frequency of female remating in the field is not known for many of the species listed in tables 6.1 and 6.3. A simple totaling of cases in which one or the other prediction is confirmed can also be misleading because of probable phylogenetic bias (Ridley 1983; Felsenstein 1985; Harvey and Pagel 1991). I will return to this point in section 6.8.

6.2 Insects

It is clear that female insects often use male seminal products to trigger their own reproductive processes (e.g., fig. 6.1). In reviewing the functions of male accessory gland secretions in insects, Raabe (1986, p. 89) noted that "in most insects, the active factor or factors inhibiting female receptivity and stimulating egg laying are generally considered to originate in the male accessory gland or in the secretory cells of the ejaculatory duct." Other reviewers concur. Chen (1984) lists the two major responses to male accessory gland products in the female as (1) elevation of oviposition rates, and (2) repression of sexual activity. Hinton (1974), Raabe (1986), and Gillott (1988) include the additional functions of triggering oviduct contractions and maturation of oocytes. A reasonably complete list (see table 6.1 at end of this chapter) includes more than sixty species and eight different female processes. In mosquitoes alone, female activities thought to be influenced by male accessory gland products include oviposition, rejection of genitalic coupling, rejection of insemination after coupling, feeding and

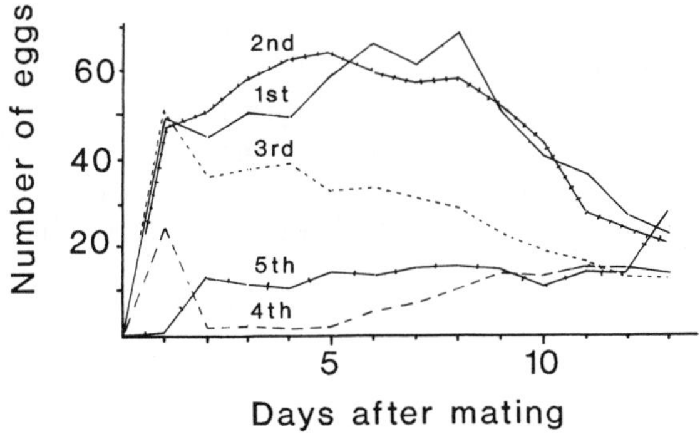

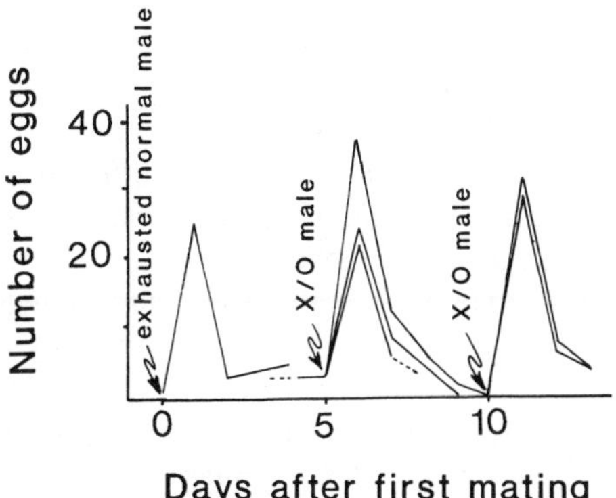

Figure 6.1 Effects of male accessory gland secretions on the female's rate of oviposition in *Drosophila melanogaster*. When a male mates with several females in rapid succession, dissections show that the contents of his accessory glands are gradually depleted but that he continues to have numerous mature sperm in his reproductive tract. *Top*: The number of eggs laid by successive females in such an experiment gradually declines, until the male's glands are completely empty (fifth female); each point is the mean of ten replicate vials with five females in each. The association between reduced egg numbers and a lack of accessory gland products rather then lack of sperm was confirmed by the results of an experiment (*bottom*). By five days after a female copulated with a normal male that had just copulated with three other females in rapid succession (first arrow), her oviposition rate had fallen almost to zero. When she was then remated on the fifth day to a mutant X/0 male (with normal accessory glands but unable to transfer sperm; second arrow), her oviposition rate shot up briefly on the following days. After it had declined again 10 days after the first mating, the female was mated with a second X/0 male (third arrow), and another burst of fertile eggs followed. (After Hihara 1981)

digestion of blood (necessary for reproduction in some species), searching for hosts, and development of oocytes and oviposition without feeding (autogeny). Since these processes may be linked physiologically, and since the studies of particular species have seldom tested more than one or two female responses (Gillott 1988), a list like this may be somewhat misleading. Nevertheless, the message is clear: male seminal products in insects have diverse, far-reaching effects on female behavior and physiology in many different species.

"Invasive" Action of Male Substances

Under natural selection, the evolution of female responses to chemical cues in male semen would presumably often arise when receptors in the female reproductive tract sensed the presence of the male product and sent this information to some coordinating site or responsive organ. An example of such a design occurs in the bug *Rhodnius prolixus*. A particular area of the wall of the female bursa senses the presence of a male accessory gland product and sends a nervous signal via the peripheral nervous system directly to the oviducts, stimulating them to contract and thus transport the sperm to the spermathecae (Davey 1958).

In contrast, sexual selection on male abilities to induce female responses could favor males able to act more directly on the female. By transferring chemical products that pass through the walls of the female's reproductive tract and enter her body cavity, the male could affect receptors in the target organs of the female (e.g., her ovaries, her nervous system). The advantage to the male of more "intrusive" action like this would be that it would be less subject to uncertain or incomplete female responses. By acting directly on the female's effector organs, the male could avoid possible incomplete responses to his stimuli in the female's receptors, her central nervous system, or other intermediate organs. In fact, the evolution of such intrusive male traits suggests that the female's responses in the past were often incomplete or unreliable.

Intrusive action of male seminal products on females seems to be common. One type of data comes from studies in which the fate of labeled male sexual products was followed in the female's body. Some male products enter the female hemolymph unaltered (Davey 1985; Gillott 1988) and arrive at the target organs themselves (table 6.1). For instance, in the migratory grasshopper *Melanoplus sanguinipes*, injection of male accessory gland products into the female causes her to increase oviposition (Pickford et al. 1969). Radioactive labeling and immunological assays demonstrated that the proteins from the male accessory gland that stimulate oviposition move rapidly from the female's reproductive tract into her hemolymph

(Friedel and Gillott 1977). At least some proteins remain unaltered by this process and rapidly accumulate in the oocytes in the ovary (Friedel and Gillott 1977). At first it was thought that these substances represented nutrient donations by the male (Friedel and Gillott 1977). But later Cheeseman and Gillott (1989) showed that the quantities involved are so minute that a nutrient function is unlikely, and they concluded that a signal function is much more probable (see sec. 6.6 for further discussion of this point). The same combination of techniques showed that in the bean weevil *Acanthoscelides obtectus* at least two male-derived, relatively large molecules or their derivatives moved into the female hemolymph between 5 and 16 hr after mating (Huignard 1983). Neither compound was incorporated into eggs.

The invasive nature of the semen of the male housefly *Musca domestica* is especially dramatic. The male accessory gland products are received in thin-walled pouches in the vagina. Shortly after they arrive, the cells in the wall of the vaginal pouch begin to break down, presumably destroyed by a component of the semen. This allows seminal products to move into the female's hemolymph (Leopold et al. 1971b). These products probably include the seminal factors known to inhibit female remating and stimulate oviposition (Riemann et al. 1967; Riemann and Thorson 1969). Radioactive labeling experiments showed that male products reached the female head and thorax only 10 min after mating began (Leopold et al. 1971b). A combination of ligation, labeling, and decapitation experiments suggested that the male substance(s) inhibiting female sexual receptivity act directly on the female's brain (Leopold et al. 1971a,b).

In another muscoid fly, the blowfly *Lucilia sericata*, injection of male accessory gland material may be accomplished by sawlike structures on the male's genitalic paraphallus, which pierce the wall of the bursa (fig. 6.2; Lewis and Pollock 1975). After mating, the female develops small dark granules in this area which appear to be scar tissue. Implantation of male accessory glands in the female reduces female receptivity to mating (Pollock 1971 in Lewis and Pollock 1975).

A second technique to determine the targets of male products involves exposing female organs to male seminal products. Results suggest that the female's nervous system is frequently involved (P. Chen 1984; Gillott 1988). Direct recordings of nervous activity in the isolated last abdominal ganglion in the silk moth *Bombyx mori* showed that a male factor (presumably the one which stimulates the female to oviposit) acts directly on this ganglion, altering the output of its neurons (Yamaoka and Hirao 1977; fig. 6.3). The last abdominal ganglion innervates both the female genitalia and her abdominal muscles. The male factor apparently alters the permeability of the neural sheath of the ganglion to ions, and the subsequent

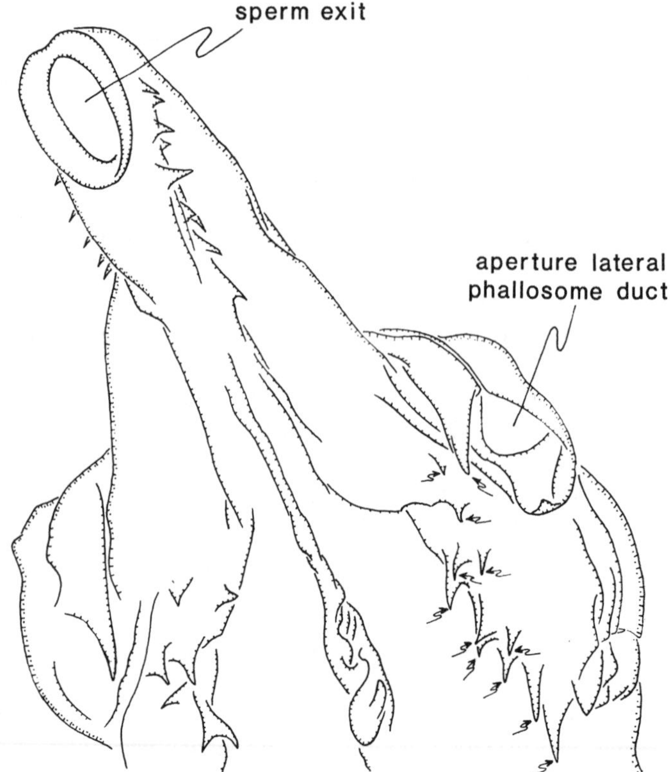

Figure 6.2 The possibly "invasive" action of the male accessory gland products in the fly *Lucilia sericata* is reflected in the morphology of the male genitalia in this drawing. There are sharp structures (arrows) which apparently produce holes in the lining of the female reproductive tract near the sites where male accessory gland substances are liberated (from the lateral phallosome duct), thus giving the male products direct access to the female's body cavity. Male seminal products inhibit female remating. (After Lewis and Pollock 1975)

change in the ionic composition bathing the interior of the ganglion results in the expression of the oviposition reflex.

In the mosquito *Aedes aegypti*, male factors inhibiting sexual receptivity also act on the female's last abdominal ganglion, which apparently controls female processes involved in genitalic coupling (Gwadz 1972). When the connection of the female's last abdominal ganglion with the rest of her nervous system was cut and the female was then injected with an extract of male accessory glands, she became highly refractory (zero of forty-five females inseminated, as compared with forty-three of forty-six similar females that were not injected; Gwadz 1972).

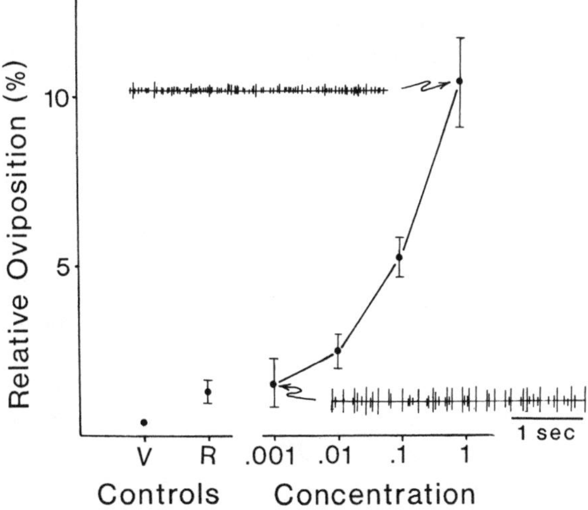

Figure 6.3 An "invasive" dosage-dependent effect on the oviposition response in virgin female silk moths *Bombyx mori*. Injections of different concentrations of extracts of male reproductive tracts into virgin females caused changes in relative oviposition activity (= number of eggs in each treatment group/average number of eggs in a group of normally mated females). The action was apparently directly on the female's last abdominal ganglion since higher concentrations of extracts of male reproductive tracts caused more frequent discharges in isolated ganglia (discharge records for concentrations of 1.0 and 0.001 are shown in the figure). Concentrations are in equivalents of a single male. (After Yamaoka and Hirao 1977)

In a different mosquito, *Culex tarsalis*, male substances that reduce female receptivity may pass unaltered into the female's body cavity; after copulation the female's body tissue acquires the ability to reduce receptivity, just as extracts of male accessory glands do, when injected into other females (Young and Downe 1987). Extracts of the heads of mated females were especially effective in reducing receptivity, and a major portion of radioactively labeled male products end up in the female's head. Thus the target organ of the male substance(s) may be the female's brain (Young and Downe 1987).

Male factors that increase both oviposition and the incorporation of labeled protein in the ovary of *D. melanogaster* apparently act on neurosecretory cells in the female's brain. By partially and completely removing the pars intercerebralis cells in the median dorsal region of the brain, Boulétreau-Merle (1976) eliminated the female's oviposition response to mating. The male sex peptide, which is thought to act hormonally to induce female oviposition, rapidly enters the female's hemolymph intact (even before copulation ends; Monsma et al. 1990), and is then transported to receptors in her brain (K. Kubli et al., unpub., cited by Spencer et al. 1992).

The unusual monomorphic configuration of the enzyme esterase-6 in those *Drosophila* species in which it is produced in large quantities in the male reproductive tract is thought to be related to the fact that its smaller size facilitates its transport across the vaginal membranes, which is necessary for its effect on oviposition and remating (Oakeshott et al. 1995). Female receptivity to remating, which is also affected by male seminal products in this species, is also under the control of certain groups of neurons in the brain (Tompkins and Hall 1983).

Less complete data suggest that male seminal products also act at sites other than the female's reproductive tract in other species. The prostaglandins produced in the spermatheca of the cricket *Teleogryllus commodus* that are derived from male seminal products are apparently released into the female's hemolymph by moving down the spermathecal duct (perhaps under female control; Ai et al. 1986). Their final target organs are unknown. Male products also act directly on female effector organs in the grasshopper *Locusta migratoria*. By bathing isolated oviducts in saline with extracts of male accessory glands, Lafon-Cazal et al. (1987) showed that the male substance probably acts on female receptors for the insect neurotransmitter octopamine (see fig. 6.6). Other substances in the male accessory glands apparently also act on other receptors in the oviduct (Lafon-Cazal et al. 1987).

Finally, there is reason to believe that male substances act on the female outside her reproductive tract in many other species in which the female target organs have not yet been determined. Most experimental demonstrations of the effects of male accessory gland substances on female oogenesis, oviposition, and receptivity in insects have involved either implantation of portions of male reproductive tract, or injections of extracts of the glands or their secretions into the female's body cavity (as contrasted with implantation or injection into her reproductive tract; forty different species in table 6.1). Riemann et al. (1967) and Loher (1984) argue that diffusion through the walls of the female reproductive tract is improbable, and that female receptors on the inner surface of her genital tract are thus unlikely to respond to substances injected into her body cavity (P. Smith et al. 1990 also argue for relative impermeability of the spermatheca of the blowfly *Lucilia cuprina*). If the female tract is indeed often relatively impermeable, then the consistent female responses in these implantation and extract injection studies imply that female receptor sites outside of the interior of her reproductive tract are very common.

Thus, while there are a few species in which male seminal substances stimulate the female within her reproductive tract (e.g., *Rhodnius prolixus* and *Helicoverpa zea* in table 6.1), the general rule appears to be more direct action within the female's body. Since selection favoring the evolution of such invasive action would depend on female responses to mating in the past not being consistently in accord with male reproductive interests, widespread invasive action thus constitutes strong evidence favoring the sexual selection over the natural selection hypothesis.

Multiplicity of Cues for the Same Female Process

If females evolved under natural selection to use male products associated with insemination as cues, then one would expect that the female would tend to use one particularly dependable cue or substance to trigger any single response (e.g., ovulation, resistance to remating). Other than advantages from possible fail-safe mechanisms, there would be no reason for extensive redundancy.

If, on the other hand, male sexual secretions evolve under sexual selection by cryptic female choice, one possible mechanism by which a male could escalate or increase the effectiveness of his signal, and offset the lack of consistent female responsiveness, would be by adding further stimulatory substances to his semen. This is the chemical equivalent of adding, for instance, an additional waving motion to a dance performed as a visual display (the "piling on" of stimuli of West-Eberhard 1983; see also Møller and Pomiankowski 1993). Again, the evolution of multiple male cues implies that cryptic female choice has been occurring due to a lack of complete female "cooperation." Thus the sexual selection hypothesis predicts that multiple rather than single substances will often, though not necessarily always, be used to trigger a given process.

Again the data fit better with the sexual selection predictions than with those from the natural selection hypothesis. Direct data on inhibition of female receptivity to further mating will be discussed first, then data on stimulation of oviposition. Keep in mind that complexity in male effect on control mechanisms is a priori less likely to have been discovered in those species that have been less intensively studied. Thus the best-studied groups (e.g., *Drosophila*, *Aedes aegypti*, *Musca domestica*, some crickets and grasshoppers) give the most powerful tests of the predictions.

Inhibition of Remating

In *Drosophila melanogaster*, multiple male stimuli switch off female remating. One seminal factor is a 36-amino-acid "sex peptide" (SP), a product of the male accessory (paragonial) gland, and it has a short-term effect (Burnet et al. 1973; Scott 1986; P. Chen et al. 1988; Aigaki et al. 1991). In transgenic females, the effects of inducing synthesis of the male sex peptide began about 1 hr after the 30 min induction period, and began to wane 7 hr later (Aigaki et al. 1991). A second stimulus that switches off female receptivity comes from the sperm or some other product of the testes acting in the female sperm storage organs. This factor has both an ephemeral, short-term effect on both female attractiveness to males (Scott and Richmond 1985; Kalb et al. 1993) and female receptivity (Hihara 1981; fig. 6.1), and a long-term effect on receptivity (Manning 1967). Short- and long-term effects seem to be independently controlled (Aigaki et al. 1991). Still another seminal substance, the enzyme esterase-6, reduces a female's receptivity but not her attractiveness to males (Gromko et al. 1984).

When females were mated with males lacking both sperm and esterase-6, females still failed to remate more often than virgin females (Gromko et al. 1984; Scott 1986). So there may be still another male stimulus (or stimuli) with a long-term effect on female receptivity.

Further cues may also be involved. Copulation must last 9–13 min to switch on the female rejection behavior involving ovipositor extension; but after only 3–4 min, females are induced to produce a substance with anti-aphrodisiac properties, and also to decrease the amount of sex attractant they produce (Tompkins and Hall 1981). P. Chen et al. (1988), in reporting on the isolation of a peptide that increases both female nonreceptivity and oviposition, suggest that other factors, such as stimuli from courtship or the act of copulation, may also affect female rejection responses in an additive or synergistic manner.

Kalb et al. (1993) suggested that perhaps the short-term SP cue was needed because sperm, which provide a long-term cue, are stored only slowly in the spermathecae. This idea does not fit well, however, with the timing of different events: storage of the maximum number of sperm occurs about 5 hr after copulation (Gilbert 1981), while short-term effects of the accessory gland product are still strong 24 hr after copulation (Kalb et al. 1993).

In sum, two different male products have short-term effects, and at least two and perhaps more have long-term effects in lowering female receptivity to male mating attempts in *D. melanogaster*. Some short-term effects are not triggered together. The male's participation in inducing the different processes involved in switching off female receptivity is clearly quite complex.

Other species also have redundant seminal trigger mechanisms. In the mosquito *Aedes aegypti* there are apparently three male stimuli that inhibit female receptivity, including two different seminal proteins, the alpha and beta fractions of "matrone." The alpha fraction alone has a weaker effect than that of the combination, while the beta fraction alone has little if any effect (Fuchs and Hiss 1970). Although the molecular weights of alpha and beta matrone given by Fuchs and Hiss were later called into question (Young and Downe 1987), their different effects on oviposition indicate that there are indeed two different components. Matrone itself has a delayed inhibitory effect on female receptivity. A third, short-term inhibition is apparently associated with filling the female bursa with seminal fluid (Gwadz et al. 1971; Gwadz 1972).

In the medfly *Ceratitis capitata*, reduction in female receptivity was triggered both by sperm and by accessory gland products (Del Rio and Cavalloro 1979). In the calliphorid fly *Lucilia cuprina*, both extracts of the accessory gland and of the testes significantly reduced female receptivity (P. Smith et al. 1989). These effects are apparently complementary, since castrating a male but leaving his accessory glands intact reduced but did not eliminate his ability to switch off female receptivity (P. Smith et al. 1990). Indirect data suggested that different products are involved. The effects of the extracts of testes on remating wore off more rapidly than did those of the extracts of accessory glands (P. Smith et al. 1989).

In the tsetse fly *Glossina morsitans* two different types of stimuli, one a chemical stimulus from the male accessory glands and the other from the physical act of copulation, reduce female receptivity (Gillott and Langley 1981). Neither is sufficient alone to induce the rapid decline in receptivity which follows a normal copulation. A similar combination of cues apparently also reduces female receptivity in *G. pallidipes* (Leegwater-van der Linden and Tiggelman 1984). Gillott and Langley (1981) note that "there is no obvious explanation for the dual mode of action," and it does not appear that one stimulus gives a short-term and the other a long-term effect. Male inhibition of female receptivity in the screw worm fly *Cochliomyia hominivorax* may also involve a two-stage mechanism (Leopold 1976).

In the housefly *M. domestica*, both the frequency and duration of switch-off effects of a male seminal substance (or substances) are also apparently supplemented by the mechanical stimuli of copulation (Riemann et al. 1967). Still further effects may result from a product of the testes (see table 6.1). Similarly, female receptivity in the Hessian fly *Mayetiola destructor* was inhibited both by the mechanical stimuli of copulation (for a short time), and by a factor from the male reproductive tract (for a longer period; Bergh et al. 1992).

Moving to other orders of insects, two male substances reduce female receptivity in the corn earworm moth *Helicoverpa zea*. Injections of a proteinaceous male factor (from his accessory gland plus an adjacent portion of his reproductive tract, the "duplex") and the receipt of a spermatophore (or a male factor derived from other, nonaccessory gland/duplex sources) both elicit cessation of the female behavior patterns of ovipositor extrusion and wing fanning that occur during pheromonal calling (Kingan et al. 1993b). Both male products are also involved, but to different extents, in the disappearance of the calling pheromone in the female's abdomen (presumably it is degraded). The component from the accessory gland plus duplex can reduce pheromone levels to those present after normal matings, while spermatophore stimuli have an approximately 5-fold weaker effect (Kingan et al. 1993b).

INDUCTION OF OVIPOSITION

In several species multiple cues from males also increase oviposition. When stimuli from genitalic contact during copulation, accessory gland products, and sperm in the female storage organs were added progressively to precopulatory courtship stimulation in *D. melanogaster*, progressive increases were apparently produced in ovarian production (Boulétreau-Merle 1974). These experiments employed males genetically lacking the ability to transfer sperm but producing normal amounts of accessory gland products; males with their internal reproductive structures (gonads, paragonial glands, ejaculatory duct) removed surgically; and males in which this removal also resulted in immobilization of the male's external genitalia. Statistical tests of significance were not reported, however, and there were no control, sham-operated males. A different study using transgenic flies in which the "main cells" of the accessory gland were inactivated indicated that both main cell secretions and sperm are needed for short-term stimulation of oviposition, while only sperm storage was correlated with long-term stimulation of oviposition (Kalb et al. 1993). Several different accessory gland products are probably involved in stimulating oviposition: the 36-amino-acid sex peptide (Chen et al. 1988; Aigaki et al. 1991); the 264-amino-acid protein *Acp 26Aa* (Monsma and Wolfner 1988; Kalb et al. 1993); and esterase-6 (Oakeshott et al. 1995).

In a congeneric species, *D. funebris*, an accessory gland product (PS-2) was associated with long-term increases in oviposition (Baumann 1974b), and other factors are probably also involved (Baumann 1974a,b). Normal mating stimulated oviposition more effectively than injections of the gland product at concentrations of up to three times the total in a single male, or of the entire gland. Both fewer eggs (maxima of 58 vs. 83 eggs/day), and a slower oviposition response (7 vs. 3 days) resulted from injections.

In the mosquitoes *Aedes aegypti* and *Culex pipiens* implants of male testes increased the percentage of females ovipositing, while implants of male accessory glands increased the percentage approximately twice as much (Leahy 1967). Products from the two sources appear to act synergistically in *A. aegypti*, since a mix of extracts in saline (half testis, half accessory gland) induced significantly more females to oviposit than a similar amount of either extract alone (data from fig. 6 of Leahy 1967; $X^2 = 6.52$, $df = 1$, $P < .01$).

Multiple cues from males of acridid grasshoppers may affect different subprocesses of oviposition. In *Melanoplus sanguinipes*, a male factor (or factors) apparently acts on the ovary to induce increased egg maturation (Friedel and Gillott 1977), while in *Locusta migratoria* a factor or factors act to induce contractions of the oviduct (Lange and Loughton 1985; Lafon-Cazal et al. 1987), presumably promoting as a result ovulation and/or oviposition. Neither species appears to have been checked directly for the responses found in the other. Nevertheless, a combination of male effects seems likely, at least in *M. sanguinipes*, where male effects on egg transport are also probable. Virgin females tended to retain mature eggs in the lateral oviducts (Friedel and Gillott 1977), and injection of male gland extracts in virgin females resulted in oviposition within 24 hr by 75% of thirty-two females that had ovulated (in controls 0% of twenty-six females that had ovulated oviposited; Friedel and Gillott 1976). Further redundancy was suggested by the fact that two antigenically different products of male accessory glands induced oviposition (Friedel and Gillott 1976). Still another stimulus, the act of mating per se, also apparently accelerates oviposition in *M. sanguinipes* (Pickford et al. 1969).

An even more complex web of oviposition induction may occur in the cricket *Acheta domesticus* (fig. 6.4). The male transfers an enzyme (or enzyme complex) that acts on the substrate arachidonic acid that accumulates in the virgin female's spermatheca, catalyzing its conversion to prostaglandin in the female's spermatheca. Experimental injections of nanogram amounts of prostaglandin into the female's body cavity cause increased

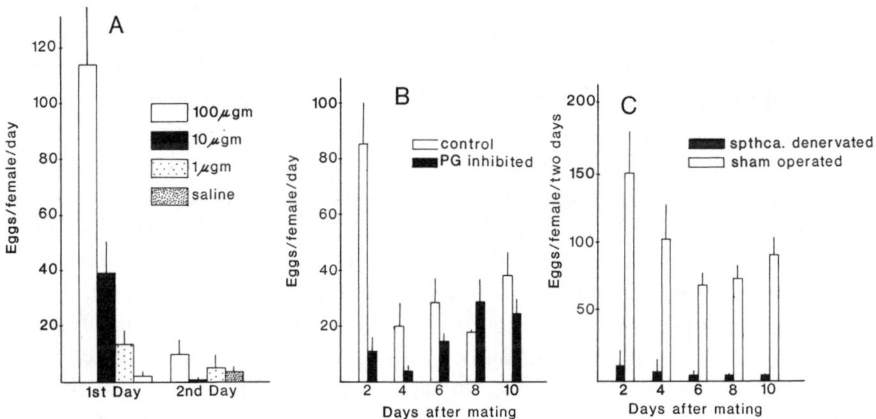

Figure 6.4 Multiple male cues influence oviposition in the cricket *Acheta domesticus*. A short-term, dose-dependent effect results from injection of prostaglandin (A). The short-term effect is eliminated by experimentally inhibiting the action of prostaglandin, but long-term effects are unaltered (B). Both short-term and long-term effects of mating on egg production depend on stimuli from the spermatheca in the female (C). (A after Destephano and Brady 1977; B after Murtaugh and Denlinger 1987; C after Murtaugh and Denlinger 1985)

oviposition (Destephano and Brady 1977; Destephano et al. 1982). Two additional factors may be involved. When the virgin female's spermatheca was surgically removed 15 days before copulation (perhaps removing the arachidonic acid), a short burst of oviposition followed during the first day after mating; but the numbers of eggs were lower than in sham-operated females, and the burst did not continue in subsequent days as in control females (Murtaugh and Denlinger 1985). An additional, long-term stimulation of oviposition apparently depends on stimulation of receptors within the spermatheca by products other than prostaglandin, apparently associated with either the sperm themselves or a factor transferred with them. Denervation of the spermatheca abolished a long-term mating-induced increase in oviposition which was independent of prostaglandins (Murtaugh and Denlinger 1985, 1987).

The prostaglandin story itself may also be complex. Males of the cricket *Teleogryllus commodus* also transfer enzymatic material to the female, and precursors for three different prostaglandins occur in the spermathecae of virgin females (Loher et al. 1981). Injections of prostaglandin E2 induced females to oviposit, but only about one-fourth as much as after normal matings (Loher et al. 1981).

In the beetle *Acanthoscelides obtectus*, egg maturation and oviposition were enhanced primarily by an accessory gland secretion, and less strongly by sperm in the female's genital ducts (Huignard 1970). Huignard suggested that males affected egg production by contributing nutrients, but quantitative considerations of the amount transferred throw doubt on the nutritional significance of the male contributions (Kaulenas 1992; see sec. 6.7). Similarly, both a male factor (or factors) as well as viable sperm were needed to trigger maximum oviposition rates in the lepidopterans *Zeiraphera diniana* (Benz 1969) and *Trichoplusia ni* (Karpenko and North 1973). Removal of the male's gonads but not his accessory glands in *T. ni* reduced but did not eliminate his ability to stimulate the female to oviposit. This species showed the additional twist that only one of the two kinds of sperm that are produced by males raised oviposition rates above those resulting from copulations in which females did not receive any sperm.

In sum, duplications of triggering functions for both switching off female receptivity and oviposition occur in various groups, as predicted by the sexual selection hypothesis. In fact, multiple male cues for a given female response appear to be the rule rather than the exception. Redundancy in male effects on female response mechanisms might sometimes evolve under natural selection as a fail-safe device to control a short-term response independently of a long-term response. However, the evolution of the use of multiple cues *from the male*, instead of a single cue from the male followed by others from the female herself seems less likely, because the female would be better able to control both production and release. On balance, the data fit with the predictions of the sexual selection hypothesis but may not give strong reason to reject the alternative, natural selection hypothesis.

Graded Female Responses to Quantitative Variation in Male Cues

If females use chemical cues produced by males to signal that mating has occurred, one would expect natural selection to tend to produce all-or-none switches, rather than dose-dependent, variable responses. This is because a female has either mated or she has not, and she is not usually expected to respond partially or halfheartedly to a normal mating cue that signals "mating has occurred." Stated in different terms, the threshold for maximum female response is expected usually to be set by natural selection at or below the lowest amount of signaling substance transferred in mating with fully fertile males (fig. 6.5).

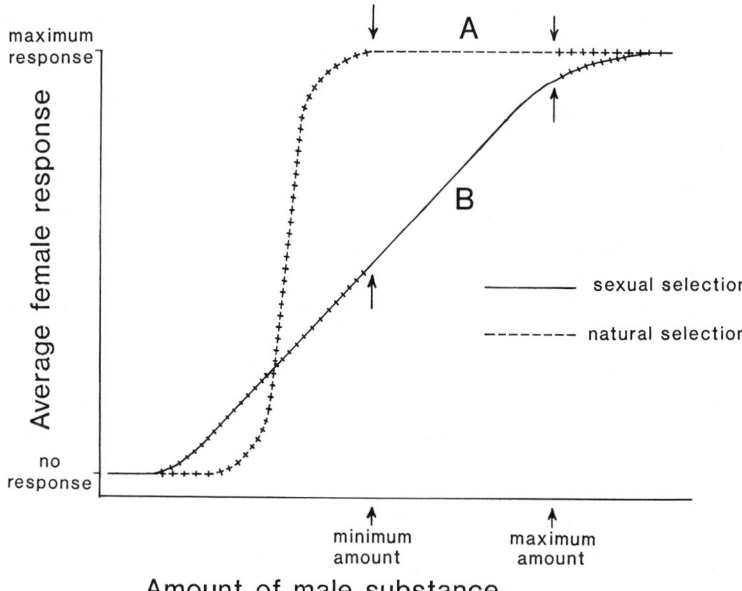

Figure 6.5 Contrasting predictions regarding female responses to different amounts of male accessory gland substances (cross-hatched portions of the lines could have other forms). Under natural selection to sense that a successful mating has occurred, a relatively complete, all-or-none response is expected to occur, with maximum responses throughout the range of minimum to maximum amounts transferred by normally fertile males (A). Under sexual selection a similar response curve could occur, but an increase in female response with increased amounts from the male (B) could sometimes be a mechanism by which females selected among males.

Only if reduced amounts of the male cue were correlated with substandard numbers of sperm or some other male product crucial to egg production might partial female responses be favored. This seems unlikely to generally be the case. Sperm numbers are often far in excess of those needed by females (e.g., Parker 1970c; see chap. 8.1). In addition, the volume of sperm is not necessarily correlated with the volume of other seminal products. The degree of dilution of sperm may vary, and in at least some groups sperm are transferred more or less separately from other seminal products (e.g., the housefly *Musca domestica*: Murvosh et al. 1964 and Riemann et al. 1967; many mammals: Mann and Lutwak-Mann 1981). Another possible reason that females might respond submaximally to some ejaculates would be if male seminal products were important nutritionally to the female; but nutritional effects of accessory gland substances may be relatively rare (see sec. 6.7). Thus the correlation between ejaculate characteristics that directly affect production of fertile eggs and the quantity of male signaling substance is not necessarily common.

Sexual selection, on the other hand, may often favor increases in the intensity of male chemical signals. If female response thresholds are appropriate (fig. 6.5), males could be required to compete on the basis of amounts transferred to the female. This would be similar, in a vocal display, to singing louder. Females might benefit from accentuating the degree of dose dependence of their responses, for instance because a steeper response curve or a curve displaced toward higher amounts of seminal product would allow them to discriminate more effectively in favor of males delivering larger amounts. Larger quantities could signal greater male size or vigor (good viability genes), or greater ability to induce female responses (good attractiveness genes). Thus quantitative, dose-dependent female responses whose maxima fall within or above the range of normal amounts transferred during copulation are predicted to occur often (though not always) by the sexual selection hypothesis (fig. 6.5); they are less likely, though not ruled out, under the natural selection hypothesis except under certain conditions.

INDUCTION OF OVIPOSITION

A dose-dependent female response is particularly clear in the cricket *Acheta domesticus* because the active male factor has been identified. The male transfers a prostaglandin synthetase (or synthetase complex) along with his sperm, and the enzyme acts on arachidonic acid that has accumulated in the female's spermatheca to produce prostaglandin, a hormone which stimulates oviposition. When gravid females were injected with different quantities of prostaglandin, larger numbers of eggs were laid on the first day afterward by females receiving larger doses (Destephano and

Brady 1977), including doses larger than those associated with normal matings (Murtaugh and Denlinger 1982, 1985).

Manipulation of doses in the presence of other stimuli that are normally associated with copulations also indicated that the doses involved in normal matings are within the range of dose-dependent female responses. Leaving a spermatophore attached for only 2–3 min instead of 10 min sharply reduced the number of eggs laid, both on the short term and the long term (Murtaugh and Denlinger 1985; average totals for 22 days slipped about 50%, from 682 to 339 eggs). A detailed study of female behavior preceding and during oviposition (Destephano et al. 1982) showed that both the duration of oviposition and the completion of different components of a complete oviposition were apparently also influenced in dose-dependent ways by prostaglandin.

Since female crickets often remate and prostaglandin presumably has a hormonelike rather than a nutritional role in the female, the natural selection hypothesis seems unlikely to explain these results.

In view of the short- and long-term role of prostaglandin in inducing cricket oviposition, the ejaculate of males of another cricket species *Teleogryllus commodus* is impressive because it contains not only prostaglandin synthetase, but also precursors on which the enzyme can act. Males transfer physiologically significant amounts of arachidonic acid along with prostaglandin synthetase (Loher et al. 1981; Stanley-Samuelson et al. 1987). This may be a male mechanism to increase the amount of prostaglandin. There was indirect evidence of arachidonic acid depletion in *A. domesticus* (Tobe and Loher 1983), and the amounts of prostaglandin in the spermathecae of once-mated *T. commodus* females varied considerably (the coefficient of variation was 54%); some *T. commodus* spermatophores did not empty completely, even in matings with virgin females (Loher et al. 1981; Stanley-Samuelson and Loher 1983). Thus it is possible that some variation in the transfer of male seminal products may result from female effects. Another female-controlled process that may also alter the effective dosage of prostaglandin resulting from a given copulation is the transport of prostaglandins out of the spermatheca and into the hemolymph. Both the duct and the spermatheca itself are impermeable to prostaglandins (Ai et al. 1986).

When the accessory glands of a single male *Schistocerca gregaria* grasshopper were implanted in virgin females, there was a much smaller effect on eggs laid per female than when the glands of three males were implanted (av. 4 vs. 40.5 additional eggs/female; Leahy 1973). The female threshold for maximum responses thus appears to be substantially above the amount any given male can supply, giving ample margin for cryptic female choice on this character, unless implants are much less effective in transferring active male products to female target organs than are normal copulations.

The threshold for maximum female oviposition response to male accessory gland products in the cabbage fly *Hylemya brassicae* may also be near the maximum produced by some males, though the evidence is not conclusive. Of 238 control matings, 17% failed to result in eggs being laid, while 50% of the females implanted with a single accessory gland (the male has two) failed to oviposit (Swailes 1971). Similarly, implantation of a male's accessory gland tissue (the bulbous tubular median ejaculatory duct) in virgin female stable flies *Stomoxys calcitrans* resulted in less than half the normal production of eggs (av. 19.0 vs. 50.3 eggs/female; Venkatesh and Morrison 1980).

There are additional species in which a graded dose-dependent response occurs, but it is not certain whether the doses at which submaximal female responses occur are within the ranges transferred by males during normal copulations. When females of the housefly *Musca domestica* were allowed increased lengths of copulation (and thus presumably received increased amounts of the same mix of ejaculatory duct products—an untested assumption), the number of eggs laid increased (Riemann et al. 1967, Riemann and Thorson 1969). In addition, females that were allowed to copulate for longer times remained unreceptive for more days afterward (Riemann and Thorson 1969).

Injections of extracts with different concentrations of male reproductive tracts in the silkworm *Bombyx mori* also showed that the oviposition response was clearly dose dependent (Yamaoka and Hirao 1977; see fig. 6.3). There were clear dose-dependent effects of a small fraction of a single male accessory gland on the strength of contractions (up to about three times stronger), the tonus, and also frequency of rhythmic contractions in the oviduct of the grasshopper *Locusta migratoria* (Lafon-Cazal et al. 1987; fig. 6.6). Another still uncertain case is that of the beetle *Callosobruchus maculatus*. Female receptivity to remating increased when males transferred smaller amounts of sperm due to previous mating (Eady 1995); the possibility that the change resulted from seminal products other than sperm was not tested.

INHIBITION OF REMATING

Another female response showing dose-dependent responses is rejection of further mating attempts (refractory behavior). Variations in female response can occur in either the delay until the onset of refractory behavior, or in the duration of refractory behavior. In *D. funebris*, for example, the proportion of females that refused mating three days after injection with PS-1 showed a clear positive correlation with increasing concentrations of the solution, and the response thresholds could result in selection on

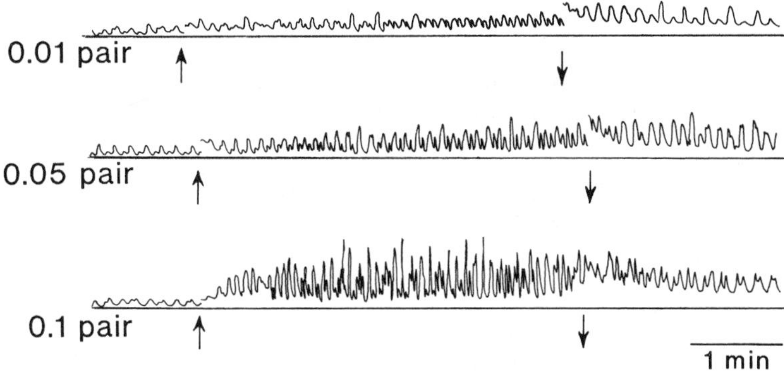

Figure 6.6 Direct stimulation of the female effector organ and possible dose-dependent effects of a male seminal product suggest the action of sexual selection in the migratory grasshopper *Locusta migratoria*. A substance in the male accessory glands causes contractions (vertical displacements in the recordings shown above) in isolated preparations of oviducts. The male substance was added at the moments indicated by the upward-pointing arrows and was washed out at each downward-pointing arrow. The substance was active in low doses (about 0.01 of the amount in a pair of opaque accessory glands) and showed a dose-dependent effect, at least in this range of concentrations; concentrations resulting from normal copulations are not known. (After Lafon-Cazal et al. 1987)

males (Baumann 1974b). The female threshold for maximum response was right at the amount (2 micrograms) thought to be present in the glands of a single normal male, and labeling experiments indicated that only about one-third of the gland's contents are transferred at each mating (Baumann 1974b).

In the blowfly *Lucilia cuprina* (P. Smith et al. 1989, 1990) the length of the delay in the beginning of the switch-off in female receptivity may be dose dependent. Switch-off after normal mating seems to be quicker (<1 hr) than in flies injected with an extract of male accessory gland, where the complete effect takes >5 hr to manifest itself (P. Smith et al. 1989). The delay in acceptance of a male (time from the first mating attempt to genitalic coupling) changed gradually (see fig. 3.11), giving evidence of what P. Smith et al. (1990) call a "graded switch-off." Large males of *Lucilia cuprina* apparently transfer greater quantities of receptivity-reducing substances (Cook 1992), so larger males may induce female refractory behavior more rapidly than small males. This seems more likely to be a biologically important advantage to larger males than their demonstrated greater ability to switch off more than five females in quick succession (Cook 1992); field data suggest that males seldom if ever have the chance for five rapid copulations (references in Cook 1992).

The size of the spermatophore introduced by the male of the butterfly *Pieris rapae* has a graded effect on the frequency of tonic afferent impulses

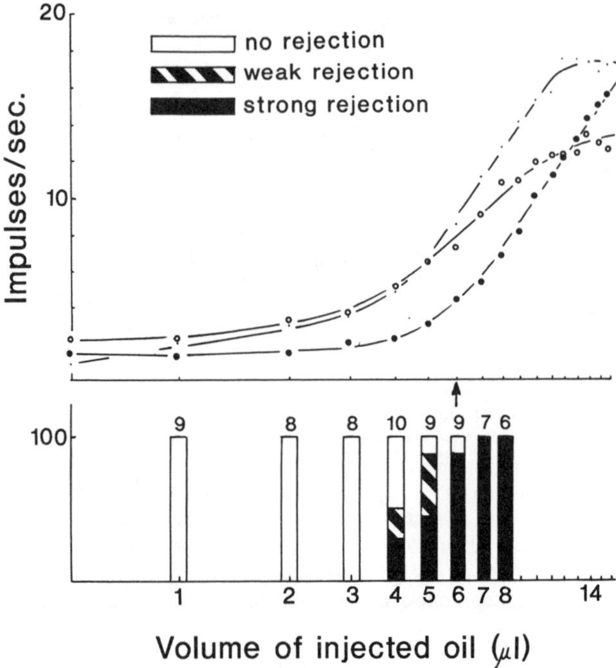

Figure 6.7 Dose-dependent response of stretch receptors in the bursa of a female *Pieris rapae* butterfly, elicited by introducing different volumes of mineral oil into the bursa (responses of three different females are shown in the upper graph; note the logarithmic scale on the *x*-axis). The volume of the spermatophore transferred into the bursa during copulation is approximately 6 microliters (arrow on *x*-axis). Variation in female rejection behavior (% of male copulation attempts rejected) is shown as a function of the volume of injected oil in the bar graphs below (sample sizes shown for each volume). The female receptor shows greatest changes in responses (steepest slope of lines in the top graph) to volumes which would represent especially large spermatophores, a design appropriate for discriminating among males on the basis of spermatophore size. Rejection is also inconsistent for volumes at and just below the average spermatophore size. A simple on-off response by the female at the lower end of the range of normal spermatophores, as predicted by the natural selection hypothesis (e.g., fig. 6.5), clearly does not occur. (After Sugawara 1979)

and the presence of silent periods from stretch receptors in the bursa, as well as on the frequency of waves of contraction in the bursa (Sugawara 1979). These responses induce female mate refusal. Thus females have graded sensory information that could be used to discriminate among spermatophores in the range of sizes normally produced by males (fig. 6.7). Experimental alterations in the volume of material in the bursa also showed that females use this information on the amount the bursa is stretched in triggering mate-rejection behavior (fig. 6.7; Sugawara 1979).

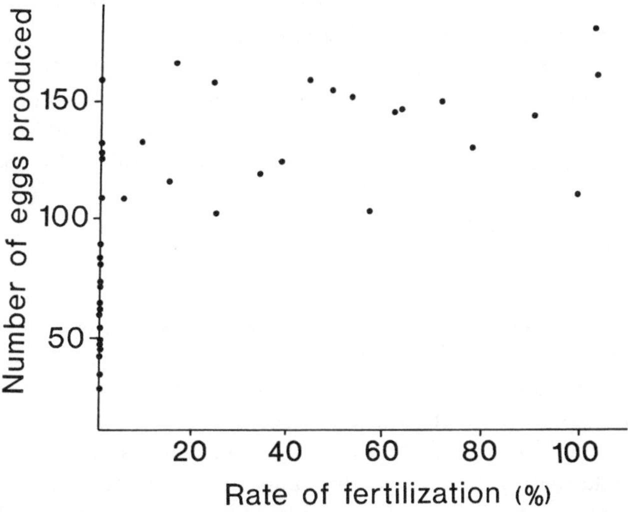

Figure 6.8 A probable all-or-none female response in egg production to a male cue (viable sperm), as predicted by the natural selection hypothesis. Males of the moth *Zeiraphera diniana* that had been irradiated were mated with females. Egg production (number laid plus number in the female's reproductive tract when she died) was lower when none of her eggs were fertilized but did not increase as rates of fertilization (and presumably number of sperm transferred) increased. Such all-or-none female responses are apparently not common, suggesting that the sexual selection hypothesis may often be applicable. (After Benz 1969)

ALL-OR-NONE RESPONSES

Females of two species of moths may show the all-or-none type of response predicted by the natural selection hypothesis. In the tortricid *Zeiraphera diniana* the presence of only just enough viable sperm (or associated material) from irradiated males for 1–2% fertilization is sufficient to trigger the complete oogenesis response of the female (fig 6.8; Benz 1969). Repeated copulations did not change the number of eggs produced. Confirmation of the natural selection prediction is somewhat uncertain, however, because the total number of sperm transferred (viable plus nonviable) was not established, and sterile sperm may be equally able to trigger female responses (irradiated males of another moth, *Trichoplusia ni*, transfer reduced numbers of sperm; Holt and North 1970). Benz (1969) argues that, in contrast to egg production, triggering of oviposition in this species is dose-dependent. Oviposition is said to be cued on the concentration of spermatophore-associated substances, but I was unable to understand which data justified this claim. Benz may have been misled by using absolute numbers of eggs laid rather than proportions of eggs produced.

A second possible all-or-nothing effect occurs in the pyralid *Helico-*

verpa zea, in which injection of a male "pheromone static factor" causes attractant pheromone in the female to disappear (presumably it is resorbed or degraded) soon after mating. Kingan et al. (1993b) found that "raw data (not shown) shows that individual females may have different thresholds, above which the pheromonostatic response is often complete rather than graded." Since these thresholds were in the range of only 0.1 equivalent of a male's accessory gland plus duplex structure, all-or-nothing female responses may be the rule in this species.

"OVERPRODUCTION" BY MALES

An otherwise puzzling phenomenon that may be explained by cryptic female choice on the basis of dose-dependent effects involves cases of apparent "overproduction" of male substances. For instance, in the mosquito *Aedes aegypti*, an extract from the accessory glands of a single male is capable of inhibiting copulation for life in more than sixty females (Craig 1967). Male mating frequencies in nature are not known, but they seem unlikely to be anywhere near sixty. Several hours are required for the sterilizing effect of the male product to manifest itself, however, and females could presumably mate repeatedly during this period. The male's "overproduction" could thus result from cryptic female choice favoring greater speed in inducing female refractory behavior.

Other injection experiments gave weaker indications of overproduction. Semicastrated *Lucilia cuprina* blowfly males did not differ statistically from sham-operated males in their abilities to inhibit female receptivity seven days after copulation (P. Smith et al. 1990). Radioactive labeling studies showed that when a second mating closely followed the first, an intact male transferred more material (av. 109 counts/min vs. about 64 counts/min for a semicastrated male), but the same level of female refractory behavior was maintained both one and seven days after copulation (P. Smith et al. 1990).

Summarizing the data on dose dependence, the observations are less complete than those discussed in previous sections because fewer studies have been performed. A few species clearly show graded female responses, with a maximum response threshold higher than that predicted by the natural selection hypothesis. Data for other species suggest, though less conclusively, similar traits. A few species appear to have an all-or-nothing response, as predicted by the natural selection hypothesis.

Interspecific Diversity of Male Products

Under natural selection there is no reason to expect chemical cues in the male semen to evolve particularly rapidly or divergently. In fact, messenger molecules for such crucial processes might be expected to be relatively conservative. Internal messenger substances are isolated from most effects of habitat change and act after all precopulatory species-isolating mechanisms that reduce hybridization (due to natural selection against the production of hybrids) have exercised their effects. Many hormonal substances have been extraordinarily conservative over evolutionary time (see the end of this section).

The prediction under sexual selection is different, since rapid, divergent evolution is common in sexually selected characters (Darwin 1871; West-Eberhard 1983, 1984; Eberhard 1985). The combination of the inexorable competition between males for access to female gametes, the evolutionarily ephemeral nature of advantages in this competition, and the advantages that females can derive from adjusting their criteria to favor particular males would continue to favor new products or variations of old products in the male semen that induced favorable female responses.

This evolution may result in different patterns of divergence. Because males can benefit from adding items to their arsenals of charms, and because a large number of stimuli within the female could possibly increase female responses, males may often evolve to add new substances to the mix of products already present in their semen. Such a change in males could lead to an increase in female response without necessarily eliminating responses to previous products. A second possible type of male evolution would be alteration of existing seminal products. This type of divergence is less likely, however, if males are mimicking female messenger molecules. Probably because female messenger molecules often control a number of different processes, and thus have multiple, pleiotropic effects in her body, such molecules and the female responses to them tend to be very conservative evolutionarily (see below).

These considerations suggest relative conservatism in the basic control mechanisms of many female processes, though not necessarily in their details. At the same time, diversity is expected in male products and the mixtures of such products that function to affect these female processes. One practical consequence for the present discussion is that tests of cross-specific stimulation of females may give less consistent results than direct assays for divergence in male seminal products.

There are multiple taxonomic levels at which one can check for diversity. The contrast between the predictions of natural selection and cryptic female choice is sharpest in comparisons between closely related species, in which there has not been enough time for the slower changes expected under natural selection, but sexually selected characters have often had sufficient time to diverge.

As already noted by P. Chen (1984), data on the accessory gland substances of the best-studied genus, *Drosophila*, indicate very rapid divergence indeed, and thus clearly support the sexual selection hypothesis. The most direct comparisons are in those species in which the male products have been characterized. A male substance that increases female fecundity in *D. melanogaster* is a 36-amino-acid peptide (Chen et al. 1988), while one that has the same effect in *D. funebris* is apparently a glycine-carbohydrate derivative (Baumann 1974a). The accessory gland of a third species, *D. nigromelanica*, completely lacks a sex peptide but has a very large amount of glutamic acid (which may function as a neurotransmitter; Chen and Oechslin 1976). The *Acp26Aa* and *Acp26Ab* genes that code for accessory gland proteins in four species of the *D. melanogaster* subgroup show an unusually high level of amino acid replacement that is apparently not neutral (Aguadé et al. 1992).

A second possible group of well-characterized molecules that permits comparisons is the esterases. In *D. melanogaster* esterase-6 (EST-6) in the semen influences sperm transfer, sperm precedence patterns, loss of stored sperm, and female oviposition and remating (Brady and Richmond 1990; Oakeshott et al. 1995). High levels of expression of the esterase *EST-6* gene in the male ejaculatory duct occurs only in the three sibling species of the *D. melanogaster* subgroup; in some but not other species in this genus, additional copies of this gene occur and are also expressed at high levels in the ejaculatory bulb, but their products lack esterase activity (Oakeshott et al. 1995).

Comparison of all the major protein components in the male accessory glands of related species also reveals striking divergences. The major protein components of accessory gland secretions in eight *Drosophila* species differed in molecular weight and electrical charge (fig. 6.9), as well as in isoelectric points. There were even clear differences among the three sibling species *D. melanogaster*, *simulans*, and *mauritiana* (Chen and Oechsin 1976; Chen 1984; Chen et al. 1985; Stumm-Zollinger and Chen 1988). These differences are especially dramatic, since females of these three species are nearly indistinguishable on the basis of their external morphology, and male identification depends on small differences in the genitalia.

Bownes and Partridge (1987) found a similar diversity in electrophoretic analyses of the accessory gland proteins of *D. melanogaster* and *D. pseudoobscura*. Not a single one of the eight major polypeptides of one species was present in those of the other. Accessory gland proteins differ more among different species of *Drosophila* than do testis-specific proteins (Coulhart and Singh 1988). Electrophoresis of the contents of the female spermathecal duct after mating in two species of *Nezara* stinkbugs also showed species-specific differences in large proteins (>64,000 Da; Kon et al. 1993). The results of cross-specific matings confirmed that these products came from males rather than females.

The electrophoretic mobilities of beta esterases in the male ejaculatory

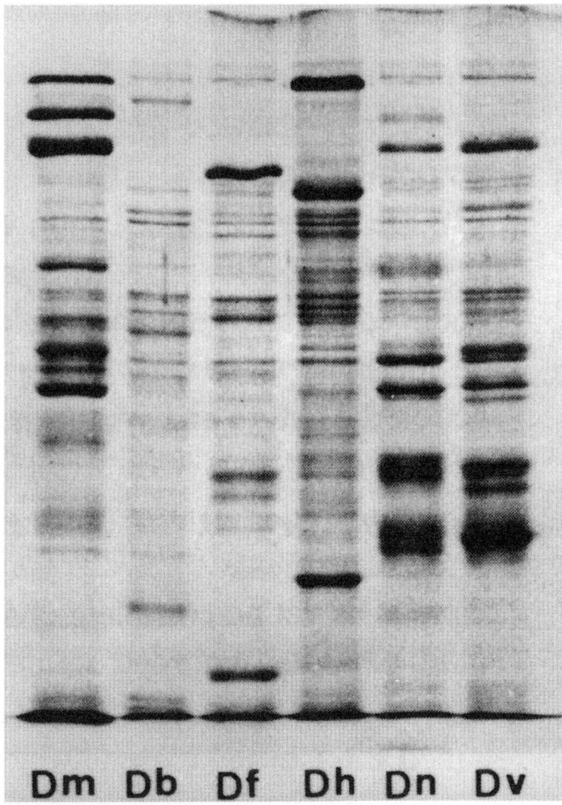

Figure 6.9 Electrophoretic separation of proteins from the male accessory glands of six species of *Drosophila* (Dm = *D. melanogaster*, Db = *D. busckii*; Df = *D. funebris*; Dh = *D. hydei*; Dn = *D. nigromelanica*; Dv = *D. virilis*). Species differences in the arrays of proteins are striking, despite the fact that the glands are morphologically quite similar (the glands also contain many nonproteinaceous substances not stained in this preparation). Some proteins in male semen are known to manipulate female reproductive processes to the male's advantage. The interspecific differences illustrated here suggest rapid divergent evolution in male accessory gland products, as expected to occur often if they are under sexual selection. (From Chen 1984)

bulb also varied widely among species of *Drosophila* (Johnson and Bealle 1968). While the functions of most of these proteins are not known, it seems likely that many are transferred to the female during copulation or are associated with such substances (e.g., are precursors).

It is important to compare the divergence of supposedly sexually selected characters with that of other, presumably not sexually selected traits of the same species (West-Eberhard 1983). Comparisons of this sort, which are lacking in some of the studies just cited, were attempted directly in an electrophoretic study of scarab beetles. As predicted by the sexual selection hypothesis, major proteins in the accessory glands of eight species in the genus *Phyllophaga* and four in the genus *Macrodactylus* showed greater diversity than did proteins in the digestive tract and muscles (Eberhard and Vargas, in prep.).

A second type of evidence concerning the diversity of male seminal products in related species comes from tests of cross-reactivity. When the females of one species receive male products from a different species, changes in their responses can give a conservative indication of differences among male products. Although there were early claims that male products and female responsiveness did not differ among groups (based partly on qualitative rather than quantitative tests; e.g., Leahy 1970), the pattern that has emerged subsequently is that it is common for male products to fail to elicit as strong a response in females of congeneric or closely related species as in conspecific females (table 6.2). In fifteen species pairs, products from conspecific males elicited larger responses than those from males from another congeneric species; in six the effects of heterospecific males were the same; in one case they were greater; and in one the responses were asymmetrical.

One should keep in mind that the results of many of the cross-reactivity studies in table 6.2 were given as simple yes-no answers, and may have missed fine differences in reactions. In this sense, the table is biased against finding differences among species. Some species are included in more than one comparison, introducing taxonomic biases.

It is noteworthy that even the very closely related sibling species *D. melanogaster*, *D. simulans*, and *D. mauritiana* showed differences between the effects of cross-injections as compared with conspecific injections on both ovulation and female receptivity (Stumm-Zollinger and Chen 1988). The pair of species with asymmetrical effects (*D. pulchrella* and *D. suzukii*) are also very closely related and can produce fertile hybrids (Fuyama 1983).

Cross-specific implantations of the male accessory glands of *Aedes* mosquitoes stimulated egg development less effectively than similar transplants in conspecific females (Klowden and Chambers 1991). The differences in relative accessory gland size and effects on oviposition behavior (and lack of it) in virgin females after feeding also suggest interspecific variation in control within one subgenus of *Anopheles* mosquitoes (Lounibos 1994). Additional types of evidence also suggest intrageneric diversity. Similar molecules in different species of *Drosophila* have different functions. A polypeptide of 27 amino acid residues reduced female receptivity in *D. funebris*, while a similar molecule stimulated oviposition in *D. melanogaster* (Chen 1984).

Differences in the morphology of the accessory glands of closely related species also suggest possible diversity in male seminal products. Gillott (1988) noted "considerable differences in the form of the glands" in several groups, and cites a key for the identification of nineteen species of bruchid beetles based on the number, shape, and structure of these glands. In three congeneric mosquito species, *Aedes triseriatus*, *A. aegypti*, and *A. taeniorhynchus*, gland structure and sites of matrone production also differ (Lum 1961; Ramalingam and Craig 1976). Anterior secretory cells in the pear-shaped accessory gland of *A. aegypti* produce substances that both inhibit remating and stimulate oviposition. In *A. triseratus* the same gland is elongate, differentiated into anterior and posterior portions, and the substance or substances that inhibit female remating are produced only in the posterior region (Ramalingham and Craig 1978).

The significance of the rapid divergence in male products just discussed is emphasized by comparison with the extreme conservatism in female messenger molecules, and in hormones in general. For example, estradiol functions as a hormone in starfish (Schoenmakers 1980) and a variety of vertebrates (e.g., Crews and Silver 1985) (and it also occurs, with as yet undetermined functions, in flatworms, molluscs, and insects; Sandor 1980). Other examples are ecdysteroids (in insects, crustaceans, spiders, and ticks; Hagedorn 1980; Bonaric 1987; Connat et al. 1986), juvenile hormone and similar compounds in insects and ticks (Engelmann 1970; Connat et al. 1986), and prostaglandins E and F (in corals, fish, mammals, ticks, birds, and insects; Lange 1984; Stacey 1984; Destephano et al. 1982; Murtaugh and Denlinger 1982).

A possible link between a male seminal product and an extremely conservative molecule of this sort occurs in *D. melanogaster*. Monsma and Wolfner (1988) determined the amino acid sequences of two proteins (coded by tightly linked genes) in the accessory glands of *D. melanogaster*. The function(s) of the proteins are not known, and their relations (if any) with the biologically active substances isolated by other workers (e.g., Baumann 1974a,b in *D. funebris*; Chen et al. 1988 in *D. melanogaster*) are not clear. But one stretch of 17 amino acids in the 264-amino-acid molecule *msP 355a* (later *Acp26Aa*) shows 11 identical amino acids (allowing one gap in each sequence) when compared with a portion of the 36-amino-acid molecule ELH of the very distantly related mollusc, *Aplysia californica* (fig. 6.10). This protein, along with two other similar proteins (the three coding genes share 90% sequence homology), elicits egg-laying behavior in the mollusc (Scheller et al. 1983). Males of *D. melanogaster* transfer *Acp26Aa* to the female during copulation. If it functions to elicit oviposition in *D. melanogaster*, and if the similarity with ELH is not due to chance (Monsma and Wolfner calculate the probability of this

Table 6.2.
Tests of Differences in the Effects of Male Seminal Products on Conspecific and Heterospecific Female Flies When Male Accessory Glands (or Extracts of Them) Were Introduced into the Female

Male	*Female*	*Product*	*Heterospecific Male Gives Less, Equal, or Greater Effect?*	*References*
Aedes aegypti	*A. albopictus*	acc. gl.	less (ovip.),[a] equal (ovip.)[b]	Leahy and Craig 1965; Leahy 1970
A. aegypti	*A. atropalpus*	acc. gl.	less (remate)[c]	Craig 1967
A. aegypti	*A. triceratus*	acc. gl.	less (remate)[a]	Craig 1967
A. albopictus	*A. aegypti*	acc. gl.	less (ovip.)	Leahy 1970; Leahy and Craig 1967
A. aegypti	*C. pipiens*	acc. gl.	equal (ovip.)	Leahy 1967
		acc. gl.	less (remate)[a]	Craig 1967
A. aegypti	*A. scutellaris*	acc. gl.	less (remate)[a]	Craig 1967
D. melanogaster	*C. pipiens*	acc. gl.	equal (ovip.)	Leahy 1967
D. melanogaster	*A. aegypti*	acc. gl.	less (ovip.)	Leahy 1967
A. aegypti	*D. melanogaster*	acc. gl.	less (ovip.)	Leahy 1967
			less (remate)[a]	Craig 1967
A. sollicitans	*A. taeniorhynchus*	acc. gl.	greater (oogen.)[d]	O'Meara and Evans 1977
A. triseriatus	*A. aegypti*	acc. gl.	less (ovip.)[a] less (remate)[a,b]	Craig 1967; Ramalingam and Craig 1976
A. triseriatus	*A. atropalpus*	acc. gl.	less (remate)[a,b]	Craig 1967; Ramalingam and Craig 1976
C. pipiens	*A. albopictus*	acc. gl.	less (remate)[a,b]	Craig 1967
C. pipiens	*A. aegypti*	acc. gl.	less remate[a]	Craig 1967
			equal (ovip.)[e]	Leahy 1970
C. pipiens	*D. melanogaster*	acc. gl.	equal (ovip.)[e]	Leahy 1970
Stomoxys calcitrans	*Musca domestica*	ej. duct	less (remate)	Morrison et al. 1982
S. calcitrans	*Phormia regina*	ej. duct	less (remate)[a]	Morrison et al. 1982
S. calcitrans	*Sarcophaga bullata*	ej. duct	less (remate)[a]	Morrison et al. 1982
P. regina	*M. domestica*	acc. gl.	less (remate)[a]	Nelson and Adams unpub. in Riemann and Thorson 1969
Drosophila funebris	*D. melanogaster*	acc. gl.	less (remate)[a]	Baumann 1974b
D. suzukii	*D. pulchella*	acc. gl.	less (ovul.)	Fuyama 1983
D. pulchella	*D. suzukii*	acc. gl.	equal (ovul.)	Fuyama 1983
D. busckii	*D. melanogaster*	acc. gl.	less (ovul.)[a]	Chen et al. 1985
D. funebris	*D. melanogaster*	acc. gl.	less (ovul.)[a]	Chen et al. 1985
D. hydei	*D. melanogaster*	acc. gl.	less (ovul.)[a]	Chen et al. 1985
D. nigromelanica	*D. melanogaster*	acc. gl.	less (ovul.)[a]	Chen et al. 1985
D. virilis	*D. melanogaster*	acc. gl.	less (ovul.)	Chen et al. 1985

Table 6.2 (*cont.*)

Male	*Female*	*Product*	*Heterospecific Male Gives Less, Equal, or Greater Effect?*	*References*
D. melanogaster	*D. simulans*	acc. gl.	equal (ovip., remate)[e]	Chen et al. 1988
D. melanogaster	*D. mauritiana*	acc. gl.	equal (ovip., remate)[e]	Chen et al. 1988
D. melanogaster	*D. seychellia*	acc. gl.	equal (ovip., remate)[e]	Chen et al. 1988
D. melanogaster	*D. funebris*	acc. gl.	equal (ovip., remate)[e]	Chen et al. 1988

[a] No or "almost no" effect in cross-specific pairs.

[b] Differences were relatively slight and possibly not statistically significant (no tests were reported).

[c] Cross-specific male products affected the female, but lack of quantitative data precludes comparison with the effects of those of conspecific males.

[d] This result is especially surprising, since *A. sollicitans* is essentially never autogenous.

[e] No data were given other than qualitative statements.

Figure 6.10 A possible case of extreme conservatism in a product that may alter female behavior is illustrated by comparing a male accessory gland substance in *Drosophila melanogaster* with a female hormone in the nudibranch mollusc *Aplysia californica*. A protein consisting of 155 amino acids produced by the male *D. melanogaster* has a region of striking similarity (11 aa identical in a stretch of 17 aa, indicated by boxes in the figure) with a smaller peptide (36 aa long) from the nudibranch (the chances that the similarity is coincidental are about 3.1×10^{-6}). The *Aplysia* molecule acts as a hormone, releasing oviposition behavior. Presumably the male *Drosophila* protein stimulates oviposition behavior in conspecific females, though proof is lacking. (After Monsma and Wolfner 1988)

as 3.1×10^{-6}), then females of these extremely distant groups have retained a striking similarity in cues used to trigger oviposition.

In summary, the data from the best-studied genus, *Drosophila*, clearly indicate substantial intrageneric differences, giving seemingly rather spectacular confirmation of the prediction of the sexual selection hypothesis. Differences in patterns of protein banding with electrophoresis, differences in the chemical nature of signaling substances, and lack of cross-reactivity in both induction of ovulation and switching off of female receptivity also point to intrageneric diversity of male seminal products affecting female reproduction. Less complete data from mosquitoes and other orders seem to follow the same trends. Thus, at least for the few genera for which there are data, the prediction of intrageneric differences made by the sexual selection by cryptic female choice hypothesis seems to be generally confirmed.

A final word of caution is in order. Striking *intra*specific variation occurs in seven of the fourteen major accessory gland proteins in *Drosophila melanogaster* and includes variation among animals from the same geographic sites (Whalen and Wilson 1986). "Numerous" electrophoretic types of beta esterases occurred in different *D. aldrichi* males (Johnson and Bealle 1968). Cooke et al. (1987) found ten allozymic classes at the esterase-6 locus in a population of *D. melanogaster*. Intraspecific variation of this sort could be due to a lack of selective importance for these substances. The question of whether or not the bands that differ in these electrophoretic analyses (and those of Chen et al. 1985; Stumm-Zollinger and Chen 1988; Bownes and Partridge 1987 on other *Drosophila*; Kon et al. 1993 on stinkbugs; Eberhard and Vargas, in prep., on beetles) correspond to proteins with important physiological effects on females remains to be solved.

Triggering of Appropriate Female Responses

If male chemical cues and female responses to them evolved under natural selection, then some of the female responses could be either unfavorable to males or have no effect on their reproduction. If, on the other hand, male chemical cues have been under sexual selection by cryptic female choice, then the only female responses expected are those which would increase the male's reproduction. This test is weak because, as noted above, many of the female responses favored by natural selection are also favorable to male reproductive interests, and *natural* selection on males could also act against any seminal products having negative effects on the male's reproduction. In light of the wide variety of female responses affected by male sexual products (table 6.1), however, it is worth checking to see if male reproductive interests are always favored.

Nearly all the data tell the same story: the effects of male accessory gland products transferred to the female favor the male's reproduction either directly or indirectly (Chen 1984). In only two cases do aspects of the female's response appear to reduce rather than favor the male's reproductive interests, and in at least one the total effect seems nevertheless to favor the male. Esterase-6 in the semen of *D. melanogaster* works against the male's reproduction by increasing the long-term probability that the female will remate, probably due to sperm leakage from female storage organs (Gilbert et al. 1981a). In this case, however, the same substance also both reduces female remating on the short term (Scott 1986), and strongly increases the male's sperm precedence (Gilbert et al. 1981a). On balance it is thought that the effect of esterase-6 is favorable to the male (Gilbert et al. 1981a).

In the beetle *Acanthoscelides obtectus* one of the two components (B) of the active fraction from the male spermatophore (which as a whole stimulates oogenesis) inhibits oogenesis in low amounts (0.08 μg), perhaps by antagonizing the oogenesis-stimulating effect of the other component; it is also toxic to the female at higher amounts (0.3 μg; Huignard et al. 1977). The significance of these contradictory effects is unclear. To evaluate the effect of component B in a biologically realistic situation, one would need to test the effect on oogenesis when the spermatophore contains B and when it lacks B.

In sum, the trend predicted by the sexual selection hypothesis clearly occurs. The predictions of the natural selection hypothesis are not sufficiently distinct, however, to permit discrimination.

Mimicry of Female Messenger Molecules

Under the natural selection hypothesis, the origin of female use of male products as reproductive cues would occur when the female began to sense and respond to a substance in the ejaculate that had some other functional significance (e.g., provide nutrition for sperm, lubricate their passage, buffer adverse conditions within the female ducts). It seems unlikely that such male substances with such different roles would tend to resemble molecules such as hormones or neurotransmitters used by the female as messengers within her own body. Such hormone-like substances have been available as possible candidates when females evolved to respond to male seminal products only if such female messenger molecules were present in the male (perhaps a common situation), and if they "leaked" into his ejaculate in large enough quantities to be sensed by the female, or if they also happened to have some direct positive effect on sperm nutrition, survival, or movement (presumably relatively uncommon conditions).

In contrast, if males are under selection to influence and manipulate female responses for their own interests, molecules used as messengers within the female body could be especially powerful tools. By introducing such a substance in a female, the male could take advantage of her built-in responses to the same or similar molecules. He might also reduce or avoid the risk of having his message filtered or possibly ignored by the female. Use of such substances could thus be an especially effective "sensory trap" (West-Eberhard 1983; Christy 1995; see chap. 2.6). This is another "invasive" or aggressive trait like those discussed above, and resembles similar manipulations practiced by parasitoid wasps on their hosts (e.g., Holler et al. 1994).

The predictions of the two hypotheses are only approximate. The natural selection hypothesis predicts that mimicry will be absent or very rare; sexual selection predicts that mimicry will be more common, but possibly only sporadic. Chemical mimicry would not necessarily evolve under sexual selection in all groups, since it would depend on imponderables such as the occurrence of appropriate mutations or the preexistence of similar molecules in males. The data are scarce, as the chemical nature of male accessory gland products is known for only a few species of insects (Gillott 1988). Even when a male product has been isolated and characterized, lack of demonstration that it resembles a female molecule may simply mean that no one has attempted to compare it with female molecules. Nevertheless, some striking resemblances between male seminal cues and female messenger molecules do occur.

INDUCTION OF OVIPOSITION

Perhaps the clearest cases are the crickets *Acheta domesticus* and *Teleogryllus commodus*, in which, as already mentioned, males transfer an enzyme or enzyme complex which, in the spermatheca of the female, acts on a substrate to produce prostaglandin, a hormone with a variety of physiological effects (Destephano and Brady 1977; Loher et al. 1981). In gravid females of both species, prostaglandin injections cause sharp increases in the numbers of eggs oviposited (Destephano and Brady 1977; Loher 1979). The seminal fluid of the silkworm *Bombyx mori* also contains prostaglandin (Setty and Ramaiah 1979, 1980). Females of both *T. commodus* (Ai et al. 1986) and *B. mori* (Setty and Ramaiah 1980) produce prostaglandins of their own, and the widespread occurrence of these substances and their varied physiological effects (Brady 1983; Loher 1984) make it very probable that the males of all these species are "mimicking" female hormone molecules.

Males of the grasshopper *Locusta migratoria* transfer a substance that is apparently similar to the neurotransmitter octopamine-2A of the female. It appears to promote ovulation and/or oviposition by acting on the receptors for this transmitter in the female oviduct (Lafon-Cazal et al. 1987).

In *Drosophila melanogaster*, a portion of what is probably a female fertility-enhancing protein, *Acp26Aa*, is produced in the male accessory gland and transferred to the female reproductive tract during mating, and some of it rapidly enters her hemolymph (Monsma et al. 1990; after only 10 min of the 20 min copulation, the protein is already in her hemolymph). As mentioned previously, this protein is highly similar to a protein used in the female organs of the distantly related mollusc *Aplysia* to release oviposition behavior (fig. 6.10; Monsma and Wolfner 1988). If female *Drosophila* utilize a protein similar to that used by female *Aplysia*, the male substance may be a mimic of that of the female.

INHIBITION OF REMATING

In discussing evidence of the effects of receptivity-inhibiting substances in flies (*Musca* in particular), Gillott (1988) favored the hypothesis that these substances compete with juvenile hormone for the same receptor sites in the female, since in many species juvenile hormone enhances willingness to mate (e.g., Engelmann 1970). Gillott noted that the apparently dose-dependent effects (e.g., duration of copulation correlated with length of refractory period; Riemann and Thorson 1969) are in accord with this receptor-block model of male receptivity-inhibiting substances.

Further suggestions that males use juvenile hormone (JH) or analogues in their semen come from studies of moths and roaches. Males of several species of saturniid moths produce and store large quantities of JH in the abdomen (references in Webster and Cardé 1984). Males of *Hyalophora cecropia* accumulate large quantities of JH in their accessory glands and transfer them to females during copulation (Shirk et al. 1980). After mating, female calling behavior is reduced. Webster and Cardé (1984) showed that JH induced both termination of calling behavior and the gradual elimination of attractant pheromone from the female's glands in another moth, *Platynota stultana* (in a different family). They speculated that the function of JH transfer in the seminal fluid of some moths may be to induce the switch from virgin to mated behavior in the female. Female moths also synthesize their own juvenile hormone, which can affect pheromone release activities (Cusson and McNeil 1989). In the roach *Blabera fusca* topical applications of acetone extracts of the spermatophore induce the same increases in vitellogenesis as result from similar applications of juvenile hormone mimics (Brousse-Gaury and Goudey-Perriere 1983).

INDUCTION OF SPERM TRANSPORT

Finally, the male seminal factor in the bug *Rhodnius prolixus* that induces female contractions which result in sperm transport to the female's spermathecae is an indolakyl amine, with properties similar to serotonin (Mann

1984). The accessory gland of the roach *Periplaneta americana* also contains a pharmacologically active material (Davey 1960a).

In summary, the chemical identity is not known for enough male factors to permit widespread tests of the mimicry predictions. There are several cases, however, in which indirect evidence suggests that male mimicry of female hormonal substances may occur, as predicted by the sexual selection hypothesis. Mimicry of this sort is unlikely under the natural selection hypothesis.

Sperm vs. Other Seminal Products as Signals

Under natural selection on the female to use a male product as a signal that a successful copulation has occurred, usually the most important information is whether or not there are sufficient healthy sperm in the female's storage organ or elsewhere in her reproductive tract (e.g., Drummond 1984). There would be strong natural selection against triggering responses such as oviposition and reduced receptivity following copulations in which potent sperm failed to arrive in female storage organs even though other male seminal products were transferred. Other male seminal products, such as accessory gland substances, can give only indirect indications of the basic "I have received sperm" message.

The selective importance to the female of using other seminal products as opposed to sperm as cues would depend on the frequency of matings in which accessory gland products but not potent sperm are transferred in normal amounts. Sterile copulations per se appear to be common in many groups (see chap. 9.1), and transfer of incomplete sperm complements are presumably even more common. But it is not clear how strong the correlation between failure to transfer sperm and failure to transfer other components of the semen is. The more frequently such correlations fail to occur, the greater the disadvantage to the female of using an indirect indicator of insemination, such as accessory gland products rather than sperm. In sum, one would predict a general though inconsistent trend for females to use sperm rather than other substances in the semen as signals.

The sexual selection hypothesis, on the other hand, makes a quite different prediction. Male signaling substances independent of sperm would be able to escape the confines of the female's reproductive tract (where the sperm must remain in order to fertilize eggs) and move "invasively" to the very organs whose activity the male is attempting to influence. In addition, the diversity of chemicals incorporated in sperm themselves would presumably be more limited than that of other substances in the semen, because of the other important functions of sperm, such as fertilizing eggs.

The absolute quantities of male products included in a given sperm cell would also be more limited.

The data in table 6.1 suggest that insects tend to utilize accessory gland substances rather than sperm as signals, which is predicted by the sexual selection hypothesis. This trend may be the result of a bias in the experimental techniques used, however. The most common experimental techniques (injections of extracts and organ implants in the female body cavity) are not likely to give adequate simulations of the stimuli associated with the sperm themselves, which would presumably be sensed inside the female reproductive tract. Thus sperm effects may be underestimated. A further possible bias is that many studies (especially those involving implantation) report on only the effects of one type of tissue—accessory glands, *or* testes, *or* sperm, but not all three. Larger numbers of cases involving accessory glands may simply be due to researchers more often testing accessory glands than testes—possibly a vicious cycle started by the bias in the potential of results from injection and implantation techniques.

In sum, the effects of male accessory glands certainly outweigh those of sperm in the studies cited in table 6.1, and this trend follows the prediction of the sexual selection hypothesis rather than that of the natural selection hypothesis. The data, however, may be heavily biased in some respects. The confirmation of the sexual selection hypothesis is not convincing. If the data had shown the opposite trend, however, they would have given strong evidence against the sexual selection hypothesis.

6.3 Ticks

Ticks are highly derived arachnids living as ectoparasites on vertebrates. They are only distantly related to insects, and presumably many aspects of their reproductive biology have evolved independently from those of insects. The importance of ticks as disease vectors, combined with their moderate size (permitting surgical manipulations), have led to a considerable literature on their reproductive physiology (table 6.3). Small sample sizes characterize many of these studies, however, and one paper with particularly wide-ranging conclusions (Galun and Warburg 1967) included almost no data. I have assumed in the discussions below that the trends described by the authors are real.

A brief summary of tick reproductive behavior will make the descriptions that follow more intelligible. Male ticks are apparently often attracted to and recognize receptive females chemically (in some groups the females are already attached to the host and are immobile; Feldman-Muhsam 1986). Mating apparently always involves preliminary insertion of the

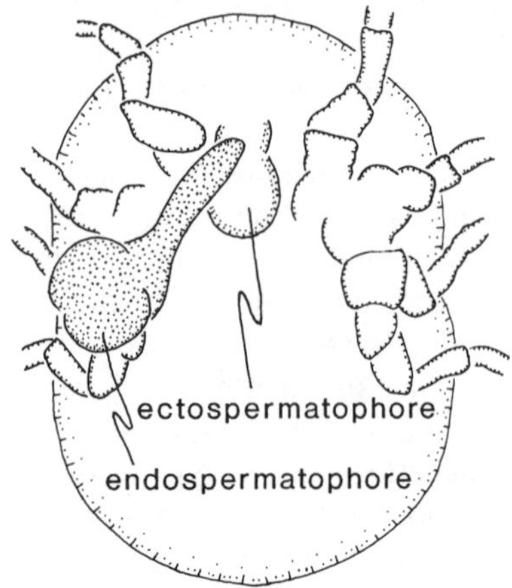

Figure 6.11 A male *Ornithodoros savignyi* tick holds a grenadelike spermatophore in his mouthparts. The ectospermatophore, a smaller, harder structure, is normally placed on the outer surface of the female, with its tip inserted in her vagina. Insertion is followed by a dramatic expansion of an internal structure, the endospermatophore (stippled), which is driven deeper into the female's body by generation of CO_2 gas in the ectospermatophore. Endospermatophore expansion occurred in this spermatophore after the male was removed from the female. (After Feldman-Muhsam 1986)

male's mouthparts into the female's genital opening (Feldman-Muhsam 1986). In some species the male inserts and withdraws his mouthparts repeatedly (as in genitalic thrusting; see chap. 5.2); he sometimes also repeatedly distends and contracts the anterior portion of his body, as if pumping, but it is not clear whether any substances are taken up or transferred (e.g., see Oliver et al. 1984 on *Ornithodoros parkeri*).

The male then emits a spermatophore from his own genital opening, grasps the tip of the spermatophore with his mouthparts (in at least *Argas persicus* he removes the tip), and inserts the tip into the female's genital opening. Processes within the spermatophore, which in *O. savignyi* include generation of CO_2 gas, and an increase in the internal pressure that causes the quiescent sperm, which are contained in one or more capsules, to be projected deeper into the female, into her uterus (fig. 6.11). In some species relatively complex structures are associated with the sperm capsules (Feldman-Muhsam and Borut 1983). Sperm apparently often remain within the capsule in the female until they are used to fertilize eggs, up to two years later. Prior to fertilization sperm are capacitated, a complex process that includes an approximate doubling of length (they have no flagella), and acquisition of the ability to penetrate female tissues (Alberti 1991). Male secretions that may be transferred to the female include saliva produced during the preliminary mouthpart insertions and during spermatophore insertion, and the contents of the spermatophore itself. Female processes that must occur if the female is to reproduce, and which are often triggered by mating, include feeding (engorgement), digestive activity in her midgut cells, vitellogenesis, ovulation, oviposition, and an increase in metabolic rate (Diehl et al. 1982; Oliver 1986; Connat et al. 1986).

Several aspects of tick natural history need emphasis. The probability of remating may often be influenced by production (or lack of production) of attractant chemicals by the female (Khalil et al. 1983). Several blood proteins are involved in egg development, and when some but not others are not in appropriate concentrations, oocyte development is sometimes aborted before reaching completion (Connat et al. 1986). Oviposition is also complex and includes ovulation, movement of eggs to the uterus, and laying of eggs.

These processes are not necessarily controlled by the same cues. For instance, injection of testes extracts caused some mature eggs in *Argas persicus* to migrate to the uterus, but they were not laid (Khalil and Shanbaky 1975). Subprocesses were generally not distinguished in the studies cited below, and are subsumed in the categories in the tables.

The data discussed below are set out in table 6.3 and summarized in table 6.4. The discussions focus on egg maturation and oviposition. The predictions of the natural selection and sexual selection hypotheses are the same as those described in the corresponding sections on insects.

Invasive Action of Male Substances

In *Ornithodoros parkeri* the site of action of male products is probably within the female's hemocoel, as predicted by the cryptic female choice hypothesis, rather than in the lining of the female reproductive tract. Extracts of male reproductive organs (accessory glands, testes, and the entire reproductive system of the male) were inserted in the female uterus (where male products are normally placed), and identical extracts were injected into the hemocoel of other females. Both treatments increased oviposition, but there was a greater response in all cases to hemocoelic injections (both in percentage of females ovipositing, and in numbers of eggs laid; Oliver et al. 1984). This confirmation of the sexual selection prediction over the natural selection prediction also supports the idea that the female reproductive tract may be at least somewhat impermeable for male seminal products that affect female reproductive responses. There may also be a greater response to hemocoelic than intrauterine injections of salivary gland extracts in *O. parkeri* (Oliver et al. 1984), but the sample sizes were too small to permit analyses.[1]

Table 6.3
Effects of Products in Male Semen on Reproductive Processes of Female Ticks

Taxon	*Male Organ*	*Product*	*Effect on Female*	*Techniques*	*References*
Ixodidae					
1 *Dermacentor variabilis*	saliva or stim. with mpts.	?	feeding[a,b]	Seal male gonopore, irradiate males	Pappas and Oliver 1972
	sptop. or acc. gl. (not sperm)	?	feeding[a]		
2 *Amblyomma americanum*	acc. gl.	sptop. (mech. stim.)	not affect oogen.[a]	Inject extract, insert sand into female uterus	Oliver et al. 1975
Argasidae					
3 *Ornithodoros moubata*	testes, acc. gl.	sperm, proteins, 2 proteins, sptop. (mech.)	ovip., oogen., digest blood	Inject washed sptop., inject intact and homog. sptop., inject homog. testes, acc. gl., and distended sem. ves. of male (contains immat. sptop.), insert metal balls in female uterus, inject supernatants	Aeschlimann and Grandjean 1973a,b; Germond and Aeschlimann 1977; Diehl et al. 1982; Sahli et al. 1985; Connat et al. 1986
4 *O. tholozani*	testes, acc. gl.	sperm, sptop. (mech.), DOPA	ovip.[c], oogen.[c]	Irradiate males (no sperm), insert plastic beads and saline in female uterus, inject DOPA	Galun and Warburg 1967
5 *O. parkeri*[a]	acc. gl., testes, sal. gl.	L-DOPA	ovip., oogen.	Inject extracts, DOPA	Oliver et al. 1984; Oliver 1986
6 *Argas persicus*	acc. gl., testes	sptop. (mech.), sperm	oogen., sperm transport(?)[d] ovip.[e]	Insert plastic beads in uterus, inject extracts, insert spermless sptop., bathe isol. cockroach heart	Tatchell 1962; Leahy and Galun 1972
7 *A. arboreus*	acc. gl., testes	?, sptop. (mech.)	oogen., ovul.[b]	Inject extracts, needle through vaginal wall	Khalil and Shanbaky 1975

NOTES: All effects on remating were inhibitions; all other effects were increases.
[a] Sample size so small that confident conclusions cannot be drawn. [b] Response less than normal mating.
[c] No data given. [d] Rate of contraction increased from 75 to 83/min (no statistical test of significance presented).
[e] Assume that all other substances are transferred in normal amounts when male transfers a spermiophore without sperm.

Multiplicity of Cues for the Same Female Process

In *Ornithodoros moubata*, vitellogenesis is partially but not completely stimulated by mechanical stimuli (produced by small metal balls in the uterus) simulating those produced by a sperm capsule (Connat et al. 1986). Vitellogenesis is also triggered by chemical stimuli from males, as females responded to injections of both a homogenate of spermatophores taken from recently mated females, and a homogenate of distended seminal vesicles (Connat et al. 1986). In addition, females responded to one or more proteins associated with sperm as they completed their maturation within the female 12–24 hr after mating (Sahli et al. 1985). Two molecules, with molecular weights of $1.6–1.9 \times 10^6$ daltons and $0.1–0.2 \times 10^6$ daltons (from sperm and spermatophores, respectively), were implicated in inducing vitellogenesis. It is possible that several subunits form a larger active compound (Connat et al. 1986). There thus appear to be at least three cues of male origin in this species.

In *O. parkeri*, introduction of 2–4 male equivalents of the accessory gland, testes, and the entire male reproductive system into the female uterus resulted in only partial induction of the oviposition response that normally follows mating (Oliver et al. 1984). These authors speculated that the incomplete response may have been due to lack of complementary action of other stimuli (male mouthparts and saliva in the female vagina) normally associated with mating.

In virgin female *Argas persicus*, mechanical stimuli simulating spermatophores (plastic beads placed in the female's uterus; Leahy and Galun 1972) triggered the production of some mature oocytes. Chemical stimuli also influence females, as injection of an extract of male accessory glands and seminal vesicles into virgin females resulted in increased oogenesis.

Graded Female Responses to Quantitative Variation in Male Cues

When small plastic beads were inserted in the uterus of female *Argas persicus*, larger numbers of mature oocytes resulted from larger numbers of beads (Leahy and Galun 1972). In contrast, controlled manipulation of doses in injections of DOPA in this species failed to show dose-dependent effects on vitellogenesis in the range of 10^{-2} M to 10^{-4} M (Oliver 1986).

Yolk deposition in *O. tholozani* was slightly more frequent when greater concentrations of DOPA were injected (Galun and Warburg 1967), but small sample sizes preclude confident conclusions. The demonstration of invasive male action cited above, in which greater numbers of females oviposited, and greater numbers of eggs were laid by *O. parkeri* after hemocoelic injection of extracts of male reproductive organs than after insertion of similar amounts in the uterus (Oliver et al. 1984), also suggests dose-dependent effects of male products.

Interspecific Diversity of Male Products

Different male stimuli appear to trigger female reproductive responses in closely related species of argasid ticks (table 6.4). Tactile stimuli from the uterus (from beads introduced into the uterus) induced vitellogenesis in female *Argas persicus* (Leahy and Galun 1972), while a similar treatment in a congener, *A. arboreus*, did not induce egg maturation; but injection of male accessory gland and/or testes extracts in *A. arboreus* did (Khalil and Shanbaky 1975). When males of *A. persicus* were mated to female *A. arboreus*, they introduced spermatophores and sperm into the female but did not trigger vitellogenesis. Thus the ovary-stimulating products of male *A. arboreus* are apparently not the same as those of male *A. persicus* (Khalil and Shanbaky 1975).

Another pair of congeneric species in the same family differ in similar ways. Virgin female *Ornithodoros tholozani* females failed to respond to mechanical stimulation of the uterus by insertion of a plastic bead (Galun and Warburg 1967), while those of *O. moubata* not only developed eggs but also oviposited when small metal beads were introduced into the uterus (Connat et al. 1986). There was partial cross-reactivity between species in vitellogenesis but not in oviposition. When female *O. tholozani* were mated to male *O. moubata*, egg development showed "almost the same rate" (no data were given) as when females mated normally with conspecific males; but these females did not oviposit (Galun and Warburg 1967). Virgin females of *O. parkeri* which received an injection of male accessory glands from male *O. moubata* began egg development but did not oviposit as they did after receiving similar injections from conspecific males (Aeschlimann and Grandjean 1973a,b; Oliver et al. 1984). Injections of spermatophores of *O. tholozani* and *O. tartakowsky* did, however, induce oviposition in virgin females of *O. moubata* (Connat et al. 1986).

Mature sperm of *O. tholozani* and *O. tartakowskyi* also induced oocyte development and oviposition in female *O. moubata* (Diehl et al. 1982; no data were given to allow checks for different degrees of stimulation). Conspecific sperm stimulated female *O. tholozani* to oviposit, but those of *O. moubata* apparently did not (Galun and Warburg 1967).

As noted by Oliver (1986), the intrageneric diversity of female responses to male stimuli in *Ornithodoros* and *Argas* is striking in view of their ecological similarity as nest inhabitors that mate before or after beginning to feed (Feldman-Muhsam 1986) and undergo several trophic cycles.

Table 6.4
Summary of Data on Effects of Male Seminal Products on Female Reproductive Processes in Argasid Ticks

	Taxon				
	Ornithodoros			*Argas*	
Female Process and Male Mechanism That Affects It	*moubata*	*tholozani*	*parkeri*	*persicus*	*arboreus*
Vitellogenesis					
Mechanical	yes	no	?	yes	no
Acc. gland	(no)[a]	yes	yes	yes	yes
Testes	no	?	yes	?	yes
Saliva	?	?	(yes)[b]	?	?
Sperm	yes	?	?	?	?
Oviposition					
Mechanical	yes	(no)[c]	?	no	?
Acc. gland	no	?	yes	no	no
Testes	?	?	yes	?	?
Saliva	?	?	?	?	?
Sperm	yes	yes	?	(prob.)	(prob.)

NOTES: See Table 6.3 for techniques and references.
[a] The material of the spermiophore from the seminal vesicles, but not sperm, induces vitellogenesis.
[b] Small samples.
[c] Uncertain since vitellogenesis did not occur.

In contrast to the pattern of diversity in argasids, the male seminal signals of ixodid ticks seem relatively nonspecific. Mating between *Amblyomma maculatum* and *A. americanum* resulted in normal engorgement weights, rapidity of engorgement, preoviposition periods, and egg mass weights (Gladney and Dawkins 1973). Cross-specific matings among a total of eight different congeneric species in three genera resulted in apparently normal engorgement and oviposition (Diehl et al. 1982; no quantitative data were given for most species, however).

This lack of specificity may constitute further support for the sexual selection hypothesis because ixodid females are generally thought to mate only once, in contrast to argasid females (Feldman-Muhsam 1986). It is thus possible that the apparent difference between the families is the result of different mating systems. If females mate only once, there will be no sexual selection by cryptic female choice and thus the sexual selection hypothesis would predict the observed relative lack of diversity. Nevertheless, data on mating frequencies in ticks are relatively sparse. In light of the many other groups of animals in which additional data have shown that apparent female monogamy was illusory (see chap. 9.4), this confirmation is only tentative.

Just as in insects, there is also one piece of evidence in ticks of remarkable long-range conservatism in chemical signals. "Several" female *O. tholozani* that received an extract of male accessory glands of the cockroach *Periplaneta americana* in the uterus carried out oogenesis (no quantitative data were given; Galun and Warburg 1967).

Mimicry of Female Messenger Molecules

Virgin female *Ornithodoros moubata* were induced to develop eggs but not to oviposit by injection of both a homogenate of the male accessory gland, and by a homogenate of the nervous system of a mated and engorged female (Aeschlimann 1968; Aeschlimann and Grandjean 1973a). The similar combination of effects is suggestive, but inconclusive evidence that male and female products are the same.

6.4 Mammals

Most studies of mammalian semen have differed from those of arthropod semen in emphasizing lists of the substances that are present. There has unfortunately been much less emphasis on the effects of experimental introduction of different components of semen into females. This emphasis has resulted in a relatively detailed understanding of the chemical composition of semen (especially that of humans), but surprisingly poor experimental documentation of the functions of different components. Statements regarding function are usually based on the chemical properties of the molecules or on their physiological effects in other contexts, rather than on direct experimental determinations of responses. Conclusions regarding their possible effects when transferred must thus be tentative. Unless otherwise noted, the discussion that follows employs data on human semen, which seems to be typical of that of other species of mammals in many respects (Mann and Lutwak-Mann 1981).

Mammalian semen is chemically very complex (Price and Williams-Ashman 1961; Mann and Lutwak-Mann 1981; Harrison and Lewis 1986). Several components of human semen probably function to keep sperm alive inside the female, including energy sources (such as fructose, glycerylphosphorylcholine, and carnitine), osmotic buffers (sorbitol), lubricants (mucus), defenses against oxidizing agents (ergothioneine), and coagulating factors (Mann and Lutwak-Mann 1981). Some components, however, probably function to trigger female responses that favor the male's reproductive interests.

Induce Muscle Contraction (Transport)

Human semen contains a variety of compounds known to affect the activity of smooth muscles. These include epinephrin, norepinephrin, histamine, and at least seventeen different prostaglandins (Mann and Lutwak-Mann 1981; some prostaglandin molecules are unstable and may spontaneously give rise to others or be breakdown products of them, so this number may be an overestimate; see Kelly 1981). Other mammals also have appreciable amounts of prostaglandin and serotonin (5-HT) in their semen (Mann and Lutwak-Mann 1981), which also induces contractions in smooth muscle. Prostaglandins are especially effective contraction inducers, and have, in fact, been used medically to induce contractions of the uterus.[2] Seminal plasma has the highest concentration of prostaglandins of any body fluid (Mann and Lutwak-Mann 1981).

Given their pharmacological properties, it seems likely that some or all of these substances function to increase the chances of contractions in the female reproductive tract which will result in sperm transport to fertilization sites (some may also have other effects, such as Ca ion transport or storage, which could be important for sperm). Limited evidence supports the transport-induction function. When prostaglandin F-2 alpha was added to ram semen before artificial insemination in the cervix, higher numbers of sperm reached the ewe's oviducts (Edquist et al. 1975 in Mann and Lutwak-Mann 1981). Men with low seminal prostaglandin levels are infertile (Albone 1984). Peristaltic contractions of the female reproductive tract are probably the rule during sperm transport in mammals. For instance, Yamanaka and Soderwall (1960) observed "large masses of semen moved anteriorly by vigorous antiperistaltic actions" of the uterus of newly mated female hamsters (p. 473; see also chap. 7.1).

Prostaglandins also have other possible effects on females that could favor males. Both estrus and ovulation can be induced in anestrous mares by introducing stallion semen into the uterus, where it is normally deposited during copulation, and there are indications that prostaglandin is the male factor responsible (Mann and Lutwak-Mann 1981). Prostaglandin E2 increases female lordosis responses to male mating attempts in the rat, perhaps through inducing the release of luteinizing hormone-releasing hormone (LH-RH) (Dudley and Moss 1976). Since a male may mate repeatedly with the same female (McClintock 1984), this could be advantageous. When circulating in the female, prostaglandins can inhibit her sexual receptivity in some vertebrates (Whittier and Tokarz 1992). Radio-labeling experiments in humans showed that seminal prostaglandins pass though the vaginal wall and into the bloodstream (Albone 1984), and they "may reach the ovarian surface" (Jöchle 1975, p. 186). Addition of human seminal plasma can induce contraction of human ovaries in vitro (Palti and

Freund 1972), and contraction of the ovary can result in ovulation of mature oocytes (Jöchle 1975; see chap. 3.9). It is not clear, however, if the quantities of seminal prostaglandins are large enough to affect either female receptivity or ovulation.

Gangliosides are still another type of spasmogenic compounds that appear to occur in human seminal plasma. The major seminal component responsible for female uterine contractions in the rat has the chemical and physical properties of a ganglioside (Ventura and Freund 1973).

Hormones

An impressive array of female hormone molecules has been discovered in mammalian seminal plasma by immunodetection techniques (Mann and Lutwak-Mann 1981). These include follicle-stimulating hormone (FSH), which females produce and use to stimulate the egg follicles; luteinizing hormone (LH), which females use to trigger ovulation, among other things (Milligan 1982) (a mixture of FSH and LH is required for implantation of the embryo in the uterus of the mouse: Marchlewska-Koj 1983); chorionic gonadotrophin, which induces oocyte maturation and ovulation, among other things (Donahue 1972); and prolactin, which maintains corpora lutea activity and thus prepares the uterus for implantation (Marchlewska-Koj 1983). Female steroid hormones, including estradiol-17 beta or E2, which trigger the surge of luteinizing hormone, among other things (Fink 1979), estriol, and estrone also occur in the seminal plasma at levels higher than those in the blood. Estradiol is essential for maintaining the increased pituitary sensitivity to LH-releasing hormone before and during preovulatory LH surge in humans, rabbits (where dosage of the steroids may be critical and seminal contributions thus likely to affect females), and rats (Yen et al. 1975; Fink and Henderson 1977a,b; Dufy-Barbe et al. 1975). Small amounts of LH and E2 could have a "priming" effect on the secretion and release of the female's own LH from her pituitary or on gonadotropin-releasing hormones in her brain, leading to the surge of LH that triggers ovulation (Fink 1979). Positive feedback effects involving E2 may also occur in so-called spontaneously ovulating mammals such as rats, sheep, and humans (Fink 1979; Milligan 1982). The measured amounts of LH in human seminal plasma vary by a factor of about forty, and range up to about 0.48 μg in an ejaculate (assuming a 3 ml ejaculate; Mann and Lutwak-Mann 1981). This is much less than the mg amounts that produce significant physiological effects when injected in females (Yen et al. 1975), but the possible importance of the fact that semen is deposited relatively close to the female reproductive organs remains to be evaluated.

Other mammals have similar compounds in their semen. Ram and rat semen contain prolactin (Mann and Lutwak-Mann 1981). An unknown component of rat semen is thought to induce female quiescence, which may reduce the chances of rapidly encountering and mating with another male, and thus favor the transport of the semen through her reproductive tract before the next mating (McClintock 1984; see also chap. 3.5). As predicted by the sexual selection hypothesis, there are wide species variations in the chemistry of accessory gland secretions: some anatomically homologous structures secrete quite different substances, and some substances are secreted by different structures in different species (Price and Williams-Ashman 1961; Mann and Lutwak-Mann 1981).

These data must be viewed cautiously. As already mentioned, experimental determinations of the effects of these substances on male reproductive success in quantities and contexts similar to those transferred in normal matings are generally lacking. Many of these molecules also have additional effects to those just mentioned. For instance, FSH also stimulates spermatogenesis in the male (Lofts 1987). Estradiol can affect folliculogenesis, anovulation, ovulation, the ovulatory cycle, and reproductive senesence (Dukelow and Erwin 1986). Nonetheless, the occurrence of so many different "female" hormones in seminal plasma strongly suggests that the functional "mimicry" predicted by the sexual selection hypothesis occurs in mammals. Males apparently often include the same molecules in their semen which the female uses for hormonal communication within her own body.

Induce Reproductive Readiness

Although they constitute precopulatory stimuli and thus fall outside the scope of cryptic female choice, it is worth mentioning that some male substances in mammals, especially in the urine, have reproduction-enhancing effects on a female's physiological readiness to breed and may also be under sexual selection by female choice (Andersson 1994). One male effect on females, hastening of estrus due to the presence of the male or his urine, occurs in the mouse, several species of voles (*Microtus*), the prairie deermouse, guinea pigs, the lab rat, and the mink, as well as in domestic animals such as sheep, goats, cattle, and pigs (Milligan 1980, 1982; Albone 1984; Lyons and Getz 1993). These effects are different from the effects on abortion discussed in chapter 4.5.

Female sexual maturity is also sometimes hastened by contact with male pheromones (Drickhamer 1985). In at least the mouse, it appears there may be a conflict of reproductive interests between males and females. Females with accelerated puberty conceive their first and sometimes several litters prior to completion of normal development, and they suffer higher mortality during subsequent pregnancies and lactation (Drickhamer 1985). Young females actively avoid the male pheromones (Drickhamer 1985), also suggesting that males have evolved to manipulate females.

The same reasoning used in the discussions of male accessory gland substances suggests that differences in male abilities to elicit female reproductive readiness may lead to sexual selection on males. For instance, responses of mouse females differ depending on the male's social status (Vandenbergh and Coppola 1986). Concentration of the acceleratory substance is higher in the urine of dominant males, and the female puberty acceleration response to dominant males is greater and would thus favor their reproduction (Drickhamer 1985; Vandenbergh 1987). The possibility that sexual selection by direct male-male interactions has influenced the evolution of male abilities to influence other males chemically has been mentioned previously (Jennett 1985). But, aside from a brief mention by Andersson (1994), the possible connection between stimulatory male effects on female reproductive physiology and sexual selection by female choice seems to have gone unnoticed to date.

Prairie voles offer another possible example. Females in seminatural conditions responded with accelerated estrus more consistently to stimuli from unfamiliar males when the male was unpaired than when he was paired (Lyons and Getz 1993). In neither case, however, was the female response complete (52% and 18%), and longer exposure to a given male had a stronger effect. Thus males with more potent or perhaps greater quantities of pheromone, or that were better able to distribute their odors, could be favored. Since females sometimes mate with more than one male during a single estrus, a male paired to another female could perhaps benefit reproductively by inducing estrus in a neighboring female.

Fragmentary evidence suggests a possible similar effect in humans on reproductive maturity which may be associated with copulation per se. The development of secondary sexual characters in girls under eight years old is associated with a history of sexual abuse, and the association is strongest in girls with evidence of vaginal penetration (Herman-Giddens et al. 1988). The cause and effect relations are unknown, however (abuse because of sexual characters? sexual characters because of abuse?).

There is "some" cross-specific activity, as male rat urine can accelerate puberty in female mice (Albone 1984). More quantitative studies are needed on this point.

These ideas can be extended to competitive interactions among females that involve the physiological control of reproductive processes of other females. The prediction would be that the traits discussed earlier (invasive action, multiplicity of cues, etc.) will also characterize the female stimuli that reduce or inhibit the fertility of other females in the same social group (as occurs in many mammals, including mice, elephants, wolves, wild dogs, dwarf mongoose, naked mole rats, marmosets, and many others; see Wasser and Barash 1983 for a list and references). This is essentially the same prediction as that made by West-Eberhard (1984) regarding the mechanisms by which queen bees repress reproduction by workers.

6.5 Other Animals

There are scattered indications that substances in the semen affect female reproductive processes in other groups. Perhaps the best evidence comes from barnacles in the genus *Balanus*, which are sessile hermaphrodites (Barnes et al. 1977). An individual acting as a male uses its long, mobile penis to search for an individual nearby that is prepared to act as a female. A receptive female allows the penis to enter her mantle cavity, where it deposits a substantial quantity of semen in about 40 sec and then withdraws. After several minutes, the male again extends his penis toward the female and into her mantle cavity and deposits more semen. Usually 6–8 such copulations occur over 30–90 min, and a total of about half of the male's body weight is transferred to the female. During the approximately 20 min after copulation, eggs emerge from the female's oviduct and move into the egg sac in her mantle cavity, where they are fertilized by sperm that have entered the sac via small pores (Klepal et al. 1977).

A substance in the semen induces muscle contractions in the female reproductive tract and/or mantle cavity, causing the eggs to be moved first into the oviduct (ovulation) and then into the egg sac (oviposition) (Barnes et al. 1977; Klepal et al. 1977). A direct contraction-inducing effect was demonstrated in *B. balanus* by applying semen, from which sperm had been removed by centrifugation, to isolated retractor muscles of the opercular membrane (Barnes et al. 1977). Since up to six males may contact a ripe female with their penes at the same time (Barnes et al. 1977), it is clear that a greater ability to induce ovulation could give a male a reproductive advantage if he arrived first.

Oogenesis in the nematode *Coenorhabditis elegans* is also apparently induced by a substance in either the sperm or the seminal fluid (Ward and Carrel 1979). Mutant males with rudimentary gonads and no vasa deferentia copulated normally but failed to elicit the increase in oogenesis that normally follows copulation. Some sterile males that transferred infertile sperm elicited oocyte production, while others did not.

In the spiny dogfish *Squalus acanthias*, as much as 6.25% of the total dry weight of the male's siphon accessory secretion is serotonin (5-HT), a substance renowned for its contraction-inducing effects on muscles (Mann

1984). When a small amount of diluted fluid from the siphon was applied to the isolated uterine horns of a female, powerful contractions occurred immediately and persisted for hours afterward. Serotonin is also present in spermatophore glands of *Octopus vulgaris*. Mann (1984) argued that the serotonin in this species functions to move the spermatophore within the male, but it could also be used to stimulate sperm transport within the female; peristaltic contractions of the octopus oviduct carry the sperm to the spermatheca just after the spermatophore is inserted (Mann 1984). In several prosobranch gastropods a substance from the male gonad also stimulates spawning in the female (Webber 1977).

In the European adder *Vipera berus*, male semen produces a sort of physiological plug of the female's cloaca. It stimulates the posterior portion of the oviduct to constrict following mating, probably making it difficult for other males, which are commonly present, to insert and inflate their hemipenes (Andren and Nilson 1987; see fig. 3.6 for a similar female contraction in a beetle). The constriction of the female snake's cloaca may also reduce the chances that the female will eject the male's semen (Stille et al. 1986).

Male seminal products might also be expected to competitively affect female reproductive processes in groups with external fertilization. I have not searched the literature to test this prediction. A hint of such an effect occurs in the herring *Clupea harengus*, where spawning is released by contact with the suspended semen (milt) of a male (Crews and Silver 1985).

I was not able to find data on the effects of male seminal products on female reproductive processes in some large animal groups, such as birds and crustaceans. This lack may be an artifact, however, as the possibility of such effects seems not to have occurred to many workers (see comments of Adiyodi and Adiyodi 1988 to this effect on crustaceans).

6.6 A Third Hypothesis: Nutritional Effects on Females

The Importance of Distinctions

Some of the "triggering" properties of male seminal products listed in tables 6.1 and 6.3 and discussed above may occur because of nutrients they supply to the female. For instance (as has often been argued in studies of behavioral ecology), male substances could increase egg production by providing the female with more raw materials from which she could synthesize eggs. In the male active–female passive tradition, the female is often taken to be a passive, egg-making apparatus whose output is strictly determined by the male's input.

Alternatively (as has often been supposed in experimental studies of the physiological effects of male accessory glands), the nutritional value of male products may be relatively insignificant, and their action may stem from hormonelike triggering effects on the female. The distinction between nutritional and nonnutritional effects may seem academic in the context of this book, since both nutrient and hormonelike activities could result in sexual selection on males via cryptic female choice. Nevertheless, the difference could have important consequences for the arguments of this chapter.

First, male nutritional effects on females may often be under natural selection rather than sexual selection (see chap. 1; also Pitnick et al. 1991). If a male's reproduction is strongly affected by his ability to transfer nutrients or defensive compounds to the female, then his reproduction may be limited by his ability to accumulate such compounds. His reproduction could be limited by his ability to accumulate resources rather than by more direct competition with other males for access to females and their gametes.

Second, possible female benefits from discrimination among nutrient effects are different. Females could gain directly from favoring males that make larger donations via the positive effect on their own reproductive output. They could also gain indirectly, via the incorporation of good quality viability genes from the male in their offspring (see Cordero 1995).

Hormonelike, nonnutritional male products are, in contrast, unlikely to be associated with better male abilities to accrue resources (unless, for some reason, hormones or other signal molecules tend to be especially complex or otherwise costly). In fact, hormones are apparently usually not especially costly (see below) and are often active in quite small quantities. Male reproduction would thus more often be limited by competition with other males to sire the female's offspring. The payoff to a female for discriminating among males on the basis of such signal molecules would only be indirect, through increased abilities in her sons to manipulate females of the succeeding generation (i.e., possession of good attractiveness genes). Direct benefits to the female's reproduction (more offspring, better survivorship) are not expected from nonnutrient male cues, except if females come to depend on male cues as triggers (see chap. 2.1). In fact, nonnutrient male effects could result in direct losses to females. A particular female reproductive behavior that is in the male's best interests (e.g., massive oviposition soon after copulation and before mating with other males) would not necessarily be in the female's best interests (e.g., greater dispersion of eggs in space and time, retention of reserves for future contingencies). Thus it is important to attempt to evaluate the likelihood of hormonal versus nutritional effects for more than just their intrinsic interest from the point of view of proximate mechanisms.

There is a body of recent literature, mostly concerning the behavioral ecology of orthopterans and lepidopterans, that tests the possible nutritional effects of male seminal products (mostly spermatophores) on female reproduction (e.g., fig. 1.5). Male nutritional donations to females have been termed male parental effort or nonpromiscuous mating effort, while nonnutritional donations are designated as male mating effort or promiscuous mating effort (e.g., Thornhill and Alcock 1983; Gwynne 1986; Oberhauser 1992). A second, largely separate set of studies, published mostly in physiology journals, has documented apparently nonnutritional, hormonal effects of male seminal products on female reproduction in a large variety of insects (especially dipterans). The data presented in previous sections of this chapter are nearly all from this second, physiological literature. One objective in this section is to draw these two bodies of knowledge closer together.

The discussion is focused on the species in tables 6.1 and 6.3, because they are the ones in which male seminal products have been thought to act as hormonal triggers for female reproductive processes. I will show that the available evidence, while incomplete for many species, is in much better accord with the nonnutritional than the nutritional hypothesis. This agrees with the emphasis placed on nonnutritional effects by most authors in this field (e.g., reviews of A. Chen 1984; Raabe 1986; Gillott 1988; Kaulenas 1992), and in the previous sections of this chapter.

Most of the studies cited in previous sections employed the technique of implanting organs or injecting extracts from one individual into another. Unfortunately, it was usually only presumed, either overtly or tacitly, that the male substances were hormonelike in their action. In most cases the possibility that nutritional contributions from males may trigger female responses only indirectly was not considered. There are several exceptions, however, in which authors deduced a nutrient function after finding that radioactively labeled seminal products move out of the female's reproductive ducts and into her ovaries, where they are incorporated into developing eggs: Friedel and Gillott (1977) on the grasshopper *Melanoplus sanguinipes*; Huignard (1983) on the beetle *Acanthoscelides obtectus*; Sivinski and Smittle (1987) on the fly *Anastrepha suspensa*; Bownes and Partridge (1987) and Markow and Ankney (1988) on several species of *Drosophila*; and Bowen et al. (1984) on the katydid *Requena verticalis*. Subsequent analyses have shown, however, that in at least the first four species, the amounts of male material are so small that nutrient effects seem unlikely (Sivinski and Smittle 1987; Cheeseman and Gillott 1989; Kaulenas 1992). For instance, in *M. sanguinipes* the male contributes about 5 micrograms of protein to the eggs, as compared to 1,400,000 micrograms of protein from the female.

Difficulties in Making Distinctions

Two crucial distinctions for the arguments made here are the following:

1. Nutritional effects involve direct incorporation of male products into eggs; triggering effects, in contrast, will often occur elsewhere in the female's body rather than the egg.
2. Nutritional effects involve amounts or types of materials that are relatively costly in terms of the energy in the product itself, or the time and energy needed to acquire it; triggering effects will tend to involve investments that are energetically cheap for the male.[3]

The distinction between nutritive and nonnutritive male contributions is sometimes difficult (fig. 6.12). In fact, some of the commonly used criteria by which male nutritional effects have been distinguished from nonnutritional effects in both the physiology and the behavioral ecology literature can be misleading. And in certain cases in which females come to depend on male cueing substances, the differences between nutritional and nonnutritional effects may disappear.

A common test of the nourishment hypothesis, used especially in katydids, crickets, and butterflies, in which males transfer spermatophores, is to ascertain whether the effect of male donations on female fecundity varies with the female's nutritional status. When male stimulatory effects on oviposition or refractory periods following mating are accentuated when the female is malnourished (e.g., Boucher and Huignard 1987 on a bruchid beetle; Gwynne 1990, Gwynne and Simmons 1990 on a katydid; W. Brown in prep. b on a cricket), the deduction is that the male has a nutritional effect on the female.

Such results, however, are inconclusive. It is possible that the male's semen provides nutrition to the female, but that this is an incidental side effect of a selectively more important triggering effect. For instance, the male ejaculate of the monarch butterfly *Danaus plexippus* increases female egg production somewhat when the female is poorly fed. Nevertheless, the male's spermatophore is only degraded slowly inside the female; a major portion of a large spermatophore is still intact when the female remates (Oberhauser 1992). Assuming at least some degree of last-male sperm precedence (as is typical in Lepidoptera; Drummond 1984; Gwynne 1984c), and similar nutritional values of the different portions of the spermatophore that are digested at different times after mating, Oberhauser concluded that a male that made a large spermatophore derived more reproductive benefit from the resulting prolongation of the female's refractory period than from the use of nutrients he supplied to her in the spermatophore to make additional eggs.

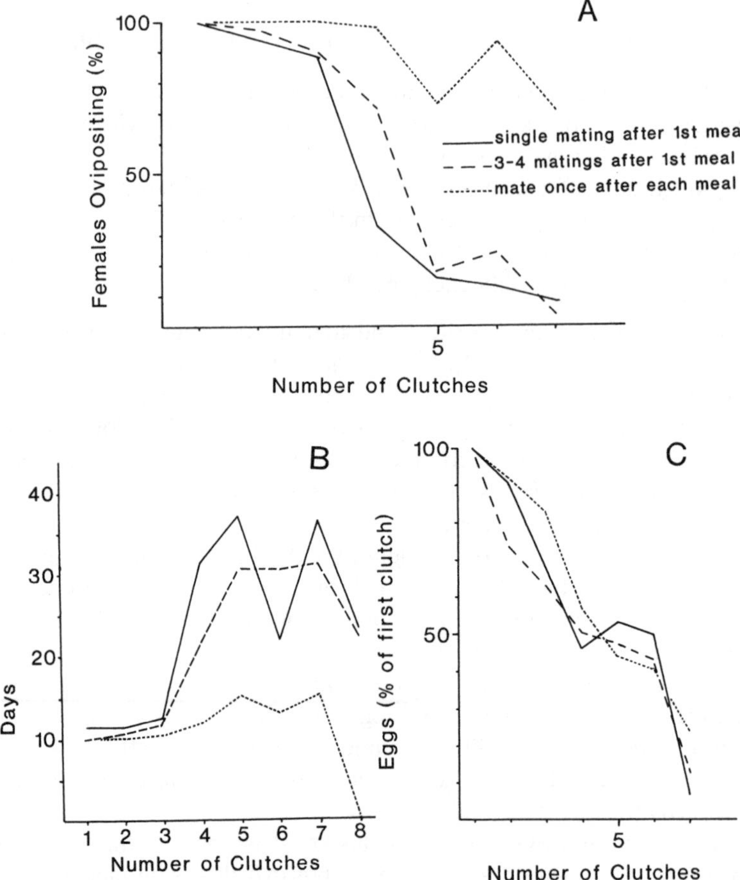

Figure 6.12 Distinguishing between a male's effects on his mate's reproduction via nutritional contributions as opposed to hormonal triggers can be difficult. In the tick *Ornithodoros moubata*, females often lay a single clutch of eggs after each blood meal. When the female mates after each meal, the probability that she will oviposit goes up (A), and the delay between feeding and oviposition goes down (B), as compared with females that mated multiply after the first blood meal and females which mated only a single time, following the first blood meal. This suggests a possible nutritional contribution of the male. Further data, however, including the numbers of eggs in each clutch (C), indicate that the male contribution is a trigger of some sort rather than just nutritional, since repeated mating has no effect. (After Aeschlimann and Grandjean 1973b)

Conversely, even if poorly fed females fail to respond more to male donations, one cannot conclude that male nutritional effects are not important. It is possible that the male is donating a "special" nutrient that cannot be produced by the female, and thus limits her reproduction whether she is well nourished or not (Gwynne 1988a,b). A special nutrient of this sort could conceivably even explain cases in which the effects of female nutrition are inverted, with better-fed females responding *less* strongly to male products (e.g., Fox 1993 on the bruchid beetle *Callosobruchus maculatus*).

Similarly, increases in the rate of oviposition are not necessarily evidence of nutritional effects. For example, repeated mating in the cotton stainer bug *Dysdercus cardinalis* results in a higher rate of oviposition, but this may result from mobilization of female reserves rather than from nutritional donations of the male. Females that remated more often and produced more eggs per day soon after copulation lived for shorter periods, despite being kept with abundant food (Kasule 1986). Singly- and multiply-mated females produced similar lifetime totals of eggs. This could be a case in which male and female interests with respect to oviposit rate are in conflict (see chaps. 1.6, 3.7), although it is also possible that those females that remated gained a reproductive advantage from producing eggs more rapidly (Kasule 1986).

Even what would seem to be the most direct test of all—radioactively labeling male seminal products and tracing their fate within the female—can give inconclusive results. This technique is so sensitive that quantities so small as to be probably of no nutritional significance can be mistaken as nutrients (see above; Cheeseman and Gillott 1989; Kaulenas 1992). As noted above, the quantities that elicit female responses in *Melanoplus* are extremely small. In other species they are even smaller: nanogram quantities of male substance induce oviposition in the cricket *Teleogryllus commodus* (Loher et al. 1981); trace contaminants from syringes that had been cleaned after injection of microgram amounts stimulated oviposition in *Acheta domesticus* (Destephano et al. 1982). Finding labeled male products in the ovary may also not be reproductively significant for the male if they also occur in the rest of the female's body (e.g., Sivinski and Smittle 1987 on the fly *Anastrepha suspensa*; Markow and Ankney 1988 and Pitnick et al. 1991 on *Drosophila*).

It must be noted, however, that arguments based on the quantity of male substance are also open to doubt. For instance, Cheeseman and Gillott (1989) and Kaulenas (1992) argue that the amount of marked male material that arrived in the ovary in *Melanoplus sanguinipes* was too small to be nutritionally significant. Nevertheless it is possible (though not demonstrated in this case, nor for that matter in any other) that the material transferred by the male is a "special" nutrient that the female is unable to produce (Gwynne 1988a; see also Bownes and Partridge 1987). A special nutrient of this sort could be nearly indistinguishable in some respects from a hormonal triggering substance (see below).

The inverse argument, that a relatively large quantity of ejaculate is an indication of nutritional effects (e.g., Pitnick et al. 1991), is also inconclusive. For instance, male *Gryllus supplicans*, *Decticus verrucivorus*, and *Poecilimon veluchianus* (crickets and katydids) produce huge male spermatophores, which average, respectively, about 3%, 9%, and 25% of the male's body weight (Sakaluk 1985; Wedell 1993; Reinhold and Heller 1993) and are eaten by the female. But experimental manipulations of the amounts consumed by females revealed no perceptible nutritional benefits (no increases in egg size, number of eggs laid before next mating, lifetime fecundity, female longevity, no decrease in number of days until first egg laid; Will and Sakaluk 1994; Wedell and Arak 1989; Wedell 1991; Reinhold and Heller 1993). Even when the protein intake of female *D. verrucivorus* was reduced, no male nutritional effects were evident. A similar lack of correlation between relative size of ejaculates and their effects on female egg production occurs in some butterflies (Svard and Wiklund 1991) and beetles (Eady 1995). Perhaps the most extreme example is the barnacle *Balanus* mentioned above. The male transfers about half his body weight, but this transfer probably has no nutritional effect. Much of the male material is rapidly discarded, and the female's eggs for that year are in any case already mature and about to be ovulated (Barnes et al. 1977).

Possible Criteria for Distinctions

One simple criterion for disentangling paternal investment from male mating effort that can be used in species that produce spermatophores is to observe how females treat spermatophores. Sometimes it is assumed that spermatophores in general provide significant nutrition to the female (e.g., Ridley 1988), but this is apparently not the case in many species. Females of some insects and mites routinely discard spermatophores apparently unaltered (and not infrequently, with sperm still inside them), indicating that spermatophores are not nutritionally important. Spermatophore dumping occurs in the grasshoppers *Syrbula fuscovittata*, *Taeniopoda eques*, *Gomphocerus rufus*, *Melanoplus sanguinipes*, and *Locusta migratoria* (Gregory 1965; Whitman and Loher 1984; Cheeseman and Gillott 1989); the weta *Hemideina femorata* (another orthopteran; only two of thirty-five spermatophores consumed; Field and Sandlant 1983); the diopsid fly *Cyrtodiopsis whitei* (Lorch et al. 1993); the midge *Culicoides melleus* (spermatophores were fresh and often contained substantial numbers of sperm when females were nonvirgins; Linley and Hinds 1975b); the darkling beetle *Onymacris unguicularis* (DeVilliers and Hanrahan 1991; early stages of spermatophore removal were performed by the male, the final stages by the female, and the spermatophore was partially degraded); the plant beetle *Macrohaltica jamaicensis* (intact, often with sperm; Eberhard and Kariko, in press); the tiger beetles *Pseudoxychila tarsalis* and *P. bipustulosa* (intact, often with sperm; R. Rodriguez and Eberhard, in prep.); the soldier beetle *Chauliognathus* sp. (intact; Eberhard, unpub.); the lygaeid bugs *Ozophora baranowskii* and *O.* sp. near *parva* (intact, sometimes with sperm; R. Rodriguez in prep.); an unidentified gelastocorid bug (intact, sometimes with sperm; R. Rodriguez and Eberhard, unpub.); and the mite *Aculus cornutus* (Oldfield and Newell 1973). Females of the grasshopper *M. sanguinipes* also discard other proteins that are abundant in the ejaculate; accumulations are cleaned from the ovipositor valves by the hind legs but not ingested (Cheeseman and Gillott 1989). Similarly, in some other species without spermatophores females routinely discard semen, indicating that it is of no significant nutritional benefit (tables 3.1, 3.2; chap. 3.2, 3.3). Semen ejection seems to be the rule, for example, in mammals (Austin 1975).

The inverse argument—that when females do not discard spermatophores or semen they probably do gain nutritionally—is not correct. The female may fail to discard male seminal products because she is unable to do so, or because it is not worth the effort. If, on the other hand, females have special characteristics that apparently function to ensure spermatophore digestion, then a likely nutritional benefit can be deduced. For instance, female *Macrodactylus* beetles have a special pocket into which spermatophores are transferred after sperm is transferred to the spermathecal duct, and where they are then digested (fig. 6.13; sometimes with sperm still inside; Eberhard 1993a). Analogous structures occur in the mite *Acarus siro* (Witalinski et al. 1990) and triclad flatworms (Sluys 1989). Structures of this sort are not known, to my knowledge, in any of the species in tables 6.1 or 6.3.

There are several types of evidence that suggest nonnutritional effects:

a. *When the active male substance directly affects a female effector organ instead of being incorporated into eggs or female food reserves.* This occurs in the stimulation of a female abdominal ganglia of *Bombyx mori* to induce oviposition (fig. 6.3; Yamaoka and Hirao 1977), and *Aedes aegypti* to induce resistance to genitalic coupling (Gwadz 1972); the stimulation of contraction of muscles in the oviduct of *Locusta migratoria* (Lafon-Cazal et al. 1987) to induce oviposition, the female bursa of *Rhodnius prolixus* to induce sperm transport (Davey 1958, 1965b), and the vagina or uterus in several mammals (Jöchle 1975; Mann and Lutwak-Mann 1981); and in the probable stimulation of the brain of *Musca domes-*

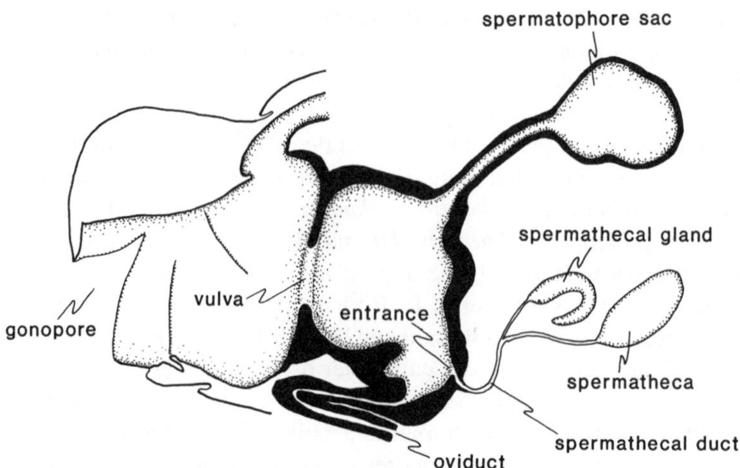

Figure 6.13 The female reproductive tract of the scarab beetle *Macrodactylus costulatus* includes a special structure, the spermatophore sac, in which spermatophores are digested, suggesting the deduction that females benefit from resorbing the spermatophore. During copulation the male deposits a spermatophore at the entrance to the spermathecal duct. The female then moves it to the spermatophore sac (sometimes before all its sperm have been transferred) where it is degraded. (After Eberhard 1993a)

tica (Leopold et al. 1971a,b) and *Drosophila melanogaster* (Baumannn 1974a,b; Boulétreau-Merle 1976). Exclusively nutritional effects are eliminated directly in these cases.

b. *When the active male substance also exists in the female and acts as a hormone or other messenger substance, or is an enzyme or a precursor leading to the production of such a substance.* Examples are prostaglandin synthetase in the crickets *Acheta domesticus* (Destephano and Brady 1977) and *Teleogryllus commodus* (Loher 1979; Loher et al. 1981); prostaglandin in *Bombyx mori* (Setty and Ramaiah 1979, 1980); esterase-6 in *Drosophila melanogaster* (Gilbert et al. 1981a,b; Scott 1986); an octopamine mimic in *Locusta migratoria* (Lafon-Cazal et al. 1987); perhaps juvenile hormone or juvenile hormone mimics in *Musca domestica* (Gillott 1988) and *Hyalophora cecropia* (Shirk et al. 1980); a probable oviposition-inducing hormone in *D. melanogaster* (Monsma and Wolfner 1988); and a variety of neurogenic, spasmogenic, and hormonal molecules in the seminal plasma of many mammals and a few other animals (Mann and Lutwak-Mann 1981). Nutritional arguments are extremely improbable in these cases. The male nutrient would also coincidentally have to have hormonal properties. "Special nutrient" arguments invoking a female inability to produce the male factors (e.g., Gwynne 1988a) are ruled out.

c. *When nonnutritional mimics of stimuli produced by male products also induce female responses.* Examples include the mechanical stimulation of the female reproductive tract by objects mimicking spermatophores that increase oogenesis in the roaches *Nauphoeta*, *Leucophaea*, and *Diploptera* (Engelmann 1970), and the ticks *Ornithodoros moubata* (Connat et al. 1986) and *Argas persicus* (Leahy and Galun 1972). Direct nutrient arguments are again eliminated. The nutrition hypothesis could be saved, at least partially, by arguing that even though the male cue for the female response is nonnutritional, the female response evolved due to nutritional effects because the nonnutritional male cue is always associated with nutrition. In this case, however, the cue itself would be expected to evolve under sexual selection because variations would be especially likely to give rise to increased male abilities to trigger female responses (see chap. 2.1). I know of no examples of such combinations of nutritional and signal functions.

d. *When alteration of the portion of the female nervous system on which the male product is thought to act abolishes the female response, even though she receives the same amount of nutrition from the male.* This occurs in long-term oviposition induction in *Acheta domesticus* (Murtaugh and Denlinger 1985); sperm transport in *Pieris brassicae* (Tschudi-Rein and Benz 1990); inhibition of pheromone release in *Epiphyas postvittana* (Foster 1993); and inhibition of female receptivity in *Musca domestica* (Leopold et al. 1971a) and *Aedes aegypti* (Gwadz 1972). As in the previous case, the nutrition hypothesis could be saved by arguing that a nonnutritious, related cue is signaling the receipt of a different, nutritionally significant donation from the male. The preceding comments regarding the predisposition of such cues to sexual selection are also applicable here.

e. *When the male products of one species do not have an effect on females of a closely related species similar to the effect they produce on conspecific females.* Such specificity occurs in oviposition induction in some *Aedes* mosquitoes (Leahy and Craig 1965; O'Meara and Evans 1977), and *Drosophila* (Fuyama 1983; Chen et al. 1988); and in inhibition of remating by female muscoid flies in the genera *Stomoxys*, *Phormia*, *Musca*, and *Scathophaga* (Morrison et al. 1982), some *Drosophila* (Baumann 1974b; Chen et al. 1988), and *Aedes* and *Culex* (Craig 1967; Ramalingam and Craig 1976). The nutritional hypothesis could be saved with the ad hoc argument that closely related species need quite different "special nutrients," or that different chemical cues are used to signal nutrient transfer in different species, so that heterospecific females do not sense that they have received nutrients. I know of no data to test this idea. The comments in (c) regarding the probable predisposition to sexual selection of male cues of this sort also apply here.

f. *When male supplies of the substance are apparently not limiting.* If the male is able to transfer enough seminal products to elicit normal, complete female responses when he copulates with several different females in rapid succession or transfers apparently exorbitant amounts to a given female, models involving costly, limited nutrients seem less likely (such limitation has been adduced to favor a nutrient function; e.g., Gwynne 1993). In several species, male substances that inhibit female receptivity do not appear to be costly or limiting: *Aedes aegypti* males have more than sixty times the amount needed to render a female unreceptive (Craig 1967); *Lucilia cuprina* males can render up to ten virgin females unreceptive when mating in rapid succession, despite a very low likelihood of encountering even two in succession in the field (Cook 1992). *Mayetiola destructor* males can generally mate with 15–20 females in rapid succession before the ability to induce oviposition begins to decline, and they can still reduce female receptivity after mating 45 different females in a single morning; this is probably far beyond a male's likely matings in nature (Bergh et al. 1990, 1992). Some fecundity-enhancing substances are also apparently not costly. When male *Glossina pallidipes* flies copulated in captivity four times in rapid succession (an average of 10 min between copulations), the lifetime fecundity of the fourth female was the same as that of the first, even though such rapid encounters with virgin females are again unlikey in nature (Jaenson 1979, 1980; Leegwater-van der Linden and Tiggelman 1984). The nutrient hypothesis could be saved in such cases by supposing that the nutrient is somehow difficult or impossible to acquire for the female but not the male. The converse argument—that if the male has only enough substance available to affect a single female (as is also common in nature), then one can deduce he has important nutritional effects on the female—is not justified. The advantages of simple economy (refrain from dedicating resources to semen until they are needed), or of chemical purity (only use freshly synthesized and thus more potent accessory gland products) could limit reserves even if male contributions were nonnutritional.

g. *When the effect on the female is too rapid to be due to nutrition.* Short-term increases in oviposition begin as soon as 6 min after injection and amount to increases from 2 (control) to an average of 114 eggs per female in 24 hr in the cricket *Acheta domesticus* (Destephano and Brady 1977; Destephano et al. 1982). In many (most?) species studied the male effects on female receptivity are apparently very rapid (usually <1 hr) and seem unlikely to result from the nutritional effects per se. The nutritional hypothesis could accommodate these data if the male nutrient material also functions as a cue. For instance, if unlaid eggs represent a source of nutrients that the female can resorb in times of need (e.g., Kaulenas 1992), then after receiving nutrients from the male the female may begin to lay the eggs she has already formed, because she now has the nutrients necessary to prolong her own survival and perhaps to synthesize further eggs. As noted above in (c), a signal function of this sort, which was consistently associated with nutrients, could result in selection on males favoring increased signaling (triggering) abilities per se; the combination of nutritional and triggering effects, if it occurs, could be relatively subtle.

h. *When the effect of the male substance lasts for much longer than the substance itself in the female.* In *D. melanogaster* the male protein *msP 355a* is thought to increase oviposition, an effect that can last for weeks. But the male protein is detectable in the female, using the sensitive technique of monoclonal antibodies, only until about 4–6 hr after copulation (Monsma et al. 1990). Direct nutritional effects of this substance, if it is indeed responsible for the long-term effects, seem to be ruled out. Again, the nutrition hypothesis could be saved by supposing that this male product is a cue that is consistently associated with nutrition (see "c" above).

In sum, there are numerous species in tables 6.1 and 6.3 in which a nutritional role for the male seems unlikely. In addition, there is not a single species in which there is good evidence favoring nutritional effects. It must be noted that the types of observation which might demonstrate nutritional effects have seldom been made on these species. But the lack of clear indications favoring the hypothesis suggests that nutritional effects may be uncommon among these species.

There are probably other species in which nutritional effects of male seminal products on females are selectively important. There is relatively convincing (though still incomplete) documentation in a pierid butterfly (Rutowski et al. 1987), three katydids (Gwynne 1984b, 1988a,b; Gwynne and Simmons 1990; Simmons 1990), and a stinkbug (McLain et al. 1990 and Kon et al. 1993). Less complete evidence points in the same direction in some other insects, such as the beetles *Melolontha melolontha* (Landa 1960), *Phyllophaga* spp. (Eberhard 1993b), and *Macrodactylus* spp. (Eberhard 1993a). There are also cases in which males may transfer defensive substances to females, such as poisons or repellent substances (e.g., Gilbert 1976 on *Heliconius* butterflies; Doussourd et al. 1988 on the moth *Utethesia ornatrix*), which protect the female herself and/or her eggs from predation, and in which selection may favor males that transfer larger amounts (LaMunyon and Eisner 1994).

6.7 Taking Stock

This chapter attempts to establish a bridge between two poorly connected fields—the study of physiological effects of products in male semen on females, and sexual selection (see Stephenson and Bertin 1983; Queller

1987, 1994 for a similar link between sexual selection and reproductive physiology in plants). Previous work on the physiological mechanisms of animal reproduction has been carried out with little or no reference to sexual selection theory. And evolutionary biologists interested in mating behavior have not exploited the rich literature on reproductive physiology. This lack of communication may result in part from the efforts by evolutionary biologists to distinguish between proximate mechanisms and ultimate causation, and a preoccupation of physiologists with mechanisms alone. The material in this chapter shows that physiology and evolutionary theory can be mutually illuminating.

The central argument is that the multiple effects of male seminal products on female reproductive behavior and physiology have often evolved under sexual selection by cryptic female choice. I am proposing that male accessory gland products often act as "chemical genitalia," inducing the female to use the male's sperm to fertilize her eggs. Such effects on female reproductive processes are clearly very widespread, at least in insects and ticks, and they also occur in some other groups. Data of several types consistently, though sometimes only weakly, suggest an important role for sexual selection. One major problem is that some of the questions of particular interest have not often been asked in previous studies. For instance, male-female conflict under natural selection (Chapman et al. 1995; R. Alexander et al., in press) could result in the same trends predicted by the sexual selection hypothesis. I hope this book will excite physiologists with the possibility of connecting their work with evolutionary theory and encourage them to make further, more powerful tests of the various sexual selection and natural selection hypotheses.

It is important to keep in mind that the discussions focused on the *origins* of different male traits (e.g., invasive action, differences between species, etc.). It is possible that natural selection has contributed to the subsequent maintenance of traits that originally spread due to sexual selection. Females can come to depend completely on male cues. For instance, females of the bug *Rhodnius prolixus* are apparently unable to move sperm from the bursa into the spermatheca unless a male accessory gland factor is also present (Davey 1958). Thus disentangling naturally selected and sexually selected aspects of maintenance may sometimes be difficult.

Female dependence on male cues could have important evolutionary consequences. Females may continue to use male cues even during periods when there is no genetic variation among males, and females obtain no advantage from making discriminations among them. Evolved female dependence on male substances could also blur the line between hormonelike and nutrient effects. If the male provides a particular hormonelike triggering substance that the female subsequently ceases to make (for reasons of efficiency), is the male substance now a critical nutrient supplied by the male? Female dependence on male cues may also reduce the likelihood that asexual reproduction will arise.

Interpretation of the data is also complicated by the data's undoubted phylogenetic bias. Most involve insects, and within insects the Diptera are strongly overrepresented. Data from some major groups of animals such as crustaceans and birds are essentially absent. I suspect that the underrepresentation of some major groups is due to researchers not having asked the appropriate questions, rather than to a lack of the predicted effects (see Adiyodi and Adiyodi 1988 on crustaceans). Only further studies will tell if this guess is correct. I know of no a priori reason to suppose that the groups that are best studied are atypical with respect to the effects of male seminal products on females.

One powerful test of the ideas presented in this chapter would be to contrast species in which females mate with only a single male (no sexual selection expected by cryptic female choice), and in closely related species in which females mate with different males. The cryptic female choice hypothesis would predict less invasive action, less male specificity of effects, less divergence of mechanisms, etc., in the seminal products of groups in which females are strictly monandrous. Possible pairs of groups for such comparisons include termites (monandrous females) versus cockroaches, ixodid ticks (possibly monandrous females) versus agasid ticks (polyandrous females), different subgenera of *Heliconius* butterfly species with monandrous and polyandrous females (K. Brown 1981; Eberhard 1985), and primate species whose social behavior dictates that females normally mate with only a single male each estrous cycle versus those in which females may mate with several males (Dixson 1987). Termites, and perhaps also the ticks, would be especially interesting because of their probably very long history of monogamy, which would reduce the possible effects of phylogenetic inertia. In groups with shorter histories of female monogamy (e.g., different *Heliconius* species), female dependence on male cues might result in extended maintenance of male cues with sexually selected traits. I know of no data presently available with which to make such comparisons.

6.8 Summary

Male seminal products often elicit female responses such as sperm transport, oviposition, and resistance to remating in the groups of animals (insects, ticks, and to a lesser extent mammals) in which male effects have been studied. Variations in such female responses can result in cryptic female choice on males. This chapter presents data that give preliminary confirmation of the possibility that the male seminal products with these effects have often evolved under sexual selection by cryptic female choice. The evolution of special male adaptations to induce increased female responses suggests that partial or inconsistent female responses occurred in the past, and probably continue in the present. Two alternative explanations of the male effects on females are also discussed: natural selection on females to use male seminal cues to signal that mating has occurred; and natural or sexual selection on males to nourish females with seminal products.

Various trends in the details of the ways in which male seminal products function are more in accord with predictions from the cryptic female choice hypothesis than with alternative hypotheses. Male seminal products frequently work in a hormonelike manner, eliciting rapid female responses with relatively tiny amounts. Instead of being sensed as signals in the female's reproductive tract, male products often seem to move into the female's body cavity and act directly on female effector organs, such as her ovaries or ganglia in her nervous system that control oviposition, or directly on the muscles in her reproductive tract that move sperm or eggs. In a number of cases the active male product is a female messenger molecule, or a mimic of such a molecule. Such mimicry may represent male use of a "sensory trap" within the female's body to manipulate her behavior and physiology. Male seminal products may evolve rapidly and divergently compared with many other characters, a common pattern in sexually selected traits. Multiple male products often affect the same female response, as would be expected under sexual selection on males to add additional stimuli in order to obtain favorable female responses. In some cases female responses increase gradually with larger amounts of male substance in ways that suggest cryptic female choice favoring males transferring larger quantities of particular seminal products.

Current data are generally not sufficient to convincingly rule out in particular species all explanations other than cryptic female choice. But the trends suggest that cryptic female choice will be found to be an important factor in the evolution of male seminal products. Further work is needed to test the predictions of cryptic female choice more rigorously, and in a wider variety of animal groups.

Notes

1. Oliver et al. (1984) argue that the presence of prostaglandin in the saliva supports the possibility that male saliva functions to release female reproductive responses in *O. parkeri*. Prostaglandins also occur, however, in the saliva of females, where they may serve to increase the permeability of the host's capillaries (Dickinson et al. 1976).

2. Peristaltic contractions of the uterus can travel upstream (toward the ovaries) or downstream, depending on factors such as hormone concentrations; presumably contractions induced during or soon after copulation travel upstream.

3. Possible genetic "paternal" effects on offspring ability to obtain resources from the female constitute a third, nonnutritional and nonhormonal class of male effects. They are not discussed here for lack of evidence, and because they would presumably be independent of the experimental effects listed in tables 6.1 and 6.3 discussed in this chapter.

Table 6.1
Effects of substances in Male Semen on Female Reproductive Processes in Insects

Family, Species	*Production Site in Male*	*Name of Product*	*Effect on Female*	*Site Where Act in Female*	*Techniques*	*References*
Orthoptera						
Acrididae						
1 *Melanoplus sanguinipes*	acc. gl.	?	ovip.	?	Implant glands, inject extracts	Gillott 1976, 1977; Friedel et al. 1969
2 *Locusta migratoria*	acc. gl. (opalescent)	prostaglandin 51000, 35000MW[b]	ovip.[a]	?	Inject prostaglandin, inject extracts gland, spermatophore, bathe oviduct in extracts gland	Lange 1984; Lange and Loughton 1985; Lafon-Cazal et al. 1987; Kaulenas 1992
3 *Schistocerca gregaria*	acc. gl. (not testes)	13000MW	oogen., ovip.[c]	?	Implant glands, spthca., sem. ves.	Leahy 1973
4 *Gomphocerus rufus*	acc. gl.	sptop.	remate	spthca.	?	Hartmann unpub. in Hartmann and Loher 1974
Gryllidae						
5 *Teleogryllus commodus*	testes	prostaglandin synthetase	ovip., oogen.	oviduct muscles?[d]	Remove testes, inject prostaglandin assay precursor of synthetase, inhibit synthetase, denervate spthca., cut spthca. duct, remove spthca., implant spthca., crush spthca., chromotogr. separation	Loher and Edson 1973; Loher 1979; Loher et al. 1981; Tobe and Loher 1983; Ai et al. 1986

6 *Acheta domesticus*	testes	prostaglandin synthetase	ovip.	spthca.	Elim. sperm (irrad. male), inject prostaglandin, remove sptop.	Destephano and Brady 1977; Destephano et al. 1982; Murtaugh and Denlinger 1985, 1987
	testes	sperm[e]	ovip. (long term)	spthca.		
7 *Plebeiogryllus guttiventris*	testes	?	ovip.,	?	Castrate male	Bentur and Mathad 1975
	not testes		oogen.	?		
8 *Gryllus bimaculatus*[f]	?		ovip.[g]	spthca.[h]	Remove spthca., remove corpora alata, inject neurosecretory block, remove sptop. prematurely, alter filling spthca., apply extract to duct	Bentur et al. 1977; Orshan and Pener 1991; Loher et al. 1993, Kimura et al. 1988
9 *Gryllodes sigillatus*	testes (not acc. gl.)	?	ovip.	spthca.	Remove testes, inject acc. gl. fluid, vary number of matings, remove spthca.	Itgi et al. 1982
Tettigoniidae						
10 *Kawanaphila nartee*[f]	?	ampulla sptop.	remate	?	Vary amount entering female	Simmons and Gwynne 1991
Blaberidae						
11 *Blabera fusca*	acc. gl.[i]	JH?	oogen.	?	Topical appl. extract of sptop.	Brousse-Gaury and Goudey-Perriere 1983
12 *Nauphoeta cinerea*	acc. gl.	sptop.(mech.)	oogen.	bursa	Implant artif. sptop. in bursa	Engelmann 1970
13 *Leucophaea maderae*	acc. gl.	sptop.(mech.)	oogen.	bursa	Implant artif. sptop. in bursa	Engelmann 1970
Diplopteridae						
14 *Diploptera punctata*	acc. gl.	sptop.(mech.)	oogen.	bursa	Implant artif. sptop. in bursa	Engelmann 1970
Homoptera						
Delphacidae						
15 *Prokelisia dolus*	acc. gl., testes+ sem. ves.	?	remate[j]	?	Inject homogenates tissue	Heady 1993

Table 6.1 (*cont.*)

Family, Species	*Production Site in Male*	*Name of Product*	*Effect on Female*	*Site Where Act in Female*	*Techniques*	*References*
Hemiptera						
Reduviidae						
16 *Rhodnius prolixus*	testes? acc. gl. (opaque)	sperm?	oogen., ovip., transp. sperm	spthca., bursa[k]	Implant spthca. mated, virgin females, remove spthca., remove semin. ves., acc. gl., kill sperm, apply gland contents to female ducts, paralyze female muscles with N2	Davey 1958; Davey 1965a,b, 1967; Ruegg 1981 in Gillott 1988
Cimicidae						
17 *Cimex lectularius*	testes	sperm	oogen.	upper reprod. tract	Remove sperm from semen, inactivate sperm, remove and implant conceptacula seminis, artif. insem.	Davis 1965a,b
Lepidoptera						
Pyralidae						
18 *Helicoverpa virescens*	testes, acc. gl.?	eupyrene sperm[l]	pheromone prod., remate[m]	?	Aspermic hybrid males, amt. sperm in spthca., attract males with females in traps	Proshold and LaChance 1974; Raulston et al. 1975; Raina 1989; Raina and Stadelbacher 1990; Ramaswamy and Cohen 1992

19 *H. zea*	testes, ejac. duct, acc. gl.+ duplex ejac. duct	eupyrene, sperm,[n] protein	pheromone prod. transp. sperm, deplete pher. calling behav., recept. mating attempts	spthca.,? muscles in repr. tract	Aspermic (irrad.) males, remove testes, inject juv. hormone, inject juv. hormone antag., inject extract acc. gl.+duplex, decapitate female, remove acc. gl.+duplex, ligate simplex, obs. contractions female ducts	Snow et al. 1972; Raina 1989; Ramaswamy and Cohen 1992; Kingan et al. 1993a,b; Callahan and Cascio 1963
20 *Anagasta* (=*Ephestia*) *kuhniella*	testes?	sperm?	calling behavior	spthca.?	Spermathecal filling	Norris 1933
Lymantriidae						
21 *Lymantria dispar*	testes[n]	?	deplete pheromone, remate	?	Artif. stim. genitalia, cut bursal nerves, remove spthca., inject hemolymph mated female	Giebultowicz et al. 1991a,b
Tortricidae						
22 *Zeiraphera diniana*	testes (not acc.gl.)	sperm	oogen., ovip.[o]	?	Irradiate males	Benz 1969
23 *Epiphyas postvittana*	?	sptop. or sperm	pheromone	?[p]	Inject extracts, interrupt cops., transect nerves different times after cop.	Foster 1993
Noctuidae						
24 *Trichoplusia ni*	acc. gl., testes	?, sperm?[q]	ovip.	?	Remove testes, irradiate males, check number sperm in female, remove acc. gl.	Karpenko and North 1973; Leopold and Derugillier in Leopold 1976

Table 6.1 (*cont.*)

Family, Species	Production Site in Male	Name of Product	Effect on Female	Site Where Act in Female	Techniques	References
Plutellidae						
25 *Acrolepiopsis assectella*	testes	eupyrene, sperm,^{m} sptop.	oogen., ovip., remate	spthca.,?	Remove spthca., bursa sham operations	Thibout 1976
Saturniidae						
26 *Bombyx mori*	? reprod. tract	prostaglandin	ovip.	?, abd. gang.	Inject substances, inhib. male prod. subst., add extracts to muscle, preparations, inject extracts, record activity of isol. gangl.	Setty and Ramaiah 1979, 1980; Yamaoka and Hirao 1977
Yponomeutidae						
27 *Atteva punctella*	?	sperm	remate	spthca.	Spermathecal filling	Taylor 1967
Pieridae						
28 *Pieris brassicae*	acc. gl.	?	transp. sperm^{r}	oviduct	Paralyze female, obs. contract. female ducts when contacted by male product	Tschudi-Rein and Benz 1990
29 *Pieris rapae*	acc. gl.	sptop.	remate	bursa	Artif. inflate bursa	Sugawara 1979
Coleoptera						
Bruchidae						
30 *Acanthoscelides obtectus*	acc. gl.+ testes, acc. gl.	?	ovip., oogen.	?	Implant empty, full sptop., inject extract	Huignard 1969; Huignard et al. 1977

Diptera						
Calliphoridae						
31 *Lucilia cuprina*	acc. gl.+ testes[s]	?	remate, ovip.	?	Inject extracts, vary length cops.	Smith et al. 1989; Barton-Browne et al. 1990
32 *L. sericata*	acc. gl.?	?	remate	?	Implant acc. gl.	Pollock 1971 in Lewis and Pollock 1975
33 *Cochliomyia hominivorax*	?	<3000 MW	remate	?	Inject extracts in *M. domestica*	Nelson et al. 1969
34 *Phormia regina*	?	?	remate	?	Inject extracts in *M. domestica*	Nelson et al. 1969
Muscidae						
35 *Stomoxys calcitrans*	med. ejac. duct	?	remate, ovip.[t]	?	Implant glands, inject extracts glands	Morrison et al. 1982
36 *Glossina morsitans*	acc. gl.	peptide?[u]	remate	?	Implant glands, inject extracts glands	Gillott and Langley 1981
37 *G. pallidipes*	?	?	remate	?	Mate with depleted male	Leegwater-van der Linden and Tiggelman 1984
38 *Musca domestica*	ejac. duct, testes[w]	?, <3000 MW	remate, ovip.[x]	brain?[v]	Inject extracts, deplete male, irradiate male al. (aspermic), castrate males, implant ejac. duct, decapitate female	Adams and Nelson 1968; Riemann et al. 1967; Nelson et al. 1969; Riemann and Thorson 1969; Leopold et al. 1971a
Anthomyiidae						
39 *Delia antiqua*	paragon. gl.	peptide	ovip., remate	?	Inject extracts	Spencer et al. 1992
40 *Hylemya brassicae*	acc. gl. (not testes)	?	ovip., remate	?	Inject extracts, implant glands, testes	Swailes 1971

Table 6.1 (*cont.*)

Family, Species	Production Site in Male	Name of Product	Effect on Female	Site Where Act in Female	Techniques	References
Tephritidae						
41 *Ceratitis capitata*	acc. gl. (not testes)	?	remate	?	Implant testes, acc. gl., remove spthca., testes	Saul and McCombs 1993
	testes	sperm[y]				
42 *Bactrocera cucurbitae*	acc. gl.[z]	?	ovip., remate	?	Aspermic males, interr. cop. before sperm transfer	Tsubaki and Sokei 1988; Kuba and Soemori 1988 in Itô and Yamagishi 1989; Yamagishi and Tsubaki 1990; Kuba and Itô 1993
Drosophilidae						
43 *Drosophila melanogaster*	acc. gl., testes	esterase-6, sperm, peptide protein	remate (short and long term),[b1] attractive-ness[c1] ovip.,[d1] larger eggs	?[a1] ? ?	Implant acc. gl., testes, males with nonmotile sperm, mutant males, interr. second cop., inject extracts, deplete males, gen. engineer females that produce sex peptide, transgenically eliminate cells from acc. gl.	Manning 1967; Merle 1968; Burnet et al. 1973; Fowler 1973; Gilbert et al. 1981; Mane et al. 1983; Gromko et al. 1984; Scott and Richmond 1985; Scott 1986; Chen et al. 1988; Aigaki et al. 1991; Kalb et al. 1993
44 *D. funebris*	acc. gl.	protein, carbohydrate	remate, ovip.[e1]	?	Inject extracts, implant acc. gl., decapitate female, incorp. labeled aa in vivo, in vitro	Baumann 1974a,b

45 *D. suzukii*	acc. gl. (not testes)	?	ovul.[f1]	?	Inject extracts, implant glands	Fuyama 1983
46 *D. pulchrella*	acc. gl. (not testes)	?	ovul.[f1]	?	Inject extracts, implant glands	Fuyama 1983
47 *D. subobscura*	testes	?	remate	?	Mutant male lacking testes	Maynard Smith 1956
Cecidomyiidae						
48 *Mayetiola destructor*	reprod. tract, mechan. stim. of cop.	?	remate,[g1] ovip.	?	Inject extracts, exhaust male (cop. 35 females in a row)	Bergh et al. 1992
Chironomidae						
49 *Chironomus riparus*	?	?	remate[h1]	?	Inject extract of body	Downe 1973
Culicidae						
50 *Aedes aegypti*	acc. gl.,[m] testes	2 proteins, tot. vol., ?	remate,[i1] feed,[j1] digest blood,[k1] seek host[m1]	abd. gangl., ovary[l1]	Vary number cops. female, remove acc. gl., cut testes, implant acc. gl., testes, deplete male, inject extracts, cut CNS connection, check insem., dir. obs. remating	Leahy and Craig 1965; Craig 1967; Leahy 1967; Judson 1967; Spielman et al. 1967; Fuchs et al. 1969; Fuchs and Hiss 1970; Gwadz et al. 1971; Gwadz 1972; Ramalingam and Craig 1976; Adlakha and Pillai 1976; Williams et al. 1978; Young and Downe 1982
51 *Aedes taeniorhynchus*	acc. gl.	?	oogen.[n1]	ovary	Inject gl. contents, inj. susp. gl., add to fat body in tissue culture	Borovsky 1985; O'Meara and Evans 1977
52 *Aedes triseriatus*	acc. gl.	?	remate ovip.[o1]	?	Implant gland	Craig 1967; Ramalingam and Craig 1976

Table 6.1 (*cont.*)

Family, Species	Production Site in Male	Name of Product	Effect on Female	Site Where Act in Female	Techniques	References
53 *Aedes albopictus*	acc. gl.	?	remate[o1]	?	Implant gland	Leahy and Craig 1965; Craig 1967
54 *Aedes atropalpus*	acc. gl.	?	remate[o1]	?	Implant gland	Craig 1967
55 *Aedes mascarensis*	acc. gl.	?	remate[o1]	?	Implant gland	Craig 1967
56 *Aedes polynesiansis*	acc. gl.	?	remate[o1]	?	Implant gland	Craig 1967
57 *Aedes scutellaris*	acc. gl.	?	remate[o1]	?	Implant gland	Craig 1967
58 *Aedes sierrensis*	acc. gl.	?	remate[o1]	?	Implant gland	Craig 1967
59 *Aedes togoi*	acc. gl.	?	remate[o1]	?	Implant gland	Craig 1967
60 *Aedes vittatus*	acc. gl.	?	remate[o1]	?	Implant gland	Craig 1967
61 *Culex pipiens*	acc. gl., testes	?	remate,[o1] feed,[m1] ovip.	?	Remove gl., cut duct testes, implant gl., testes, inj. extracts, check insem.	Craig 1967; Leahy 1967; Adlakha and Pillai 1976; Jones and Gubbins 1979
62 *Culex tarsalis*	acc. gl.	protein, 2000 MW	remate, ovip	head?, ?	Inj. extracts, check transf. labeled cmpds., vary number cops. for female	Young and Downe 1983, 1987
63 *Anopheles quadrimaculatus*	acc. gl.	?	remate[o1]	?	Implant gland	Craig 1967

Notes: In all cases, effect on female remating is to decrease it, and effect on oviposition, oogenesis, sperm transport, and pheromone depletion is to increase it.

[a] Stimulated contraction of the oviduct; apparently two types of receptors were involved.

[b] Very small amounts were effective (e.g., 1/50 of a gland, or 0.56 micrograms).

[c] Increased number of egg pods/week, number of eggs/pod, number of eggs/female, lengths of oocytes in females, decreased number of days before first oviposition, and number of oocytes resorbed. In addition, the site of oviposition changed from that typical of virgin females (eggs scattered in the cage) to that of mated females (grouped in the sand).

[d] Effect on muscles could be indirect.

[e] Other characteristics of the semen besides the sperm were influenced by irradiation, thus weakening the inference that sperm were responsible for the effect on oviposition.

[f] Female remates in nature.

[g] No effect on female receptivity.

[h] Role of spermatheca is unknown, other than the fact that it apparently does not involve neurosecretory activity, which was apparently ruled out by injections of reserpine. Presumably receptors in the spermatheca were stimulated, as sectioning the nerve cord abolished the effect. Male effect on corpora alata may be indirect rather than direct.

[i] Female was decapitated at start of the experiment. The stimulatory effect of a spermatophore inserted into the bursa, instead of the extract, was significantly larger. This could be due to additional stimulatory effects from the spermatophore in the bursa, or to greater amounts of stimulatory substance being passed to the female. Similar applications of juvenile hormone mimics gave similar reactions.

[j] Female responses were greater when mated to a virgin male than when to a male that had mated another female less than an hour before.

[k] Stimulation of contractions is apparently not due to direct action of the male product on female muscles, but rather to nervous transmission to a site farther up the oviduct where contractions originate.

[l] Some males in nature were below threshold for maximum female response.

[m] Mated females without sperm attracted significantly fewer males than those with sperm; there was also a statistically insignificant trend for numbers of sperm in mated females that attracted at least one male to be higher than the numbers in females that did not attract a male.

[n] Mechanical stimuli from genitalic coupling reduced pheromone production, but did not affect female receptivity to attempted copulations.

[o] Transfer of spermatophores was verified, but irradiated males were assumed to produce normal amounts of accessory gland products. If this species is like the moth *Trichoplusia ni*, reduced numbers of viable sperm with different doses of radiation are probably correlated with reduced absolute numbers of sperm transferred (Holt and North 1970); the fragmentary data in Benz (1969) on sperm that failed to leave the spermatophore in females with 0% fertilization suggest that this is indeed the case. Oogenesis and oviposition are controlled separately.

[p] Timing of transfer of sperm to the spermatophore within the female was not determined. The delayed onset and gradual buildup of the effect suggests that presence of sperm in the spermatheca may be the triggering stimulus.

[q] The authors did not rule out other possible products of the testes besides sperm, but did show that, of the two types of sperm normally transferred, the fertile, eupyrene sperm were responsible for increasing oviposition rates over those for copulations in which no sperm were transferred (irradiation of male results in reduced sperm transfer; Holt and North 1970); the possibility that irradiation also affected accessory gland secretion was apparently not checked.

[r] Activity of the sperm themselves may be involved (Gillott 1988), but in other lepidopterans female contractions are largely responsible for sperm transport to the spermatheca (e.g., Davey 1965b).

[s] Since effects of the testis extract wore off much more rapidly than those of the accessory gland extract, the products may be different. The degree of effect on oviposition also differed, but in the absence of additional data the two organs are grouped.

[t] Mating also induced females to ingest greater volumes of blood at each meal (Venkatesh and Morrison 1980), but the role (if any) of the accessory glands was not established. Both traits could presumably be under the same control.

[u] Evidence was not conclusive.

[v] Action on brain could be indirect; although labeled <3000MW, male ejaculatory duct substance (not known to inhibit female receptivity) was more persistent in the female head than elsewhere (Terranova et al. 1972).

Table 6.1 (*cont.*)

[w] Increased oviposition occurred in virgin females with implanted testes as compared with normal virgins, but data from sham-operated virgins were not given (Riemann and Thorson 1969).

[x] Strictly speaking, the measure of the tendency to oviposit (mg of eggs laid in cages containing many females) was inappropriate, since contributions of individual females were not determined. The effects were in general so large, however, that there was little doubt that the differences were significant. Since nearly all virgin females had eggs that were apparently ready to be oviposited (371 of 372; Riemann and Thorson 1969), the male trigger is apparently ovulation or oviposition, rather than vitellogenesis.

[y] Remating interval was reduced significantly when the first male was castrated; the number of rematings was reduced but not significantly. Possible effects of products from the testes other than sperm were not eliminated, although implantation of testes did not have an effect.

[z] Sperm and the stimuli from intromission were eliminated as possible stimuli. Other possible mechanisms of induction of refractory behavior in females, such as a possible increase in the pressure in the spermatheca from other seminal fluids, longer stimulation from copulatory courtship (if it occurs), were not eliminated.

[a1] Destruction of the pars intercerebralis region soon after eclosion mimics the effect of mating (Boulétreau-Merle 1976). The severe damage caused by this operation requires caution in interpreting these results, however (Gillott 1988).

[b1] Both short-term decrease in receptivity and long-term increase could be due to increased sperm motility causing both quicker filling and emptying of the female storage organs.

[c1] The effect on attractiveness occurred only 6–8 hrs after mating and was not present at either 4 or 10 hr after mating. Scott (1986) was unable to confirm this result.

[d1] The effect on oviposition was complex: the number of mature eggs in the ovaries was not higher than in virgin controls; but since eggs were laid at a much higher rate, rate of vitellogenesis was presumably higher. The change in oviposition rate is so rapid and large (within 4 hr the rate rose from <1 egg/hr to about 20; Manning 1967) that oviposition *sensu strictu* must be triggered by mating.

[e1] Radioactive labeling showed that the male substances P2 occurred in the head, thorax, and abdomen of the female (Baumann 1974a,b).

[f1] Author states that changes in oviposition rates were in accord with those in ovulation rates.

[g1] The switch-off of female receptivity may be under different control (or at least require greater amounts of male substances) than induction of oviposition. Only short-term suppression occurred as a result of mechanical stimulation.

[h1] Females refrain from entering swarms of males.

[i1] Since females can terminate copulation by pushing male genitalia with their legs (Spielman 1964), it must be assumed that some females were receptive to copulation but not to insemination; others rejected males by preventing genitalic coupling (Craig 1967).

[j1] See, however, Klowden 1979.

[k1] Frequency of feeding was unaffected, but size of blood meal increased.

[l1] Differences may be due to differences in amount of blood ingested (Adlakha and Pillai 1976). Chemosterilization of the ovary abolished the effect of mating, and subsequent implantations of additional ovaries reinstituted it. The effect of the ovary may be indirect (Gillott 1988).

[m1] Effect was not immediate.

[n1] Female was induced to oviposit without a blood meal instead of waiting until she had fed.

[o1] Females kept with males were not inseminated. The mechanism of female rejection was not determined.

7 Evidence That Cryptic Female Choice Is Widespread, III: Male and Female Morphology

7.1 Female Reproductive Ducts: A Tortuous Route to the Egg

Obstacles to Insemination and Female Control

It is a general rule in animals with internal fertilization that the male does not deposit his sperm directly on the female's eggs, nor at a site where fertilization will occur soon afterward (Eberhard 1985). Generally sperm must be moved (usually by the female) to storage and/or to fertilization sites (e.g., Birkhead and Møller 1992 on birds; Suarez et al. 1990 and Katz and Drobnis 1990 on mammals; Austad 1984 and Thomas and Zeh 1984 on arachnids; Davey 1965b, Engelmann 1970, Linley and Simmons 1981, Gillott 1988, Sugawara 1993, and LaMunyon and Eisner 1993 on insects; see references in Eberhard 1985 for other groups). Access to storage and insemination sites is often not easy for the male and his sperm, either because it is mechanically difficult for his genitalia to reach them or for his sperm to survive and move to them (fig. 7.1).

Consider, for instance, the obstacles mammalian sperm can face (Suarez et al. 1990; Roldan et al. 1992). Some sperm are often immediately leaked to the exterior (e.g., up to 80% of the ejaculate in the rabbit; Overstreet and Katz 1977; see also chap. 3.2), while others are phagocytized in the vagina (the site of ejaculation in many species). The surviving sperm are then called upon to survive a veritable odyssey. First the sperm cells must pass through the mucus that fills the cervix, probably at least partly under their own power. This mucus sometimes contains factors that adhere to sperm, rendering them unable to move onward (Katz and Drobnis 1990), or antibodies that agglutinate sperm (Mann and Lutwak-Mann 1981); in other cases it may be in a physicochemical state which makes sperm movement difficult (Mann and Lutwak-Mann 1981). The cervix of some species has cilia which at least sometimes create currents that quickly flush material back into the vagina (Blandau 1973). The cervical mucus apparently channels sperm toward the wall of the cervix, and the motion of the flagellum may change in close proximity to the wall (the so-called wall effect), with

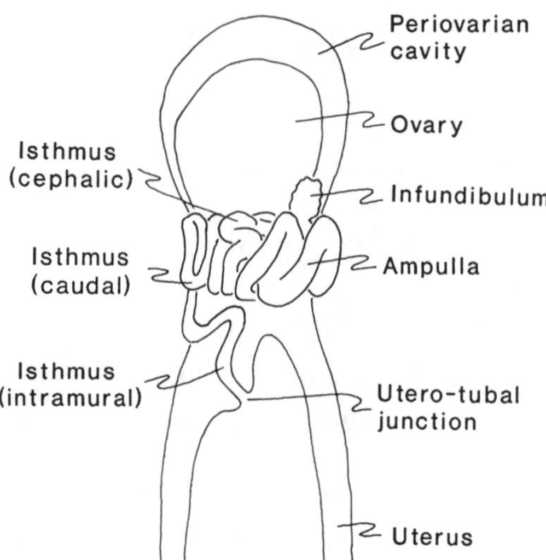

Figure 7.1 The convoluted oviduct of a hamster. The sperm must arrive at the ampulla to fertilize the eggs. Although the male deposits hundreds of millions of sperm in the female's uterus, almost none reach the ampulla. Yet fertilization rates are very near 100%, even when the entrance to the oviduct is ligated near the utero-tubal junction 1 hr after copulation. It is not clear how such incredibly fine control is achieved with millions of sperm being held back but a tiny fraction nevertheless reaching the ampulla. (After Yanagimachi and Chang 1963)

uncertain and possibly detrimental effects on sperm locomotion (Overstreet and Katz 1977). The cervix of the rat is relatively tightly shut until 5–12 hr after the beginning of estrus, possibly making successful sperm transport to the uterus less likely (Chester and Zucker 1970). Sexual arousal in the human female involves pulling the cervix *away* from the vagina, making the trip longer for the sperm (Masters and Johnson 1966). In some species the cervix may serve as a temporary reservoir (see fig. 7.2) where conditions are less hostile than in other portions of the female's reproductive tract (e.g., Morton and Glover 1974a).

Of those sperm which succeed in passing through the cervix and into the uterus, usually only a tiny fraction are then transported, by female contractions and/or cilia, to the narrow utero-tubal junction, which often has a valvelike structure (Blandau 1969). Each sperm must probably once again swim actively through the junction, possibly against a current produced by cilia in the oviduct (Suarez et al. 1990). The duct itself sometimes also *narrows* during estrus after copulation (Overstreet and Katz 1977; Suarez et al. 1990), and it may bend (rabbit), form sinuous waves (horse), or coil (rodents) (Suarez et al. 1990).

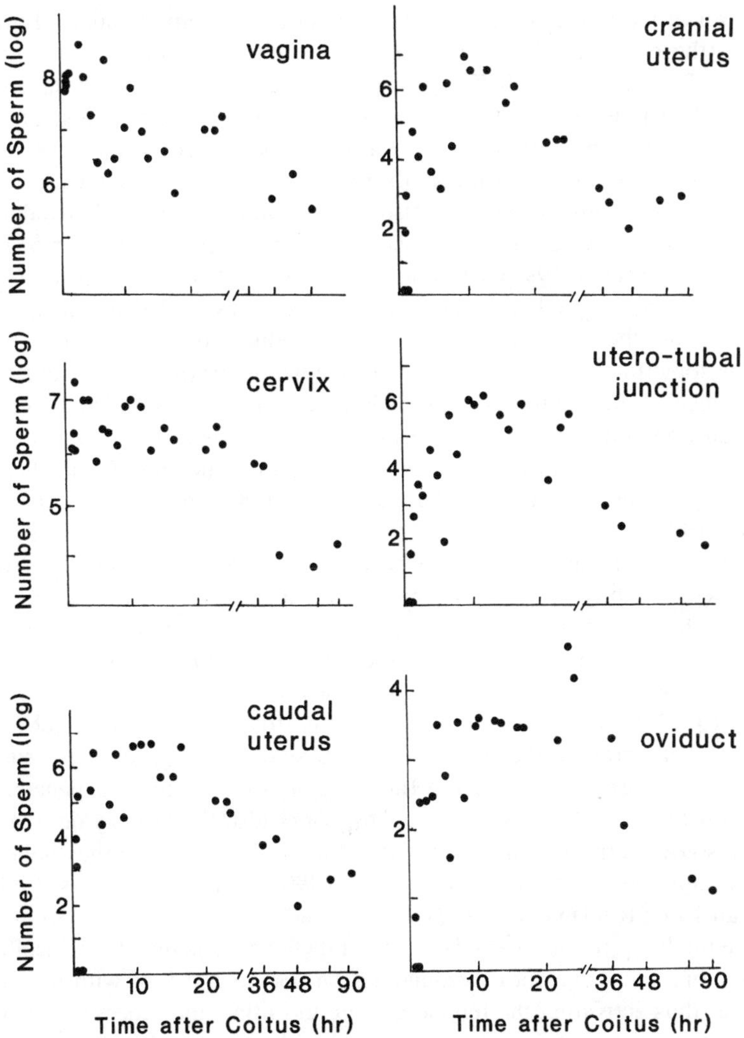

Figure 7.2 Numbers tell a story of travail and precision as sperm are moved through the reproductive tract of a female rabbit (read from top to bottom on left then on right). The number of sperm is reduced by a factor of about 10,000 from the vagina to the oviduct. Ovulation, which is induced by copulation, occurs about 10 hr afterward, and the numbers of sperm in the oviduct, where fertilization occurs, are remarkably constant at about 2000–3000 from about 10 hr until 20 hr after copulation, suggesting relative precision in control (each point is from a single female). (After Morton and Glover 1974a)

Once past the uterotubal junction, the sperm apparently must attach to the wall of the lower isthmus of the oviduct and become temporarily inactive (those which fail to attach die or lose their fertilizing capacity more rapidly; Pollard et al. 1990; T. Smith 1990). Sperm activation or "capacitation," which must occur if they are to be able to fertilize eggs, is also under female control, probably under ovarian endocrine control (Parks and Ehrenwald 1990). The sperm may disengage itself from the oviduct (or be released) and swim (or be moved) farther up the oviduct, perhaps along folds in the wall where the "downstream" current may be weaker or reversed (Suarez et al. 1990). Blandau (1969) judged that the downward current normally observed in the rabbit oviduct is so strong that sperm cannot move toward the ovary unless the current is reduced or reversed. The oviduct is sometimes coiled on itself in convoluted loops (fig. 7.1). In at least some species such as the cow, pig, and rabbit, it executes peristaltic contractions, predominantly in the uterine direction (Austin 1975; Overstreet and Katz 1977), thus further lengthening this part of the journey (the direction of peristalsis sometimes changes during optimum fertility periods; Austin 1975).

There are still further problems when a mammalian sperm cell finally reaches the vicinity of an egg. It must change its behavior and use a second type of "hyperactivated" motility (which may in fact include two or more different types of movement; see Burkman et al. 1990; Young and Starke 1990) to pass through the especially viscous fluids in this region. Finally, it must push through the cumulus oophorus (follicular cells in a gel-like matrix) that surrounds the ovum, using, among other things, enzymes such as hyaluronidase (see, however, Talbot et al. 1985), and then penetrate between the cells of the corona radiata using a different enzyme. Once again, some sperm become stuck and fail to advance because they adhere to the cumulus material (Suarez et al. 1990; Katz and Drobnis 1990; Roldan 1990; Roldan et al. 1992).

The final step for the sperm is to penetrate the zona pellucida (an acellular glycoprotein envelope) surrounding the oocyte and fuse with the oolemma (thus activating the female gamete to initiate the process of swallowing the sperm and rejecting further sperm; Deuchar 1975). Again more sperm penetrate the zona pellucida than enter the egg (Longo 1987).

There is general agreement that the sperm are dependent on female help to advance at several stages of this trek (Austin 1969; Blandau 1973; Overstreet and Katz 1977; Suarez et al. 1990; Katz and Drobnis 1990). A dramatic illustration of female participation occurs in the dog *Canis familiaris*. When a female was surgically modified so the top of her uterus was open to the outside, contractions of her uterus began about one minute after copulation began and produced sufficient pressure to shoot streams of semen to a distance of 20–25 cm, like a stream ejected from a syringe

(Evans 1933). In the unmodified uterus this force would act to move sperm inward toward the oviduct.

The combination of female obstacles and active transport may be finely tuned. It is seldom mentioned that female control of sperm migration in at least some species appears to be capable of almost unbelievably precise adjustments. In the hamster *Mesocricetus auratus*, for example, the male ejaculates several hundreds of millions of sperm into the vagina in a single copulation, but only a tiny fraction ever arrives at the ampulla, the fertilization site high up the oviduct (fig. 7.2). When standard numbers of sperm were artificially introduced into the female's uterus at about the time of ovulation, they entered the ampulla only very slowly. The total numbers of living sperm did not start to exceed the number of eggs in the ampulla until after about 50% or more of the eggs had been fertilized. A relatively minor error by the female (say by a factor of one in a million) could result in the presence of hundreds or thousands more sperm when the eggs move down from the ovary, or in no sperm at all.

Similar incredibly precise reductions occur in other rodents in which nearly every sperm in the ampulla finds and enters an egg (Boatman 1990). The number of sperm in the oviduct of the rabbit (*Oryctolagus cuniculus*) also shows signs of precise regulation, remaining quite constant at 1000–2000 for many hours after copulation while ovulation occurs (Morton and Glover 1974a; fig 7.2).

Sperm numbers can be relatively constant under a surprising variety of conditions, suggesting female control. In the rabbit, artificial insemination with equal numbers of sperm in twenty times more saline resulted in nearly equal log mean numbers of sperm in the oviduct 10–11 hr later (Morton and Glover 1974b). An increase by a factor of one thousand in the number of sperm gave only a 40-fold increase in the oviduct (Morton and Glover 1974b). The fact that multiple (as compared with single) matings in rabbits have no effect on the number of tubal sperm (Morton and Glover 1974b and references) suggests the importance of active female control rather than simple diffusion or movements of different numbers of inseminated sperm. Insects may have similar flexibility. A 40% reduction in ejaculate size in *Drosophila melanogaster* also did not reduce female fertility (Kaufman and Demerec 1942 cited by Wu 1983). Thus some females may have fine control over the movement of sperm in their reproductive tracts.

The ducts leading to sperm storage sites are often especially long, twisted, or coiled. In some cases, they reach truly absurd extremes from a transportation efficiency standpoint. In the cassidine beetle *Charidotella propinqua*, for example, the female's spermathecal duct, if stretched out, would be more than twenty times the length of her body (D. Windsor, unpub.; see also fig. 1.2 and the discussion in sec. 7.2). In some other arthropods the diameter rather than the length of the spermathecal duct is "absurd." For instance, the ducts of the wasp *Dahlbominus fuscipennis* and the mite *Caloglyphus berlesi* are so narrow that only a single spermatozoan can go through at a time (Wilkes 1966; Witalinski et al. 1990). In the biting midge *Culicoides melleus*, the diameter of the narrowest portions of the spermathecal duct is 1.2 fm, enough to accommodate a maximum of only eight spermatozoa at one time if they are optimally packed (Linley 1981). The insemination duct of the spider *Clubiona pallidula* narrows to 2mm (the diameter of a single sperm cell) and is highly sclerotized only in this short region (Huber, in press). Such tiny diameters effectively rule out the possibility of penetration by any male sperm-transferring structure.

Tight fits of this sort reduce both the propulsive force the sperm can generate with their flagellae and the number of sperm that can ascend at any one time, and probably result in the sperm themselves, creating strong currents *opposing* their ascent to the spermatheca (Linley and Simmons 1981). Because of the rigid nature of the duct and the spermatheca, the liquid which spermatozoans displace as they move up the duct will have nowhere to go, except back down the duct. For example, I calculate from Linley's data that the sperm of *C. melleus* would occupy, when optimally packed and moving through the narrow portion of the duct, about 60% of the volume. Thus, movement of sperm up the duct will create a substantial downward flow in such narrow, rigid ducts, against which a sperm attempting to swim up without help from the female must have to struggle (Linley and Simmons 1981). It seems clear that there is something special about sperm transport that often favors the evolution of female traits that make passage difficult.

Physiological Barriers

In addition to the obstacles and difficulties just described, there are generally additional problems for the sperm, many of which may result from adaptations to protect the female reproductive tract from opportunistic pathogens. Mammalian sperm are ingested in the female reproductive tract by lymphocytes and leucocytes in both the uterus and the oviduct (in the rabbit, some associations with leucocytes may be reversible; Overstreet and Katz 1977), by epithelial cells of the oviduct, and by follicle cells in the mass surrounding the ovulated egg in the oviduct (Overstreet and Katz 1977; Suarez et al. 1990; Roldan et al. 1992). They are also subject to cytolysis by uterine gland tubules (Austin 1975). Transport by the female may result in damage to sperm in several species (Suarez et al. 1990) (in the rabbit the most rapidly transported sperm may not play any role at all in fertilization; Overstreet and Katz 1977). In the human female the pH within the vagina is 3.5–4.0 (Masters and Johnson 1966), which also makes sperm mobility and survival more difficult. Sperm do better at 5.5–7.0, and the buffering power of the semen usually keeps the pH above 5.5 for 3–4 hr; sometimes (with help from the female?) the pH remains high for as long as 16 hr after copulation (Masters and Johnson 1966). In some species the spermicidal properties of the female tract vary during the female's reproductive cycle (e.g., Doak et al. 1967 on the dog).

Some barriers may be especially effective against conspecific sperm, suggesting that they function primarily as sperm filters rather than defenses against infection. While conspecific sperm are often retained in the cumulus and corona radiata layers around a hamster egg (above), both layers were readily penetrated by both bullfrog and sea urchin sperm. Even a motile alga, *Chlamydomonas reinhardtii*, was able to pass through these layers and reach the zona pellucida next to the egg (Talbot et al. 1985)!

The physical and chemical environments within female invertebrates are less well known but can also pose problems for sperm. Any active sperm that are not protected by the spermatophore and leak out into the bursa of the female moth *Trichoplusia ni* are rapidly immobilized and are not transferred to the spermatheca (Holt and North 1970). Bursal sperm are also damaged relatively quickly in the bruchid beetle *Callosobruchus maculatus*—50% were damaged 2.25 hr after copulation, 90% after 5 hr, and 99% after 10 hr (Eady 1994a). In the female bug *Cimex hemipterus* the sperm must be able to perform metabolic processes that are both anaerobic (in the spermalege) and aerobic (in the haemocoel) (Ruknudin and Raghavan 1988).

Difficult barriers to sperm also occur in the barnacle *Balanus* (Klepal et al. 1977). Sperm must enter the egg sac via tiny pores which are almost too small for them. When animals were fixed for microscopy while sperm were still in the process of passing through these pores, those still in the pores were completely normal in appearance, while many of those that had passed into the interior of the sac were severely damaged, with membranes lost, the nucleus and axoneme separated, or the microtubules of the axoneme completely disorganized (Klepal et al. 1977).

Perhaps the most dramatic evidence favoring the interpretation that some female structures are designed to hinder rather than help sperm movement is the evolution of a second set of sperm storage receptacles and transport ducts that occurred in bedbugs (see chap. 2.5; Carayon 1966, 1975). Compared with related species lacking these structures, the presence of these receptacles and ducts greatly reduces the number of sperm that eventually reach fertilization sites near the ovaries. In addition, the receptacles and ducts are derived from cells and tissues that are normally used to combat invasions of pathogens and foreign materials (amoebocytes, secretions of PAS-positive membranes; Carayon 1966, 1975).

What Do Female Barriers Prevent?

Why all these impediments to successful sperm movement? Many female characters such as low pH, currents flushing material outward, phagocytic cells, and valves to prevent flow upstream may have originated due to natural selection to prevent infection of the female reproductive tract by microorganisms (above; chap. 2.1; see Profet 1993 for evidence that menstruation in mammals has a similar function). Such characters could be employed (and subsequently further elaborated) as cryptic female choice mechanisms, favoring males that could traverse the ducts by using their genitalia or by inducing the female to move sperm along the ducts (see also Eberhard 1985; Birkhead et al. 1993a). Although defense against infections may have originally been important, however, less drastic measures seem to offer sufficient protection.

Defense against microbial invasion is an unlikely general explanation, for example, of the especially tortuous nature of spermathecal ducts. In most of these groups (e.g., most insects) the female's external genital opening can be closed, and her other reproductive ducts besides the spermathecal ducts are usually much less constricted. Tortuous ducts may function to impede sperm arrival in storage, but in such a way as to allow the female to screen the abilities of the sperm themselves (Cummins 1990; Roldan et al. 1992; Birkhead et al. 1993a; Keller and Reeve, in prep.). On the other hand, it is possible that tortuous female ducts did not evolve to control of sperm *entry* into the storage organs, but rather to control sperm *exit* to fertilization sites. For instance, a tortuous spermathecal duct could reduce the chances of having too many sperm leave the spermathecal duct to fertilize any given egg. An overabundance of sperm might be a common problem because selection on sperm would tend to favor those able to make an early exit from the spermatheca. The female might die or remate before late-emerging sperm were used. In contrast, selection on the female, both to avoid polyspermy, which could result in egg loss, and to achieve efficient sperm usage (maintain a minimum number of sperm but avoid running out), would tend to favor control of sperm access to the fertilization site. A structure whose function appears to be the limitation of sperm egress from storage occurs in the crabs *Inachus phalangium* and *Pisa tetraodon*. An irislike diaphragm at the mouth of the seminal receptacle is thought to contract to a small opening during fertilization (Diesel 1990, 1991; no substantiating data were given, however).

There is a chance to examine the question of fertilization control versus insemination control using the genitalic design of entelegyne spiders, lepidopterans, some isopods, and some mites. In these animals the entrance (insemination) duct of the spermatheca is separate from the exit (fertilization) duct. If insemination ducts are consistently longer, especially when

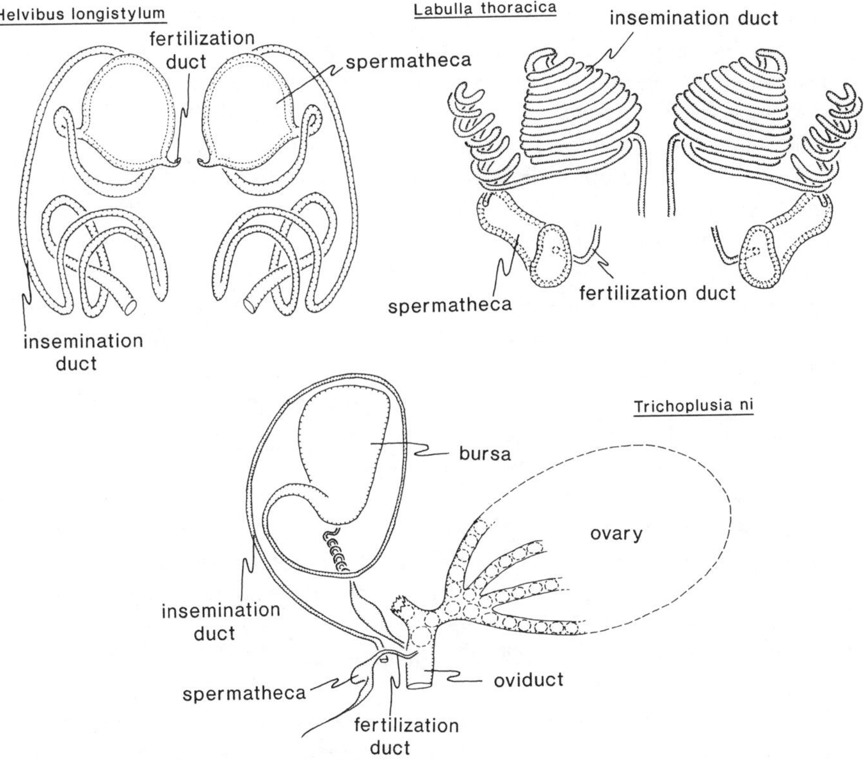

Figure 7.3 Entering is harder than leaving. *Upper left*: the spermathecae and associated ducts of the theridiid spider *Helvibus longistylum*; *upper right*: of the linyphiid spider *Labulla thoracica*; and *bottom*: of the moth *Trichoplusia ni* (drawn to different scales). Entrance (insemination) ducts are much longer and more tortuous than the exit (fertilization) ducts. In species like *Labulla*, the insemination ducts reach apparently ridiculous extremes. The consistently greater lengths of insemination ducts than fertilization ducts in spiders (see fig. 7.4) suggest that in many other groups, in which tortuous ducts serve both as entrance and exit ducts, selection favoring female control of sperm entry rather than exit was responsible for the evolution of the tortuous forms. (After Levi 1964c; Millidge 1993; and Holt and North 1970)

one of the ducts is particularly long and convoluted, then control of sperm entry (which can result in sexual selection by cryptic female choice) is a more likely function for tortuous ducts than control of sperm exit (which, as described above, could result from sperm competition and natural selection on females).

The group with the most available data is entelegyne spiders, because there are many species, and drawings of female genitalic ducts are routinely included in taxonomic descriptions (fig. 7.3). I used a map-reading tool to measure the lengths of the insemination ducts (distinguished by their usually thicker walls, wider diameters, and openings to the surface of the female) and the fertilization ducts in the drawings of female genitalia in a sample of taxonomic papers by major spider taxonomists on 111 genera in 16 families of entelegyne spiders; the measured lengths were underestimates, since the drawings gave only flattened, two-dimensional images.

The results were clear (fig. 7.4). Insemination ducts were longer than fertilization ducts in 314 species, shorter in 40, and equal in 6. There was substantial variation, both among and within genera (see standard deviations in fig. 7.4), but there was one completely consistent pattern: insemination ducts were always longer when the difference between the two ducts was substantial (i.e., when one of the ducts was relatively tortuous in design; figs. 7.3, 7.4). The great variability in lengths and forms of these ducts even among closely related species (which, of course, is why such drawings are often included in taxonomic papers), indicates that a phylogenetic bias is unlikely to explain this consistency.

Only scattered data on duct lengths are available for other groups, but they show a similar pattern. At least a few species of Lepidoptera have the spider pattern of longer insemination ducts than fertilization ducts (fig. 7.3). Similarly, the insemination duct (= "spermathecal duct") of asellotan isopods is characterized as a "long, thin-walled . . . tube" (Wilson 1991, p. 240), while the spermatheca is inside the oviduct itself, so the length of the fertilization duct is essentially zero (Wilson 1991). Fragmentary data on mites also show the same pattern (Lee 1974 on *Euepicrius filamentosus* and *Athiasella dentata*—one other rhodacarid species has approximately equal ducts; Thomas and Zeh 1984 on *Rhinophaga pongoicola* and *Tyranninyssus spinosus*; Witalinski et al. 1990 on *Acarus siro*).

Data from spiders are particularly informative because, as far as is known, spider sperm are always encapsulated and thus completely immobile until after they have been transferred (Lopez 1987). This means that the tortuous insemination ducts of females cannot be functioning to screen sperm on the basis of their abilities to move up the ducts (as may happen in at least some portions of the reproductive ducts of female mammals; see above), or to separate sperm from other, less mobile infectious organisms. Instead, the insemination ducts probably screen males on the basis of their abilities to introduce sperm deep within the female, and/or to induce her to transport them. In *Micrathena gracilis* stimuli from copulation apparently induce sperm transport to the spermatheca by the female (Bukowski and Christenson, in prep.). In some species the male intromittent organ reaches the spermatheca; in others, such as the linyphiid spiders discussed in chapter 3.5 and many others, it clearly does not (see Gering 1953; Grasshoff 1973; Huber 1993, 1994a, and 1995 for other spiders).

In sum, extensive data on separate insemination and fertilization ducts from spiders and very limited data from three other test groups suggest that

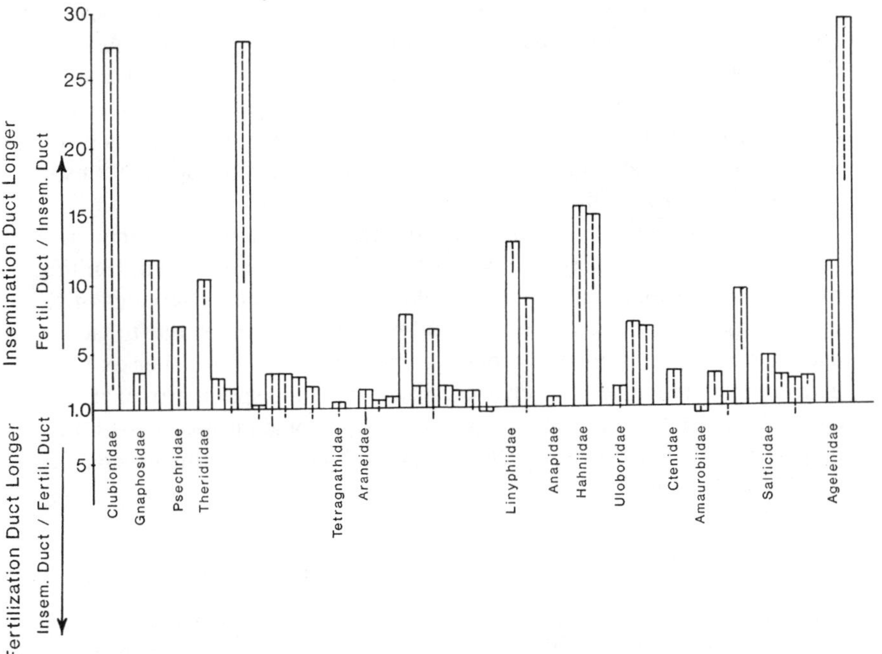

Figure 7.4 Distribution of average ratios of length of entry (insemination) duct/length of exit (fertilization) (or the inverse when the fertilization duct was longer) in forty-three genera of spiders in which three or more species were measured (measurements were directly from drawings with a map-reading instrument or curvimeter). The dashed line within each bar indicates one standard deviation of the mean. In some groups the two ducts were more or less equal; in all cases in which one duct was substantially longer than the other (averages reached nearly 30:1) the fertilization duct was the longer of the two. The genera (with number of species/genus) and the papers that were consulted, in order from left to right on the graph, were as follows: Clubionidae—*Trachelas* (16) Platnick and Shadab 1974; Gnaphosidae—*Gnaphosa* (5) Platnick and Shadab 1975; *Lygromma* (3) Platnick and Shadab 1976; Psechridae—*Psechrus* (3) Levi 1982; Theridiidae—*Achaearanea* (5) Levi et al. 1982; *Episinus* (16) Levi 1964a; *Stemmops* (3) Levi 1964b; *Helvibus* (3) Levi 1964c; *Phoroncidia* (13) Levi 1964d, 1967b; *Thymoites* (13) Levi 1964e; *Theridion* (7) Levi 1967a,b; *Anelosimus* (3) Levi 1967b; *Robertus* (3) Brignoli 1980; Tetragnathidae—*Chrysometa* (9) Levi 1986; Araneidae—*Wagneriana* (5) Levi 1991; *Nuctenea* (3) Levi 1974; *Zygiella* (3) Levi 1974a; *Larinia* (3) Levi 1975; *Acanthepeira* (4) Levi 1976; *Cyclosa* (4) Levi 1977a; *Eustala* (14) Levi 1977a; *Metepeira* (7) Levi 1977b; *Aculepeira* (5) Levi 1977b; *Micrathena* (3) Levi 1978; Linyphiidae—*Walckenaeria* (3) Millidge 1993; *Pelecopsis* (3) Millidge 1993; Anapidae—*Pseudanapis* (3) Brignoli 1980; Hahniidae—*Neoantistea* (12) Opell and Beatty 1976; *Hahnia* (4) Opell and Beatty 1976; Uloboridae—*Miagrammopes* (5) Opell 1979; *Uloborus* (7) Opell 1979; *Philoponella* (10) Opell 1979; near Ctenidae—*Machadonia* (14) Griswold 1991; Amaurobiidae—*Themacrys* (8) Griswold 1990; *Vidole* (6) Griswold 1990; *Phyxelida* (16) Griswold 1990; *Namaquarachne* (6) Griswold 1990; Salticidae—*Sitticus* (5) Galiano and Baert 1990; *Neonella* (30) Galiano 1987a; *Aillutticus* (3) Galiano 1987a; *Hisukattus* (4) Galiano 1987b; and Agelenidae—*Tegenaria* (4) Brignoli 1971; *Cicurina* (6) Brignoli 1972.

tortuous spermathecal ducts evolved to limit or control the entrance of sperm into storage, rather than the exit of sperm to fertilize eggs. Alternative hypotheses can undoubtedly be brought forward (e.g., reductions of venereal infections of the spermatheca; see above),[2] but I know of no evidence to support them. Even if an increase in the difficulty of access is due to other selective advantages, it would in any case favor male abilities to improve the access of their gametes.

As noted above, the chemical characteristics of female reproductive ducts of some groups may also be "tortuous" in the same sense of limiting male access. Cross-specific comparisons of chemical characteristics of female reproductive ducts are not currently possible, however, due to lack of data. If the interpretation of morphology made here is correct, then extravagant complexity in chemical characteristics impeding sperm arrival at storage and fertilization sites may be discovered in some groups. This in turn may help explain the diversity of sperm morphologies in some groups (see fig. 7.7; Roldan et al. 1992).

What Do Females Gain from Barriers?

What would females gain from imposing insemination barriers that screen males and their gametes? By favoring males that are better able to introduce their gametes in advantageous sites, or are better at inducing the female to transport them to such sites, the female could increase the chances her male offspring had such traits. An additional possible benefit in some groups is competition-mediated sorting among sperm themselves. This discrimination could occur in two different contexts: sperm from the same ejaculate or different ejaculates of the same male, and sperm in ejaculates from different males.

Select among Sperm from a Given Male

Possible genetic differences among sperm from a single male might alter their abilities to overcome barriers within the female. Use of sperm that had successfully passed such barriers would tend to endow the female's sons with especially successful sperm. It is possible (though not necessarily true) that such sperm would also produce zygotes with superior viability traits (e.g., Keller and Reeve 1995). These advantages to filtering sperm within an ejaculate may, however, be uncommon, due to the fact that sperm genomes are generally completely or nearly completely repressed before being engulfed in the egg (Mulcahy 1975; Poccia 1989; Cummins 1990; see also chap. 4.11).

Nevertheless, there are two possibly important nongenetic advantages to impeding sperm access to eggs for sperm from a given male. Filtering could prevent malformed, aged, or otherwise phenotypically less competent sperm from reaching the egg. If such phenotypic problems are associated with reduced fertilization or developmental capacities, then filtering could benefit the female (as well as the male). Decreased access could also avoid danger to eggs and zygotes from overaggressive sperm (e.g., polyspermy). Chromosomal aberrations in sperm are relatively common in at least some species. For instance, in human sperm from four different individuals, an average 1.9% were hyperhaploid, 3.8% aneuploid, and 2.8% had other abnormalities (Estop et al. 1990). Abnormalities in general average about 10% with wide variance among different individuals (Martin 1990). The frequency of abnormal sperm may be higher both with older sperm (e.g., Austin 1975) and sperm from older males (e.g., Martin 1990). Similarly, twenty of twenty-eight species of felids ejaculated >30% pleiomorphic sperm, and the percentage is >60% in some species (Wildt 1990). Deformed sperm are not common in all species, however. For instance, only 2% of the sperm in the vas deferens of the bruchid beetle *Callosobruchus maculatus* were obviously defective (Eady 1994a).

The sperm of many organisms are apparently relatively fragile. Those which have been stored longer inside the female give rise to increased numbers of nonviable embryos in several mammals and birds (Lake 1975; Austin 1975). In at least some cases, mammal sperm stored longer show greater genetic damage in the form of increased frequencies of chromosome abnormalities (Munne et al. 1990). Reduced fertilizing abilities may also arise from sperm entering the egg more slowly, thus resulting in more multiple entries (polyspermy) that produce polyploid (nonviable) embryos (Gosden p. 84 in Austin 1975).

There is some evidence that these proposed filtering effects occur in some birds and mammals. When the semen from low-fertility roosters was introduced artificially into the vagina, few sperm reached the storage sites, and only occasional eggs were fertilized; but the embryos formed were mostly normal. When semen from the same males was introduced in the uterus or shell gland, considerably larger numbers of sperm reached the infundibulum (the fertilization site), and appreciable numbers of eggs were fertilized; but a high percentage of these embryos died before the egg was laid (Zavaleta and Ogasawara 1987). Abnormally high numbers of sperm in the infundibulum are known to result in exceptionally high rates of preovipositional embryonic mortality, so polyspermy could also have been responsible for the mortality (Zavaleta and Ogasawara 1987).

Sperm may be filtered in rabbits on the basis of their competitive abilities and without respect to other characteristics (summarized in Overstreet and Katz 1977). Ejaculates pooled from different males were divided, and aliquots were introduced into the vagina in one group of females and into the oviduct in a second group. Sixteen hours later, the oviducts of both groups were flushed, their contents pooled and then reinseminated in a third group of females. The sperm that had been filtered by normal transport (inseminated in the vagina), which were strongly outnumbered by the nonfiltered sperm (ratios varied from 1:3.5 to 1:200 in different experiments), achieved a disproportionate number of fertilizations (13 of 26). Unfiltered sperm (from uterine insemination) did not produce an increase in postinsemination embryonic abnormalities.

In addition to the possible benefit of filtering sperm on the basis of quality, female barriers to sperm access could also reduce polyspermy (fig. 7.5; Cummins 1990; Cherr and Dulcibella 1990). Entry of too many sperm leads to abnormal development in many invertebrates, fish, anurans, birds, and mammals (Austin 1975; Davey 1965b; Longo 1987). A barrier to polyspermy could be nonselective, in the sense that the traits of given sperm cells (and/or the seminal plasma accompanying them) would have no effect on the likelihood that the sperm would ever arrive near an egg and have a chance to fertilize it. Several egg traits indicate that limiting sperm entry is advantageous: rapid electrical depolarization of the egg membrane, which is probably associated with a block to further sperm, occurs after sperm attachment in several groups, including echinoderms, annelids, tunicates, molluscs, frogs, and possibly mammals; a complex, slower block to sperm entry is associated with exocytosis of the cortical granules from the egg in many groups; the elevation and hardening of a fertilization membrane around eggs also impedes sperm entry; and supernumerary sperm are physically eliminated in some species (Longo 1987). In some animals, however, particularly those with large eggs such as urodeles, reptiles, and birds, polyspermy is common (Longo 1987).

Even in these cases too many sperm can result in failure of zygote development, so limits on the number of sperm entering are advantageous. Lake (1975) summarized evidence that too many sperm in the vicinity of the egg in the chicken interferes with normal development of the blastodisc. About sixty sperm normally penetrate into an ovum, and abnormal embryonic development occurs if more than about two hundred enter. This control must again be relatively precise, since fertility decreases rapidly once the density of sperm trapped on the egg membrane drops below a "certain threshold" (Wishart 1987).

Females are thus often in a delicate position. Premature triggering of powerful blocking mechanisms such as cortical granule discharge is likely to result in loss of the egg due to lack of fertilization. But delay in response to sperm entry can result in egg loss due to polyspermy. And selection on sperm will continue to favor more aggressive traits that enable them to be the ones that overcome female blocking mechanisms if sperm from more than a single male are in position to fertilize given eggs (see chap. 8.4 regarding possible consequences of these considerations for egg "inviability").

One prediction from these ideas is that species in which females are monandrous should have relatively modest but especially effective female sperm barriers. If females mate with only a single male, there would be no conflict of interest between male and female in winnowing out defective sperm and avoiding polyspermy. Reproduction of both male and female is increased if defective sperm are not used, and if none of the female's eggs are lost to polyspermy. Males would thus not be expected to evolve traits to overcome the filters and barriers imposed by females, and the barriers would be expected to be relatively simple (i.e., not like those in figs. 1.2, 7.1, 7.3). Unless the morphology and behavior of individual sperm from a given male differ according to genetic differences among them (unlikely if haploid transcription is rare), such filtering processes should be efficient and effective. Even at the level of individual sperm cells, kin selection would probably favor some degree of restraint on the part of sperm, if by being overly aggressive they were likely to seriously damage the zygote, thereby depriving a sibling of a chance to reproduce. Thus losses of embryos to excessive polyspermy, for instance, would be expected to be lower.

SELECT AMONG SPERM FROM DIFFERENT MALES

The predictions are different if the female is polyandrous (this includes species with external fertilization in which sperm from more than one male are likely to encounter a given female's eggs). Selection favoring male traits able to overcome female barriers, and thus obtain better access to fertilizable eggs, will be reinforced by selection to outdo the sperm of other males in overcoming female barriers. This could in turn lead to accentuation of female barriers, both because males would be expected to continue to escalate attempts to overcome the female barriers, and because females could obtain another advantage in addition to avoidance of fertilization by defective sperm and polyspermy: the female could obtain genes in her offspring that code for competitively superior sperm traits.

In other words, male adaptations to win in competitive interactions could sometimes turn out to be "overly aggressive," in the same way that the mating behavior of males of some toads (Davies and Halliday 1979), dung flies (Parker 1970b), wasps and bees (Evans et al. 1986; Alcock, in prep.), and ducks (McKinney et al. 1984) sometimes results in the mobbing and eventually drowning or dismembering of females with which males are trying to mate (fig. 7.5; see also chap. 8.4). This would result in selection on females to resist overly aggressive sperm characteristics (analogous, for example, to female ducks returning surreptitiously and circuitously to their nests, or hiding in vegetation; McKinney et al. 1984). Such female efforts to control sperm access to eggs could include a variety of traits, such as reduced access to fertilization sites, more rapid cortical reactions in eggs

Figure 7.5 Competition among males scrambling to mate with a female sometimes results in injury and even death for the female, exemplifying the dangers an egg may face from overaggressive sperm. Here a female *Centris pallida* bee is swarmed by males as she emerges from underground. Some female bees and wasps are literally torn to pieces by overeager males struggling to mate with them (Evans et al. 1986; Alcock, in prep.). Analogous aggressive adaptations in sperm may cause early zygote deaths and thus favor the evolution of female barriers to insemination and fertilization. These in turn could result in cryptic female choice, acting on males' abilities to overcome the barriers. (From Alcock 1989)

when sperm enter them, and more persistent or thicker coating of the point of entry (e.g., the micropyle of insects) until the egg is in a position in which sperm access can be more precisely controlled (e.g., Leopold et al. 1978 on the housefly *Musca domestica*; see below). The female could derive the additional advantage of biasing the paternity of her offspring so that her sons produced sperm better able to overcome the barriers erected by other females in the population. Female characters of this sort would of course tend to produce additional selection on sperm for improved abilities to outcompete other sperm (e.g., Roldan et al. 1992).

In such a conflict between male and female interests, extended evolutionary bouts of escalation in barriers and mechanisms to overcome them are expected to continue until limited by natural selection (e.g., selection for fertilization). Cummins (1990) predicted a balance between sperm adaptations to outcompete other sperm for access to eggs, and female-imposed barriers to reduce access and avoid polyspermy. In species with polyandrous females, barriers are expected to be relatively elaborate, and they are expected to be at least sometimes partially ineffective.

Some fragmentary data suggest such imperfection. Polyspermy is a recurrent problem in mammals, despite multiple blocking mechanisms, including cortical granule release, proteinase activity in the zona pellucida, elevation and hardening of the fertilization membrane, and changes in the membrane of the egg (Longo 1987; Cherr and Dulcibella 1990). For example, 1% of all recognized human conceptions in vivo result in triploid embryos that are aborted, and the majority of these are due to dispermic fertilization (Cherr and Dulcibella 1990). The multiple female mechanisms mentioned in the previous section which function to prevent polyspermy are also in themselves consistent with the predicted evolution of complex controls resulting from the competitive interactions just described. A further, especially dramatic example of an apparent female adaptation to prevent access of sperm other than those in the spermathecae occurs in the housefly *Musca domestica.* A plug covers the micropyle (the site of sperm entry) of the eggs as they emerge from the ovary. The plug is removed by special morphological structures and chemical products near the mouth of the spermathecal ducts (fig. 7.6; Leopold et al. 1978) just before fertilization.

In summary, female morphology, physiology, and behavior often impose substantial barriers to sperm movement within the female's body. Females may derive several different advantages from these features. Some accrue from natural selection (to avoid infection by pathogens, to avoid fertilization by defective sperm, or to avoid polyspermy). In species with polyandrous females, females may also gain by obtaining genes for their offspring to produce more competitively successful sperm.

Work on differences in sperm morphology in mammals independently led Roldan et al. (1992) to conclude that female reproductive tracts are often designed to be able to impede fertilization. Their argument is as follows. The morphology of the sperm of the many animals with external fertilization, in which sperm swim through water to encounter eggs, is relatively simple and uniform. In contrast, the morphology of the sperm of species with internal fertilization is more complex and more diverse (fig. 7.7; Baccetti and Afzelius 1976). A complex series of events must occur if a rodent's sperm is to fertilize an egg, and only a tiny fraction of the sperm in a normal ejaculate ever reaches a position in which they can fertilize an egg. This led Roldan et al. (1992) to conclude that the female tract of mammals "represents a formidable barrier for spermatozoa." Their

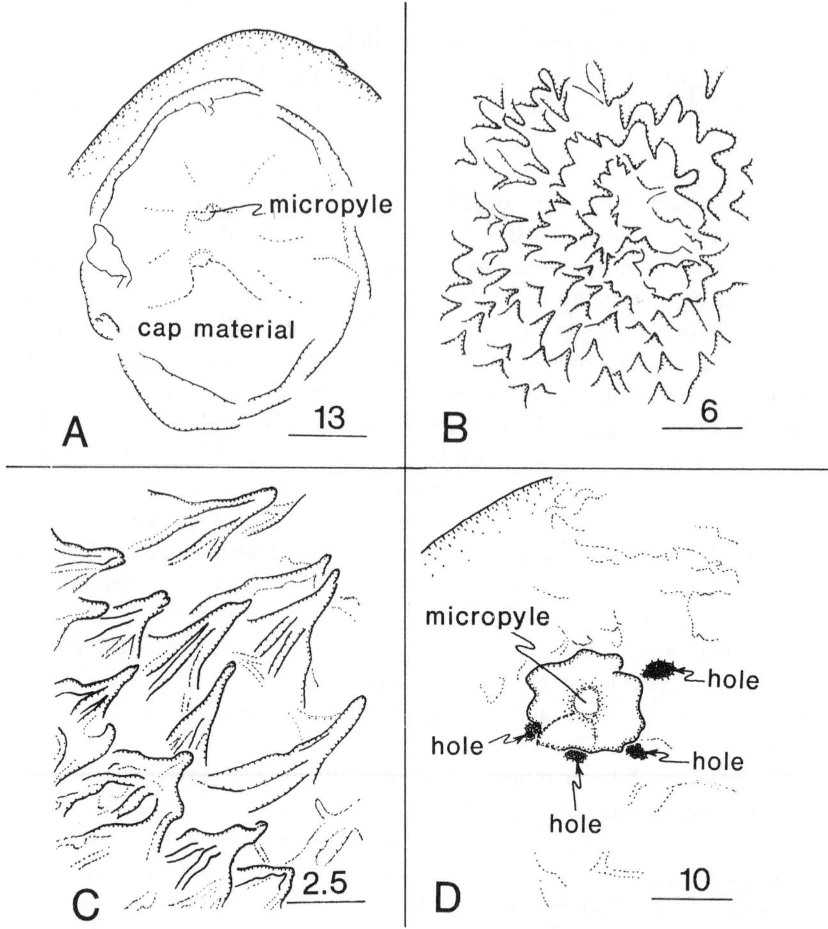

Figure 7.6 An apparent female barrier to fertilization in the housefly *Musca domestica*, which is removed by the female only just before fertilization. The micropyle (sperm entry point) of each egg is covered by cap material when the egg leaves the ovary (A). When the egg reaches the female's genital chamber and is positioned with its micropyle opposite the exit of the spermathecal duct, the cap material is removed. A special field of spines in this area (B, C) breaks up the material, and a gland product dissolves it. In a recently fertilized egg (D), much of the cap material is removed from the micropyle, and holes which apparently correspond to spines are present (numbers associated with scale lines indicate micrometers). (After Leopold et al. 1978)

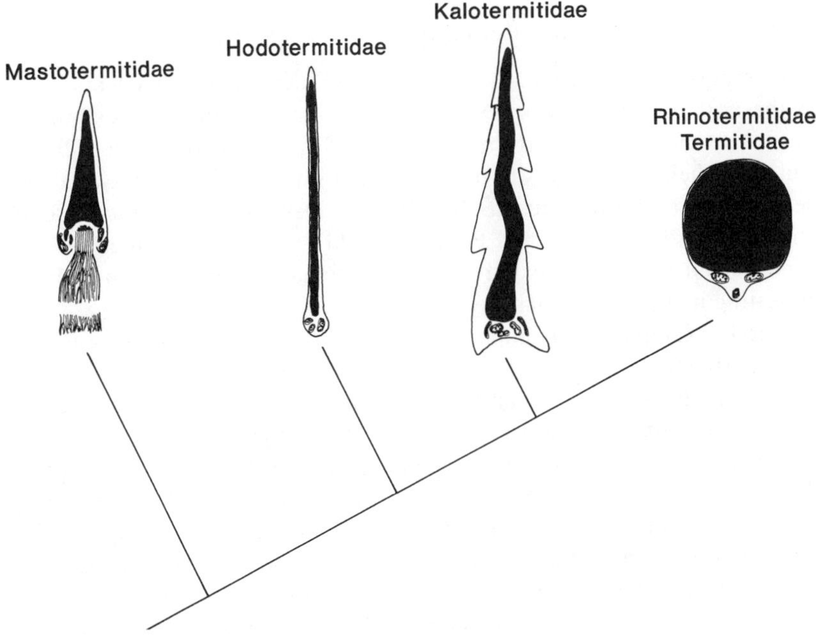

Figure 7.7 Diverse and elaborate sperm morphology in termites. The fact that female termites are generally strictly monogamous indicates that elaborate morphology is not related in this case to competition among the sperm of different males. (After Jamieson 1987)

hypothesis as to why such a barrier has evolved is similar to mine—it allows females to exercise paternity choice, both among gametes and among males.

7.2 Frequent Rapid, Divergent Evolution of Genitalia

Male intromittent genitalia and genitalic products such as spermatophores are often extremely elaborate (see fig. 1.6) and frequently differ in form among even closely related species (summaries in Eberhard 1985; Shapiro and Porter 1989). I have argued (Eberhard 1985) that cryptic female choice is responsible for this pattern of rapid divergence, which is one of the most general trends in all of animal evolution. If that argument is correct, then the evolutionary pattern of rapid divergent evolution in genitalic structures constitutes evidence that cryptic female choice itself must also be very widespread. I will thus review additional evidence and criticisms that have come to my attention since 1985 regarding this and other hypotheses concerning genitalic evolution.

The Special Problem of Asymmetric Tests

It is important to clarify at the outset that testing the sexual selection hypothesis is not always easy. Some observations constitute "tests" in the sense that they can support the hypothesis if they show one type of trend, but nevertheless the opposite trend does not give convincing reason to reject it. This asymmetry results from the intrinsic unpredictability of both the criteria (e.g., Møller and Pomiankowski 1993; Andersson 1994) and the mechanisms used in sexual selection by female choice, and is a problem in testing many evolutionary hypotheses.

For an example of this unpredictability, consider the possible results of an experiment designed to determine whether experimentally modifying a male's genitalia affects the amount of sperm from former mates which he induces the female to discard. A positive effect (modified males induce less sperm to be discarded than sham-operated males) would constitute support for the sexual selection hypothesis. But the lack of an effect, or even a negative effect (more sperm discarded with modified male), would not give grounds for rejecting the sexual selection hypothesis in general but only with respect to this particular cryptic female choice mechanism. The same genitalic structure might influence other female processes which were not measured (storage of present male's sperm, oviposition, ovulation, egg size, etc.; see chaps. 3 and 4). Thus confident rejection using this type of experiment would be possible only after consistently negative effects and lack of effects were obtained in an exhaustive series of studies of possible criteria and mechanisms of cryptic female choice. Even then, one could argue that all possible effects had not been tested.

This is not to say that the sexual selection hypothesis cannot be refuted. If, for instance, a group (e.g., a genus, a subfamily) of animals is found in which species-specific differences occur in male genitalia, and within which females of different species never make genitalic contact with more than one male, the hypothesis must be rejected for that group (Eberhard 1985) unless relatively special conditions occur. Huber (1993) has argued that even in species in which females mate only once, selection to produce rapid divergent evolution can also be expected on the basis of more offspring from males that transfer sperm more efficiently. This could occur only if the reproductive disadvantage to variant females, with criteria that result in decreased reproduction when they mate with some types of males, was more than compensated for by other benefits such as, say, avoidance of infection of the reproductive tract by pathogens or having sons capable of fertilizing more females (Birkhead et al. 1993a). Whether such imbalances ever occur in nature is not known.

Similarly, if species of major groups in which females mate with multiple males tended to show less genitalic differentiation than those of related groups in which females mated with only one, it would constitute strong evidence against the sexual selection hypothesis (especially if possible differences in movements of the genitalia could be ruled out). In fact, the available data on this point show a trend in the opposite direction (Eberhard 1985; see also discussion of primates and bees below).

Experimental Tests

I know of only three direct, experimental tests of the effects of male genitalic morphology on male reproductive success. All three tested the relatively simple hypothesis that variation in size correlates with differences in insemination or fertilization success. They thus constituted asymmetrical tests, as negative results would not contradict the theory. Positive results, on the other hand, would clearly support it. Two of the three gave positive results.

In the spider *Nephila clavipes* (see fig. 4.12), larger males had both larger genitalia and more sperm and transferred larger numbers of sperm to the female (only first matings were tested). There was no relation, however, between the size of a male's intromittent organs and the proportion of his sperm that reached the female's spermathecae (Cohn 1990). There was also no correlation with several measures of male copulatory behavior, other than that longer time spent in copulation (extrapolated from a sample of 1 hr; mating lasts up to 48 hours; see chap. 4.8) was associated with greater transfer. When the tips of the conductors (and presumably also the tips of the intromittent organs inside them) were cut off, the male failed to couple successfully. These two structures apparently mesh with different female structures (Schult and Sellenschlo 1983; see, however, Huber 1993 for evidence that the technique of artificial expansion of male genitalia used in this study can lead to mistaken interpretations). Females showed no signs of overt rejection of modified males and did not push them away more often as they attempted to mate.

In contrast, in the chrysomelid beetle *Chelymorpha alternans*, the size of a male genitalic structure did have a significant effect on male paternity (V. Rodriguez 1994a; V. Rodriguez et al., subm.). Each of thirty-two virgin females was mated to a series of two or three males carrying different genetic markers for body color, and paternity was determined by raising large numbers of the female's offspring to adulthood (av. offspring/female = 179 ± 74). The proportion of the offspring sired by each male was significantly correlated with the length of the male's flagellum (a long genitalic structure that is threaded up the female's spermathecal duct, at least occasionally reaching her spermatheca; fig. 7.8). No other measured male character (pronotum width, elytron length, body length, color, order of mating, duration of copulation) correlated consistently with male paternity

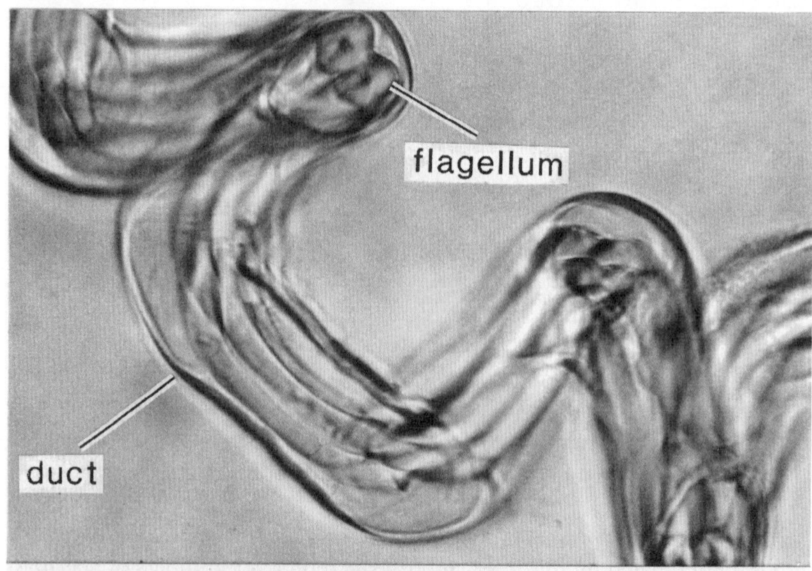

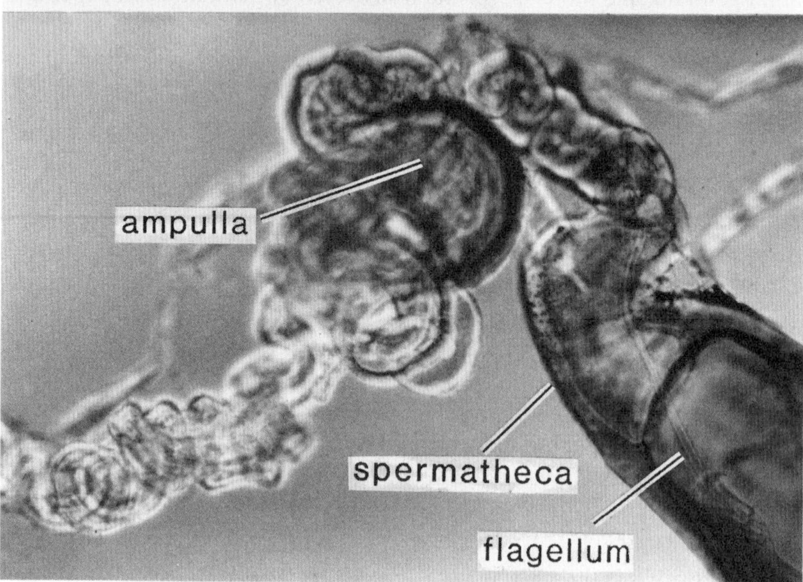

Figure 7.8 The highly tortuous female spermathecal duct of a beetle in which cryptic female choice favors males with longer genitalia. The male's genitalic flagellum in the tortoise beetle *Chelymorpha alternans* is threaded up the highly coiled spermathecal duct of the female (*top*). Usually it extends only to a widening of the duct (the ampulla), just short of the spermatheca. Rarely (*bottom*), it coils in the ampulla (out of focus) and extends the rest of the way to the spermatheca itself. The sperm of males that had a longer flagellum were discarded less often by the female and were transferred in greater numbers to the female's spermatheca, and these males had a greater probability of fathering her offpsring (V. Rodriguez 1994a; V. Rodriguez et al., subm.).

success. For each female taken separately, the male with the most offspring had a longer flagellum than the one with the least, in both two- and three-male experiments. If one accepts the likely assumption that survivorship of offspring into adulthood was not different for offspring of males with different flagellum lengths, then flagellum length, or some other trait associated with it, apparently influences a male's chances of paternity.

The cryptic female choice mechanism by which some males were favored over others in these beetles may be sperm dumping. During a virgin female's first copulation, she sometimes ejected a drop of sperm (13.1% of 84 matings; see fig. 3.1). In those copulations in which a drop was emitted, the number of sperm that were stored in the female's spermatheca was reduced by about 35%. The length of the flagellum of twenty-three males whose virgin female mates had ejected a drop was shorter than that of thirty-one males for which no drops were emitted. Active female participation in control over sperm emission was suggested by an increase in the frequency of emission in virgin females when the female's spermathecal muscle was cut (from 13.6% in sham-operated females to 44%; V. Rodriguez 1994b).

Experimental manipulation of the length of the male's flagellum indicated that the flagellum itself rather than some associated character is important. When the distal portion (av. 24.8%) of the flagellum was cut off in thirty-two males, and these males were then allowed to mate with virgin females, the percentage of females that emitted a drop of sperm jumped from 13.1% in controls to 87.5%.[3] Since sperm precedence in this species is apparently determined largely by relative numbers of sperm rather than by their positions in the spermatheca, and since females appear to mate multiply in the field (V. Rodriguez 1994a; V. Rodriguez et al., subm.), it is likely that reduction in sperm numbers in the spermatheca has important reproductive consequences for the male. Whether longer, more natural delays between matings by different males also give correlations between flagellum lengths and paternity remains to be determined, however.

In a third study, the effects of the sharp, basally directed spines on the glans of the penis of male rats (*Rattus norvegicus*) on the induction of the luteal cycle in females was studied by modifying the spines with hormone treatments (O'Hanlon and Sachs 1986). Adult males were castrated and then subjected to treatments with testosterone or estradiol. Some treatments resulted in reduction or elimination of the spines, and these males were less able to induce the female luteal cycle, as would be expected if the spines function as courtship devices. The treatments also resulted, however, in changes in penis responses and in reduced likelihood of ejaculation. It is also possible that they modified male accessory gland secretions in the semen. After taking into account the results of other studies, the authors nevertheless concluded tentatively that there was a possible stimu-

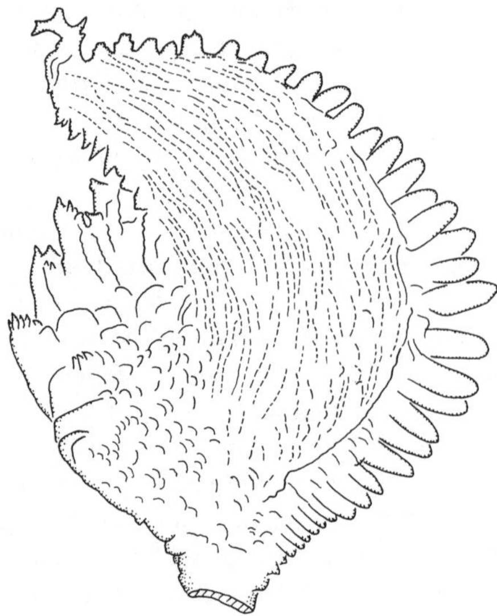

Figure 7.9 Dorsal view of the appendage on the second antenna of a male *Eubranchipus serratus* fairy shrimp which is used to clasp the female prior to copulation. When these species-specific structures were removed, operated males continued to be able to clasp females as successfully as unoperated males, but they were only able to induce females to allow intromission about 25% as often. (After Belk 1984)

latory effect of spines, or some condition of spines, on luteal induction in females (O'Hanlon and Sachs 1986).

If male genitalia are especially prone to rapid divergent evolution because they are in contact with (and stimulating) the female in sexual contexts, as postulated by the sexual selection hypothesis, then there are two clear predictions. Those nongenitalic portions of the male which consistently contact the female and become specialized for such contact should tend to evolve like genitalia and diverge rapidly; they should also often function as courtship devices to influence female choice.

The first prediction for "nongenitalic contact structures" was confirmed for a variety of groups (Eberhard 1985). The second remained untested, except for some dragonflies (Loibl 1958; Krieger and Krieger-Loibl 1958). The results of experimental modifications of the second antennae of males of the fairy shrimp *Eubranchipus serratus* constitute a second confirmation (Belk 1984, 1991). The male antennal appendages (fig. 7.9) lie along the female's dorsum when the male clasps the female with his second antennae prior to mating. When the appendages were cut off, operated males were no less proficient in clasping females (250 vs. 253 clasps in trials with five operated males, five control males, and five females). Unreceptive females often struggled after being seized, and operated males were also no less able to maintain their hold on such females (64.5% of 237 rejections of operated males occurred within 6 sec, vs. 63.5% of 200 rejections of control males). But unoperated, intact males induced females to copulate significantly more often (20.9% of 253 clasps) than operated males (5.2% of 250 clasps).

Other Evidence

Several other studies give indirect support for the sexual selection hypothesis. One prediction of the sexual selection theory (both by female choice or by direct male manipulation of other males' sperm or their chances of insemination—e.g., by sperm removal or plugs) is that when taxa in which the females of some species mate with only a single male are compared with those in which females mate with multiple males, the male genitalia of species in the second group should show more rapid, divergent evolution. Sexual selection by cryptic female choice is possible if females mate with more than one male but not possible if they mate only with a single male. This prediction does not logically follow from the other hypotheses to explain the evolution of genitalia. I showed previously that this prediction was met in two groups: termites compared with their close relatives, the roaches; and different species groups of butterflies in the genus *Heliconius* (Eberhard 1985). Two subsequent tests in bees (Roig-Alsina 1993) and in primates (Dixson 1987) have also demonstrated the predicted pattern.

The analysis in bees was relatively simple. The morphology of the male endophallus was classified as simple or complex, and the mating system as monandrous or polyandrous. The results were clear: each of seven monandrous species (in as many genera in three different families) had a simple endophallus; thirteen of fourteen species with polyandrous females (in six genera in four families) had a complex endophallus (Roig-Alsina 1993) (fig. 7.10; the endophallus of the fourteenth species had an intermediate complexity). Insofar as the literature on mating systems is accurate,[4] this is a striking confirmation of the prediction.

A survey of 130 species of primates showed that the following traits related to penis morphology and behavior are more complex in species in which females are more likely to mate with more than a single male during each estrous cycle: the size of the spines or papillae on the distal portion of the penis; the relative length of the baculum (penis bone); the relative length of the flaccid pars libera; the complexity of the distal region in terms of overall shape; and the complexity of copulation behavior (multiple intromissions, or prolonged single intromissions; Dixson 1987). Although Dixson used each species as a separate data point and thus did not take into account the possibly confounding effects of common heritage (Ridley 1983; Felsenstein 1985; Harvey and Pagel 1991), a reanalysis of his data

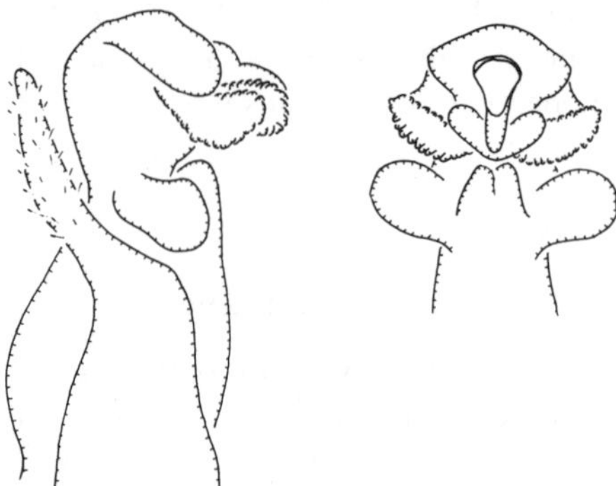

Figure 7.10 The complex inflated internal sac of the male genitalia of the andrenid bee *Arhysosage flava* is in accord with the sexual selection by cryptic female choice for the evolution of genitalia (lateral view on *left*, dorsal view on *right*). Females of this species mate repeatedly. There is an association between complex internal sac morphology and polyandrous mating systems in bees, in accord with predictions of the theory. (After Roig-Alsina 1993)

using independent derivations gave generally similar results (Verrell 1992), conforming to the predictions of the sexual selection hypothesis.

This test, as with the one involving bees, has the weakness of implicitly assuming that "more is always better" (e.g., greater length, longer spines, etc., will always be favored by female discrimination). This may not be correct, as it is easy to imagine situations in which such a bias would not occur (e.g., female receptors located nearer the external opening of her reproductive tract might be stimulated by males with shorter genitalia; see also L. Brown et al. 1995 on the need to take female morphology into account, and the discussion of the unpredictability of sexual selection in sec. 7.2 above). Nevertheless, the sample was large, and the data on primate mating systems may be relatively good compared with those in many other groups.

Harcourt and Gardiner (1994) confirmed Dixson's findings that most of the genitalic characters he discussed are most developed in species with multi-male mating systems. But they questioned his conclusion regarding penile spines. After performing additional analyses by carefully taking into account possible phylogenetic biases, they argued that stimulation by the male may function to coordinate male and female reproductive responses such as ovulation in the female rather than as a mechanism of female choice. This interpretation was based on doubts regarding opportunities for sexual selection by cryptic female choice in species with "dispersed" mating systems (see Dixson 1987 for a different view). In any case, their hypothesis is clearly not necessarily *alternative* to the hypothesis of sexual selection by cryptic female choice. As noted in chapters 3.9 and 6, male ability to induce ovulation is one of the mechanisms of male competition by cryptic female choice.

Behavioral and morphological data from five species of sphaerocerid flies in the genus *Coproica* were used to test the sexual selection hypothesis as an explanation for their species-specific male genitalia (Lachmann 1994). They supported the hypothesis in several respects: females probably control sperm transfer to the spermathecae, where there are muscles and valves that may also enable the female to manipulate sperm already received; females probably can sense the differences among the genitalia of different individual males; males make sustained stimulatory movements with their genitalia prior to ejaculation; there are interspecific differences in these movements; and females remate. Data from these species failed to fit two predictions of an alternative hypothesis (lock and key; Lachmann 1994).

A different type of data comes from hybridization of sibling species of *Drosophila* (Coyne 1993). When female *D. simulans* mated with male *D. mauritiana*, females struggled vigorously, and copulations were abnormally short. The two species differ morphologically only in the shape of the male genitalia and the number of genes involved, and the relative effects of genes on different chromosomes were the same for copulation duration as for the genitalic differences. This led Coyne to suspect that shortened copulations were due to the female's perception of species-specific differences in male genitalia. Such perceptions and responses are in accord with the sexual selection by cryptic female choice hypothesis. Since there were only four chromosomes involved, however, this similarity may be only coincidental.

Another way to examine the female choice hypothesis is to test alternative hypotheses. The most commonly cited alternative explanations involve possible selection on genitalia to avoid cross-specific pairings, either by mechanical exclusion or via stimulation of the female (Shapiro and Porter 1989; Edwards 1993). Thus demonstrations of rapid, divergent evolution of male genitalia in groups in which, because of their ecology or geographic ranges, females never come into contact with males of closely related species constitute strong evidence against species isolation hypotheses.

In addition to the cases cited previously (Eberhard 1985), an especially clear case has recently been described in ten new species of the nematode genus *Parasitodiplogaster* (Poinar and Herre 1991). These nematodes pass part of their lives in fig wasps of the family Agaonidae that have evolved as mutualistic pollinators in intimate association with fig trees (*Ficus* spp.). All three groups of organisms are highly specific in their associations: the closed inflorescence (syconium) of each fig species always harbors only a single species of wasp; each wasp species inhabits only a single species

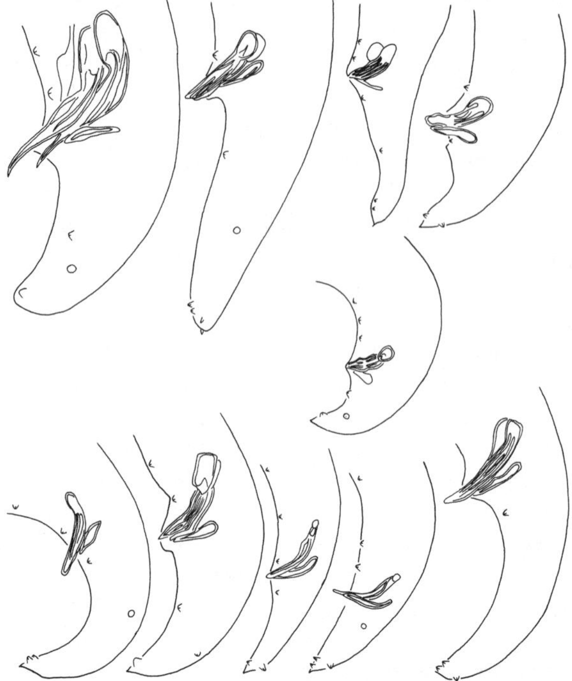

Figure 7.11 Posterior portions of the males of ten species of *Parasitodiplogaster* nematodes give clear evidence against species isolation hypotheses of genitalic evolution. Both spicules and genital papillae are species-specific in form and location, despite the nematodes' highly specific host relations with fig wasps, which are in turn highly host-specific in the fig species they inhabit. Cross-specific encounters between *Parasitodiplogaster* nematodes probably occur extremely rarely, if ever. (After Poinar and Herre 1991)

of fig; each species of wasp is attacked by a single, different species of *Parasitodiplogaster* nematode; and each species of nematode attacks only a single species of wasp. The nematodes reproduce only within the confines of a *Ficus* syconium, and they have never been found in any other habitat (though the possibility of future discoveries cannot, of course, be excluded).

Thus the members of a given species of *Parisitodiplogaster* nematode apparently seldom (and possibly never) contact members of other congeneric species. Confirmation that these are long-term, intimate evolutionary associations among wasps, nematodes, and figs comes from both the fine adjustments of the degree of virulence of different nematode species to the likelihood of more than one wasp entering a given syconium (Herre 1993), and the congruence of phylogenetic trees of wasps and figs deduced from molecular data (A. Herre, pers. comm.). Nevertheless, the intromit-

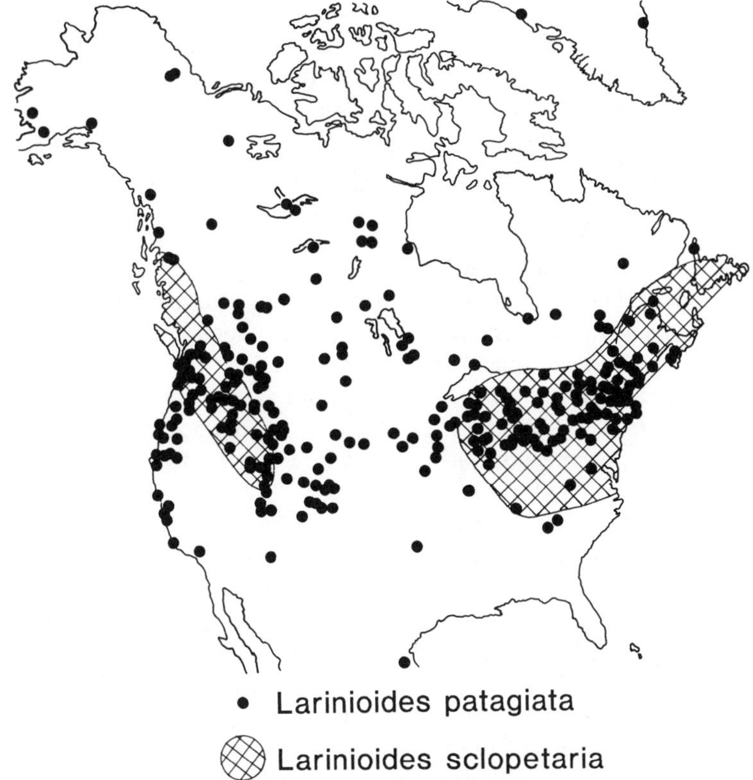

Figure 7.12 Evidence against species isolation hypotheses for genitalic evolution from a pair of spider species with partially overlapping ranges. Although the forms of the male genitalia of both species varied geographically, the genitalia of *Larinioides patagiata* were not more distinct from those of *L. sclopetaria* in the area where their ranges overlap than in other parts of their range. The failure of the species isolation hypothesis in this carefully selected test case strengthens the argument that genitalia have evolved under sexual selection by cryptic female choice. (After Ware and Opell 1989)

tent genitalia (spicules) of these nematodes are elaborate and species-specific in form, and the male genital papillae on the body surface near the gonopore also differ (fig 7.11; Poinar and Herre 1991). These data clearly contradict the predictions of the species isolation hypotheses.

An additional test that produced data not in agreement with the species isolation hypotheses was made by Ware and Opell (1989). They studied the geographic pattern of variation of male genitalic morphology in a pair of closely related species of the spider genus *Larinioides* that have partially overlapping geographic ranges (fig. 7.12). The species isolation hypotheses predict that if male genitalia vary, differences between species should be greater in areas of overlap (character displacement). The case was carefully chosen to maximize the chances of being able to find the predicted character displacement in genitalia. The taxonomy of the genus had been recently revised, reducing the probability of misidentifications; the size and basic structure of the genitalia of the two species is similar, and their breeding seasons overlap substantially, so cross-specific matings were feasible; male genitalic characters were somewhat variable in both species; no other closely related species that might have produced additional patterns of character displacement were present; and sizable collections were available for study. The ranges of the two species have apparently not changed substantially during the last one hundred years. Despite these favorable conditions, character displacement did not occur. In neither species did variation in the male genitalia correlate with zones of overlap.

A similar deduction that species isolation has not caused rapid divergence in genitalia will follow if genitalia diverge without preventing hybridization. In pierid butterflies, *Tatochila mercedis* hybridizes readily with two closely related species, *T. sterodice* and *T. vanvolxemii*, in nature and captivity, despite clear differences in the male genitalia (Porter and Shapiro 1990). Similarly, several species of the fairy shrimp genus *Streptocephalus* and the damselfly genus *Ischnura* form cross-specific pairs despite clear differences in the shapes of male clasping organs (Wiman 1979a, 1979b; Leong and Hafernik 1992). There was also no sign of the predicted character displacement in sympatric populations of *Ischnura* (Leong and Hafernik 1992). Study of the behavior and morphology of copulation in three species of the scarab chafer beetle genus *Macrodactylus* also showed that selection for species isolation was unlikely to be able to account for relatively rapid divergence in at least some genitalic structures (Eberhard 1992a, 1993a).

Another line of approach is to examine the evolutionary sequence of events in the evolution of male and female genitalia. A careful determination of the order of events in ground beetles of the genus *Platynus* by cladistic analysis also failed to fit species-isolation hypotheses (Liebherr 1992). Male and female genitalia did not change hand-in-hand, as expected with the species isolation hypothesis. Instead, changes in female genitalic characters (a dorsal pouch on the bursa and its subsequent sclerotization and narrowing) preceded these in the associated genitalic characters of the male (various modifications of the tip of the median lobe). This sequence is also difficult to understand under other hypotheses, however, including the sexual selection hypothesis. Further data (e.g., possible movements of male structures within the female, the function of the dorsal pouch) will be needed to clarify the possible impact of this case on the competing theories.

Finally, further data have illuminated a few cases I was unable to understand in 1985. One group of structures that seemed inexplicable was the elaborate, species-specific ovigerous legs of male pycnogonids. It now appears that they are used as courtship devices. Males stroke the female as she lays eggs, perhaps inducing her to lay more eggs (and thus allow him to fertilize more; Nakamura and Sekiguchi 1980). Further studies have also revealed that additional species-specific male structures can be included in my list of nongenitalic contact courtship devices: dermal papillae, hooks and stylets on the heads of onycophorans (Tait and Briscoe 1990); abdominal setae of zorapterans (Choe, 1995); sculptured abdominal sternites of *Phyllophaga* beetles (Eberhard 1993b); adornments on the chelicerae of pholcid spiders (Huber 1994a; Uhl et al. 1994); ventral spines and sculptured front legs of *Macrodactylus* beetles (Eberhard 1993c). As predicted, all contact females in ways that could stimulate the female in sexual contexts.

There are also scattered studies that claim to have tested and confirmed, or to have tested and rejected, the sexual selection hypothesis for particular groups, but which I believe do not provide convincing data for either conclusion. Edwards (1993), for instance, claims that nonmechanical species isolation mechanism hypotheses were insufficiently considered, apparently having overlooked the extensive evidence against "sensory (nonmechanical) lock and key" mechanisms in a variety of taxa (Eberhard 1985). In concluding that the arguments favoring a sexual selection interpretation were "proposed wholly within the entomological context," he seems to have also missed the extended discussions and numerous examples of fish, molluscs, nematodes, mammals, and many arthropods other than insects.

Zunino (1988) noted that in some scarabaeoid beetles female genital morphology often changes in step with that of the male, and proposed that male and female morphology evolve under a process of reciprocal sexual selection. As Zunino himself notes, such parallel evolution of male and female is not, however, contrary to expectations of the sexual selection hypothesis.

Mikkola (1992) described many "lock and key" mechanisms in the internal genitalia of noctuid moths, involving parallel coadjustments in male and female genitalic morphology of different species. But the abilities of different female designs to exclude heterospecific males, as would be required by the lock-and-key species isolation hypothesis, were not demonstrated. Females of many species have relatively small pockets or enlargements of the bursal duct, or larger diverticula in the soft tissue of the bursa itself; in some cases mechanical exclusion of other species seems improbable. Also, as noted by Mikkola (1992), parallel coadjustments are compatible with other hypotheses, including that of sexual selection by cryptic female choice. Mikkola's argument that sexual selection is improbable because the female is already in copula when these male structures are brought into play, and that they are "stable" (apparently meaning that they are not easily forced out of the female), misses the point that many post-intromission cryptic female choice mechanisms are available to a female, even if she does not break away from the male.

The most taxonomically useful portion of the male genitalia of the butterfly *Maniola jurtina* do not contact the female during copulation, arguing against sexual selection by female choice (Goulson 1993). The possibility that the male genitalia move and touch the female during or preceding copulation was not considered (observations were of frozen pairs). Given the extensive movements of male genitalia in butterflies (Lorkovic 1952) and other insects (Alexander 1959; Eberhard 1985, 1994, and in prep. a; see chap. 5.2), further observations are needed.

McAlpine (1988) pointed out two possible cases of genitalic character displacement in different families of flies that conflict with the sexual selection hypothesis. He also noted, however, a general absence of this pattern: "Despite such examples as this, I find a detailed uniformity in genitalic characters over a considerable geographic range to be usual in the dipterous groups I have studied" (p. 75).

R. Alexander et al. (in press) argue that male-female conflict is a more likely general explanation of genitalic evolution than sexual selection. Their misunderstandings of sexual selection and mechanisms of female choice, and their failure to deal with several types of contradictory data (see Eberhard, in press), call their arguments into doubt. As noted in chapter 2.2, my rejection of sperm removal as a widespread genitalic function on the basis of the fact that male genitalia often do not reach to "the" female's sperm storage organ (Eberhard 1985) was probably premature. The ultimate impact of these considerations on the sexual selection hypothesis remains to be determined. An inverse underestimate has also occurred. Rejection of the possibility that genitalic devices that remove sperm also function to stimulate the female (e.g., Siva-Jothy 1987a) will not be convincing until possible stimulation effects are tested (see discussion in chap. 5.2 and fig. 5.9).

New Ideas

Some new interpretations have been proposed regarding how selection could affect the complex genitalic fit between male and female in groups in which female genitalia are both complex and relatively rigid. When mating pairs of the nesticid spider *Nesticus cellulanus* were frozen in liquid nitrogen and then sectioned, the mechanical mesh was extremely complex, with eleven discernible points of locking and contact (fig. 7.13; Huber 1993a). The portions of the male genitalia that made contact with the female were, as might be expected with the sexual selection hypothesis,[5] those portions which most often differ among closely related species (i.e., which evolve especially rapidly; Huber 1993). The same correlation could occur, however, with other types of selection on genitalia, such as lock and key.

Examination with the scanning electron microscope and serial sections

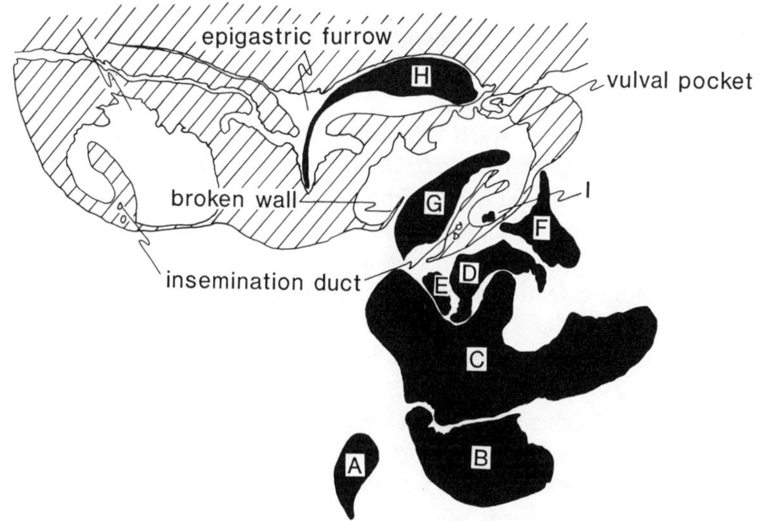

Figure 7.13 A section through the genitalia of a pair of spiders (*Nesticus cellulanus*) frozen *in copula* rules out one possible stimulatory function for the male genitalia but suggests a pair of others. The male genitalia (black) look like a jigsaw puzzle with many complex pieces, some of which brace only against each other (B,C) or brace both against each other and the female (D-F); one (G) breaks through a thin wall of the female, and another (H) is driven deep into a furrow. Sperm delivery occurs through still another sclerite (I). The portions of the male that touch the female are the portions of his genitalia that vary most clearly among species, as predicted by the sexual selection hypothesis (and some others). But the portions of the female (at least those on the external surface of her genitalia) that he touches do not have touch receptors, so simple rubbing or pressing to stimulate her is ruled out. The tip of the deeply intromittent structure H apparently presses very firmly on a membranous portion of the female (note distortion in the wall of the epigastric furrow), and it is possible that the complex array of male structures is a sort of Rube Goldberg apparatus to facilitate this deep penetration and possible stimulation of the female. The male rhythmically expands and collapses a large membrane at the base of the array of sclerites (about 1/sec early in the av. 10.3 min insertion, later slowing to about 0.1/sec), that may produce a thrusting movement of sclerite G or H. Another nonexclusive possibility is that in the process of achieving the strong clamping alignment or while expanding and collapsing his genital membranes, the male twists or displaces the female's abdomen, and she senses this. (After Huber 1993a)

revealed no obvious sensory structures, such as lyriform organs or setae, on these same contact surfaces of the female. So if sexual selection does act on the male genitalia of *N. cellulanus*, it must be on the basis of mechanical fit between male and female or of stimulation deeper within the female (perhaps by the male's median apophysis, which pierces the thin cuticular wall of the female vulval pocket, or the male paracymbium, which hooks into the epigastric furrow; fig. 7.13). Stimulation in these sites might also occur on the basis of overall mechanical fit, perhaps signaled by pressing or twisting the entire epigynum or the female's entire abdomen during successful insertion (which would involve an appropriate mesh of the contact surfaces on male and female genitalia).

A further possibility, which is feasible in some other spiders such as *Agelenopsis* (Gering 1953) and *Histopona* (Huber 1994b), but perhaps not in *N. cellulanus*, is that the male intromittent structures stretch parts of the female's internal genitalia. Such stimulation would depend on the existence of proprioceptive mechanisms in the female; such structures have apparently not been searched for in the membranous portions of the female genitalia of spiders. They do exist associated with membranes in their legs, and also in membranous internal structures in some insects (e.g., Sugawara 1979; Taylor 1989; and references therein).

The idea that females may sometimes use only unspecialized receptors to sense the male receives support from studies of the cricket *Gryllus bimaculatus*. Experimental manipulations (chap. 5.4) showed that a female response that was necessary for copulation to proceed was triggered by stimulation from the male epiphallus. Nevertheless, the female had no specialized sensillae for sensing this structure (Sakai et al. 1991). Even so, the striking lack of sense organs on many of the portions of female spider genitalia that are contacted by those portions of male genitalia which differ among closely related species (and are thus presumably subject to female choice) supports the argument that mechanical mesh rather than stimulation is important (Huber 1993b).

Data on another spider, *Neriene litigiosa*, may be relevant to the question of how female spiders could detect differences in male genitalia. Increased frequency of male failure in insertion attempts ("flubs"), which may stem from inappropriate morphological mesh between male and female genitalia, was part of the measure of reduced male "copulatory vigor," which correlated with lower paternity probabilities for second-mating males (Watson 1991). Failed insertion attempts also occur in other spiders (e.g., Huber 1993a on *Nesticus*; Prenter et al. 1994 on *Metellina*). The frequency of flubs could be one of the mechanisms postulated by Huber (1993b) to evaluate male genitalic variants for mechanical fit, and scraping movements that may represent exploratory attempts to achieve mechanical locks (or also stimulation of the female) are also apparently widespread (Huber 1995). It has not been established, however, whether improper male morphology or improper positioning of the male genitalia prior to insertion (or both) was responsible in any of these cases. In summary, mechanical mesh per se of genitalia seems to be important in some species of spiders, though the mechanisms which translate it to changes in male reproductive success are not yet clear.

Summary of Evidence on Genitalic Evolution

The case for the hypothesis that sexual selection by cryptic female choice has acted on male genitalia has been strengthened rather than weakened by evidence that has accumulated since 1985. Although there are studies of genitalic function and evolution which give data that also allow alternative explanations of rapid, divergent genitalic evolution (e.g., Siva-Jothy 1987a; Proctor et al., in press), I know of no studies before or after 1985 which rule out sexual selection by female choice as an explanation for rapid divergent genitalic evolution. On the contrary, there are several new confirmations in particular cases (*Chelymorpha* beetles, bees, primates).

7.3 Summary

This chapter discusses two topics related to sexual morphology: the widespread occurrence of features in female reproductive tracts that impede or restrict the access of males and their gametes to fertilization and/or to sperm storage sites; and the widespread pattern of rapid divergent evolution of the morphology of male genitalia. Overall trends in the data on both topics are in agreement with the hypothesis of cryptic female choice.

A widespread feature of the morphology and physiology of female reproductive tracts is that access to and from the sites where sperm are stored and are used to fertilize eggs is relatively difficult for the male and his gametes. Comparisons of duct lengths in spiders in which the entrance and exit ducts associated with sperm storage sites are separate showed clearly that difficulty of entry rather than of exit has been the selective factor favoring the tortuous nature of these ducts. Since spider sperm are immobile when transferred, it is apparently access of the male's genitalia, rather than access of swimming sperm, which has been under selection. These interpretations are reinforced by observations in other groups in which female behavior impedes rather than facilitates sperm access to fertilization or storage sites. Additional support comes from the evolution of a second system of sperm transport and storage structures in cimicid bugs, which is derived from cells whose normal role is to *kill* invading organisms. The most logical explanation for the widespread occurrence of difficult access seems to be that the female structures serve to filter males and/or their gametes after intromission has begun.

An even more general trend in animal evolution is for the morphology of male genitalia to diverge relatively rapidly when compared with that of other structures. I have argued previously that sexual selection by cryptic female choice is largely responsible for this pattern (Eberhard 1985). Evidence for or against this theory of genitalic evolution would give additional independent reasons to accept or to doubt the generality of cryptic female choice. I thus reviewed additional evidence that has come to my notice since 1985.

In general, no strongly conflicting data have appeared since 1985, though it is important to note that the nature of the hypothesis and its predictions makes it relatively difficult to obtain convincing negative data. There are, on the other hand, some striking confirmations from direct manipulations of male genitalia and from comparisons of genitalic evolution in groups with monandrous versus polyandrous females.

Notes

1. I am informed by G. Hormiga that there are occasional exceptions such as *Linyphia* and *Neriene*, which did not happen to be included in the survey, in which the fertilization duct is also tortuous.
2. Entelegyne spiders are unusual in having rigid, permanently open and exposed insemination apertures. These could conceivably make them more susceptible to infections of the spermathecae. However, I know of no observations of foreign organisms in spider spermathecae.
3. Lack of a control treatment, however, means that other factors, such as the loss of the tip of the flagellum or the stress of the operation, rather than the reduction on length per se, cannot be eliminated as possible causes of this change.
4. There is reason for caution on this point, as female tendencies to remate are easily underestimated (see chap. 9.4), and the observations of the bees' mating behavior are not particularly exhaustive. Even so, they may be sufficient to indicate *relative* differences in female mating behavior.
5. This prediction is weak, however. The form of a noncontact area of the male (e.g., B,C in fig. 7.13) could be under sexual selection because it modified the way in which another area (e.g., D,E in fig. 7.13) contacted the female.

8 Related Topics

This chapter presents data and ideas that represent extensions of the hypothesis of cryptic female choice to related topics. It attempts to show that insights derived from cryptic female choice can be used to improve understanding in other fields. The chapter is written in the spirit of exploring possible limits and consequences of cryptic female choice and associated phenomena rather than on conclusive proofs. Given our present ignorance of the evolutionary biology of complex internal events during mating, it is useful to push the basic ideas to their limits in at least one portion of this book rather than to be overly conservative. If nothing else, perhaps some rash statements here will motivate researchers to perform the critical observations to show that I am wrong.

8.1 Significance of Variation in Volumes of Ejaculates and Sperm Storage Organs

In many species the female sperm storage organs are rigid and have fixed volumes. Comparisons of male ejaculate volumes with female spermathecal volumes in such species offer a chance to understand the evolution of male investments in ejaculates (table 8.1). Current explanations of ejaculate size focus on the balance between the benefits of more sperm (in terms of increased likelihood of fertilizations) and their costs (energy and materials which could otherwise be used to achieve other copulations and fertilizations). As is often the case, most attention has been directed to the male side of this interaction. A common expectation, based on the relatively low cost of male gametes, is that males will usually transfer more sperm than needed to fill up the female (e.g., Parker 1970c; Eberhard 1985). Comparisons of the relative volume or number of sperm in a single ejaculate with the capacity of the sperm storage organ of the female reveal, however, a surprising diversity of relationships (table 8.1). In some species males produce the expected "oversized" ejaculates, which are substantially larger than the female's capacity, so that only a fraction of the male's sperm ends up in female storage organs. In some of these species a single ejaculate fills the female storage organ, but in others substantial space is left for further sperm. In other species the male's ejaculate is substantially smaller than the female's storage capacity.

Table 8.1
A Sample of Species with Fixed-Volume Sperm Storage Organs (in Most Cases Highly Sclerotized and Rigid) in the Female in Which Both the Storage Capacity and the Ejaculate Size Are Known

	Ejaculate Size	*Amt. into Spthca.*	*Capacity of Spthca.*	*Sptop.?*	*Remate?*	*References*
UNDERSIZED EJACULATES						
ODONATA						
Lestes sp.	X[a]	?	"several" X	N	?	Waage 1984
Sympetrum sp.	X[a]	?	"several" X	N	?	Waage 1984
HEMIPTERA						
Hebrus pusillus	X?	X	3X	N[b]	Y[c]	Heming-van Battum and Heming 1986
ORTHOPTERA						
Taeniopoda eques	?	X	"several" X	Y	Y	Whitman and Loher 1984
COLEOPTERA						
Macrohaltica jamaicensis	X	≤X	4X[d]	Y	Y	Eberhard and Kariko, in prep.
Chelymorpha alternans	58,000[e]	22,000	80,000	Y	Y	V. Rodriguez 1994a: V. Rodriguez et al. in prep.
Labidomera clivicollis	230,000[f]	3500[g]	34,000	N	Y	Dickinson 1986, 1988
Macrodactylus spp.	X	<X	3–4X	Y	Y[h]	Eberhard 1992a, 1993a
Tetrapium sp.	X	?	>X	?	Y	Schimitschek 1929 in Ridley 1988
Anthonomus grandis	700×10^3	15×10^3	35×10^3	Y[i]	Y[j]	Villavaso 1975a; Nilakhe and Villavaso 1979
Hypera postica	?	5247	15,694	?	Y[k]	Le Cato and Pienkowski 1972
DIPTERA						
Culicoides melleus	806	804	1615	Y	Y	Linley and Hinds 1975b
Drosophila pachea	44 ± 6		304	N	Y	Pitnick and Markow 1994
D. euronotus	?	X	≥2X[l]	?	Y	Stalker 1976
D. hydei	?	X	>X	N	Y[j]	Markow 1985

Table 8.1 (*cont.*)

	Ejaculate Size	*Amt. into Spthca.*	*Capacity of Spthca.*	*Sptop.?*	*Remate?*	*References*
D. (N.) nannoptera	?	X	>X	N	Y[j]	Pitnick and Markow, subm.
D. (N.) acanthoptera	?	1023	1456[m]	N	Y[j]	Pitnick and Markow, subm.
D. (N.) "sp. W"	?	274	1240	N	Y[j]	Pitnick and Markow, subm.
Cyrtodiopsis whitei	?	160[n]	250[n]	Y	Y	Lorch et al. 1993
LEPIDOPTERA						
Helicoverpa virescens	?	25,218[o]	72,962[o]	Y	Y	Raulston et al. 1975
HYMENOPTERA						
Apis cerana	1.2×10^6	?	1.4×10^6	N	Y	Koeniger 1991
A. dorsata	$2.5\text{x}10^6$	?	3.6×10^6	N	Y	Koeniger 1991
A. florea	0.4×10^6	?	1.4×10^6	N	Y	Koeniger 1991
A. andreniformis	0.1×10^6	?	1.1×10^6	N	Y	Koeniger 1991
Dahlbominus fuscipennis	372	157	555	N	Y	Wilkes 1966
Aphytis melinus	?	?	?[p]	?	Y	Allen et al. 1994
BRACHYURA						
Inachus phalangium[q]	?	X	2–3X		Y	Diesel 1990
ISOPODA						
Paracerceis sculpta	?	X	>X[r]	N	Y	Shuster 1989
NEMATODA						
Coenorhabditis elegans	164	100	350[s]	N	Y	Ward and Carrel 1979
APPROXIMATELY EQUAL OR OVERSIZED						
ODONATA						
Mnais pruinosa	0.1–0.34 mm^2 [t]	0.16 mm^2 [u]	0.16 mm^2	N	Y	Siva-Jothy and Tsubaki 1989a
ORTHOPTERA						
Metaplastes ornatus	1.49×10^6	1.23×10^6	1.19×10^6 [v]	Y	Y[w]	von Helversen and von Helversen 1991
Locusta migratoria	?	X	X[x]	Y	Y	Gregory 1965

Table 8.1 (*cont.*)

	Ejaculate Size	*Amt. into Spthca.*	*Capacity of Spthca.*	*Sptop.?*	*Remate?*	*References*
HEMIPTERA						
Rhodnius prolixus	X	?	≤X	Y	Y	Khalifa 1950a in Ridley 1988
COLEOPTERA						
Callosobruchus maculatus	46,000	6548	6468	Y	?	Eady 1994a, 1995
DIPTERA						
Scathophaga stercoraria	>>716	717	724	N	Y	Parker 1970a; Parker et al. 1990
Lucilia cuprina	?	3019[y]	"full"[z]	N	"seldom"	Smith et al. 1988
Drosophila melanogaster	5000	1200	?1200	N	Y	Gilbert 1981; Gromko and Markow 1993
Dryomyza anilis	3–10X	X	?[a1]	N	Y	Otronen and Siva-Jothy 1991
HYMENOPTERA						
Apis mellifera	6,000,000	6,000,000[b1]	5,700,000[c1]	N	Y[d1]	Page 1986; Kerr et al. 1962
Nasonia vitripennis	?	X	<X[e1]	N	Y	Holmes 1974
Solenopsis invictus	10×10^6	7×10^6	7×10^6	?	N	Tschinkel and Porter 1988
AVES						
Prunella modularis	10^6 [f1]	?	0.7×10^6	N	Y	Birkhead et al. 1991; Davies 1992

NOTES: Numbers refer to numbers of sperm; X refers to volumes.

[a] Determined from volume of male sperm vesicles, and thus may be an overestimate if the male does not transfer all of the contents to the female.

[b] Sperm transferred both in bursa and directly into spermathecal duct.

[c] Of twenty-five females that remated with the same male, 28% mated twice, 48% three times, and 24% four times; all of these also mated the next day, presumably with other males, since males only stayed mounted for a maximum of 2 hr.

[d] Packing of the very long sperm differs (they are highly aligned in the spermatheca, tangled in the spermatophore), so the volumes may not represent strictly equivalent capacities for sperm storage.

[e] Male has more than twice as many sperm in his seminal vesicles ($130 \times 10^3 \pm 66 \times 10^3$, N = 30) as he ejaculates (V. Rodriguez 1994a).

[f] In seminal vesicle of male (not necessarily all ejaculated).

[g] Data read from fig. 3 of Dickinson 1986.

[h] Minimum 5–10 times.

[i] A large "clot" of seminal material, without apparent structure, was present in the bursa following copulation.

[j] Remate in captivity.

[k] >5 times. Both number and fertility of eggs were lower if the female mated only once.

[l] Sperm numbers were apparently determined by offspring numbers from single and double matings rather than from direct counts, though mention is made of "relatively uncrowded sperm receptacles."

[m] Seminal receptacle is elastic, and value given is maximum observed in ten females; the average was 552 sperm.

[n] Numbers from regression of sperm numbers on number of matings.

[o] Number transferred is average from twenty-two females captured copulating and not copulating in the field that had mated only once; capacity of the spermatheca is the average number of sperm in twelve females that had copulated 4–5 times in the field.

[p] Sperm loads of doubly mated females were greater than those of singly mated females, at least judging by numbers of fertilized (female) eggs.

[q] Estimated from sectioned spermathecae.

[r] Sperm masses in oviducts were appreciably smaller after one mating than after two with the same male.

[s] Estimated from all numbers transferred by male to hermaphrodite's spermatheca plus number of own (hermaphrodite's) sperm; probably an underestimate, as up to 560 sperm were transferred. "Capacity" of spermatheca may be misleading, since sperm in direct contact with the wall (a lower number) apparently have an advantage in fertilization.

[t] Areas in figures in article.

[u] Bursa; spermathecal volume is much smaller.

[v] Spermatheca is bladderlike, but probably volume changes little with the quantity of sperm (O. von Helversen, pers. comm.)

[w] Remate readily in laboratory; field observations lacking.

[x] Male transfers several spermatophores in a given copulation.

[y] Numbers transferred ranged from 0 to 10,014!

[z] Sperm counts in multiply mated females averaged the same as those from singly mated females. But sperm counts more than double the average occurred in some females after a single copulation.

[a1] Volumes of female storage organs were not given, but 70–90% of the male's ejaculate was discarded by the female soon after copulation.

[b1] Transferred not to spermatheca, but to uterus. Of these sperm, only about 10% are stored in the spermatheca and the rest are discarded.

[c1] Amount from any single male is much less.

[d1] 7–17 times.

[e1] Deduced from sperm usage patterns in double pairings with virgin and depleted males.

[f1] Ejaculate size estimated on the basis of testes size by Davies (1992). Storage capacity based on maximum observed number of sperm/storage tubule (500), multiplied by number of tubules (1400).

This section will show the possible importance of taking active female participation into account. The discussion is deliberately oversimplified in equating larger ejaculate size with greater numbers of sperm. The proportions of sperm and seminal plasma in normal copulations are not necessarily constant (they vary quite dramatically interspecifically in both mammals and birds; Mann and Lutwak-Mann 1981). To date there are apparently very few intraspecific studies on this point (the only ones I know, in the katydid *Decticus verrucivorus*, found no change in proportions in different-sized ejaculates; Wedell 1992).

Possible Explanations of Oversized Ejaculates

Oversized ejaculates are probably often due to the frequently cited dilution advantage resulting from larger numbers of sperm ("raffle competition"). For example, dilution advantages probably explain the huge excess (about 10-fold) in the beetle *Callosobruchus maculatus* (Eady 1995). When a second male's ejaculate size was reduced by allowing him previous copulations (average dropped from about 56,000 sperm to about 8,700) his average percentage of paternity with females that had mated previously decreased (but not proportionally—averages declined from about 95% to 78%; Eady 1995). The female morphology and physiology of these beetles probably influence these outcomes. The male is unable to place his sperm directly in the female's spermatheca, because his thick genitalia cannot enter her long, thin spermathecal duct (Eady 1994b). Exactly how larger numbers of sperm in the bursa are translated into greater dilution in the spermatheca (whose capacity is about 6500 sperm) is not clear. Females of this species degrade sperm that do not reach the spermatheca soon after copulation. They can apparently store enough sperm from only a single mating to fertilize a lifetime of eggs (Fox 1993).

The facultative adjustments of ejaculate size in the damselfly *Mnais pruinosa* (Siva-Jothy and Tsubaki 1989a) support the idea that larger ejaculates function to dilute sperm of previous males already inside the female. The last male to mate removes at least some of the sperm from previous matings and has nearly complete sperm precedence immediately after copulation, presumably due to his sperm being better positioned in the female for fertilization, as occurs in other odonates (Waage 1984). As time passes after copulation, this positional advantage decreases, presumably due to sperm mixing. Most copulations with territorial males are immediately followed by oviposition. Thus the ejaculates of these males are not subject to dilution competition soon after copulation; in addition, a territorial male sometimes has another potential mate on his territory, and may soon need additional sperm for another copulation. It is thus consistent with the dilution hypothesis that territorial males sometimes deliver "undersized" ejaculates (four of twelve in fig. 3 of Siva-Jothy and Tsubaki 1989a).

Nonterritorial males, in contrast, deliver larger ejaculates. These males do not generally have other possible mates immediately available; in addition, their mates do not usually oviposit immediately, so sperm mixing has a chance to occur. Sometimes oviposition may occur several days later. Thus nonterritorial males may obtain fertilizations by winning "raffle" competitions. The female's behavior (whether or not she remates just before ovipositing—in at least 37% of the observations she did not) can also have a strong effect on the success of these ejaculation tactics.

Facultative increases in ejaculate size occur in several other species with indeterminate sperm storage capacities when the likelihood of sperm from other males being present in the female is greater, again supporting the idea that dilution competition is important (Gage 1991 on a fly; Gage and Baker 1991 on a beetle; Bellis et al. 1990 on lab rats; Baker and Bellis 1989b on humans).

A further factor which may sometimes favor "oversized" ejaculates is that in some species in which females have rigid sperm storage organs, sperm can also be stored, at least for short periods, in an additional area such as the bursa or the oviduct. Dilution competition outside the spermatheca probably occurs, for example, in the oviduct of the honeybee (multiple copulations occur prior to sperm being moved into the spermatheca; e.g., Page 1986).

Other possible, perhaps less common functions of "oversized" ejaculates include swamping or leaking past female impediments to transport, or feeding the female with sperm or other male products. A feeding function can apparently be ruled out in several species where sperm are discarded by the female (chap. 3.2, 3.3, 6.6; Eberhard 1994).

Possible Explanations of Undersized Ejaculates

Explaining "undersized" ejaculates, which are too small to fill the female storage organs, is more problematic. Current theory emphasizes the probable costs of sperm to the male and the possibility of future matings (e.g., Dewsbury 1982; Parker 1984; Sakaluk 1985; Pierce et al. 1990; Møller 1991a,b; Birkhead 1991; Birkhead and Fletcher 1992; Birkhead et al., 1993b; Pitnick 1993). An additional, possibly related factor is male vigor, and the male's presumed ability to produce sperm (e.g., in the nematode *Cruznema lambdiensis* young males transfer about thirty-two sperm per copulation, while old males transfer only about twelve per copulation; Ahmad and Jairajpuri 1981). Such analyses are incomplete, however, unless they take into account both female behavioral and morphological "rules of the game." For instance, there may be no increase in payoffs to males for larger ejaculates if females have "acceptance limits" and reject larger proportions of sperm in large ejaculates.

LIMITED SPERM RESERVES IN THE MALE

If females remate frequently and if there is last-male sperm precedence, it will not be advantageous for males to transfer more sperm than the female is likely to use in the best possible circumstances (e.g., Ridley 1988). Sperm limitation in the male (e.g., Dewsbury 1982) could accentuate the advantage of transferring reduced amounts of sperm (Pierce et al. 1990). This combination of circumstances appears to explain the undersized ejaculate of *Drosophila pachea,* which is about one-seventh of the female's spermatheca capacity. The giant sperm of this species are apparently costly for the male to produce (males require 13–14 days of adult life to reach sexual maturity, females only 3; Pitnick and Markow 1994). On the average, females remate about every 14 hr (Pitnick and Markow 1994; Pitnick 1993), and pairs do not stay together after copulation (Pitnick 1993). Operational sex ratios in the field are often female-biased (Pitnick 1993). It is thus reasonable to suppose that those males which produce undersized ejaculates, and thereby retain some of their costly sperm for future mating opportunities, are favored. Similar apportioning of costly sperm among relatively common receptive females probably also occurs in *D. nannoptera* (Pitnick and Markow 1994), *D. hydei* (Markow 1985), and also in *D. wassermani*, in which sperm are not so costly (Pitnick and Markow 1994).

The parasitic wasp *Dahlbominus fuscipennis* is another species in which undersized ejaculates may be explained by this type of argument. The sex ratio is normally heavily biased toward females and mating occurs near the emergence site (Godfray 1994). Careful apportioning of ejaculates by males may be favored, because it assures adequate insemination of even the last females to emerge. The strongly biased sex ratio in this species suggests that females in nature seldom mate with males that are not their brothers and perhaps seldom remate (if inbreeding is less severe, sex ratios are expected to be more nearly 50:50), so the disadvantage of failing to fill a female's spermatheca may be relatively small. This leaves unexplained, however, the apparently overly large size of the female's spermatheca, and the strange progressive increase in sperm numbers in the spermatheca after successive copulations (after the first copulation the average is 157; an average of 47 are added after a second copulation, 33 more after a third, 42 more after a fourth; Wilkes 1966).

LIMITED VOLUME OF THE FEMALE'S RECEIVING CHAMBER

Morphological "rules of the game" may explain the undersized ejaculates of the semiaquatic bug, *Hebrus pusillus*. The volume of the site where sperm are deposited in the female (the gynantrial sac) is apparently only about one-third that of the spermatheca, and the ejaculate fills this sac (see fig. 2.9). Males obviously have more sperm available, as they often copulate up to four times with the same female in the space of only two hours. Female behavior may also be involved in limiting ejaculate sizes. Muscles attached to the walls of the sac apparently extract sperm from the male's genitalia, so the female can potentially control the amount of sperm transferred (Heming-van Battum and Heming 1986, 1989).

FEMALE REJECTION OF EXTRA SPERM

Paternity in the chrysomelid beetle *Chelymorpha alternans* is apparently determined by the relative amounts of sperm in the spermatheca (V. Rodriguez 1994a; V. Rodriguez et al., subm.), and soon after copulation sperm are virtually absent from all parts of the female's reproductive tract other than the spermathecal duct and the spermatheca; V. Rodriguez 1994b, pers. comm.). Males nevertheless transferred only about a quarter of the number of sperm that would be needed to fill the spermatheca (assuming the same observed proportion of ejaculated sperm that arrive in the spermatheca; see table 8.1). Males are clearly not sperm limited. They had more sperm in reserve in the seminal vesicle than were transferred in a single ejaculate (averages of about 120,000 in the vesicle vs. about 58,000 ejaculated), and they also replaced sperm rapidly (males frozen 10–20 sec after concluding the second of two consecutive copulations, with 1–5 min between copulations, had an average of 118,000 sperm in the seminal vesicle; V. Rodriguez 1994a). In addition, there was no effect of mating order in *C. alternans*, at least when matings were closely spaced (V. Rodriguez 1994a; V. Rodriguez et al., subm.). Males in nature are seldom found with more than one female at a time, and generally leave the female after a single copulation (V. Rodriguez, pers. comm.), so saving sperm for subsequent copulations soon afterward is an unlikely explanation.

There is thus no sign that males are sperm limited or that restrictions on ejaculate size serve to take advantage of future mating opportunities. The most likely explanation for undersized ejaculates in this species seems to be due to "pure" female control. Presumably by limiting the maximum amount she will ever transfer to her spermatheca from a single copulation, the female may set an upper limit to the adaptive size of the male's ejaculate. Females often emit sperm during copulation, and on average only about 38% of the male's ejaculate is transferred to the spermatheca. Even when no sperm are ejected during copulation, the spermatheca is not filled, so sperm are probably also digested or otherwise eliminated internally. Experimental removal of the spermathecal muscle resulted in changes in both the number of sperm stored in the spermatheca and the frequency with which females discard sperm during copulation, so the female does indeed seem to have at least partial control of sperm transfer (V. Rodriguez 1994b).

Another species with undersized ejaculates is the cricket *Gryllus bimaculatus*. This species was not included in table 8.1 because the female spermatheca is flexible and can expand to accommodate over thirty ejaculates (Simmons 1986). The male attaches a spermatophore to the female's genitalia, and sperm transfer is slow and gradual, taking about 60 min (Simmons 1986; see also chap. 4.4). Males can transfer more than one spermatophore in relatively rapid succession (Simmons 1986).

Since sperm apparently mix completely in the spermatheca, larger numbers of sperm should translate directly to greater reproductive success for a male via raffle competition (Simmons 1987a). The female role in limiting ejaculate size is especially clear in *G. bimaculatus*. The female consumes the spermatophore after the male has inserted it into her genitalic opening (see chap. 4). She usually waits for 1–80 min after copulation ends before removing a spermatophore (forty of forty-nine were removed in less than 60 min; Simmons 1986). Thus there is an upper limit on the amount of sperm it is advantageous for a male to attempt to transfer; quantities larger than this will be eaten by the female before they enter her reproductive tract.

The biting midge *Culicoides melleus* may have a similar explanation for undersized ejaculates, but interpretation is less certain in this case. Males produce ejaculates that contain only about half as many sperm as can fit into the spermatheca (Linley and Hinds 1975b). Females of this species routinely eject spermatophores before they are emptied of all their sperm (Linley and Hinds 1975b; see fig. 3.14), so female limitation of sperm transfer may be involved. The situation may be more complex, however. Males show some signs of being sperm limited (they reduce the number of sperm ejaculated when copulating with less receptive females (Linley and Hinds 1975a; see chap. 3.8), even though a male's chances of immediately encountering additional females may not be high (mating occurs at sites where males and females are emerging) (Linley and Adams 1972). Many copulations with virgin females resulted in complete transfer of the sperm from the spermatophore (86 of 120 cases; Linley and Hinds 1975b), not as would be expected if female limitation of sperm transfer were responsible for the undersized ejaculates (no attempts were made to check ejaculate size against the likelihood that sperm would be discarded). In sum, data for this species are equivocal.

INTERNAL RESISTANCE OR LACK OF TRANSPORT IN THE FEMALE

The details of copulation in another chrysomelid, *Labidomera clivicollis* (fig. 8.1), are quite different, but they lead to a similar conclusion regarding female-imposed rules (Dickinson 1986, 1988). In successful copulations, the male apparently places his virga (the tip of his intromittent organ) at the

Figure 8.1 A pair of *Labidomera clivicollis* beetles engage in a copulation which will result in a paradoxically small amount of sperm being transferred. Although a male normally has on the order of 175,000 sperm available and is able to rapidly produce more, and the female's spermatheca can hold up to about 30,000–40,000, only about 3000–4000 sperm are transferred during the average of about 2.5 copulations that occur in an hour. By a process of elimination of other possible explanations, it appears that in this species, and several others, males may refrain from transferring larger numbers of sperm because females will not accept them (Dickinson, in press). (Photo courtesy of J. Dickinson)

entrance or basal end of the spermathecal duct, and his semen move directly into the spermatheca from there (Dickinson 1988 and in press). Single copulations lasted an average of only about 15 min, but males stayed mounted on females for up to 42 hr (av. = 18 hr) and copulated on average about 2.5 times per hour (nearly half of these copulations were "partial intromissions," often following full intromissions; partial intromissions may have involved incomplete extension of the male's genitalia or improper alignment within the female; see chap. 3.4, fig. 3.6). A pair of beetles thus usually copulates several tens of times before separating. In matings with virgin females, males transferred about 3,000 to 4,000 sperm in the first hour of pairing. This was only about one-tenth of the apparent capacity of the spermatheca and about one-fiftieth of the number of active sperm left in the male's seminal vesicles following a single copulation.

Dickinson (1986) noted that it is puzzling that males inseminate females repeatedly with undersized ejaculates. The male's reproductive success (both as first and as second male) was apparently determined by relative sperm numbers rather than by the order of copulations, and a mounted male was sometimes displaced by another male or by the female. Thus more rapid ejaculation of larger numbers of sperm would seem favorable for the male. Dickinson suggested that female refractory behavior, such as kicking, which occurs throughout mounted courtship and copulation, may make it difficult for the male to maintain his genitalia in contact with her spermathecal duct. An additional possible female role is that female cooperation may well be necessary to allow sperm to enter the spermatheca. Active transport of sperm by absorption of spermathecal liquids seems likely, as has been documented for other insects with rigid spermathecae and a thin spermathecal duct similar to those of *L. clivicollis* (Linley and Simmons 1981). As Dickinson (1988) notes, changes in female uptake behavior could explain the otherwise puzzling increase in second male predominance in paternity when two 15 hr matings are separated by 5 days instead of occurring directly one after the other.

Females of this species may benefit directly from resisting insemination and restricting sperm uptake in addition to the indirect benefits of discrimination among potential fathers of their offspring. By delaying or prolonging insemination, females oblige males to remain on their backs, and may gain protection from predators or parasites (the beetles feed on heavily defended plants in the genus *Asclepias* and are aposematically colored).

A somewhat similar situation may occur in the dunnock *Prunella modularis*, though the data are incomplete. The female bird has about 1400 sperm storage tubules, and the maximum number of sperm observed in a tubule was about 500 (Birkhead et al. 1991). Although the female storage organs are not rigid, her total sperm capacity is probably between 7×10^5 (estimated from maximum numbers of sperm in any single storage tubule; Davies 1992) and 1.4×10^6 (estimated from volume of sperm and storage tubules; Birkhead et al. 1991; fig. 8.2). The size of an ejaculate has not been determined directly. Birkhead et al. (1991), using the relationship found between testes weight and ejaculate size in other birds, estimated it to be about the same (1.9×10^6) as the female's storage capacity (Birkhead et al. 1991). A male dunnock has available, however, enough sperm (1×10^9) for about one thousand such ejaculations in his cloacal protuberance, so probably ejaculates are larger. This plus the fact that females mate very frequently (on the order of up to twenty times per day for about ten days; Davies 1992) and the likelihood of female control of sperm storage (fig.

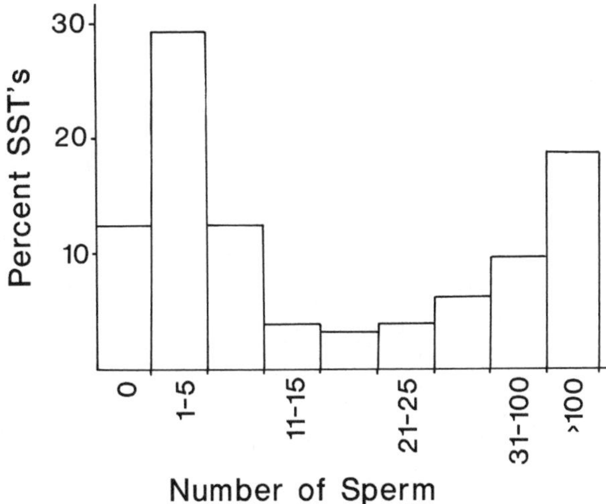

Figure 8.2 Bimodal distribution of number of sperm in a sample of the approximately 1400 sperm storage tubules in a female bird, the dunnock *Prunella modularis*. This female had already laid four eggs and had probably copulated 10–20 times/day during the week before she was killed (Davies 1992). Since some full tubules were adjacent to empty ones, large-scale differences in the location of tubules (e.g., nearer or farther from the cloaca) are not likely to explain the bimodality. Apparently either sperm arrival or exit from storage tubules (or both) is nonrandom, suggesting possible active female roles. (After Birkhead et al. 1991)

8.2) make it likely that (1) the overwhelming majority of the sperm from the average copulation do not end up in the female's storage organs; (2) females are probably responsible for some portion of this failure to be stored; and (3) males withhold some available sperm when they copulate.

OTHER FEMALE-IMPOSED BEHAVIORAL RULES

Another striking case of undersized ejaculates occurs in some honeybees (*Apis*). Four of five species (the exception is the domestic honeybee *A. mellifera*) have undersized ejaculates (Koeniger 1991; table 8.1). The smallest ejaculates (9.1% of the spermathecal capacity in *A. adreniformis*, 29% in *A. florea*) occur in the only two species in which the male apparently succeeds in positioning the tip of his endophallus deep enough in the female to reach the mouth of her spermathecal duct (in other species, the male deposits his sperm in the oviduct; fig. 8.3; Koeniger 1991). Males strongly outnumber females at mating sites in *Apis*, and the male dies as a result of copulating; he becomes paralyzed as he everts his genitalia into the female, and the terminal portion of his body breaks off there, forming a plug which is apparently occasionally successful in preventing insemina-

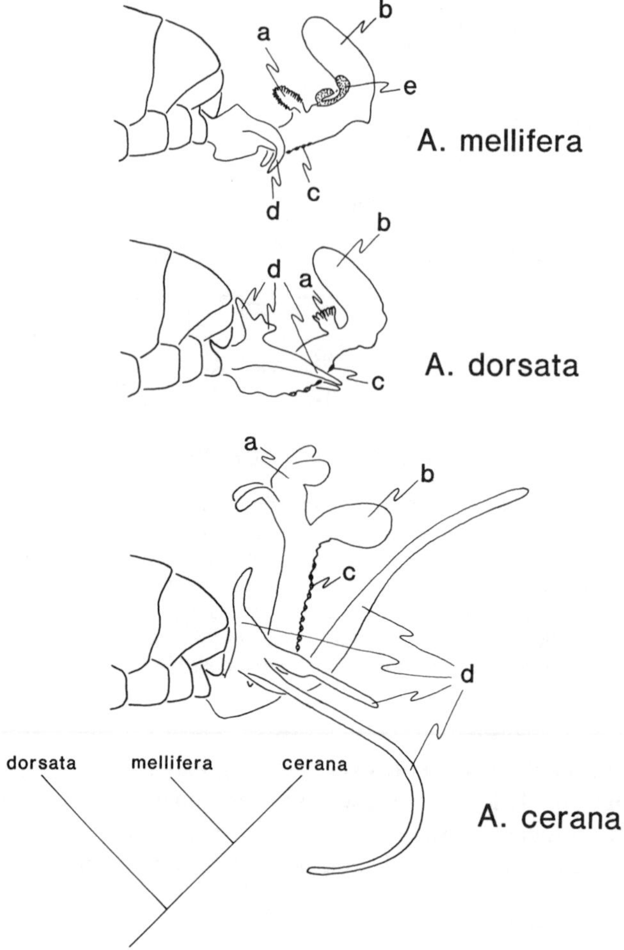

Figure 8.3 The everted male genitalia of three species of *Apis* honeybees (letters indicate homologous structures). In four of five species (including *cerana* and *dorsata*), the male ejaculate size is smaller than the capacity of the female's spermatheca. Since mating is suicidal for *Apis* males, the small ejaculates are not explicable with the usual argument that reduced ejaculate size is an adaptation to conserve limited supplies for future matings. Those *Apis* species in which the male genitalic structures permit deposition of sperm closest to the entrance of the female's spermathecal duct have the most extremely undersized ejaculates (as low as 9.1% of the spermathecal volume). Female-imposed limitations on the amount of sperm accepted from any given male may explain undersized ejaculates. (After Simpson 1970; cladogram after B. Alexander 1991)

tion by subsequent males (Page 1986). There is thus no chance that males are saving sperm for future copulations.

Substantial amounts of seminal products (mucus) are also transferred in some species, but investments in mucus are probably not responsible for undersized ejaculates of sperm. The largest amount of mucus (estimated at 5% of the male's body weight) occurs in the species with the largest relative numbers of sperm, *A. mellifera,* while the species with the most undersized ejaculate, *A. florea,* produces very little mucus. Sperm from different males mix relatively uniformly in the spermatheca, with only moderate clumping, at least in the best-studied species, *A. mellifera* (Page 1986; see fig. 4.9) so sperm numbers are important determinants of a male's reproductive success.

Thus by elimination of other possible explanations, it seems likely that undersized ejaculates of *Apis* are related to female-imposed limitations. Starr (1984) proposes the female-imposed "rules of the game" hypothesis that this limitation is due to rapid female flight at mating sites, so that presumably only males carrying relatively small testes will be able to win out over other speedy males. Why the female flies so rapidly was not discussed. This argument is not entirely convincing, since larger drones would presumably be faster fliers and carry greater payloads of sperm. Drones are nevertheless smaller than queens, suggesting selection favoring larger drone size has not been important. The best design for a drone would depend on the balance of benefits from larger drones (faster flight, larger ejaculates) and the costs of producing them (lower numbers). More data on the trade-offs are needed to evaluate Starr's argument.

A second, nonexclusive alternative is that the queen will only accept a certain maximum amount of sperm from any given male. Queen bees could benefit from such limitations, because there are several possible payoffs they could derive from multiple paternity (Starr 1984; Page 1986). There is fine female control of the next step after copulation—the passage of sperm from the oviducts into the spermatheca; muscles in the female's ducts, fluid from her spermathecal gland, and even the number of worker bees attending her affect the entry of sperm into the spermatheca (Koeniger 1986).

POSSIBLE TESTS

One possible test of the idea that rejection of sperm is responsible for undersized ejaculates would be to determine whether, in species with undersized ejaculates in which females eject or fail to store some sperm, larger proportions of larger ejaculates do not reach female storage organs. The only possibly related data I have found do not fit. In five species of *Rhabditis* nematodes there was no relation between the number of sperm transferred and the probability of loss via vaginal discharge (Hass 1990). It is not clear, however, whether the simple spermatheca of these organisms is rigid and whether it can hold more than a single ejaculate. It should be relatively easy to manipulate a male's ejaculate size in at least some species by controlling the number of his preceding copulations.

It is also interesting to ask whether species with undersized ejaculates are as common as suggested by the data in table 8.1. I think it is too early yet to obtain a clear answer. Other variables may be correlated with ejaculate sizes, such as frequency and timing of remating by females, sperm precedence patterns, and efficiency of sperm usage, so separating causes and effects may be difficult (Pitnick and Markow 1994). One thing that *is* already clear is that even in closely related species male ejaculation tactics can vary substantially (e.g., species in the *nannoptera* species group of *Drosophila*; Pitnick and Markow 1994; see table 8.1).

What Determines the Sizes of Sperm Storage Organs?

It is possible to push this analysis one step back, and ask what factors have favored different volumes of sperm storage organs and amounts of sperm accepted by females (e.g., Walker 1980). I will refrain from straining my imagination and the readers' patience with a list of unsupported speculations, other than to remind them that the advantages to females of multiple paternity seem to be sizable. This might favor larger storage organs. One of the basic lessons from the discovery of widespread extra-pair copulations in purportedly monogamous birds is that females must benefit more from insemination by multiple mates than had generally been thought previously.

I know of only one extensive interspecific comparative study on spermathecal volumes (or at least the contents of spermathecae in females that had just finished their final mating and had not begun oviposition). By comparing the number of sperm against the maximum expected number of female offspring in twenty-five species of fifteen genera of ants (male offspring come from unfertilized eggs), Tschinkel (1987) concluded that the spermathecae of ants are as small as they can be and still hold enough sperm to produce all the fertilized eggs that a successful queen is likely to lay in a lifetime. Tschinkel argued that such a design minimizes the female's sperm maintenance costs (see Parker 1970c). The limited volume is particularly dramatic in some species, such as the fire ant *Solenopsis invicta*, because the reproductive life of a queen with a large colony sometimes ends prematurely due to limited sperm supplies (when her sperm run out, the queen is replaced by a fully inseminated female; Tschinkel and Porter 1988). Presumably the advantages of smaller volumes earlier in the queen's life (most queens perish attempting to found a colony) must outweigh these occasional tragedies (from the queen's point of view).

Ants may be a special case, since the early mating and subsequent long

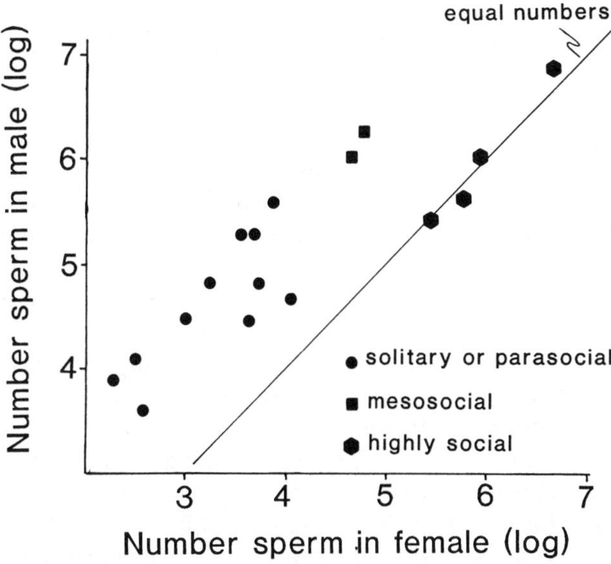

Figure 8.4 Correlation in 17 different species of solitary and social bees between the number of sperm available in the male and the number of sperm in the female (line marks equal numbers). Reproductive females of highly social species, which lay more eggs, have higher numbers of sperm while their males have relatively smaller numbers of sperm available (not, of course, necessarily equal to the numbers transferred in a single copulation). The species at the upper right is the only *Apis* included, *A. mellifera*. (After Garófalo 1980)

reproductive life of a queen mean that sperm maintenance may be unusually costly (a newly mated queen of *S. invicta* carries about 7 million sperm and can live up to about 6.8 yr; Tschinkel and Porter 1988). Nevertheless, minimum spermatheca sizes occurred even in ant species with colonies of fewer than one hundred workers in which sperm storage costs are presumably much smaller.

A similar survey showed a positive correlation between sperm numbers in female spermathecae and in the testes and seminal vesicles of 17 species of solitary and social bees. Unfortunately there was no control over the female's prior mating and oviposition history: since females of at least some species may mate repeatedly (Roig-Alsina 1993), spermathecal sperm counts may have been substantially lower than capacities. In fourteen species the male had at least three times more sperm than females, in two species the numbers were about equal, and in one (the highly social *Melipona marginata*) the male had about half the sperm in the female (fig. 8.4; Garófolo 1980). In accord with the data on ants, queens of highly social species, which have more populous nests, had higher numbers of sperm.

8.2 Intraspecific Variation in Cryptic Female Choice Criteria and Sexual Selection Theory

If male seminal products are often under sexual selection by female choice (chap. 6; Cordero 1995; Eberhard and Cordero 1995), then data on the responses of females when they are exposed to different concentrations of male products can throw light on variability in female choice criteria and male display characters. This is because both male seminal stimuli and female reception of them, which were discussed in chapter 6, were in some cases unusually well controlled, in comparison with more usual tests of sexual selection by female choice. The females in many of these tests were carefully standardized with respect to variables such as age, feeding, and reproductive status. The "stimuli" provided by males were chemical; their intensities could thus be standardized by pooling them and controlling dilutions. The stimulating factors may be relatively simple, and they may exercise their effects without the intervention of much of the female's central nervous system (or at least not primarily through higher centers of her central nervous system; see chap. 6). Variability arising from differences in the status and activity of the central nervous systems of different females may thus have had relatively minor effect on female responses to males.

Compare, for instance, a test of classic, overt female choice acting on a rooster's comb size in the chicken (Zuk et al. 1992) with a test of the effect of male accessory gland substance on ovulation in *Drosophila melanogaster* (fig. 8.5; Chen et al. 1988). Differences in a rooster's behavior from one trial to the next will unavoidably introduce differences in the stimuli received by the female. In addition, differences in the behavior of the female herself (e.g., the degree of distraction by other stimuli, disturbance resulting from being placed in the test situation) will also lead to variation in the stimuli received by her central nervous system. And the CNS itself is probably very variable from one moment to the next.

In contrast, the physiology experiment with the fruit flies may be much simpler. The male stimulus, a standard quantity of a common extract, is presumably close to perfectly uniform for different females, especially if the exact site at which it is injected into the female's body does not alter its effects. This seems to be the general assumption in these studies; I know of no data to test it, other than one gross comparison of inside versus outside the female's reproductive tract in a tick (Oliver et al. 1984). The female's reception of the stimulus from an injection is probably also largely free of the vagaries of the activity of her central nervous system. Despite this probably unusual simplicity and uniformity, the clear pattern in studies of both remating inhibition and oviposition responses was for a substantial intraspecific variation in female responses.

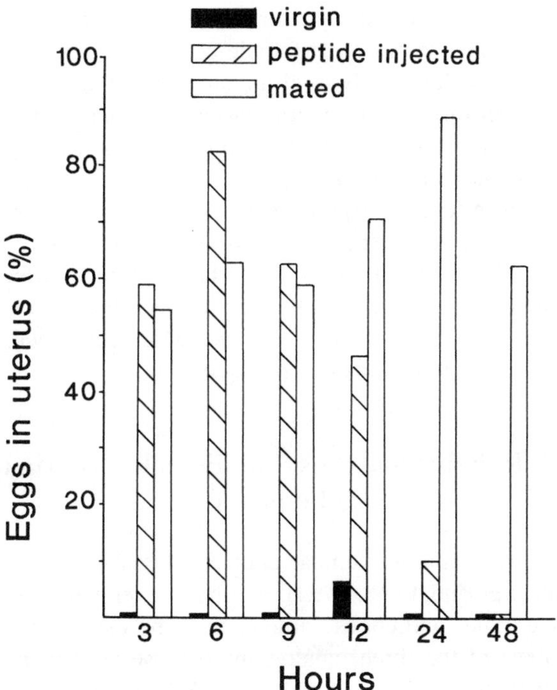

Figure 8.5 Variation in the ovulation response of female *Drosophila melanogaster* to an injection of a standard amount of purified peptide material from male accessory glands. Despite the highly standardized stimulus, female responses were variable. The lack of accord was especially pronounced during the first 12 hr, with females being divided about 50-50. Variability of this sort under standardized treatments suggests that selection on male stimuli may be less consistent than some theoretical models have supposed. (After P. Chen et al. 1988)

These variations raise an important point regarding the evolution of female responses to male signals, because they challenge the idea that females of a species all have the same criterion for choosing males. Perhaps the thresholds at which different females respond to the male product differ. Or maybe females differ from one another with respect to the number and intensity of other stimuli (from the male, from the environment, from within their own bodies), which affect their responses. The important point is that even when female reception of male stimuli was unusually constant, the response criteria of different females appeared to vary substantially (fig. 8.5).

If such a lack of fine-tuning in female criteria is a general phenomenon, it could have important theoretical consequences. Theoretical discussions of sexual selection by female choice are usually strongly typological ("the" male signal, "the" female criterion for choice; see, for instance, my own discussion in the next section!). If female criteria are not consistent, then selection on males is less likely to eliminate genetic variance in male display characters. In turn, such genetic variance among males in sexually selected characters means that females can gain reproductive payoffs from choosing males via increased display abilities in their offspring (though lower uniformity in female responsiveness would also reduce the chances that any particular male design would be best at obtaining positive responses from all females). Intraspecific variation in genes affecting female criteria for classic, precopulatory choice have been found in some species (Andersson 1994; Wagner et al. 1995). What selective factors maintain such female variability remains an open question.

8.3 Relative and Absolute Female Criteria in Cryptic Female Choice

As in classic precopulatory female choice (Maynard Smith 1987; Real 1991; Downhower and Lank 1994), a female exercising cryptic choice could compare males and choose among them by using three basic techniques (or mixes of the three) analogous to those used in precopulatory courtship. All three would presumably result from characteristics of the female's nervous system and/or sensory receptors, and/or the morphology and/or physiology of her genitalia or other portions of her body that influence the abilities of males to father her offspring. Previous experiences (e.g., observations of her parents) as well as genetic influences could be involved in establishing particular female criteria.

A. Before entering into a sexual interaction, the female might compare the male against some predetermined "absolute" image or ideal ("absolute standard"; e.g., accept any male larger than size *x*, or any male whose call intensity exceeds *y* decibels, etc.).

B. The standard used to judge males could be a "sliding" one and adjusted with respect to the female herself ("self-reference sliding standard"; e.g., accept any male larger than *x* times her own size).

C. The standard could be "sliding" with respect to other males the female had encountered ("cumulative sliding standard"; e.g., accept any male larger than the others seen previously).

The second and third possibilities have been modeled mathematically (Maynard Smith 1987).

All three techniques could presumably be used in both classic, precopulatory female choice, and in cryptic female choice. They could also be mixed (e.g., compare on a sliding scale only those males that have already passed some predetermined absolute standard), and they could involve several criteria (e.g., Trail and Adams 1989). The same criteria need not even necessarily be applied equally to all males. For instance, the criteria for a female's first mating might be more lax, in order to avoid going uninseminated (Jackson 1981).

Is there any evidence that any of these mechanisms are especially likely to be used to guide cryptic female choice? The use of absolute and self-reference standards rather than standards involving comparisons with other males appear likely to be more common. Decisions made during and following a first copulation can only be based on predetermined standards (absolute or sliding self-reference criteria), since the female has not yet experienced copulation with any other males. Many important cryptic female choice decisions are apparently made during or following a first copulation (for instance, to emit or not emit a drop of the male's sperm in the beetle *Chelymorpha alternans*; mature or not mature eggs and oviposit in many insects and ticks: tables 6.1, 6.3; reject or not reject additional males in many insects: table 6.1). Presumably at least part of the reason for such immediate responses is the disadvantage to the female of delaying reproductive responses following copulation. An observational bias may have also contributed to this apparent trend. Decisions made on the basis of one copulation are easier to determine, as compared with differences in the effects of several different copulations. Nevertheless, the frequent immediate responses of females of many species following a single copulation indicate that comparisons of males against predetermined standards may be common in cryptic female choice.

The seldom-asked question of whether such predetermined criteria are absolute or sliding with reference to the female's own characters can be broached in species in which female choice may be made on the basis of mechanical mesh with male genitalic or nongenitalic morphology (the nongenitalic contact courtship structures of males; Eberhard 1985). Both reduced variance in the sizes of the portions of the female's body with which the male must mesh, and a lower slope when the size of such an area is regressed on female's body size would indicate a relatively fixed, absolute female criterion. Data of this sort on the rigid female genitalic plates with which male genitalia must mesh in the spiders *Argiope trifasciata* and *Philoponella vicina* (Eberhard et al., in prep.) show such low slopes and coefficients of variation when compared with other female structures such as cephalothorax and leg length. In other words, the females of these species have relatively standard-sized genitalia. This suggests relatively fixed absolute standards and is in accord with similarly standard-sized male genitalia in these same species (Eberhard et al., in prep.). Relative constancy in the size of male genitalia, as compared with other body parts, may be a widespread trend in many other species (Byers 1976, 1983, 1990;

McAlpine 1988; Eberhard 1985; Coyle 1985a; Wheeler et al. 1993; Eberhard et al., in prep.). Interestingly, the genitalia of males of several species in which female genitalic structures are not rigid, and do not impose problems of mechanical mesh, are also relatively invariable in size (Eberhard et al., in prep., and refs.). This suggests that cryptic female choice that involves stimulation as well as mechanical fit may also be based on relatively fixed absolute standards.

Perhaps in some cases there is a "lock and key" relationship, not in the original sense of a mechanism to isolate one species from another (Shapiro and Porter 1989), but to elicit more favorable responses from conspecific females. If a male's genitalia must mesh mechanically with a relatively rigid female structure (in order to stimulate receptors, or to fit against the structure), then overly large and overly small male structures may be ineffective. Similarly, extreme sizes of the female structure could be disadvantageous, making pairing with some males difficult or impossible. Even if the female structure is not rigid but has receptor organs at particular locations which the male must stimulate, or if the female otherwise senses the size of the stimulating male structures relative to her own size, then extreme size variants in male structure may be disadvantageous. Byers (1983) developed a similar argument, starting from the premise that standard size in female genitalia results from selection to lay one optimum size of egg. The two ideas are not exclusive.

Even when the female has mated with more than one male, at least some cryptic female choice discriminations apparently do not involve comparisons among males. The characteristics of the second but not the first male gave significant effects on P2 values in the spider *Neriene litigiosa* (Watson 1991). Similarly, in both the thirteen-lined ground squirrel *Spermophilus tridecimlineatus* (Schwagmeyer and Foltz 1990) the ladybird beetle *Harmonia axyridis* (Ueno 1994) and the fly *Dryomyza anilis* (Otronen 1994a,b), characteristics of the second copulation rather than the first were positively correlated with the second male's chances of paternity (the small number of litters of the ground squirrel makes the conclusion for this species somewhat tentative). The length of the first copulation but not of the second in the stinkbug *Nezara viridula* affected the proportions of offspring sired by the two males (McLain 1980).

A simple female sperm uptake mechanism in which males are apparently not compared directly may be responsible for a similar pattern in ejaculate sizes in the bruchid beetle *Callosobruchus maculatus*. Normal virgin males ejaculate about ten times more sperm into the female's bursa than can be stored in her spermatheca, and uptake into the spermatheca is probably dependent on female transport processes (Linley and Simmons 1981; Eady 1994a). Even though a male's ejaculate is strongly reduced, from about 56,000 to 8600 by several sequential copulations, reduced ejaculates were still sufficient to fill the spermatheca if the female was a virgin (Eady 1995). Such reduced ejaculates by first males had no effect on the male's paternity when the female mated twice; but sharp decreases in paternity resulted if a male with a reduced ejaculate mated second. The size of the second male's ejaculate was an important determinant of paternity not in comparison with that of the first male, but with respect to the standard ejaculate size of about 56,000 sperm.

8.4 A Possible Relationship between Infertile Eggs and Overly Aggressive Sperm

Fertilization seems often to be an especially delicate, easily disrupted event (fig. 8.6). For instance, fertilization in vitro in all well-studied mammals routinely results in large numbers of abnormal embryos (summary in Overstreet and Katz 1977). As illustrated in table 8.2, it also seems to be "a fact of life" that females of sexually reproducing organisms in a variety of taxonomic groups usually lose an appreciable portion of their eggs because they are infertile. The average egg infertility rate for the thirty-seven species in the table (using estimated median values for species for which only ranges of infertility rates are given) was 13% (range 0.2–62.5%, median 8%). While a few elevated infertility rates in table 8.2 may be artifacts of not allowing females to remate as often as they would in nature (see, for example, *Oncopeltus fasciatus*), it is nevertheless clear that egg infertility is widespread and often substantial. Data from seventeen other bird species suggested that hatching failures are not uncommon. The median was 10%, and percentages ranged up to nearly 50% in some years in boobies (*Sula dactylatra*; D. Anderson 1990). In at least some of these species most failed eggs had no visible embryo (D. Anderson 1990). Egg infertility seems also to be common in externally fertilized eggs. Except for those rare cases in which infertile eggs serve as food for other offspring (e.g., Downes 1988 on a spider), there must be strong selection on females to avoid such infertility. The substantial rates of infertility commonly found in nature can be maintained only by relatively strong counterselection. What would this be?

There are several possible explanations. Some "infertile" eggs may result from developmental failures in embryos. For instance, in the bird *Passer domesticus* an estimated one-third of the 8–14% of the eggs that failed to hatch died as embryos (Wetton and Parkin 1991). This section explores the possible importance of a second set of factors that could result in fertilization failures: failure of sperm to contact the egg; only nonviable sperm contacting the egg; or too many sperm entering the egg (polyspermy) and causing early embryo death. An extension of the arguments in chapter 7.1 suggests the possibility that such fertilization problems may often be due to

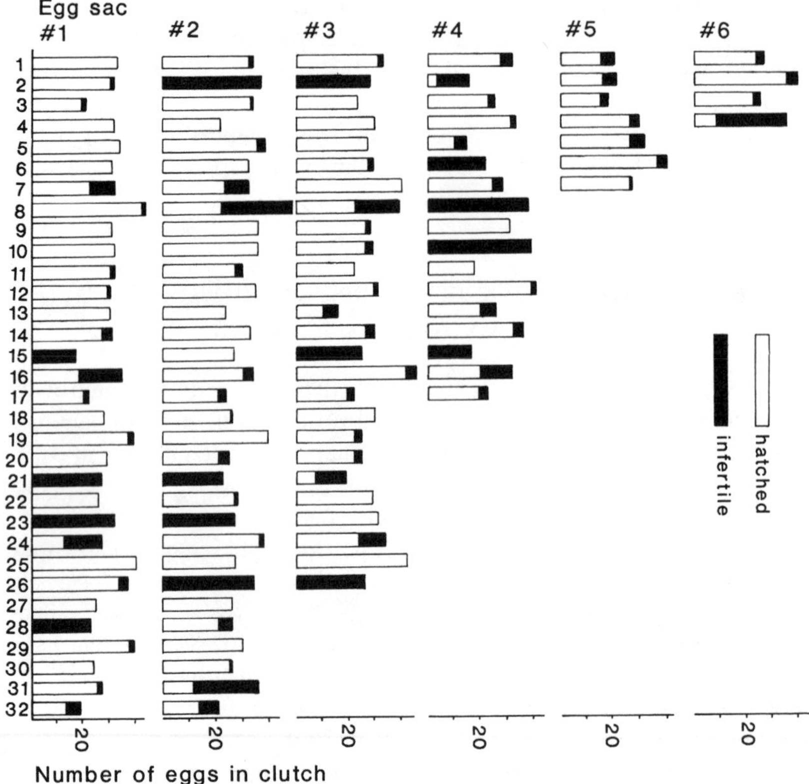

Figure 8.6 Erratic changes in egg fertility in the spider *Pholcus phalangioides* illustrate the current lack of understanding of processes involved in sperm use. The numbers of hatched and unhatched eggs in successive clutches of eggs of thirty-two different females, each of which had copulated only once, are indicated by the heights of the bars. There were substantial numbers of infertile eggs even in the first clutches of females that later laid many fertile eggs (e.g., #s 7, 16, 24); and some females (e.g., #s 2, 15, 21, 23, and 28) laid entire clutches of infertile eggs but nevertheless subsequently produced fertile eggs. Morphological studies suggest simple dynamics of sperm storage (encapsulated sperm are stored in a matrix of secretions in a widening of the lower oviduct; Uhl 1994a). The reason for such irregular infertility is thus unclear; perhaps it involves the events associated with activating the sperm. (After Uhl 1992)

Table 8.2
Percentages of Infertile Eggs (Eggs That Fail to Hatch)

Taxon	*% Infertile*	*Reference*
Pisces		
Fundulus heteroclitus	7.5[a]	M. Taylor 1986
Amphibia		
Afrixalus delicatus	6[b]	Backwell and Passmore 1990
Chiromantis xerampelina	<2	Jennions et al. 1992
Reptilia		
Vipera berus	3.9[c]	Madsen et al. 1992
Aves		
Passer domesticus	8–14[d]	Wetton and Parkin 1991
Tachycineta bicolor	0.2	Dunn et al. 1994a
Arachnida		
Araneae		
Nephila clavipes	10[e]	Christenson 1990
Agelena limbata	6[f]	Masumoto 1993
Theridion rufipes	10–20[g]	Downes 1988
Insecta		
Odonata		
Erythemis simplicicollis	5.8[h]	McVey and Smittle 1984
Diptera		
Aedes aegypti	7.1[i]	Adlakha and Pillai 1975
Culex pipiens	16.9[i]	Adlakha and Pillai 1975
Bactrocera cucurbitae	7	Tsubaki and Sokei 1988
Culicoides melleus	40[j]	Linley and Hinds 1974
Dryomyza anilis	0.9[k]	Otronen 1994
Drosophila pseudoobscura	0–2.2[l]	Maynard Smith 1956
Lucilia cuprina	8[m]	Smith et al. 1990
Hylemya brassicae	24[n]	Swailes 1971
Coleoptera		
Onymacris unguicularis	26	DeVilliers and Hanrahan 1991
Orthoptera		
Gryllus bimaculatus	20.6[o]	Simmons 1987a
Acheta domesticus	25–30[p]	Murtaugh and Denlinger 1987
Locusta migratoria	31.6[q]	Parker and Smith 1975
Allonemobius fasciatus	6.7[r]	Howard and Gregory 1993
A. socius	10.7[r]	Howard and Gregory 1993
Trichoptera		
Mystacides azurea	7[s]	Petersson 1989
Lepidoptera		
Zeiraphera diniana	13[t]	Benz 1969
Heliothis virescens	9.1[u]	Proshold and LaChance 1974
H. subflexa	21.7[v]	Proshold and LaChance 1974
Euphesia kuhniella	18	Norris 1933
Acrolepiopsis assectella	5[w]	Thibout 1975

Table 8.2 (*cont.*)

Taxon	*% Infertile*	*Reference*
Hemiptera		
Gerris remigis	7[x]	Rubenstein 1989
G. odontogaster	7	Arnqvist 1989
Nezara viridula	14.6[y]	McLain and Marsh 1990
Oncopeltus fasciatus	62.5[z]	Riemann and Thorson 1974
Abedus herberti	2.8[a1]	Smith 1979
Crustacea		
Orconectes rusticus	19.4[b1]	Snedden 1990
Uca lactea	0.35[c1]	Murai et al. 1987

[a] Standard error = 2.6%.

[b] Range = 0–100%; standard error = 3.2%, N = 32 females.

[c] Standard deviation = 8.5%, N = 34. Remating by the female did not affect the percentages of infertile eggs. Another 31.6% of the eggs produced stillborn offspring.

[d] Hatching success varied in different years at the same site. About one-third died as embryos, two-thirds as "unfertilized."

[e] Only count "productive" egg sacs; 85% of sacs early in the season are productive, while only 15% are productive late in the season.

[f] Rate for fourteen females with a complete plug; rate was 10% for three females with incomplete plugs.

[g] Fertility rate increased when female was better fed.

[h] Range = 0–41%, N = 18.

[i] Infertile = lack embryo.

[j] When the number of sperm from the male was reduced below about 600, the rate of infertility increased.

[k] Standard error = 1.2%.

[l] N = 12 females with "normal" males.

[m] Fertility judged by egg hatch.

[n] Fertility judged by egg hatch; one mating with virgin male; 13% of matings resulted in <50% hatching.

[o] Two matings with normal males.

[p] Rate varied; fertility was determined by presence of embryonic eyespots.

[q] There was great variance among individuals, but there was no difference when one or two males copulated with the female.

[r] Number of females that produced offspring after male was left with female for four days.

[s] First batch of eggs.

[t] Fertilization judged by color of egg; N = 74 females, average of 146 eggs/female.

[u] Range 2.7–23.7%; hatching rate for embryonated eggs = 94.9%.

[v] Range 10.2–100%, hatching rate for embryonated eggs = 85.9%.

[w] Combined data for females that mated once and twice (there was somewhat greater infertility when the female mated twice); N = 10,038 eggs from 122 females.

[x] Copulated once with each of two males.

[y] Range 11.8–27.6%, N = 28.

[z] Fertilized eggs judged by collapse of egg. Female copulated with only a single male; soon after mating, infertility was near 0% but gradually increased.

[a1] Eggs were cared for by males; 1431 eggs on the backs of 18 males.

[b1] Single mating; egg fertility determined by presence of embryonic eyes.

[c1] N = 10 clutches.

escalating selection for both overly aggressive male gametes and for female barriers against them.

Male gametes are probably under strong selection for "aggressive" abilities to arrive in the vicinity of eggs and to enter the egg and fuse with its nucleus. Male abilities of this sort could result in some eggs becoming overwhelmed and damaged. Although selection on males would not favor killing per se, overaggressive male competitive traits that sometimes result in female death are known in several groups. Groups of "overeager" males sometimes drown female toads (*Bufo bufo*; Davies and Halliday 1979), ducks (*Anas crecca*; McKinney et al. 1984), and dungflies (*Scathophaga stercoraria*; Parker 1970b), or literally pull a female to pieces in the sand wasp *Bembicinus quinquespinosus* (Evans et al. 1986), and the burrowing bee *Amegilla dawsoni* (Alcock, in prep.) in their frantic struggles to mate with her (see fig. 7.5). Female-imposed barriers against sperm that have "overly aggressive" adaptations of this sort could be favored. This in turn could lead to the evolution of "overly effective" barriers, which would result in some eggs going unfertilized.

Females may often be caught in the dilemma that overly rigorous barriers will prevent fertilization, while overly lax barriers against aggressive male gametes will result in early developmental failures. If sperm from different males compete for the same eggs, continuing selection for improvements in male adaptations is expected, and new male advances would necessitate readjustments in female barriers. Frequent changes in the male aggressive traits might make it difficult for females to evolve optimum barrier traits that would eliminate egg losses. A certain degree of infertility might thus be a frequent cost of sexual reproduction, due to sexual selection on male gametes.

This idea could be tested by comparing the frequencies of "infertile" eggs in species with strictly monandrous females and species with polyandrous females. In strictly monandrous species, both males and females should evolve to reduce gamete wastage in the female, and egg infertility should be reduced. Within-ejaculate competition among sperm should be weak, as long as individual sperm genotypes are not manifested phenotypically. Termites (with generally strictly monandrous females) and parasitic wasps (with strongly female-biased but facultatively adjustable sex ratios indicative of probable female monandry plus probably very precise control of sperm access to eggs; Godfray 1994) are possible test groups. At the opposite extreme, females of some bedbugs and their relatives, and other groups such as leeches with hypodermic insemination, appear to have relatively little control of male access to their eggs. All other things being equal (e.g., sperm malformations, tolerance of polyspermy in eggs; I know of no data on these points), such groups would be expected to have higher rates of infertile eggs.

8.5 "Mistimed" Matings

If female roles in sperm movement, storage, and fertilization are relatively passive, then the timing of copulations in a number of bird species is apparently inappropriate. In birds, each egg in a clutch except the first is thought to be fertilized during the approximately 1 hr after the previous egg was laid; during the rest of the 24 hr until the next egg is laid the oviduct is blocked by the descending egg. When eggs are laid early in the morning (often the case), fertilization also occurs in the early morning. Experiments using genetic markers show that the last male to mate has a greater chance of fertilizing a female's eggs (Birkhead and Møller 1992). Thus if females are passive, one might expect that (1) males are especially assiduous in copulation attempts early in the morning; (2) that they begin mating attempts soon before the female begins to lay eggs; and (3) (especially in species in which females sometimes copulate with extra-pair males) that they continue to attempt to mate with them at least until the day before the last egg is laid. These predictions are often not fulfilled.

In the relatively well-studied red-winged blackbird *Agelaius phoeniceus*, for instance, females sometimes copulate with extrapair males, and about 20% of all nestlings are of extra-pair paternity (Westneat 1993). Thus a pair male runs a real risk that a female with whom he is paired carries sperm from another male. His best strategy would seem to be to copulate assiduously just after each morning's oviposition, throughout the female's egg-laying period. Instead, both male copulation attempts and female solicitations of copulation peaked 2–3 days before the first egg was laid and declined precipitously thereafter, reaching very low levels long before the end of the female's fertile period (see fig. 1.18; Westneat 1993). Why do red-winged blackbird males make such gross "mistakes" in timing their copulations? Males lack intromittent genitalia, and both sperm storage and fertilization sites are relatively deep within the female's body, so female transport is probably important in determining the fate of a male's sperm. Perhaps female bias in sperm transport is responsible. If, for example, females fail to store sperm from copulations after day 0, or to move them to fertilization sites, there will be no advantage for males to continue to copulate with them. In other words, a female's "fertile period" may not necessarily be determined passively by the timing of her ovulation and the descent of eggs through her oviduct, as has so often been assumed in the literature on avian reproduction (summary of literature in Birkhead and Møller 1992).

An alternative explanation, that the males refrain from further copulations because they are saving ejaculates for other copulations (Westneat 1993), seems unlikely in *A. phoeniceus*, in view of the relatively high average number of copulations per clutch (about twenty). More likely, this is another "rules of the game" situation that involves female-imposed limitations in either acceptance of copulations (as also intimated by Westneat 1993, though discussed from the point of view of a male attempting to overcome female resistance), or in acceptance and use of sperm, or both. The probable general importance of female transport of sperm in birds is emphasized by the fact that male birds often do *not* tend to copulate during the female "fertilization window" following each oviposition (Birkhead and Møller 1993b), when the empty oviduct would give direct male mechanisms of sperm movement (injection of sperm into the female, sperm motility) a chance to bring sperm to fertilization sites.

A cross-specific comparison of the strength of the correlation between periods of female fertility and the timing of copulations did not support several possible explanations, involving other possible partners for the males, interference with incubation, differences in offspring quality, and possible damage to unlaid eggs (Birkhead and Møller 1993b). Female promiscuity (with the resulting possibility of sperm competition) and changes in the effectiveness of copulations in producing fertilizations were possible explanations. Changes in effectiveness could be due to cryptic female choice; they could, for instance, result from changes in female tendencies to transport sperm and take them up in their sperm storage tubules. The fact that species with a greater likelihood of sperm competition showed more sustained series of copulations throughout the female's fertile period was taken as support for the sperm competition hypothesis (Birkhead and Møller 1993b). But this same pattern is expected if the timing of copulations is under sexual selection by cryptic female choice.

Such "mistakes" in timing of copulations are not limited to birds. In some marine turtles, males apparently disappear from the waters near oviposition sites weeks before the females have ceased ovipositing, and the egg fertility at the end of the season is sharply reduced perhaps because of reduced sperm stores (A. Chavez, pers. comm.). The most spectacular example of "mistiming" I know of occurs in the sepsid fly *Sepsis cynipsea*. Males patrol oviposition sites (cowpats) and mount females that arrive to oviposit. The male does not copulate, however, until the female has laid all the eggs in her ovaries and moved off the cowpat (Parker 1972)! As soon as copulation ends, the male dismounts and leaves the female. Similar sequences also occur in other species of this family (K. Schulz, pers. comm.). Again there is reason to suspect that female "rules" are involved. The first stage of oviposition in female sepsids is to move an egg from the ovary to the bursa, where she keeps it while she searches for an oviposition site. Females that have just arrived at a cowpat and have not yet begun to lay eggs sometimes already have an egg in the bursa. An egg in the bursa probably acts (perhaps inadvertently) as a "plug," as it occupies the area where the male introduces his genitalia during copulation and where the mass of sperm accumulates (see fig. 2.1). And, vice versa, the large masses of sperm left in the female by the male may be an obstacle for the descending eggs, thus favoring the usual female resistance to male attempts to copulate before she has laid all her eggs.

8.6 Summary

This chapter explores hypotheses that result from including cryptic female choice in analyses of several puzzling patterns related to sexual reproduction. The expectation from current theory regarding ejaculate sizes is that, except in species in which ejaculates are particularly costly or sperm do not mix within the female storage organs, the male will transfer enough sperm to fill up the female's sperm storage organ. Surprisingly, species with "undersized" ejaculates appear to be common, and in several of these the conventional explanation (male limitation in sperm production) is improbable. Small ejaculates may sometimes result from female limits on the quantity of sperm she will accept from a single male.

Cryptic female choice via female physiological responses to male accessory gland products are particularly interesting because females have often been standardized with respect to some important variables in published studies, and male "stimuli" and their perception by the female have also been precisely controlled. Despite this unusual simplicity for a potential female choice situation, substantial variation in female responses is clearly the rule rather than the exception. Female thresholds in general and perhaps even the criteria themselves in female choice may vary substantially, contrary to the usual discussions of "the" female choice criterion in evolutionary theory.

Several types of data suggest that cryptic female choice seems often to employ predetermined, absolute choice criteria rather than sliding and conditional criteria, perhaps because of the disadvantage to the female of postponing reproduction after she has copulated.

Females, especially in polyandrous species, may have to strike a difficult balance between overly weak barriers to counteract aggressive male adaptations that can cause zygote death, and overly strong barriers that can result in some of their eggs going unfertilized. The common occurrence of egg infertility may be a difficult-to-escape cost of sexual reproduction, due to sexual selection (often by cryptic female choice) on males and their gametes.

9 Evidence Ruling Out Cryptic Female Choice: Is It Common?

This chapter attempts to assess how frequently cryptic female choice can be ruled out for particular species. If it is common for species to have traits that eliminate the possibility of cryptic female choice, then cryptic choice may occur only infrequently. Of course the reverse is *not* true: an absence of data in a particular species justifying rejection of cryptic female choice does not prove that it occurs in that species. This chapter thus sets up a "no win" situation for the basic argument of the book (in contrast to the "no lose" situations described in chap. 7.2).

If it happens that cryptic female choice seldom can be confidently rejected, this rules out one type of evidence against the claim that cryptic female choice is widespread. Since the types of evidence and the species cited in this chapter are to a large extent different from those in previous chapters, the tests here are largely independent of previous evidence.

There are several ways in which the possible occurrence of cryptic female choice could be ruled out:

1. Lack of variation in female processes that affect the chances of paternity of a male with whom she has already begun to copulate.
2. Lack of any correlation between differences in paternity achieved by males that have achieved genitalic coupling and other characters of the male (i.e., variation in paternity is not biased with respect to any male characters).
3. Lack of intraspecific genetic variation in the male characters that are associated with paternity biases in the offspring of females with whom they have begun to copulate.
4. Strict female monandry, at least during any single fertile period.

The search for cases in which cryptic female choice can be shown not to occur confronts the same problem as the search for demonstrations that it does occur—not many species have been studied in the necessary detail.

9.1 Lack of Variation in Female Processes Determining Paternity

If females consistently perform the same reproductive processes that can affect male paternity during and following all copulations, then they are clearly not exercising choice between males. One preliminary indication of whether variation actually occurs is to measure male paternity probabilities. If all males that copulate with females under some standard set of conditions have similar chances of paternity, then no selection among these males is occurring. For instance, if all copulations result in ejaculation and sperm transfer, then no cryptic female choice is occurring via the mechanisms of premature interruptions or prevention of deeper penetration. If equal portions of all ejaculates are transported to the same sperm storage organs, then no cryptic female choice is occurring via differences in sperm transport.

This technique of checking for variation has the problem that it combines male-caused variations with female-caused variations. Suppose, for instance, there is substantial variation in sperm numbers that are stored after different copulations (as often turns out to be the case; see below). The variation could be due to variations in male characteristics such as ejaculate size rather than variations in a female-controlled process such as sperm transport. There are two reasons, however, that the technique is less limited than it seems at first. In the first place, male investments in ejaculates are often relatively minor (see chap. 8.1), and selection against male variants that reduce fertilization success after the male has successfully passed all the many precopulatory obstacles is undoubtedly intense. This gives an a priori reason, admittedly far from absolute, to suspect that females may be responsible for many of the variations noted (e.g., see fig. 5.20).

The second, more important consideration is that female "rules of the game" will often be responsible for many "male-caused" variations that are, in the end, due to female behavior, physiology, and morphology. Suppose, for instance, that larger ejaculate size is advantageous to males because females transport all sperm they receive to a storage or fertilization area that is large enough to allow sperm from different males to mix. The female design (large mixing area) and behavior (consistent transport of the entire ejaculate) are responsible, one step farther back in the chain of causation, for the "male-controlled" effect of ejaculate size; they result in selection favoring males with larger ejaculates. Different sets of female rules (for instance, "transport only a fixed number of sperm, regardless of the ejaculate size") could result in different selection on males.

This is not to say that all differences in male insemination success are due to females. There are, of course, male malfunctions that are independent of the female (e.g., failure to produce an ejaculate, production of spermatophores incapable of ejecting sperm into the female, provision of less sperm in the ejaculate than the female would have transported to storage, transfer of dead sperm). But in many respects "malfunction" is a relative term, determined to a great extent by the female. For instance, a male whose semen lacks products which stimulate the female's ducts to contract and transport sperm to storage sites may suffer reproductive failure because of the female's dependence on those products (e.g., Davey 1958; Callahan and Cascio 1963; chaps. 3.5, 6.2, 6.4). If there is variation among males with respect to their abilities to elicit female contractions, the female dependence can result in selection on the males.

Thus the fraction of the reproductive failures of copulations that is at least partially due to females may be relatively large. Lists of failures of copulations to result in paternity such as those which follow will tend to overestimate the frequency with which females vary their acceptance and use of different males' gametes; but the overestimates are not as great as might have been supposed.

Variation in Insemination Success

Selection on males is generally thought to favor relatively profligate, indiscriminate insemination tendencies—the more females a male inseminates, the greater his potential reproduction (e.g., Thornhill and Alcock 1983; Bradbury and Andersson 1987; Andersson 1994). Similarly, one would expect especially strong selection on a male to transfer sperm once he has achieved genitalic coupling. This is particularly true for species in which a given male-female pair usually copulates only once before separating. On the other hand, several possible mechanisms of cryptic female choice involve at least partial prevention of ejaculation, sperm transfer, or sperm transport (chaps. 3.2, 3.4, 3.5, 4.4, 4.8, 4.9).

Copulations which fail to result in sperm transfer to storage are quite common in different animal groups (e.g., fig. 9.1). The sample of species in table 9.1, which was compiled incidentally while searching for other information, suggests that such failures are, in fact, relatively frequent in a wide range of taxonomic groups. The average failure rate is 21.2% (range 0–80%, median 13.2%) for forty-four species in sixteen different orders; 5% or more of the copulations failed to result in any sperm being stored in the female in 75% of these species. In some species male malfunction was undoubtedly responsible for some failures but not all of them. For example, in the caddis fly *Mystacides azurea* older males (with more wing wear, and thus probably more likely to have mated previously and to have decreased reserves of seminal material) were significantly more likely to participate in the total of 28% of all copulations that were failures; but some males with little wing wear also failed to inseminate females (Petersson 1989).

Other compilations have given similar results. Birkhead and Møller (1992) noted a 30–50% rate of sperm transfer failure in three species of birds. Ridley (1988) found that in ten of eleven other insect species >9% of copulations were "impotent" (females were "infertile or had low fertility"). The average frequency of "impotent" copulations for these eleven species

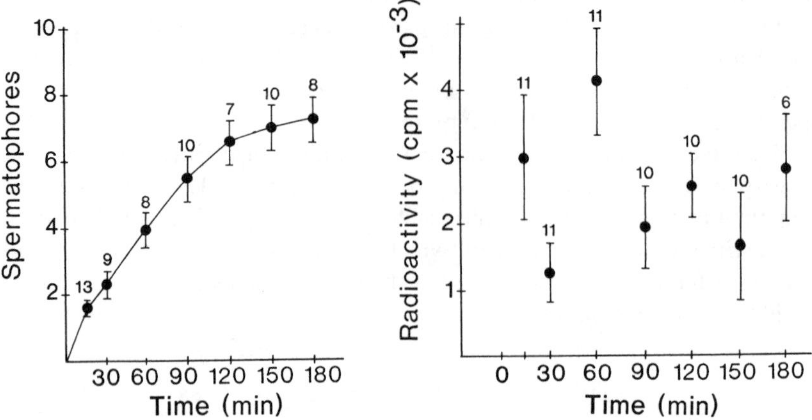

Figure 9.1 High variance in insemination success despite low variance in spermatophore transfer suggests that sperm transport may not be consistent in the grasshopper *Melanoplus sanguinipes*. Pairs usually copulate for 1–4 hr, during which time the male transfers an average of about 7 spermatophores. Despite the machinelike regularity of spermatophore transfers during a copulation (*left*) (note small standard error bars), the amount of radioactively labeled male material which accumulates in the spermatheca is highly variable (*right*; note lack of a clear trend and large standard errors). (After Cheeseman and Gillott 1989)

Table 9.1
Percentages of "Failed" Copulations Involving Normal Males and Females That Fail to Result in Sperm Transfer, or in Production of Fertile Offspring (sample sizes in parentheses)

Group	*Percentage Failed*	*Sperm Checked?*	*References*
Nematoda			
Rhabditis (2 spp.)	8	Y	Rehfeld and Sudhaus 1985
R. pellio[a]	47, 35	Y	Somers et al. 1977
Coenorhabditis elegans	≥6 (16)	Y	Ward and Carrel 1979
Panagrellus redivivus	"Occasional"	Y	Duggal 1978
Araneae			
Latrodectus mactans	29.4 (17)	N[b]	Breene and Sweet 1985
Frontinella pyramitela	8 (39)	N[c]	Suter and Parkhill 1990
Acari			
Ornithodoros tholozani	up to 100[d]	Y	Feldman-Muhsam 1986
Crustacea			
Sicyonia (2 spp.)[e]	"several"(25)	Y	Bauer 1992
Inachus phalangium	0 (3)	Y	Diesel 1990
Penaeus vannamei	80 (10)	Y[f]	Yano et al. 1988
Chionoecetes bairdi	0 (60)[g]	Y	Adams and Paul 1983

Table 9.1 (*cont.*)

Group	*Percentage Failed*	*Sperm Checked?*	*References*
Diptera			
Sabethes cyaneus	5 (20)	Y	Hancock et al. 1990
Scathophaga stercoraria	>30 (6)[h]	Y	Ward 1993
Cyrtodiopsis whitei	23.2 (56)[i]	Y	Lorch et al. 1993
Drosophila robusta	70 (10)	Y	Markow and Ankney 1988
Glossina pallidipes	14.4 (201)[j]	Y	Jaenson 1979
Hylemya brassicae	16.8 (238)[k]	N	Swailes 1971
Lucilia cuprina	4.8 (84)[l,m]	Y	Smith et al. 1988
Aedes aegypti	0 (20)	Y	Jones and Wheeler 1965
Ceratitis capitata	22.4 (48)[n]	N	Saul and McCombs 1993
	25.4 (114)[o]	Y	Seo et al. 1990
Musca domestica	35 (112)[p]	Y	Leopold and Degrugillier 1973
Coleoptera			
Diabrotica virgifera	0 (20)	Y	Lew and Ball 1980
Anthonomus grandis	8.7 (46)[q]	Y	Villavaso 1975b
Onthophagus binodis	0 (22)	N	Cook 1990
Bolithotherus cornutus	0 (22)	N[f]	Conner 1989
Callosobruchus maculatus	0.7 (134)[r]	Y	Eady 1994a,b
Labidomera clivicollis	12 (33)[s]	Y	Dickinson, in press
Lepidoptera			
Helicoverpa virescens	28.6 (213)	Y	Raulston et al. 1975
	9.8 (36)	Y	Proshold and LaChance 1974
H. subflexa	67.5 (46)	Y	Proshold and LaChance 1974
H. zea	approx. 10 (50)[t]	Y	Snow et al. 1972
Utetheisa ornatrix	21.2 (33)	N[u]	LaMunyon 1995
Carpocapsa pomonella	9.4 (265)	Y[f]	Howell et al. 1978
Apantesis phalerata	42 (57)	N	Bacheler and Habeck 1974
A. radians	30 (20)	N	Bacheler and Habeck 1974
Atteva punctella	34 (100)	N[v]	Taylor 1967
Ephestia kuhniella	3.7 (54)[w]	Y	Norris 1933
Trichoptera			
Mystacides azurea	28 (49)	Y	Petersson 1989
Orthoptera			
Eyprepocnemis plorans	0 (10)	Y[x]	Lopez-Leon et al. 1994
Zoraptera			
Zorotypus barberi	15 (20)	Y	J. Choe 1995
Hemiptera			
Nezara viridula	5.3 (19)	Y	Kon et al. 1993
N. antennata	4.8 (21)	Y	Kon et al. 1993
Pisces			
Lebistes reticulatus	67.8 (28)[y]	Y	Clark and Aronson 1951

Table 9.1 (*cont.*)

Group	Percentage Failed	Sperm Checked?	References
Amphibia			
Desmognathus fuscus	29.4 (34)[z]	Y	Verrell 1994
Rodentia			
Meriones unguiculatus	30 (13)[a1]	N	Agren 1990
Rattus norvegicus	5[b1]	N	Chester and Zucker 1970
Primates			
Cercopithecus aethiops	12 (33)[c1]	N	Gartlan 1969
Aves			
Streptopelia risoria	70 (11)[d1]	N	Cheng et al. 1981

NOTES: Partial insemination failures, in which some but less than a normal amount of sperm are transferred, are counted as successful. Data for an additional 13 species of Lepidoptera can be found in Drummond 1984, and 8 additional species in different orders of insects in Ridley 1988.

[a] 47% of the females failed to produce offspring; in 35% no sperm was transferred. Some males that failed in one copulation later succeeded in another (of an average of 21.3 copulations/male, only an average of 11.8 resulted in insemination of the female).

[b] Percentage of egg clutches that were infertile.

[c] 8% of copulations did not last the 15 min necessary for complete insemination.

[d] Males less than two weeks old copulate up to five times in this period but never transfer sperm; thereafter they apparently transfer sperm relatively consistently (but not always) unless starved.

[e] In seven pairs the male copulated on one side that received no sperm, giving a failure frequency of 14%.

[f] Checked only for the presence of a spermatophore, so may be an underestimate of failure to get sperm into the spermatheca.

[g] After a single clutch of eggs (numbers varied by a factor of about 5) the number of sperm in the female varied by a factor of about 100.

[h] Data read from graph (fig. 2) for females that copulated more than 20 min and laid eggs soon thereafter.

[i] Possibly an underestimate, as all males that failed to transfer sperm in two matings were eliminated from consideration. Data were read from a graph (fig. 3) for all copulations lasting more than 40 sec (all uninterrupted).

[j] Includes both copulations with no jerking behavior (which were shorter—av. 19 ± 31 min) and copulations with jerking behavior (av. 69 ± 21 min).

[k] Percentage of females that did not lay eggs.

[l] Males were virgins

[m] Only 0–200 sperm were transferred, compared with normal average of 3019; 34 of 84 males transferred sperm numbers gave a significant reduction in the frequency of fertilization of eggs (Smith et al. 1988, 1990).

[n] Data from males with two different genetic markers combined. The number is probably an underestimate, as no copulations for less than 15 min were included.

[o] Sperm in spermathecae but not in the ventral receptacle were counted, so failures may be overestimated. Especially long and especially short copulations were more likely to fail to result in sperm transfer.

[p] Sham-operated females.

[q] In 6.9% of twenty-nine females sham-operated when they were less than 24 hrs old, and 11.8% of seventeen untreated females.

[r] Huge variation in numbers—range in twenty-six females was about 8000 to 81,000 sperm in a single copulation.

[s] Pairs had probably copulated 2–3 times, as females were examined after an hour, and they averaged 2.5 copulations/hr. The same failure rate occurred in ten pairs dissected after 24 hr together.

[t] In another 20%, only sterile, apyrene sperm reached the spermatheca.

[u] Proportion of the field-collected females in which no offspring were sired by the last male to mate; in many other females the last male sired only a relatively small proportion of the offspring.

[v] Percentage of females that failed to lay fertile eggs after a single copulation.

[w] At room temperature; at higher temperatures failures were much more common.

was 27% ± 16%. Keep in mind that these are probably severe underestimates of *partial* failures of sperm transfer, since transfer of any moderate amount of sperm was classified as "success" (see, for example, Birkhead and Fletcher 1992 for >100-fold variation in the number of sperm transferred in successful copulations in the zebra finch *Taeniopygia guttata*; further examples are given in table 3.3). In sum, it is clearly common for a substantial proportion of copulations to fail to result in successful sperm transfer.

Variation in Sperm Precedence

There is a substantial literature on the effects of mating order on paternity, when a female mates with more than one male (usually only two; e.g., reviews by Ridley 1988, 1989a, 1989b on insects; Austad 1984 and Eberhard et al. 1993b on spiders; Ginsberg and Huck 1989 on mammals; Birkhead and Møller 1992 on birds). The favored position for a male (first, last) varies, and usually both males father some portion of the offspring (e.g., figs. 1.12, 9.2). If females that mate with several males fail to discriminate among these males on the basis of other traits, then the paternity success of a male may be determined largely by his order of mating with the female. Conversely, if females discriminate via cryptic female choice on other male characters, then substantial variance in P2 values is expected for different males (note, however, that lack of variation in P2 values does not rule out some choice mechanisms, such as variation in oviposition, remating, or investment in offspring). As discussed above, such variation could be due to male as well as female causes.

Again, there is a clear trend, noted by Lewis and Austad (1990) in insects and also present in other taxonomic groups: P2 values tend to show large *intra*specific variation (table 9.2). The average coefficient of variation (standard deviation/mean) was 78.8% (range 0.5–180%, median 77%). In some species, P2 values range from 0 to 100% despite only small samples, such as in the spider *Physocyclus globosus* ($N = 12$), the sand lizard *Lacerta agilis* ($N = 20$), the thirteen-lined ground squirrel *Spermophilus*

---

[x] Embryo chromosomes were checked to be sure parthenogenesis had not occurred.

[y] Data are from "long" copulations (1.3–2.4 sec); a similar percentage of insemination failures occurs in short copulations (at least 67% of eighteen copulations).

[z] Number of females with no sperm mass in the cloaca the next morning/number of females for which a spermatophore was deposited by the male.

[a1] Single ejaculate series.

[b1] Each female received two ejaculations.

[c1] Percentage lack of ejaculatory pause at end of copulation; three of the four copulations occurred out of season, all with nulliparous females.

[d1] All eggs in the nest were infertile; pairs were allowed only one or two copulations.

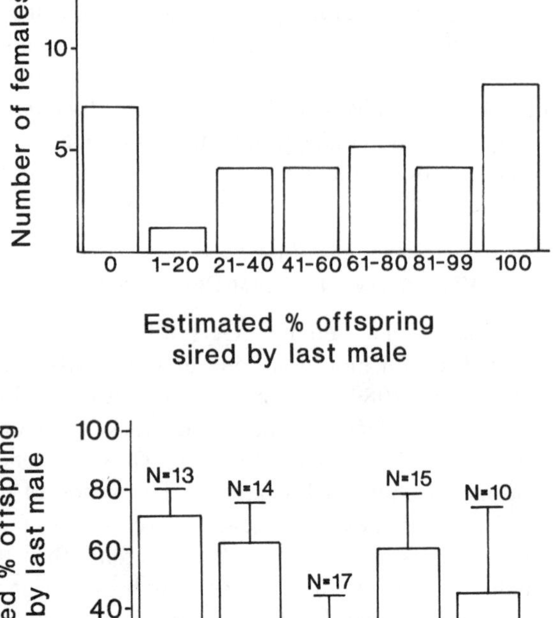

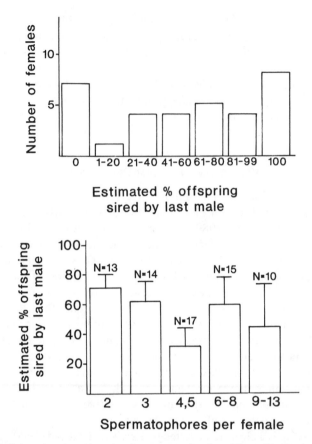

Figure 9.2 Extreme variance in paternity success of males of the moth *Utetheisa ornatrix* in the field. Females were collected in copula, and the proportion of each female's offspring sired by her partner (the "last male") was determined using electrophoretic markers in her offspring. Many females had mated several times prior to being captured. Some last males apparently failed to sire a single offspring, while others apparently sired them all; still others had intermediate success (*top graph*). The previous mating history of the female had little effect on male paternity success (*bottom*). Males failed to transfer spermatophores in only 2.4% and 8.6% of copulations in two series of observations in captivity (N = 82, 58; LaMunyon and Eisner 1993, 1994), thus much of the variation in male success may be due to variation in female responses (such as sperm transport), rather than male failures to transfer spermatophores. (After LaMunyon 1994)

Table 9.2
Intraspecific Variance in an Incomplete Sampler of P2 or Sperm Precedence Values (Percentage of Offspring Fathered by Second of Two Males to Mate with the Same Female) in Animals Other Than Insects

Taxon	*Average P2 ± S.D.*	*Range*	*N*	*Coeff. of Var. (%)*	*References*
Araneae					
Nephila clavipes	21 ± 33.6	0–99	32	160.0	Christenson and Cohn 1988
Phidippus johnsoni	34 ± 45	0–100	22	132.3	Jackson 1980
Physocyclus globosus	38 ± 30	0–100	12	78.9	Eberhard et al. 1993b
Acari					
Argas persicus	58 ± 38[a]	0–99	18	65.5	Sternberg et al. 1973
	14 ± 18[a]	0–69	20	128.6	
Ixodes dammini		0–>76	27		Yuval and Spielman 1990
		0–>76	29		
Macrocheles muscaedomesticae	0.3		410[b]		Yasui 1988
Crustacea					
Orconectes rusticus	95 ± 15[c]	60–100	15	15.8	Snedden 1990
	87 ± 0.5[c]	86–88	?	0.5	
Artemia salina	100[d]	?	?		Bowen 1962
Inachus phalangium	100[d]	?	3		Diesel 1990
Scopiomera globosa	87 ± 16[d]	?	7	18.4	Koga et al. 1993
Nematoda					
Coenorhabditis elegans	?	10–80	?	?	Ward and Carrel 1979
Rodentia					
Mus musculus	27 ± 20[e]	9–86	14	74.0	Levine 1967
	5 ± 9[e]	0–33	14	180.0	
Spermophilus tridecimlineatus	?	0–100	13	?	Schwagmeyer and Foltz 1990
Meriones unguiculatus	79.5 ± 10[f]		13	12.6	Agren 1990
Reptilia					
Lacerta agilis	48 ± 38	0–100	20	79.2	Olsson et al. 1994

NOTES: For many data on insects, which show equally wide intraspecific variations, see Lewis and Austad 1990. N = number of females or number of clutches unless otherwise specified.

[a] Percentages of eggs that hatched when one sterile male and then one fertile male mated, when five sterile males and then one fertile male mated. All clutches had >50 eggs (av. were 91.3 and 106.8 eggs/clutch).

[b] Number of offspring.

[c] Depends on whether sterilized male was first or second.

[d] Female moults after each oviposition, so spermathecal contents may be discarded.

[e] Depends on which genetic marker was first.

[f] Both first and second males were allowed a single ejaculatory series.

tredecimlineatus ($N = 13$), and the hermaphroditic snail *Arianta arbustorum* (Baur 1994), and the insects *Chelymorpha alternans* ($N = 11$; V. Rodriguez 1994a), *Utethesisa ornatrix* ($N = 33$; fig. 9.2; LaMunyon 1994), and *Choristoneura fumiferana* ($N = 15$; Retnakaran 1974; see also Svard and McNeil 1994 on the moth *Pseudaletia unipuncta*, and references in LaMunyon 1994). The causes of this variance remain to be determined in most cases, and it is possible that processes other than female choice are involved. Nevertheless, a lack of variation, which could signal a lack of female discrimination, seems not to be common.

9.2 Lack of Correlation between Paternity and Other Male Characters

Inspection of the possible examples of cryptic female choice in chapters 3 and 4 suggests that often there may be no correlation between male fertilization success and other male characters. If fertilization success varies, but the variation occurs randomly with respect to other male characters, it will not result in selection on the males. For instance, variation in insemination success and P2 values is apparently random with respect to male size in both *Chelymorpha* and *Labidomera* beetles (V. Rodriguez 1994a; V. Rodriguez et al., subm.; Dickinson 1986), and thus will not produce selection on male size. If variation in paternity is not linked to any male character, it will not result in cryptic female choice.

Further consideration, however, shows that the lack of correlations may be an illusion. Most studies did not attempt to find any correlation with male characters, and at best usually tested only one or two relatively crude, easily measured characters (e.g., the size or age of the male). Until further, more detailed tests are made (as in the spider *Neriene litigiosa*, where a combination of preinsemination copulation time, rate of intromissions, and frequency of intromittent failures was correlated with the paternity success of males that mated with nonvirgin females; Watson 1991), it is clearly premature to reject the possibility of correlation with male characters.

In fact, this method for rejecting the hypothesis of cryptic female choice is relatively weak. To show that no male characters are correlated with paternity success, it would be necessary to test all possible male characters. It will thus always be possible to argue that the list of male characters examined is not complete and that the truly important characters have not been examined yet.

A reasonable way to deal with this problem would be first to determine which male traits vary among species, since such divergent traits are likely to have been under sexual selection. If tests that focus on these male traits fail to show the predicted correlations with female responses, they will give strong (though not conclusive) reason to doubt the cryptic female choice hypothesis. I do not know of any such searches. The most reasonable conclusion is that lack of correlation with other male characters is a logically reasonable criterion, but it is difficult to apply in practice. The currently available data, which suggest that this is an important reason to reject the cryptic female choice argument, are not convincing.

9.3 Lack of Intraspecific Genetic Differences

If there is no genetic variation among males in the traits that affect cryptic choice responses of females, then variation in female responses cannot influence the evolution of the males until some genetic variation occurs. Lack of variation in males would not, however, rule out the possibility that cryptic female choice operated in the past; it could, in fact, be responsible for the lack of variation in the present.

One type of data relevant to this question comes from studies of genetic differences in fertilization abilities of males. Two techniques have been widely used: allowing females to copulate with males of different genetic races; and performing artificial insemination with mixed semen from males of different races. In contrast to an earlier conclusion (Walker 1980), I think that the basic pattern from both types of study is that differences among different races are common (figs. 1.12, 9.3; table 9.3). Intraspecific differences associated with different genotypes have been found in at least twenty species of birds, mammals, and insects. In most species, unfortunately, only male differences were tested, without testing for variation in females.

Differences in the species most intensively studied in this regard, *Drosophila melanogaster*, are particularly dramatic. Different genetic strains from nature show differences among males in the number of sperm stored by the female (Gilbert 1981), the rapidity with which they are stored (Gilbert 1981), the survival of sperm once they are stored (DeVries 1964), the ability to gain precedence over sperm already in the female from previous matings (Clark et al. 1995), and the ability to resist displacement by sperm from subsequent matings (Clark et al. 1995). There are polymorphic variants in nearly one-half of the accessory gland proteins, including differences among stocks of flies recently derived from natural populations (Whalen and Wilson 1986). At least some accessory gland proteins have strong effects on female oviposition and remating (see chap. 6).

Table 9.3 might be thought to be biased in favor of genetic differences, however, since a lack of differences may be more likely to go unreported

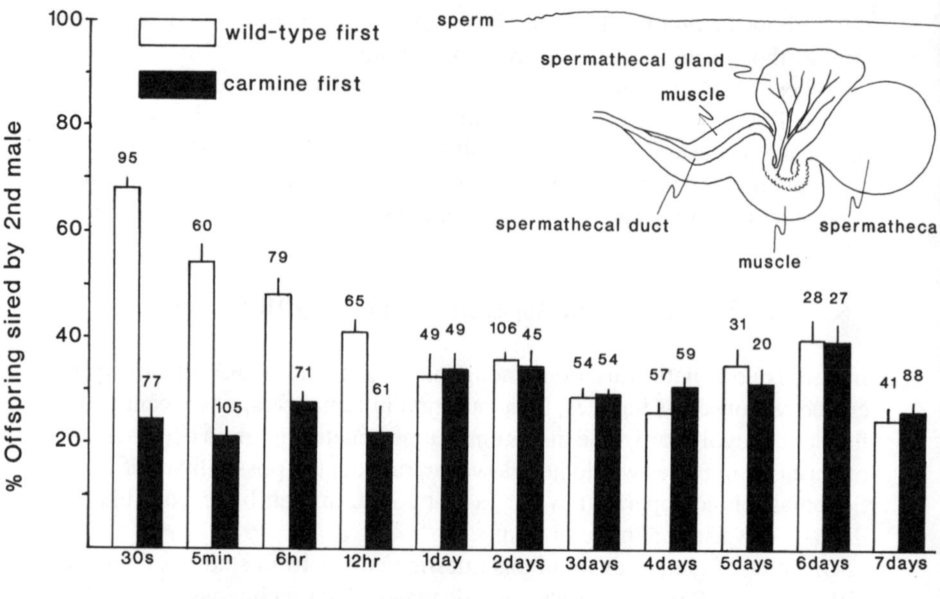

Figure 9.3 Males of the parasitic wasp *Dahlbominus fuscipennis* differ in their competitive abilities. Massive sample sizes (numbers at tops of standard error lines are numbers of females) showed that paternity success with twice-mated females varied dramatically for genetically different males when the two copulations were less than a day apart but did not differ when copulations were separated by a day or more. Drawings at the top (not to the same scale) represent the helical-headed sperm and the female reproductive tract. The muscles along the spermathecal duct are thought to control the sharp bend in the duct that can block sperm movement in or out. (After Wilkes 1965, 1966)

in the literature. Even taking this bias into account, however, the overall bias is probably in the opposite direction, that is, of underestimating male variation. This is because, in all species studied, differences in male fertilizing ability were discovered "accidentally," only as a result of their incidental association with one of a small number of other phenotypic traits (usually alternative eye or coat colors). Focused searches for genetic differences in fertilizing abilities per se would almost certainly reveal even greater degrees of difference in a larger number of species. The only search of this sort I have found (Clark et al. 1995) revealed four alleles at different loci affecting accessory gland proteins that influenced the ability of a male's sperm to resist displacement by subsequent sperm, and showed no correlation between this "defensive" ability and genetic variations in the "offensive" ability to displace resident sperm.

Second, the list of possible cryptic female choice mechanisms checked, even for a well-studied species like *D. melanogaster*, is far from complete.

Table 9.3
A Sample of Species in Which Male Genotype Effects on Sperm Precedence Were Tested When Sperm from More than One Male Are Present in the Same Female

	Tests of Genotype Were Performed by Varying the					
Species	*Male*	*Female*	*Phenotypic Marker*	*Insem. Technique*	*No. of Matings Certain?*	*References*
I. *Genetic effect occurs*						
DIPTERA						
Drosophila melanogaster	Y	Y and N	eye, body color	nat.	Y and N	Prout and Bundgaard 1977; Childress and Hartl 1972; Gromko et al. 1984; Letsinger and Gromko 1985
D. hydei	Y	N	eye color	nat.	Y	Markow 1985
D. pseudoobscura	Y	N	eye color	nat.	?	Dejianne et al. 1978
Ceratitis capitata	Y	N	pupal color	nat.	Y	Saul and McCombs 1993
HYMENOPTERA						
Dahbominus fuscipennis	Y	N	eye color[a]	nat.	Y	Wilkes 1966
Nasonia vitripennis	Y	Y	eye color	nat.	Y	Holmes 1974
HEMIPTERA						
Abedus herberti	Y	N	body color	nat.	Y	Smith 1979
COLEOPTERA						
Tribolium castaneum	Y	Y	body color	nat.	N	Lewis and Austad 1990
T. confusum	Y	Y	body color	nat.	N	Vardell and Brower 1978
ORTHOPTERA						
Blatella germanica	Y	N	eye color	nat.	N	Cochran 1979
RODENTIA						
Mus musculus	Y	N	hair color	AI and nat.	Y	Levine 1967
Mesocricetus auratus	Y	N	hair color	nat.	Y	Oglesby et al. 1981; Ginsberg and Huck 1989; Huck et al. 1985; Lanier et al. 1979
Peromyscus maniculatus	Y	N	hair color	nat.	Y	Dewsbury and Baumgardner 1981

Table 9.3 (*cont.*)

Species	*Tests of Genotype Were Performed by Varying the* Male	Female	*Phenotypic Marker*	*Insem. Technique*	*No. of Matings Certain?*	*References*
Microtus ochrogaster	Y	N	hair color	nat.	Y	Dewsbury and Baumgardner 1981
Rattus norvegicus	Y	N	hair color	nat.	Y	Lanier et al. 1979
II. *No genetic effect occurs*						
COLEOPTERA						
Chelymorpha alternans	Y	N	body color	nat.	Y	V. Rodriguez 1994a; V. Rodriguez et al., subm.
Tribolium castaneum	Y	N	body color	nat.	N	Schlager 1960
HEMIPTERA						
Nezara viridula	Y	N	body color	nat.	Y	McLain 1980
HYMENOPTERA						
Apis mellifera	Y	N	body color	AI, nat.	Y/N	Page 1986; Raphael de Almeida, unpub. (see fig. 4.9)
RODENTIA						
Meriones unguiculatus[b]	Y	N	hair color	nat.	Y	Agren 1990

NOTES: AI = artificial insemination; nat = natural matings. In those cases in which the number of matings was not certain (N), it is possible that differences stemmed from one genotype copulating more frequently than another with same female. Unless otherwise specified, only two alleles or strains were compared.

[a] Advantage of mutant males is thought to be associated with increased ability of sperm to enter the spermatheca soon after copulation.

[b] Apparently test was not extensive, as author was attempting to avoid possible genetic effect.

For instance, in order to be confident that genetic differences in the abilities of male *D. melanogaster* to influence cryptic female choice did not exist, one would have to check, in addition to the data in table 9.3, abilities to avoid premature interruption of copulations, and to induce more rapid oviposition, greater numbers of eggs laid, longer refractory periods, larger eggs, and to bias P2 values favorably when the female mates multiply at different time intervals.

Thus, at least in the small set of species that has been studied, genetic variation among males with respect to characters potentially related to cryptic female choice appears to be common, even when male abilities to influence only a fraction of cryptic female choice mechanisms are considered.

9.4 Female Monandry

If a female copulates with only a single male, she will be unable to exercise cryptic choice. The only theoretical exceptions I know of would be female choice among sperm from a single male (chap. 4.6), or a facultatively parthenogenetic female that could use or refrain from using sperm from a single male (W. Brown, in prep. a). I know of no concrete examples of either.

Despite many confident statements in the literature regarding the monandry or polyandry[1] of the females of many species, definitive data on this point are both scarce and relatively difficult to obtain. It is generally impossible to follow undisturbed females in the field for their entire reproductive lives to determine whether they mate with more than a single male. A relatively low frequency of multiple matings could be enough to result in strong selection on males. If, for instance, males competed to induce females to refrain from remating and if second males achieved strong fertilization precedence, even a low rate of female polyandry could exercise appreciable selection in a male's ability to reduce subsequent receptivity. Even genetic tests of the offspring can only give underestimates, because copulations that did not result in fertilizations will not be noted.

Observations in captivity both for and against monandry are unconvincing because of the possible effects of unnatural conditions on encounter frequencies and female receptivity. The difficulty of determining mating frequency in the field is illustrated in several bird species by the abandonment of former confident assertions that females are monandrous in light of further data proving polyandry (Birkhead and Møller 1992). Similar revisions have occurred in a number of other relatively well-studied species: Young and Downe (1982, 1983) on the mosquitoes *Aedes aegypti* and *Culex tarsalis*; Bergh et al. (1992) on the Hessian fly *Mayetiola destructor*; Allen et al. (1994) on the parasitic wasp *Aphytis melinus*; Waddy and Aiken (1991) on the American lobster *Homarus americanus* (fig. 9.4); Beninger et al. (1993) on the snow crab *Chionoecetes opilio*; Paul (1984) on the tanner crab *C. bairdi*; Schwartz et al. (1989) on the garter snake *Thamnophis sirtalis*; Palombit (1994) and Reichard (1995) on the gibbons *Hylobates syndactylus* and *H. lar*; Chism and Rowell (1986) and Rowell (1988) on *Cercopithecus* and *Erythrocebus* monkeys; Sussman and Garber (1987) on tamarins and marmosets. Even a species such as *T. sirtalis*, in which males leave a copulatory plug and apply antiaphrodisiac substances to the female and the female rapidly leaves the area where mating occurs, most females produced multiply-sired broods (an estimated 72%; Schwartz et al. 1989); and this, of course, represents a lower estimate for the frequency of remating, given the probable occurrence of some infertile matings.

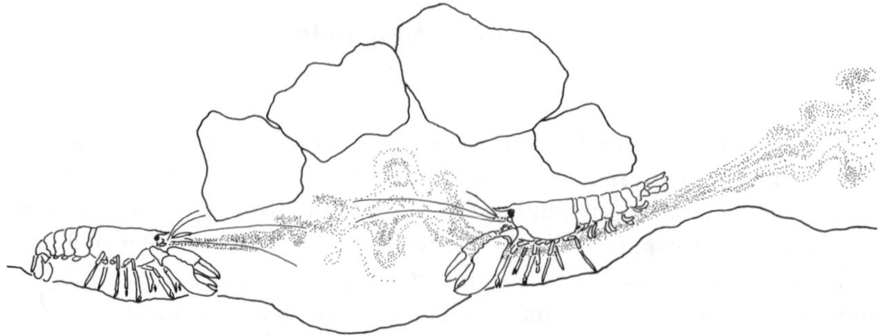

Figure 9.4 A relatively well-studied species, the American lobster *Homarus americanus*, in which former statements that females mate only once were disproven by later evidence. Similar abandonment of earlier claims of female monandry has occurred in a number of other relatively well known species. Female polyandry seems often to result from females actively seeking additional matings rather than having them forced on them by aggressive males. Here the female lobster (*left*) begins an interaction by releasing pheromones that are carried to the male in his retreat by the current he produces with his pleopods. (After Atema 1986)

Overly categorical statements have clearly been a problem in the literature on remating. For instance, researchers repeatedly stated that female *Aedes aegypti* mosquitoes are monandrous (Craig 1967; Fuchs et al. 1969; Gwadz et al. 1971), despite the fact that the same authors and others had shown that females consistently refused to remate only if sufficient time had elapsed after the first mating (Spielman et al. 1967; Craig 1967). Later discoveries showed that remating also often occurs after prolonged delays if oviposition is allowed to occur (Young and Downe 1982; it had been prevented in previous studies by the rearing conditions).

In general, I believe that for most animals the question of how often females remate in the field is still undecided (see discussion in Eberhard 1985). As noted by Drummond (1984), the difficulty of documenting remating in the field can lead to underestimates due to ignorance.

Of the data available, there are two types which give the best estimates of the lower limits of the frequency of female remating. In some species, successful copulation results in the female acquiring an object inside her body (generally a piece of a spermatophore, or of the male's own genitalia) or a mark or scar on the outside. By counting such signs of mating on field-collected females, one can estimate the number of times they have mated. A second technique involves paternity exclusion by genetic analysis of a female's offspring. Isozyme and DNA analyses are the two most commonly utilized techniques of this type.

Both of these techniques can give only underestimates of the numbers of males with which a female has mated. When physical markers are used, any mating which did not result in the production of an enduring mark will not be counted: if no spermatophore was produced, the spermatophore was digested quickly or possibly removed by subsequent mates, the male genitalia did not break off, or they had already broken off as a result of a previous copulation (e.g., Breene and Sweet 1985 on the black-widow spider *Latrodectus mactans*). In addition, some fraction of the females that are examined are likely to be younger individuals that would have remated later if they had been given the chance (Drummond 1984).

Similarly, genetic techniques are unable to sense any mating that did not result in the production of at least one sampled offspring; and female rematings (at least in species with continuous egg production) subsequent to the sampling of her offspring will be missed. Genetic studies are also often limited by cost to relatively small samples. A variety of ecological factors can also affect mating frequency, so relatively extensive studies may be needed before complete lack of remating can be assured. Drummond (1984) mentions demonstrations that remating by female butterflies and moths can be affected by seasonality, migratory status, maturity of food plants, temperature, humidity, sunny versus cloudy conditions, female feeding status, male access to plants with pheromone precursors, timing and placement of traps, and sex ratio.

Keeping these biases and limitations in mind, the available evidence seems to favor the idea that strict female monogamy is surprisingly rare. For instance, of thirty-two supposedly monogamous bird species in which >15 broods were checked for paternity, 78% proved to have at least occasional eggs fertilized by nonpair males as well as by the pair male (Birkhead and Møller 1992). A summary of data on spermatophore numbers in seventy-nine species of field-collected moths and butterflies failed to reveal a single species in which females were strictly monogamous (Drummond 1984). Even the relatively small broods of the pseudoscorpion *Cordylochernes scorpioides* usually included offspring of more than one father (74% of 23; D. Zeh and J. Zeh, in prep.).

I think it is also fair to say that recent reviews of behavioral observations generally show substantial numbers of species in which females mate with more than a single male—e.g., Leuthold 1977 on African ungulates; Dixson 1987; Sussman and Garber 1987; Rowell 1988 and Palombit 1994 on primates; Dewsbury 1984 on rodents; Roldan et al. 1992 on other mammals; McKinney et al. 1984 and Birkhead and Møller 1992 on birds; Devine 1984 and Slip and Shine 1988 on reptiles; Halliday and Verrell 1984 and Houck et al. 1985 on salamanders; Constanz 1984 on internally fertilizing poeciliid fish; Austad 1984 on spiders; Feldman-Muhsam 1986 on ticks; Thomas and Zeh 1984 on other arachnids; Starr 1984 on hymenopterans; Page 1986 on social insects; Drummond 1984 on lepidopterans; Arnqvist, in press, on water striders; Wedell 1993 on katydids; Ridley 1988, 1989a on insects in general.

It is important to note that the discovery of widespread female polyandry in some groups also has implications for the discovery of polyandry in other as yet unstudied groups. There are several reasons to suppose that courtship and copulation are costly for females (increased risk of predation, increased risk of infection by parasite and disease organisms, loss of time and energy; Daly 1978). The existence of polyandry in many well-studied species argues that it is common that one or more of several possible benefits of polyandry (e.g., Walker 1980; Andersson 1994) is large enough to counterbalance these costs. Presumably, similar benefits of polyandry also occur in many as yet unstudied groups.

This is not to say that there are no species in which females mate with only a single male. Clearly, female monandry occurs in nature. In some species such as the gorilla, a single male is able to defend a receptive female from all other males during any given period of estrus. In others, such as many parasitic wasps, some mites, and other insects, strongly female-biased sex ratios indicate that females usually mate with only a single male (Hamilton 1967; Godfray 1994). Female monogamy also occurs throughout some major groups, such as in the termites, where mated females are physically sequestered in sealed chambers, so strict female monandry is probably nearly universal. In these groups the possibility of cryptic female choice can be confidently rejected. It seems equally clear, however, that female monandry is not the general rule in animals with internal fertilization.

In sum, female monandry does occur in some species and is sufficient reason to rule out nearly all types of cryptic female choice. It is not easy to document, however. It appears to be less prevalent than was previously believed.

9.5 Summary

This chapter utilizes four types of data, largely independent of those in other parts of this book, to assess how often cryptic female choice can be ruled out for a given species or group of species. If such causes for rejection were common, they would give reason to doubt the basic argument of this book that cryptic female choice is a widespread and important phenomenon. The converse is not true: lack of rejection in a particular species does not constitute reason to accept cryptic female choice in that species.

Several tests (variation in insemination success in copulations, variation in sperm precedence values, genetic differences among males in fertilization abilities) failed to show that conditions for rejecting certain types of cryptic female choice are common. These tests were weak, in the sense that they applied only to certain mechanisms of cryptic female choice; even if females showed little or no variance in P2 values, for example, they could still exercise cryptic female choice by ovipositing more, or remating more after some copulations than after others.

Another possible type of data that would justify rejection of cryptic female choice seems at first glance to be common among the species discussed in chapters 3 and 4. A consistent bias in cryptic female choice mechanisms favoring particular types of males has been demonstrated relatively infrequently. Very few tests of this possibility have been made, however. This criterion is probably of limited usefulness, since rejection of the cryptic female choice hypothesis would require that the null hypothesis be proven for all male characters, an impractical undertaking.

Probably the most powerful criterion for rejecting cryptic female choice is female monandry. It is clear that females are monandrous in a substantial number of species. Nearly all types of cryptic female choice can be confidently ruled out in these species (although the possibility of cryptic choice among sperm from a given male remains). The frequency of female monandry in the field is unknown for most species, however, and it is difficult to determine under natural conditions. Many confident statements must be treated with skepticism, as testified by the number of relatively well studied species in which previous characterizations of females as monandrous have later had to be retracted. Current techniques of paternity determination have revealed that polyandry is much more prevalent than previously suspected in several groups. These data suggest that the advantages a female gains from mating with more than one male are greater than previously suspected, which in turn gives reason to expect that female polyandry also occurs in many as yet unstudied groups.

In sum, there is clear evidence to reject cryptic female choice for some species. There is no indication, however, that it can be ruled out on a large scale. The idea that cryptic female choice is common thus survives, though slightly battered, the data presented in this "no win" chapter.

Notes

1. Different authors have used the terms *monandry* and *polyandry* in different ways (Andersson 1994). I use them here as in Thornhill and Alcock (1983): *monandry*—the female copulates with only a single male during any given fertile period or estrus; *polyandry*—the female copulates with more than one male.

10 Summary and Conclusions

Most of this book reduces to a single argument—that cryptic female choice is a common feature of evolution. The likely effects of many possible mechanisms of cryptic female choice on male reproduction have already been recognized by others. For instance, the possibility that females may favor some males over others by refusing (or failing to refuse) remating has undoubtedly been recognized for a long time. The general possibility of cryptic female choice has also already been discussed (Thornhill 1983). The idea that cryptic female choice is widespread is not new either. Indeed, it formed a central thread in my argument that male genitalia are under sexual selection by female choice (Eberhard 1985), and has been discussed at length for plants (e.g., Willson 1979, 1990, 1994; Willson and Burley 1983; Stephenson and Bertin 1983; Queller 1987, 1994). Yet the importance of cryptic female choice for sexual selection has not been generally appreciated, either by those describing the mechanisms or by those attempting to discuss sexual selection.

Some studies of animal reproduction (especially in birds and mammals, e.g., Birkhead and Møller 1992; Ginsberg and Huck 1989) have emphasized that all copulations are not equally likely to result in offspring, and that in particular the timing of copulations with respect to the female's estrous or fertility cycle may have an effect on the male's chances of paternity. But even here the "myth of the passive female" (Batten 1992) dies hard. The assumption has often been that the closer the male's copulation is to the female's "window" of fertility, the greater his chances of fertilization. The possibility of female selectivity via, for example, induced variations in gamete transport or ovulation has only recently begun to be considered seriously (e.g., Lifjeld and Robertson 1992; Lifjeld et al. 1993; Birkhead et al. 1993a).

The idea of cryptic female choice has had little impact on general discussions of sexual selection in animals since its first careful formulation more than ten years ago (Thornhill 1983). Cryptic female choice does not even appear in the index of the massive summary of Andersson (1994), and is not mentioned in many other recent treatments of sexual selection by female choice (e.g., Maynard Smith 1987, 1991; Kirkpatrick 1987; Kirkpatrick and Ryan 1991). When it does appear, its predictions tend to be misrepresented or oversimplified, even by leaders in the field (Thornhill and Sauer 1991; Hunter et al. 1993; Alcock 1994). What has been lacking, and what this book attempts to supply, is a pulling together of the surprising amount of previously dispersed evidence on cryptic female choice.

10.1 Overview of the Arguments

More than twenty mechanisms by which females can exercise cryptic choice were compiled, and at least preliminary evidence that cryptic female choice actually occurs in nature was found in more than one hundred species. Some mechanisms are actually composites of several different processes that may have at least partially independent controls. The list of independent female processes which occur after copulation begins and which could have effects on male reproduction probably numbers more than thirty. The species involved belong to a wide variety of taxa, including nematodes, scorpions, snakes, spiders, rats, fruit flies, and humans. Biased operation of such mechanisms, favoring the reproduction of those conspecific males with some behavioral or morphological characters but not others, has been demonstrated in at least a preliminary way for sixteen of these mechanisms. This assemblage of diverse data reveals both a previously unsuspected variety of female mechanisms, and a striking array of animals in which cryptic female choice may occur.

A second objective was to answer the more difficult question of whether cryptic female choice is a rare biological curiosity, or whether it is a widespread phenomenon with far-ranging consequences. There are theoretical arguments that suggest that cryptic female choice may be relatively likely to evolve and persist. Several different major categories of data also argue that cryptic female choice is widespread. The apparently very common occurrence of male courtship behavior after copulation has begun suggests that there is selection on many males to induce female responses that improve the male's chances of paternity. Incomplete or inconsistent female execution of possible cryptic female choice processes also seems the most likely explanation for the frequently invasive and manipulative action of male seminal products on female reproductive processes, such as induction of oviposition and inhibition of remating. Such effects are widespread in animals such as insects and ticks. The fact that morphological and physiological features of female reproductive tracts often greatly impede access of sperm to storage and fertilization sites, sometimes in quite elaborate ways, suggests selection on females to exercise selective control over paternity. The pervasive pattern of relatively rapid divergent evolution of male genitalia is most easily explained if the genitalia often function as internal courtship organs to influence cryptic female choice.

Much of the evidence on both the existence of different female choice mechanisms and the widespread occurrence of cryptic female choice is suggestive rather than conclusive, however. There is also a strong taxonomic bias in much of it favoring insects and mammals. This book does not by any means provide final proof that cryptic female choice is a common, important evolutionary phenomenon. Such proof can only come from further studies appropriately designed to test for cryptic female choice. Rather than being a definitive summary, this book is an attempt to convince biologists to change the set of hypotheses that they usually include when studying reproductive physiology, behavior, and morphology. I am arguing that cryptic female choice is sufficiently likely to be an important phenomenon that heretofore neglected hypotheses involving cryptic choice should be taken more seriously. If this book stimulates further, better tests of cryptic female choice, it will have largely accomplished its objective. An increased recognition of the importance of both passive and active female roles in biasing paternity may lead to fundamental changes in how we look at animal reproduction and the evolution of reproductive processes.

10.2 Consequences for Sexual Selection Theory

The idea that female choice occurs after copulation has already begun necessitates changes in some basic concepts and habits of thought related to sexual selection. Both courtship behavior and the morphological traits associated with courtship have generally been thought to function to induce the female to allow copulation, i.e., to influence what I have called classic, "overt," or precopulatory female choice. The concept of courtship needs to be expanded. All male behavior prior to, during, and following copulation which serves to induce the female to perform both pre- and postintromission responses that increase the chances that his sperm, rather than those of another male, will fertilize her eggs can reasonably be considered as courtship.

The classic Darwinian criterion for female choice, which is based on the male's ability to induce the female to allow him to copulate, should also be extended to include male abilities to induce critical postintromission reproductive processes in the female. Sexual selection itself should probably be redefined to emphasize that males are competing not for access to females, but for access to their eggs and the opportunity to use them to produce offspring.

Studies that attempt to quantify male reproductive success should abandon (or at least carefully test) the common use of copulation frequency as a measure of male success, as if all copulations are equally likely to produce offspring. Most studies of male reproductive success still assume that copulation is equivalent to fertilization, and that male reproductive success can be measured using male copulation success, despite repeated warnings to the contrary (e.g., Thornhill 1983; Arnold and Wade 1984; Shively and Smith 1985; Waage 1986; Simmons 1987a; Rowell 1988; Lewis and Austad 1990; Lifjeld et al. 1993). For example, a survey of recent issues of *Animal Behaviour* (vols. 42–44) showed twenty-seven studies in which arguments relating to selection on particular male behavioral or morphological characters or on female selectivity were based on male copulation success, and only three in which paternity was used. The possibility cannot be ignored that some males are particularly good at obtaining copulations but nevertheless poor at siring offspring, or that some are only average at obtaining copulations but are especially good at siring offspring. Some rejections of the possibility of sexual selection (e.g., Hafernik and Garrison 1986) are premature until possible cryptic female choice has been checked. A similar need for attention to offspring rather than matings has also been emphasized in discussions of sexual selection in plants (Stephenson and Bertin 1983).

Of course, mating success is much easier to determine than paternity, and mating success is undoubtedly often strongly correlated with paternity success. But the payoffs for attention to the possibility of cryptic female choice may be considerable. It can provide explanations for otherwise puzzling traits in behavior (such as copulatory courtship), physiology (such as invasive action of male accessory gland substances on female reproductive physiology), and morphology (such as the extreme elongation of ducts giving access to sperm storage sites in the female).

The existence of cryptic female choice neither negates nor diminishes the importance of either classical, precopulatory female choice, or of sexual selection by direct male-male conflicts. Rather, these different aspects of sexual selection may often work in concert. A male may often be "tested" or compete sequentially with respect to his ability to physically (or sneakily) overcome other males and gain access to females; then for his ability to induce females to allow him to initiate copulation; and then for his ability to induce females to respond behaviorally and physiologically in ways that increase his likelihood of paternity (see fig. 1.6).

Cryptic female choice links evolutionary theory with a large body of studies on reproductive physiology which has heretofore been largely unconnected to evolutionary theory. This link (see also Gomendio and Roldan 1993) promises to foster the growth of an "evolutionary physiology" of male-female interactions, similar to other synthetic fields such as ecological physiology and evolutionary medicine (Williams and Nesse 1991; Ewald 1994). Such approaches may have important practical consequences for attempts to manipulate female reproductive physiology such as artificial insemination of humans and other animals (Jöchle 1975; Eberhard 1991b).

The hypothesis of cryptic female choice makes predictions which may fruitfully orient future studies in reproductive physiology. I predict, for example, that experimental injection of male accessory glands, their extracts, and other male seminal products will probably often result in dramatic effects on female reproduction in groups such as spiders, crustaceans, birds, and mammals, just as they do in insects and ticks. I also predict that seminal products will often prove to differ among closely related species.

If it is true that copulatory courtship functions to influence cryptic female choice, then the kinds of behavior that males perform as copulatory courtship strongly suggest that indirect female benefits from discrimination among males are often obtained through acquisition of "good attractiveness genes" for their male offspring. This conclusion is in accord with data on genitalic evolution (Eberhard 1993d) but differs from the conclusions of several recent publications (e.g., Kodric-Brown and Brown 1984; Kirkpatrick and Ryan 1991; Maynard Smith 1991). One possibly important factor is that many (probably most) previous studies of precopulatory female choice have involved "public" male signals that operate over relatively long distances, such as frog calls or peacock tails (see summary in Andersson 1994). The male signals that have been associated with cryptic female choice are, in contrast, relatively "private" or "intimate," operating at relatively short distances, such as the "silent songs" of some insects (Henry 1994). Private signals are likely to be subject to somewhat different selection. For instance, the benefits of greater intensity, and thus of communication at greater distances (e.g., Ryan and Keddy-Hector 1992), may be smaller, while the costs to the male from increased predation and interference from other males are likely to be smaller.

The frequent assumption that males have active roles while females are passive during male-female interactions (Batten's "myth of the passive female") also needs to be discarded in studies of the behavioral ecology, physiology, and morphology of reproduction. Both the planning and the execution of observations, and also the development of theory for their interpretation, should abandon preconceptions of female passivity. The idea of cryptic female choice focuses attention on female-imposed "rules of the game," and raises the possibility that these rules have taken certain forms and not others because they result in particular biases in male-female interactions which are advantageous to females.

Abandoning the idea that females are morphologically and behaviorally passive and inflexible in male-female interactions promises to give a more complete understanding of sexual selection (e.g., Lifjeld et al. 1993; Birkhead et al. 1993a). The recent flood of evidence that female choice occurs in many different groups (e.g., Andersson 1994) makes one forget that its very existence was considered controversial less than twenty years ago

(Maynard Smith 1976; Taylor and Williams 1982; Searcy 1982; Thornhill 1983). Recognition of the importance of cryptic female choice is one more step in the continuing widespread reevaluation of the possibility of active female roles in a number of fields, including anthropology and primatology (Hrdy 1981; Batten 1992).

References

Abele, L. G., and S. Gilchrist. 1977. Homosexual rape and sexual selection in acanthocephalan worms. *Science* 197:81–83.

Adams, A. E., and A. J. Paul. 1983. Male parent size, sperm storage and egg production in the crab *Chionoecetes bairdi* (Decapoda, Majidae). *Int. J. Invert. Reprod.* 6:181–187.

Adams, T. S., and D. R. Nelson. 1968. Bioassay of crude extracts for the factor that prevents second matings in female *Musca domestica. Ann. Ent. Soc. Am.* 61:112–116.

Adiyodi, K. G. 1988. Annelida. In K. G. and R. G. Adiyodi, eds., *Reproductive Biology of Invertebrates*, vol. 3, *Accessory Sex Glands*, 189. Wiley, New York.

Adiyodi, K. G., and R. G. Adiyodi. 1988. Crustacea. In K. G. and R. G. Adiyodi, eds., *Reproductive Biology of Invertebrates*, 204–213. Wiley, New York.

Adlakha, V., and M.K.K. Pillai. 1975. Involvement of male accessory gland substance in the fertility of mosquitoes. *J. Insect Physiol.* 21:1453–1455.

Adlakha, V., and M.K.K. Pillai. 1976. Role of male accessory gland substance in the regulation of blood intake by mosquitoes. *J. Insect Physiol.* 22:1441–1442.

Adler, N. T. 1969. Effects of the male's copulatory behavior on successful pregnancy of the female rat. *J. Comp. Physiol. Psychol.* 69:613–622.

Aeschlimann, A. 1968. La ponte chez *Ornithodorus moubata* Murray (Ixodoidea, Argasidae). *Rev. Suisse Zool.* 75:1033–39.

Aeschlimann, A., and O. Grandjean. 1973a. Influence of natural and "artificial" mating on feeding, digestion, vitellogenesis and oviposition in ticks (Ixodoidea). *Folia Parasitologica* (Praha) 20:67–74.

Aeschlimann, A., and O. Grandjean. 1973b. Observations on fecundity in *Ornithodorus moubata*, Murray (Ixodoidea: Argasidae). Relationships between mating and oviposition. *Acarologia* 15:206–217.

Agren, G. 1990. Sperm competition, pregnancy initiation and litter size: Influence of the amount of copulatory behaviour in Mongolian gerbils, *Meriones unguiculatus. Anim. Behav.* 40:417–427.

Aguadé, M., N. Miyashita, and C. H. Langley. 1992. Polymorphism and divergence in the *Mst26A* male accessory gland gene region in *Drosophila. Genetics* 132: 755–770.

Ahmad, I., and M. S. Jairajpuri. 1981. The copulatory behaviour of *Cruznema lambdiensis* (Nematoda: Rhabditidae). *Rev. Nematol.* 4:151–156.

Ai, N., S. Komatsu, I. Kubo, W. Loher. 1986. Manipulation of prostaglandin-mediated oviposition after mating in *Teleogryllus commodus. Intl. J. Invert. Reprod. Devel.* 10:33–42.

Aigaki, T., I. Fleischmann, P.-S. Chen, and E. Kubli. 1991. Ectopic expression of sex peptide alters reproductive behavior of female *D. melanogaster. Neuron* 7:557–563.

Aiken, R. B. 1992. The mating behaviour of a boreal water beetle, *Dytiscus alaskanus* (Coleoptera Dytiscidae). *Ethology* 4:245–254.

Alberti, G. 1991. Spermatology in the Acari: Systematic and functional implications. In R. Schuster and P. W. Murphy, eds., *The Acari—Reproduction, Development and Life-History Strategies*, 77–105. Chapman and Hall, London.

Albone, E. S. 1984. *Mammalian Semiochemistry*. Wiley, New York.

Alcock, J. 1979. The evolution of intraspecific diversity in male reproductive strategies in some bees and wasps. In M. S. Blum and N. Blum, eds., *Sexual Selection and Reproductive Competition in Insects*, 381–402. Academic Press, New York.

Alcock, J. 1989. Size variation in the anthophorid bee *Centris pallida*: New evidence on its long-term maintenance. *J. Kans. Ent. Soc.* 62:484–489.

Alcock, J. 1994. Postinsemination associations between males and females in insects: The mate-guarding hypothesis. *Ann. Rev. Ent.* 39:1–21.

Alcock, J. In prep. The relation between male body size and mating success in Dawson's burrowing bee, *Amegilla dawsoni* (Apidae, Apinae, Anthophorini). *J. Zool.*

Alcock, J., and S. L. Buchmann. 1985. The significance of post-insemination display by male *Centris pallida* (Hymenoptera: Anthophoridae). *Z. Tierpsychol.* 68:231–243.

Alcock, J., C. E. Jones, and S. L. Buchmann. 1976. Location before emergence of the female bee *Centris pallida* by its male (Hymenoptera: Anthophoridae). *J. Zool.* 179:189–199.

Alexander, B. 1991. A cladistic analysis of the genus *Apis*. In D. R. Smith, ed., *Diversity in the Genus Apis*, 1–28. Westview Press, Boulder, Colo.

Alexander, R. D. 1959. The courtship and copulation of *Pasimachus punctulatus* Haldemann (Coleoptera: Carabidae). *Ann. Ent. Soc. Amer.* 52:485.

Alexander, R. D. 1964. The evolution of mating behaviour in arthropods. *Symp. Roy. Ent. Soc. Lond.* 2:78–94.

Alexander, R. D., D. Marshall, and J. Cooley. In press. Evolutionary perspectives on insect mating. In J. Choe and B. Crespi, eds., *Social Competition and Cooperation in Insects and Arachnids. I. Evolution of Mating Systems*. Cambridge University Press, Cambridge, U.K.

Allen, G. R., D. L. Kazmer, and R. L. Luck. 1994. Post-copulatory male behaviour, sperm precedence and multiple mating in a solitary parasitoid wasp. *Anim. Behav.* 48:635–644.

Altmann, J., G. Hausfater, and S. A. Altmann. 1988. Determinants of reproductive success in Savannah baboons, *Papio cynocephalus*. In T. H. Clutton-Brock, ed., *Reproductive Success—Studies of Individual Variation in Contrasting Breeding Systems*, 403–418. University of Chicago Press, Chicago.

Alvarado, G. 1992. Comportamiento reproductivo en trés especies de garzas: *Cochlearius cochlearius* (Linnaeus), *Agamia agamia* (Gmelin) y *Butorides striatus* (Linnaeus) (Ardeidae) en Westfalia—Limon, Costa Rica, 1988–1989. Master's thesis, University of Costa Rica.

Anderson, D. J. 1990. Evolution of obligate siblicide in boobies. I. A test of the insurance-egg hypothesis. *Am. Nat.* 135:334–350.

Andersson, M. 1982. Female choice selects for extreme tail length in a widowbird. *Nature* 299:818–820.

Andersson, M. 1994. *Sexual Selection*. Princeton University Press, Princeton.

Andren, C., and G. Nilson. 1987. The copulatory plug of the adder, *Vipera berus*: Does it keep sperm in or out? *Oikos* 49:230–233.

Arnold, S. J. 1972. The evolution of courtship behavior in salamanders. Ph.D. diss., University of Michigan, Ann Arbor.

Arnold, S. J. 1994. Is there a unifying concept of sexual selection that applies to both plants and animals? *Am. Nat.* 144 (suppl.):S1–S12.

Arnold, S. J., and M. J. Wade. 1984. On the measurement of natural and sexual selection: Applications. *Evolution* 38:720–734.

Arnqvist, G. 1989. Sexual selection in a water strider: The function, mechanism of selection and heritability of a male grasping apparatus. *Oikos* 56:344–350.

Arnqvist, G. 1994. The cost of male secondary sexual traits: developmental constraints during ontogeny in a sexually dimorphic water strider. *Am. Nat.* 144:119–132.

Arnqvist, G. In press. The evolution of water strider mating systems: Causes and consequences of sexual conflicts. In J. C. Choe and B. J. Crespi, eds., *Social Competition and Cooperation in Insects and Arachnids. I. Evolution of Mating Systems*. Cambridge University Press, Cambridge, U.K.

Arnqvist, G., and L. Rowe. 1995. Sexual conflict and arms races between the sexes: A morphological adaptation for control of mating in a female insect. *Proc. Roy. Soc. Lond. B* 261:123–127.

Aron, C., G. Asch, and J. Roos. 1966. Triggering of ovulation by coitus in the rat. *Int. Rev. Cytol.* 20:139–172.

Atema, J. 1986. Review of sexual selection and chemical communication in the lobster, *Homarus americanus*. *Can. J. Fish. Aquat. Sci.* 43:2283–2290.

Austad, S. 1982. First male sperm priority in the bowl and doily spider (*Frontinella pyramitela*). *Evolution* 36:777–785.

Austad, S. 1984. Evolution of sperm priority patterns in spiders. In R. L. Smith, ed., *Sperm Competition and the Evolution of Animal Mating Systems*, 223–250. Academic Press, New York.

Austin, C. R. 1969. Fertilization and development of the egg. In H. H. Cole and P. T. Cupps, eds., *Reproduction in Domestic Animals*, 355–384. Academic Press, New York.

Austin, C. R. 1975. Sperm fertility, viability and persistence in the female tract. *J. Reprod. Fert.* (suppl.) 22:75–89.

Baccetti, B. 1991. *Comparative Spermatology 20 Years After*. Serono Symposia Publications from Raven Press, vol. 75. New York.

Baccetti, B., and B. A. Afzelius. 1976. *The Biology of the Sperm Cell*. Karger, New York.

Bacheler, J. S., and D. H. Habeck. 1974. Biology and hybridization of *Apantesis phalerata* and *A. radians* (Lepidoptera: Arctiidae). *Ann. Ent. Soc. Am.* 67:971–975. *Ann. Ent. Soc. Am.* 67:971–975.

Backwell, P., and N. I. Passmore. 1990. Polyandry in the leaf-folding frog, *Afrixalus delicatus*. *Herpetologica* 46:7–10.

Bagnara, J. T., L. Iela, F. Morrisett, and R. K. Rastogi. 1986. Reproduction in the Mexican leaf frog (*Pachymedusa dacnicolor*). I. Behavioral and morphological aspects. *Occ. Pap. Mus. Nat. Hist., Univ. Kansas* 121:1–31.

Baker, R. R., and M. A. Bellis. 1988. "Kamikaze" sperm in mammals? *Anim. Behav.* 36:936–939.

Baker, R. R., and M. A. Bellis. 1989a. Elaboration of the kamikaze sperm hypothesis: A reply to Harcourt. *Anim. Behav.* 37:865–867.

Baker, R. R., and M. A. Bellis. 1989b. Number of sperm in human ejaculates varies in accordance with sperm competition theory. *Anim. Behav.* 37:867–869.

Baker, R. R., and M. A. Bellis. 1993a. Human sperm competition: Ejaculate manipulation by females and a function for the female orgasm. *Anim. Behav.* 46:887–909.

Baker, R. R., and M. A. Bellis. 1993b. Human sperm competition: Ejaculate adjustment by males and the function of masturbation. *Anim. Behav.* 46:861–885

Barclay, M. R., and D. W. Thomas. 1979. Copulation call of *Myotis lucifugus*: A discrete situation-specific communication signal. *J. Mammal.* 60(3):632–634.

Barker, D. M. 1994. Copulatory plugs and paternity assurance in the nematode *Caenorhabditis elegans*. *Anim. Behav.* 48:147–156.

Barnes, M. 1992. The reproductive periods and condition of the penis in several species of common cirripedes. *Oceanogr. Mar. Biol. Ann. Rev.* 30:483–525.

Barnes, H., M. Barnes, and W. Klepal. 1977. Studies on the reproduction of cirripedes. I. Introduction: Copulation, release of oocytes, and formation of the egg lamellae. *J. Exp. Mar. Biol. Ecol.* 27:195–218.

Barnett, M., and S. R. Telford. 1994. The timing of insemination and its implications for sperm competition in a millipede with prolonged copulation. *Anim. Behav.* 48:482–484.

Barton-Browne, L., P. H. Smith, A.C.M. van Gerwen, and C. Gillott. 1990. Quantitative aspects of the effect of mating on readiness to lay in the Australian sheep blowfly, *Lucilia cuprina*. *J. Insect Behav.* 3:637.

Bateson, P., ed. 1983. *Mate Choice*. Cambridge University Press, Cambridge, U.K.

Batten, M. 1992. *Sexual Strategies*. Putnam's, New York.

Battin, T. J. 1993a. The odonate mating system, communication, and sexual selection. *Boll. Zool.* 60:353–360.

Battin, T. J. 1993b. Revision of the *puella* group of the genus *Coenagrion* Kirby, 1890 (Odonata, Zygoptera), with emphasis on morphologies contributing to reproductive isolation. *Hydrobiologia* 262:13–29.

Bauer, R. T. 1992. Repetitive copulation and variable success of insemination in the marine shrimp *Sicyonia dorsalis* (Decapoda: Penaeoidea). *J. Crust. Biol.* 12(2): 153–160.

Bauer, R. T., and J. W. Martin. 1991. *Crustacean Sexual Biology*, Columbia University Press, New York.

Baumann, H. 1974a. Isolation, partial characterization and biosynthesis of the paragonial substances, PS-1 and PS-2 of *Drosophila funebris*. *J. Insect Physiol.* 20:2181–2194.

Baumann, H. 1974b. Biological effects of paragonial substances PS-1 and PS-2, in females of *Drosophila funebris*. *J. Insect Physiol.* 20:2347–2362.

Baur, B. 1994. Multiple paternity and individual variation in sperm precedence in the simultaneously hermaphroditic land snail *Arianta arbustorum*. *Behav. Ecol. Sociobiol.* 35:413–421.

Beatty, R. A., and N. S. Sidhu. 1969. Polymegaly of spermatozoan length and its genetic control in *Drosophila* species. *Proc. Roy. Soc. Edinburgh* (B)71:14–28.

Belk, D. 1984. Antennal appendages and reproductive success in the Anostraca. *J. Crust. Biol.* 4:66–71.

Belk, D. 1991. Anostracan mating behavior: A case of scramble-competition polygyny. In R. T. Bauer and J. W. Martin, eds., *Crustacean Sexual Biology*, 111–125. Columbia University Press, New York.

Bell, P. D. 1980. Multimodal communication by the black-horned tree cricket, *Oecanthus nigricornis* (Walker) (Orthoptera: Gryllidae). *Can. J. Zool.* 58:1861–1868.

Bella, J. L., R. K. Butlin, C. Ferris, and G. M. Hewitt. 1992. Asymmetrical homogamy and unequal sex ratio from reciprocal mating-order crosses between *Chorthippus parallelus* subspecies. *Heredity* 68:345–352.

Bellis, M. A., R. R. Baker, and M.J.G. Gage. 1990. Variation in rat ejaculates consistent with the kamikaze-sperm hypothesis. *J. Mammal.* 71:479–480.

Bellve, A., K. H. Thomas, R. Chandrika, T. M. Wilkie, and M. I. Simon. 1987. Mammalian spermiogenesis: Assembly and fate of perinuclear theca constituents. In D. W. Hamilton and G.M.H. Waites, eds., *Cellular and Molecular Events in Spermiogenesis*, 217–240. Cambridge University Press, New York.

Beninger, P. G., C. Lanteigne, and R. W. Elner. 1993. Reproductive processes revealed by spermatophore dehiscence experiments and by histology, ultrastructure, and histochemistry of the female reproductive system in the snow crab *Chionoecetes opilio* (O. Fabricius). *J. Crust. Biol.* 13:1–16.

Bentur, J. S., and S. B. Mathad. 1975. Dual role of mating in egg production and survival in the cricket, *Plebeiogryllus guttiventris* Walker. *Experientia* 31:539–540.

Bentur, J. S., K. Dakshayani, and S. B. Mathad. 1977. Mating induced oviposition and egg production in the crickets, *Gryllus bimaculatus* De Geer and *Plebeiogryllus guttiventris*. *Z. ang. Ent.* 84:129–135.

Benz, G. 1969. Influence of mating, insemination and other factors on oogenesis and oviposition in the moth *Zeiraphera diniana*. *J. Insect Physiol.* 15:55–71.

Bergh, J. C., M. O. Harris, and S. Rose. 1990. Temporal patterns of emergence and reproductive behavior of the Hessian fly (Diptera: Cecidomyiidae). *Ann. Ent. Soc. Am.* 83:998–1004.

Bergh, J. C., M. O. Harris, and S. Rose. 1992. Factors inducing mated behavior in female Hessian flies (Diptera: Cecidomyiidae). *Ann. Ent. Soc. Am.* 85(2):224–233.

Berglund, A. 1993. Risky sex: Male pipefishes mate at random in the presence of a predator. *Anim. Behav.* 46:169–75.

Bertram, B.C.R. 1975. Social factors influencing reproduction in wild lions. *J. Zool.* 177:463–482.

Birch, M. C., D. Lucas, and P. R. White. 1989. The courtship behavior of the cabbage moth, *Mamestra brassicae* (Lepidoptera: Noctuidae), and the role of male hair pencils. *J. Insect Behav.* 2:227–239.

Birdsall, D. A., and D. Nash. 1973. Occurrence of successful insemination of females in natural populations of deer mice (*Peromyscus maniculatus*). *Evolution* 27:106–110.

Birkhead, T. R. 1991. Sperm depletion in the Bengalese finch (*Lonchura striata*). *Behav. Ecol.* 4:267–275.

Birkhead, T. R. 1993. Female control of paternity. Abstr. *23rd Int. Congr. Ethol.*, 72.

Birkhead, T. R. 1994. Enduring sperm competition. *J. Avian Biol.* 25:167–170.

Birkhead, T. R., and F. Fletcher. 1992. Sperm to spare? Sperm allocation by male zebra finches. *Anim. Behav.* 43:1053–1055.

Birkhead, T. R., and F. M. Hunter. 1990. Numbers of sperm-storage tubules in the zebra finch (*Poephila guttata*) and Bengalese finch (*Lonchura striata*). *Auk* 107:193–197.

Birkhead, T. R., and C. M. Lessells. 1988. Copulation behaviour of the osprey *Pandion haliaetus*. *Anim. Behav.* 36:1672–1682.

Birkhead, T. R., and A. P. Møller. 1992. Numbers and size of sperm storage tubules and the duration of sperm storage in birds: A comparative study. *Biol. J. Linn. Soc.* 45:363–372.

Birkhead, T. R., and A. P. Møller. 1993a. Female control of paternity. *TREE* 8:100–104.

Birkhead, T. R., and A. P. Møller. 1993b. Why do male birds stop copulating while their partners are still fertile? *Anim. Behav.* 45:105–118.

Birkhead, T. R., and A. P. Møller. 1995. Extra-pair copulation and extra-pair paternity in birds. *Anim. Behav.* 49:843–848

Birkhead, T. R., L. Atkin, and A. P. Møller. 1987. Copulation behaviour of birds. *Behaviour* 101:101–138.

Birkhead, T. R., J. V. Briskie, and A. P. Møller. 1993b. Male sperm reserves and copulation frequency in birds. *Behav. Ecol. Sociobiol.* 32:85–93.

Birkhead, T. R., K. Clarkson, and R. Zann. 1988. Extra-pair courtship, copulation and mate guarding in wild zebra finches *Taeniopygia guttata*. *Anim. Behav.* 35: 1853–1855.

Birkhead, T. R., B. J. Hatchwell, and N. B. Davies. 1991. Sperm competition and the reproductive organs of the male and female dunnock *Prunella modularis*. *Ibis* 133:306–311.

Birkhead, T. R., F. M. Hunter, and J. E. Pellatt. 1989. Sperm competition in the zebra finch, *Taeniopygia guttata*. *Anim. Behav.* 38:935–950.

Birkhead, T. R., A. P. Møller, and W. J. Sutherland. 1993a. Why do females make it so difficult for males to fertilize their eggs? *J. theor. Biol.* 161:51–60.

Birkhead, T. R., M. T. Stanback, and R. E. Simmons. 1993b. The phalloid organ of buffalo weavers *Bubalornis*. *Ibis* 135:326–331.

Blades, P. I. 1977. Mating behavior of *Centropages typicus* (Copepoda: Calanoida). *Mar. Biol.* 40:57–64.

Blades, P. I., and M. J. Youngbluth. 1979. Mating behavior of *Labidocera aestiva* (Copepoda: Calanoida). *Mar. Biol.* 51:339–355.

Blades-Eckelbarger, P. I. 1991. Functional morphology of spermatophores and sperm transfer in calanoid copepods. In R. T. Bauer and J. W. Martin, eds., *Crustacean Sexual Biology*, 246–270. Columbia University Press, New York.

Blandau, R. J. 1969. Gamete transport—comparative aspects. In E.S.E. Hafez and R. J. Blandau, eds., *The Mammalian Oviduct*, 129–162. University of Chicago Press, Chicago.

Blandau, R. J. 1973. Sperm transport through the mammalian cervix: Comparative aspects. In R. J. Blandau and K. Moghissi, eds., *The Biology of the Cervix*, 285–327. University of Chicago Press, Chicago.

Blaney, W. M. 1970. Some observations on the sperm tail of *D. melanogaster*. *Drosoph. Inv. Serv.* 45:125–127.

Blest, A. D., and G. Pomeroy. 1978. The sexual behaviour and genital mechanics of three species of *Mynoglenes* (Araneae: Linyphiidae). *J. Zool. Lond.* 185:319–340.

Blommers-Schlosser, R.M.A. 1975. A unique case of mating behaviour in a Malagasy tree frog, *Gephyromantis liber* (Peracca, 1893), with observations on the larval development (Amphibia, Ranidae). *Beaufortia, Zool. Mus. Univ. Amster.* 296(23):17–21.

Blommers-Schlosser, R.M.A., and L.H.M. Blommers. 1984. The amphibians in Madagascar. In A. Jolly, P. Oberle and R. Albignac, eds., *Key Environments*, 89–104. Pergamon Press, New York.

Blurton-Jones, N. G., and J. Trollope. 1968. Social behaviour of stump-tailed macaques in captivity. *Primates* 9:365–394.

Boatman, D. E. 1990. Oviductal modulators of sperm fertilizing ability. In B. D. Bavister, J. Cummins, and E.R.S. Roldan, eds., *Fertilization in Mammals*, 223–238. Serono Symposia, Norwell, Mass.

Bockwinkel, G., and K. P. Sauer. 1994. Resource dependence of male mating tactics in the scorpionfly, *Panorpa vulgaris* (Mecoptera, Panorpidae). *Anim. Behav.* 47:203–209.

Bodnaryk, R. P. 1978. Structure and function of insect peptides. *Adv. Ins. Physiol.* 13:69–132.

Bonaric, J.-C. 1987. Moulting hormones. In W. Nentwig, ed., *Ecophysiology of Spiders*, 111–118. Springer-Verlag, New York.

Boonstra, R., X. Xia, and L. Pavone. 1993. Mating system of the meadow vole, *Microtus pennsylvanicus*. *Behav. Ecol.* 4:83–89.

Borovsky, D. 1985. The role of the male accessory gland fluid in stimulating vitellogenesis in *Aedes taeniorhynchus*. *Arch. Insect Biochem. Physiol.* 2:405–413.

Boucher, L., and J. Huignard. 1987. Transfer of male secretions from the spermatophore to the female insect in *Caryedon serratus* (Ol.): Analysis of the possible trophic role of these secretions. *J. Insect Physiol.* 33:949–957.

Boulétreau-Merle, J. 1974. Importance relative des stimulations de l'accouplement: Parade, copulation et insémination sur la production ovarienne de *Drosophila melanogaster*. *Bull. Biol. Fr. Belg.* 108:61–70.

Boulétreau-Merle, J. 1976. Destruction de la pars intercérébralis chez *Drosophila melanogaster*: Effet sur la fecondité et sur sa stimulation par l'accouplement. *J. Insect Physiol.* 22:933–940.

Bourne, G. R. 1993. Proximate costs and benefits of mate acquisition at leks of the frog *Ololygon rubra*. *Anim. Behav.* 45:1051–1059.

Bowen, B. J., C. G. Codd, and D. T. Gwynne. 1984. The katydid spermatophore (Orthoptera: Tettigoniidae): Male nutritional investment and its fate in the mated female. *Aust. J. Zool.* 32:23–31.

Bowen, S. T. 1962. The genetics of *Artemia salina*. I. the reproductive cycle. *Biol. Bull.* 122:25–32.

Bownes, M., and L. Partridge. 1987. Transfer of molecules from ejaculate to females in *Drosophila melanogaster* and *Drosophila pseudoobscura*. *J. Insect Physiol.* 33:941–947.

Bradbury, J. W., and M. B. Andersson, eds. 1987. *Sexual Selection: Testing the Alternatives*. Wiley, New York.

Braden, A.W.H., and S. Gluecksohn-Waelsch. 1958. Further studies of the effect of the T locus in the house mouse on male fertility. *J. Exp. Zool.* 138:431–452.

Brady, J. P., and R. C. Richmond. 1990. Molecular analysis of evolutionary changes in the expression of *Drosophila* esterases. *Proc. Natl. Acad. Sci. USA* 87:8217–8221.

Brady, U. E. 1983. Review: Prostaglandins in insects. *Insect Biochem.* 13(5):443–451.

Breene, R. G., and M. H. Sweet. 1985. Evidence of insemination of multiple females by the male black widow spider, *Latrodectus mactans* (Araneae, Theridiidae). *J. Arachnol.* 13:331–335.

Briceño, R. D., and W. G. Eberhard. 1995. The functional morphology of male cerci and associated characters in 13 species of tropical earwigs (Dermaptera: Forficulidae, Labiidae, Carcinophoridae, Pygidicranidae). *Smithson. Contrib. Zool.* 555: 1–63.

Brignoli, P. M. 1971. Su alcune *Tegenaria* d'Ispagna. *Mem. Mus. Civ. St. Nat. Verona* 18:307–312.

Brignoli, P. M. 1972. Some cavernicolous spiders from Mexico (Araneae). *Accad. Naz. Lincei* 171:129–155.

Brignoli, P. M. 1980. Ragni d'Italia XXXIII. Le genere *Robertus* (Araneae, Theridiidae). *Frag. Ent.* 15:259–265.

Bronson, F. H. 1979. The reproductive ecology of the house mouse. *Quart. Rev. Biol.* 54:265–299.

Brousse-Gaury, P., and F. Goudey-Perriere. 1983. Spermatophore et vitellogenèse chez *Blabera fusca* Br. (Dictypotère, Blaberidae). *C. R. Acad. Sci. Paris* 296: 659–664.

Brown, K. S. 1981. The biology of *Heliconius* and related genera. *Ann. Rev. Ent.* 26:427–456.

Brown, L., R. W. Shumaker, and J. F. Downhower. 1995. Do primates experience sperm competition? *Am. Nat.* 146:302–306.

Brown, S. G. 1985. Mating behavior of the golden orb-weaving spider, *Nephila clavipes*. II. Sperm capacitation, sperm competition and fecundity. *J. Comp. Psychol.* 99:167–175.

Brown, W. D. 1993. Cryptic female choice in a courtship feeding tree cricket. Abstr. *23rd Int. Congr. Ethol.*, 103.

Brown, W. D. In prep. (a). Female control of paternity: Conditions favoring post-copulatory female choice and a study of tree crickets.

Brown, W. D. In prep. (b). Remating as mechanism of nutrient acquisition and postcopulatory female choice in the black-horned tree cricket, *Oecanthus nigricornis* (Orthoptera: Gryllidae: Oecanthinae).

Bruce, K. E., and D. Q. Estep. 1992. Interruption of and harassment during copulation by stumptail macaques *Macaca arctoides*. *Anim. Behav.* 44:1029–1044.

Buechner, H. K., and R. Schloeth, R. 1965. Ceremonial mating behavior in Uganda kob (*Adenota kob thomasi* Neumann). *Z. Tierpsychol.* 22:209–225.

Bukowski, T. C., and T. E. Christenson. In prep. Mating in the orbweaving spider *Micrathena gracilis*. II. Factors influencing copulation and sperm release and storage.

Bullini, L., M. Coluzzi, and A. P. Bianchi Bullini. 1976. Biochemical variants in the study of multiple insemination in *Culex pipiens* L. (Diptera, Culicidae). *Bull. Ent. Res.* 65:683–685.

Bunnell, B. N., B. D. Boland, and D. A. Dewsbury. 1977. Copulatory behavior of golden hamsters (*Mesocricetus auratus*). *Behaviour* 61:180–206.

Burk, T., and J. C. Webb. 1983. Effect of male size on calling propensity, song parameters, and mating success in Caribbean fruit flies, *Anastrepha suspensa* (Loew) (Diptera: Tephritidae). *Ann. Ent. Soc. Am.* 76:678–682.

Burke, H. R. 1959. Morphology of the reproductive systems of the cotton boll weevil (Coleoptera, Curculionidae). *Ann. Ent. Soc. Amer.* 52:287–294.

Burkhardt, D., I. de la Motte, and K. Lunau. 1994. Signalling fitness: Larger males sire more offspring. Studies of the stalk-eyed fly *Cyrtodiopsis whitei* (Diopsidae, Diptera). *J. Comp. Physiol. A* 174:61–64.

Burkman, L. J., N. J. Alexander, and G. D. Hodgen. 1990. Criteria for identification of human hyperactivated (HA) sperm: An introduction to visual and computer assisted methods. In B. D. Bavister, J. Cummins, and E.R.S. Roldan, eds., *Fertilization in Mammals*, 425. Serono Symposia, Norwell, Mass.

Burley, N. T., D. A. Enstrom, and L. Chitwood. 1994. Extra-pair relations in zebra finches: Differential male success results from female tactics. *Anim. Behav.* 48:1031–1041.

Burnet, B., K. Connolly, M. Kearney, and R. Cook. 1973. Effects of male paragonial gland secretion on sexual receptivity and courtship behaviour of female *Drosophila melanogaster*. *J. Insect Physiol.* 19:2421–2431.

Busse, C. D., and D. Q. Estep. 1984. Sexual arousal in male pigtailed monkeys (*Macaca nemestrina*): Effects of serial matings by two males. *J. Comp. Psychol.* 98:227–231.

Byers, G. 1961. The crane fly genus *Dolichopeza* in North America. *Univ. Kans. Sci. Bull.* 42:665–924.

Byers, G. 1976. A new Appalachian *Brachypanorpa* (Mecoptera: Panorpidae). *J. Kans. Ent. Soc.* 49:433–440.

Byers, G. 1983. The crane fly genus *Chionea* in North America. *Univ. Kans. Sci. Bull.* 52:59–195.

Byers, G. 1990. *Brachypanorpa sacajawea* n.sp. (Mecoptera: Panorpidae) from the Rocky Mountains. *J. Kans. Ent. Soc.* 63:211–217.

Cade, W. H. 1984. Genetic variation underlying sexual behavior and reproduction. *Am. Zool.* 24:355–366.

Callahan, P. S., and T. Cascio. 1963. Histology of the reproductive tracts and transmission of sperm in the corn earworm, *Heliothis zea*. *Ann. Ent. Soc. Am.* 56:535–556.

Camacho, H. 1989. Transferencia de espermatozoides en la mosca del Mediterráneo *Ceratitis capitata* Wied. (Diptera Tephritidae). Master's thesis, University of Costa Rica.

Carayon, J. 1966. Traumatic insemination and the paragenital system. In R. Usinger, ed., *Monograph of Cimicidae*. Thomas Say Foundation 7, Entomological Society of America, Philadelphia.

Carayon, J. 1975. Insémination extra-genitale traumatique. In P. P. Grasse, ed., *Traité de Zoologie*, vol. 8. Masson, Paris.

Carlson, D. A., and P. A. Langley. 1986. Tsetse alkenes: Appearance of novel sex-specific compounds as an effect of mating. *J. Insect Physiol.* 32:781–790.

Carlson, R. R., and V. J. DeFeo. 1965. Role of the pelvic nerve vs. the abdominal sympathetic nerves in the reproductive function of the female rat. *Endocrinology* 77:1014–1022.

Carré, D., and C. Sardet. 1984. Fertilization and early development in *Beroe ovata. Dev. Biol.* 105:188–195.

Carter, C. S. 1973. Stimuli contributing to the decrement in sexual receptivity of female golden hamsters (*Mesocricetus auratus*). *Anim. Behav.* 21:827–834.

Carter, C. S., and L. L. Getz. 1993. Monogamy and the prairie vole. *Sci. Am.* 268(6):70–76

Chapman, T., L. F. Liddle, J. M. Kalb, M. F. Wolfner, and L. Partridge. 1995. Cost of mating in *Drosophila melanogaster* females is mediated by male accessory gland products. *Nature* 373:241–244.

Charalambous, M., R. K. Buttin, and G. M. Hewitt. 1994. Genetic variation in male song and female song preference in the grasshopper *Chorthippus brunneus* (Orthoptera: Acrididae). *Anim. Behav.* 47:399–411.

Chardine, J. W. 1987. The influence of pair-status on the breeding behaviour of the kittiwake *Rissa tridactyla* before egg-laying. *Ibis* 129:515–526.

Chaudhury, M. F. B., and T. S. Dhadialla. 1976. Evidence of hormonal control of ovulation in tsetse flies. *Nature* 260:243–244.

Cheeseman, M. T., and C. Gillott. 1989. Long hyaline gland discharge and multiple spermatophore formation by the male grasshopper, *Melanoplus sanguinipes. Physiol. Entomol.* 14:257–264.

Chen, D.-Y. 1990. An ultrastructural study on heterofertilization in vitro between giant panda and golden hamster. In B. D. Bavister, J. Cummins, and E.R.S. Roldan, eds., *Fertilization in Mammals*, 441. Serono Symposia, Norwell, Mass.

Chen, P. S. 1984. The functional morphology and biochemistry of insect male accessory glands and their secretions. *Ann. Rev. Ent.* 29:233–255.

Chen, P. S., and R. Buhler. 1970. Paragonial substance (sex peptide) and other free ninhydrin-positive components in male and female adults of *Drosophila melanogaster. J. Insect Physiol.* 16:615–627.

Chen, P. S., and A. Oechslin. 1976. Accumulation of glutamic acid in the paragonial gland of *Drosophila nigromelanica. J. Insect Physiol.* 22:1237–1243.

Chen, P. S., E. Stumm-Zollinger, T. Aigaki, J. Balmer, M. Bienz, and P. Bohlen. 1988. A male accessory gland peptide that regulates reproductive behavior of female *D. melanogaster. Cell* 54:291–298.

Chen, P. S., E. Stumm-Zollinger, and M. Caldelari. 1985. Protein metabolism of *Drosophila* male accessory glands. II. *Insect Biochem.* 15:385–390.

Cheng, M.-F., M. Porter, and G. Ball. 1981. Do ring doves copulate more than necessary for fertilization? *Physiol. Behav.* 27:659–662.

Chenoweth, P. J. 1986. Reproductive behavior of bulls. In D. Morrow, ed., *Current Therapy in Theriogenology*, 148–152. W. B. Saunders, Philadelphia.

Cherr, G. N., and T. Dulcibella. 1990. Activation of the mammalian egg: Cortical granule distribution, exocytosis, and the block to polyspermy. In B. D. Bavister,

J. Cummins, and E. S. Roldan, eds., *Fertilization in Mammals*, 309–329. Serono Symposia, Norwell, Mass.

Chester, R. V., and I. Zucker. 1970. Influence of male copulatory behavior on sperm transport, pregnancy and pseudopregnancy in female rats. *Physiol. Behav.* 5:35–43.

Childress, D., and D. L. Hartl. 1972. Sperm preference in *Drosophila melanogaster. Genetics* 71:417–427.

Chin, D. A., and D. P. Taylor. 1961. Sexual attraction and mating patterns in *Cylindrocorpus longistoma* and *C. curzii* (Nematoda; Cylindrocorporidae). *J. Nematol.* 1:313–317.

Chism, J. B., and T. E. Rowell. 1986. Mating and residence patterns of male patas monkeys. *Ethology* 72:31–39.

Choe, J. C. 1995. Courtship feeding and repeated mating in *Zorotypus barberi* (Insecta: Zoraptera). *Anim. Behav.* 49:1511–1520.

Christenson, T. E. 1990. Natural selection and reproduction: A study of the golden orb-weaving spider, *Nephila clavipes*. In D. A. Dewsbury, ed., *Contemporary Issues in Comparative Psychology*, 149–174. Sinauer, Sunderland, Mass.

Christenson, T. E., and J. Cohn. 1988. Male advantage for egg fertilization in the golden orb-weaving spider (*Nephila clavipes*). *J. Comp. Psychol.* 102:312–318.

Christy, J. H. 1995. Mimicry, mate choice, and the sensory trap hypothesis. *Am. Nat.* 146:171–181.

Christy, J., and W. G. Eberhard. In prep. Evaluating models of sexual selection by female choice: Intrageneric comparisons.

Clark, A. G., M. Aguade, T. Prout, L. G. Harshman, and C. H. Langley. 1995. Sperm displacement and its association with accessory gland protein loci in *Drosophila melanogaster. Genetics* 139:189–201.

Clark, E., and L. R. Aronson. 1951. Sexual behavior in the guppy, *Lebiste reticulatus. Zoologica* 36:49–66.

Clark, M. M., L. Tucker, and B. G. Galef, Jr. 1992. Stud males and dud males: Intra-uterine position effects on the reproductive success of male gerbils. *Anim. Behav.* 43:215–221.

Clarke, J. R., and F. V. Clulow. 1973. The effect of successive matings upon bank vole (*Clethrionomys glareolus*) and vole (*Microtus agrestis*) ovaries. In *The Development and Maturation of the Reproductive Organs and Functions in the Female*, 160–170. Excerpta Medica International Congress Series 267, Amsterdam.

Clements, A. N. 1963. *The Physiology of Mosquitoes*. Macmillan, New York.

Clulow, F. V., and P. E. Langford. 1971. Pregnancy-block in the meadow vole, *Microtus pennsylvanicus. J. Reprod. Fertil.* 24:275–277.

Clulow, F. V., and F. F. Mallory. 1974. Ovaries of meadow voles, *Microtus pennsylvanicus*, after copulation with a series of males. *Can. J. Zool.* 52:265–267.

Clutton-Brock, T. H. 1988. *Reproductive Success: Studies of Individual Variation in Contrasting Breeding Systems*. University of Chicago Press, Chicago.

Clutton-Brock, T. H., and G. A. Parker. 1995. Sexual coercion in animal societies. *Anim. Behav.* 49:1345–1365.

Cochran. D. G. 1979. A genetic determination of insemination frequency and sperm precedence in the German cockroach. *Ent. Exp. Appl.* 26:259–266.

Cohen, A. C., and J. G. Morin. 1990. Patterns of reproduction in ostracodes: A review. *J. Crust. Biol.* 10:184–211.

Cohen, J. 1975. Gamete redundancy—wastage or selection? In D. L. Mulcahy, ed., *Gamete Competition in Plants and Animals*, 99–112. Elsevier, New York.

Cohen, J. 1991. The case for and against sperm selection. In B. Bacetti, ed., *Comparative Spermatology 20 Years After*, 759–764. Raven Press, New York.

Cohn, J. 1990. Is it the size that counts? Palp morphology, sperm storage, and egg hatching frequency in *Nephila clavipes* (Araneae, Araneidae). *J. Arachnol.* 18:59–71.

Cohn, J., F. V. Balding, and T. E. Christenson. 1988. In defense of *Nephila clavipes*: Postmate guarding by the male golden orb-weaving spider. *J. Comp. Psychol.* 102(4):319–325.

Colwell, M. A., and L. W. Oring. 1989. Extra-pair mating in the spotted sandpiper: A female mate acquisition tactic. *Anim. Behav.* 38:675–684.

Comstock, J. H. 1948. *The Spider Book*. Rev. and ed. W. J. Gertsch. Comstock, Cornell University Press, Ithaca, N.Y.

Connat, J-L., J. Ducommun, P. A. Diehl, and A. Aeschlimann. 1986. Some aspects of the control of the gonotrophic cycle in the tick *Ornithodoros moubata* (Ixodoidea, Argasidae). In J. R. Sauer and J. A. Hair, eds., *Morphology, Physiology, and Behavioral Biology of Ticks*, 194–216. Wiley, New York.

Conner, J. 1989. Older males have higher insemination success in a beetle. *Anim. Behav.* 38:503–509.

Connor, J., and D. Crews. 1980. Sperm transfer and storage in the lizard, *Anolis carolinensis*. *J. Morph.* 163:331–348.

Conoway, C. H., and M. W. Sorenson. 1966. Reproduction in tree shrews. In I. Rowlands, ed., *Comparative Biology of Reproduction in Mammals*, 471–492. Symposium Zoological Society of London 15, London.

Constanz, G. D. 1984. Sperm competition in poeciliid fishes. In R. L. Smith, ed., *Sperm Competition and the Evolution of Animal Mating Systems*, 465–485. Academic Press, New York.

Cook, D. F. 1990. Differences in courtship, mating and postcopulatory behaviour between male morphs of the dung beetle *Onthophagus binodis* Thunberg (Coleoptera: Scarabaeidae). *Anim. Behav.* 40:428–436.

Cook, D. F. 1992. The effect of male size on receptivity in female *Lucilia cuprina* (Diptera: Calliphoridae). *J. Insect Behav.* 5(3):365.

Cooke, P. H., R. C. Richmond, and J. G. Oakeshott. 1987. High resolution electrophoretic variation at the esterase-6 locus in a natural population of *Drosophila melanogaster*. *Heredity* 59:259–264.

Cordero, C. 1995. Ejaculate substances that affect female insect reproductive physiology and behavior: Honest or arbitrary traits? *J. theor. Biol.*174: 453–461.

Cordero, A., S. Santolamazza-Carbone, and C. Utzeri. 1995. Male disturbance, repeated insemination and sperm competition in the damselfly *Coenagrion scitulum* (Zygoptera: Coenagrionidae). *Anim. Behav.* 49:437–449.

Cosmides, L. M., and J. Tooby. 1981. Cytoplasmic inheritance and intragenomic conflict. *J. theor. Biol.* 89:83–129.

Costa, F. G., and J. R. Sotelo. 1994. Stereotypy and versatility of the copulatory pattern of *Lycosa malitiosa* (Araneae, Lycosidae) at cool *versus* warm temperatures. *J. Arachnol.* 22:200–204.

Coulhart, M. B., and R. S. Singh. 1988. Differing amounts of genetic polymorphism in testes and male accessory glands of *Drosophila melanogaster* and *D. simulans*. *Biochem. Genet.* 26:153–164.

Cowan, D. 1986. Sexual behavior of eumenid wasps (Hymenoptera: Eumenidae). *Proc. Ent. Soc. Wash.* 88:531–541.

Cox, C. R., and B. J. LeBoeuf. 1977. Female incitation of male competition: A mechanism in sexual selection. *Am. Nat.* 111:317–335.

Coyle, F. A. 1985a. Two-year life cycle and low palpal character variance in a Great Smoky Mountain population of the lamp-shade spider (Araneae, Hypochilidae, *Hypochilus*). *J. Arachnol.* 13:211–218.

Coyle, F. A. 1985b. Observations on the mating behaviour of the tiny mygalomorph spider, *Microhexura montivaga* (Araneae, Dipluridae) Crosby & Bishop. *Bull. Brit. Arachnol. Soc.* 6:328–330.

Coyle, F. A. 1986. Courtship, mating and the function of male-specific leg structures in the mygalomorph spider genus *Euagrus* (Araneae, Dipluridae). In W. Eberhard, Y. Lubin, and B. Robinson, eds., *Proceedings of the Ninth International Congress of Arachnology, Panama 1983*, 33–38. Smithsonian Institution Press, Washington, D.C.

Coyle, F. A., and T. C. O'Shields. 1990. Courtship and mating behavior of *Thelechoris karschi* (Araneae, Dipluridae), an African funnelweb spider. *J. Arachnol.* 18:281–296.

Coyne, J. A. 1993. The genetics of an isolating mechanism between two sibling species of *Drosophila*. *Evolution* 47:778–788.

Crabtree, C. 1941. The structure of Bowman's capsule as an index of age and sex variations in normal mice. *Anat. Rec.* 79:395–413.

Craig, Jr., G. B. 1967. Mosquitoes: Female monogamy induced by male accessory gland substance. *Science* 156:1499–1501.

Crews, D. 1973. Coition-induced inhibition of sexual receptivity in female lizards (*Anolis carolinensis*). *Physiol. Behav.* 11:463–468.

Crews, D. 1980. Interrelationships among ecological, behavioral, and neuroendocrine processes in the reproductive cycle of *Anolis carolinensis* and other reptiles. *Adv. Stud. Behav.* 11:15–74.

Crews, D., and W. R. Garstka. 1982. The ecological physiology of a garter snake. *Sci. Am.* 247(5):159–168.

Crews, D., and R. Silver. 1985. Reproductive physiology and behavior interactions in nonmammalian vertebrates. In N. Adler, D. Pfaff, and R. Goy, eds., *Handbook of Behavioral Neurobiology*, vol. 7, 101–182. Plenum, New York.

Cronin, H. 1991. *The Ant and the Peacock—Altruism and Sexual Selection from Darwin to Today*. Cambridge University Press, New York.

Cronin, J. E., R. Cann, and V. M. Sarich. 1980. Molecular evolution and systematics of the genus *Macaca*. In D. G. Lindburg, ed., *The Macaques: Studies in Ecology, Behavior and Evolution*, 31–51. Van Nostrand-Reinhold, New York.

Cruzan, M. B., and S.C.H. Barrett. 1993. Contribution of cryptic incompatibility to the mating system of *Eichhornia paniculata* (Pontederiaceae). *Evolution* 47:925–934.

Cuellar, O. 1966. Oviductal anatomy and sperm storage structures in lizards. *J. Morphol.* 119:7–20.

Cummins, J. W. 1990. Evolution of sperm form: Levels of control and competition. In B. D. Bavister, J. Cummins, and E.R.S. Roldan, eds., *Fertilization in Mammals*, 51–64. Serono Symposia, Norwell, Mass.

Curie-Cohen, M., D. Yoshihara, L. Luttrell, K. Benforado, J. W. MacCluer, and W. H. Stone. 1983. The effects of dominance on mating behavior and paternity in a captive troop of rhesus monkeys (*Macaca mulatta*). *Am. J. Primatol.* 5:127–138.

Cusson, M., and J. N. McNeil. 1989. Involvement of juvenile hormone in the regulation of pheromone release activities in a moth. *Science* 24:210–212.

Daly, M. 1978. The cost of mating. *Am. Nat.* 112:771–774.

Dane, B., and W. van der Kloot. 1962. An analysis of the display of the goldeneye duck (*Bucephala clangula* [L.]). *Behaviour* 22:282–328.

Darwin, C. 1871. *The Descent of Man and Selection in Relation to Sex.* Reprinted. Modern Library, New York.

Davey, K. G. 1958. The migration of spermatozoa in the female of *Rhodnius prolixus* Stal. *J. Exp. Biol.* 35:694–701.

Davey, K. G. 1960a. A pharmacologically active agent in the reproductive system of insects. *Can. J. Zool.* 38:39–45.

Davey, K. G. 1960b. The evolution of spermatophores in insects. *Proc. Roy. Ent. Soc. Lond.* A35:107–113.

Davey, K. G. 1965a. Copulation and egg-production in *Rhodnius prolixus*: The role of the spermathecae. *J. Exp. Biol.* 42:373–378.

Davey, K. G. 1965b. *Reproduction in Insects.* Oliver and Boyd, London.

Davey, K. G. 1965c. The physiology of reproduction in insects. *Ann. Ent. Soc. Quebec*, 13–25.

Davey, K. G. 1967. Some consequences of copulation in *Rhodnius prolixus. J. Insect Physiol.* 13:1629–1636.

Davey, K. G. 1985. The female reproductive tract. In G. A. Kerkut and L. Gilbert, eds., *Comprehensive Insect Physiological Biochemistry and Pharmacology*, vol. 1, 15–36. Pergamon Press, New York.

Davies, N. 1983. Polyandry, cloaca-pecking and sperm competition in dunnocks. *Nature* 302:334–336.

Davies, N. B. 1985. Cooperation and conflict among dunnocks, *Prunella modularis*, in a variable mating system. *Anim. Behav.* 33:628–648.

Davies, N. 1992. *Dunnock Behaviour and Social Evolution.* Oxford University Press, Oxford.

Davies, N. B., and T. R. Halliday. 1979. Competitive mate searching in male common toads, *Bufo bufo*. *Anim. Behav.* 27:1253–1267.

Davies, N. B., I. R. Hartley, B. J. Hatchwell, A. Desrochers, J. Skeer, and D. Nebel. 1995. The polygynandrous mating system of the alpine accentor, *Prunella collaris*. I. Ecological causes and reproductive conflicts. *Anim. Behav.* 49:769–788.

Davis, H. N., Jr., G. D. Gray, and D. A. Dewsbury. 1977. Maternal age and male behavior in relation to successful reproduction by female rats (*Rattus norvegicus*). *J. Comp. Physiol. Psychol.* 91:281–289.

Davis, H. N., Jr., G. D. Gray, M. Zerylnick, and D. A. Dewsbury. 1974. Ovulation and implantation in montane voles (*Microtus montanus*) as a function of varying amounts of copulatory stimulation. *Hormones Behav.* 5:383–388.

Davis, N. T. 1965a. Studies of the reproductive physiology of Cimicidae (Hemiptera). II. Artificial insemination and the function of the seminal fluid. *J. Insect Physiol.* 11:355–366.

Davis, N. T. 1965b. Studies of the reproductive physiology of Cimicidae (Hemiptera). III. The seminal stimulus. *J. Insect Physiol.* 11:1199–1211.

de Benedictis, T. 1973. The behavior of young primates during adult copulation: Observations of a *Macaca irus* colony. *Am. Anthropol.* 75:1469–1484.

De Feo, V. J. 1966. Vaginal-cervical vibration: A simple and effective method for the induction of pseudopregnancy in the rat. *Endocrinology* 79:440–442.

Degrugillier, M. E., and R. A. Leopold. 1972. Abdominal peripheral nervous system of the adult female house fly and its role in mating behavior and insemination. *Ann. Ent. Soc. Am.* 65:689–695.

Dejianne, D., A. Pruzan, and S. H. Faro. 1978. Sperm competition in *Drosophila pseudoobscura*. *Behav. Genet.* 8:544.

Del Rio, G., and R. Cavalloro. 1979. Influenza dell'accoppiamento sulla recettivita sessuale e sull'ovideposizione in femmene di *Ceratitis capitata* Wiedmann. *Entomologica, Bari* 15:127–143.

Denell, R. E., and B. H. Judd. 1969. Segregation distorter in *D. melanogaster* males: An effect of female genotype on recovery. *Molec. Gen.* 105:262–274.

Deputte, B., and M. Goustard. 1980. Copulatory vocalizations of female macaques (*Macaca fascicularis*): Variability factors analysis. *Primates* 21:83–99.

Destephano, D. B., and U. E. Brady. 1977. Prostaglandin and prostaglandin synthetase in the cricket, *Acheta domesticus*. *J. Insect Physiol.* 23:905–911.

Destephano, D. B., U. E. Brady, and C. A. Farr. 1982. Factors influencing oviposition behavior in the cricket, *Acheta domesticus*. *Ent. Soc. Am.* 75:111–114.

Dethier, V. G. 1988. Induction and aversion-learning in polyphagous arctiid larvae (Lepidoptera) in an ecological setting. *Can. Ent.* 120:125–131.

Deuchar, E. M. 1975. *Cellular Interactions in Animal Development*. Chapman and Hall, London.

DeVilliers, P. S., and S. A. Hanrahan. 1991. Sperm competition in the Namib desert beetle, *Onymacris unguicularis*. *J. Insect Physiol.* 37:1–8.

Devine, M. C. 1975. Copulatory plugs in snakes: Enforced chastity. *Science* 187: 844–845.

Devine, M. C. 1977. Copulatory plugs, restricted mating opportunities and reproductive competition among male garter snakes. *Nature* 267:345–346.

Devine, M. C. 1984. Potential for sperm competition in reptiles: Behavioral and physiological consequences. In R. L. Smith, ed., *Sperm Competition and the Evolution of Animal Mating Systems*, 509–521. Academic Press, New York.

DeVries, J. K. 1964. Insemination and sperm storage in *Drosophila melanogaster*. *Evolution* 18:271–282.

Dewsbury, D. A. 1972. Patterns of copulatory behavior in male mammals. *Quart. Rev. Biol.* 47:1–33.

Dewsbury, D. A. 1975. Diversity and adaptation in rodent copulatory behavior. *Science* 190:947–954.

Dewsbury, D. A. 1982. Ejaculate cost and mate choice. *Am. Nat.* 119:601–610.

Dewsbury, D. A. 1984. Sperm competition in muroid rodents. In R. L. Smith, ed., *Sperm Competition and the Evolution of Animal Mating Systems*, 547–571. Academic Press, New York.

Dewsbury, D. A. 1988. Copulatory behavior as courtship communication. *Ethology* 79:218–234.

Dewsbury, D. A., and D. Baumgardner. 1981. Studies of sperm competition in two species of muroid rodents. *Behav. Ecol. Sociobiol.* 9:121–133.

Dewsbury, D. A., and D. Q. Estep. 1975. Pregnancy in cactus mice: Effects of prolonged copulation. *Science* 187:552–553.

Dewsbury, D. A., and J. D. Pierce, Jr. 1989. Copulatory patterns of primates as viewed in broad mammalian perspective. *Am. J. Primatol.* 17:51–72.

Diakow, C. 1973. Male-female interactions and the organization of mammalian mating patterns. *Adv. Stud. Behav.* 5:227–268.

Diamond, M. 1970. Intromission pattern and species vaginal code in relation to pseudopregnancy. *Science* 169:995–97.

Diamond, M. 1972. Vaginal stimulation and progesterone in relation to pregnancy and parturition. *Biol. Reprod.* 6:281–287.

Dickinson, J. L. 1986. Prolonged mating in the milkweed leaf beetle *Labidomera clivicollis clivicollis* (Coleoptera: Chrysomelidae): A test of the "sperm-loading" hypothesis. *Behav. Ecol. Sociobiol.* 18:331–338.

Dickinson, J. L. 1988. Determinants of paternity in the milkweed leaf beetle. *Behav. Ecol. Sociobiol.* 23:9–19.

Dickinson, J. In press. Multiple mating, sperm competition, and cryptic female choice in the leaf beetles (Coleoptera; Chrysomelidae). In J. C. Choe and B. J. Crespi, eds., *Social Competition and Cooperation in Insects and Arachnids. I. Evolution of Mating Systems*. Cambridge University Press, Cambrige, U.K.

Dickinson, R. G., J. E. O'Hagan, M. Schotz, K. C. Binnington, and M. P. Hegarty. 1976. Prostaglandin in the saliva of the cattle tick *Boophilus microplus*. *Aust. J. Expt. Biol. Med. Sci.* 54(5):475–486.

Diehl, P. A., A. Aeschlimann, and F. D. Obenchain. 1982. Tick reproduction: Oogenesis and oviposition. In F. D. Obenchain and R. Galun, eds., *Physiology of Ticks*, 277–350. Pergamon Press, New York.

Diesel, R. 1990. Sperm competition and reproductive success in the decapod *Inachus phalangium* (Majidae): A male ghost spider crab that seals off rivals' sperm. *J. Zool. Lond.* 220:213–223.

Diesel, R. 1991. Sperm competition and the evolution of mating behavior in Brachyura, with special reference to spider crabs (Decapoda, Majidae). In R. T. Bauer and J. W. Martin, eds., *Crustacean Sexual Biology*, 145–163. Columbia University Press, New York.

Dilley, W. G., and N. T. Adler. 1968. Postcopulatory mammary gland secretion in rats. *Proc. Soc. Exptl. Biol. Med.* (NY) 129:964–966.

Dixson, A. F. 1977. Observations on the displays, menstrual cycles and sexual behaviour of the "black ape" of Celebes (*Macaca nigra*). *J. Zool. Lond.* 182:63–84.

Dixson, A. F. 1987. Observations on the evolution of the genitalia and copulatory behaviour in male primates. *J. Zool. Lond.* 213:423–443.

Dixson, A. F. 1991a. Penile spines affect copulatory behaviour in a primate (*Callithrix jacchus*). *Physiol. Behav.* 49:557–562.

Dixson, A. F. 1991b. Sexual selection, natural selection and copulatory patterns in male primates. *Folia Primatol.* 57:96–101.

Doak, R. L., A. Hall, and H. E. Dale. 1967. Longevity of spermatozoa in the reproductive tract of the bitch. *J. Reprod. Fertil.* 13:51–58.

Dobzhansky, T. 1970. *Genetics of the Evolutionary Process.* Columbia University Press, New York.

Dodson, G. 1978. Morphology of the reproductive system in *Anastrepha suspensa* (Loew) and notes on related species. *Fla. Ent.* 61:231–239.

Donahue, R. P. 1972. The relation of oocyte maturation to ovulation in mammals. In J. D. Biggers and A. W. Schuetz, eds., *Oogenesis*, 413–438. University Park Press, Baltimore.

Donaldson, W. E., and A. E. Adams. 1989. Ethogram of behavior with emphasis on mating for the tanner crab *Chionoecetes bairdi* rathbun. *J. Crust. Biology* 9(1): 37–53.

Doussard, D. E., K. Ubik, C. Harvis, J. Resch, J. Meinwald, and T. Eisner. 1988. Biparental defensive endowment of eggs with acquired plant alkaloid in the moth *Utethesia ornatrix. Proc. Natl. Acad. Sci. USA* 85:5992–5996.

Downe, A.E.R. 1973. Some factors influencing insemination in laboratory swarms of *Chironomus riparius* (Diptera: Chironomidae). *Can. Ent.* 291–298.

Downes, M. F. 1988. The production and use of inviable eggs by *Theridion rufipes* Lucas (Arancae: Theridiidae). In A. D. Austin and N. W. Heather, eds., *Australian Arachnology*, 127–135. Australian Entomol. Soc., Brisbane.

Downhower, J. F., and D. B. Lank. 1994. Effect of previous experience on mate choice by female mottled sculpins. *Anim. Behav.* 47:369–372.

Drew, G. 1911. Sexual activities of the squid *Loligo pealei* (Les). *J. Morphol.* 22:327–360.

Drickhamer, L. C. 1985. Puberty-inducing chemo-signals in house mice: Ecological and evolutionary considerations. In D. Duvall, D. Muller-Schwarze, and R. M. Silverstein, eds., *Chemical Signals in Vertebrates*, vol. 4, 441–456. Plenum, New York.

Drickhamer, L. C. 1989. Pregnancy block in wild stock house mice, *Mus domesticus*: Olfactory preferences of females during gestation. *Anim. Behav.* 37:690–698.

Drummond, B. A. 1984. Multiple mating and sperm competition in the Lepidoptera. In R. L. Smith, ed., *Sperm Competition and the Evolution of Animal Mating Systems*, 291–371. Academic Press, New York.

Dudley, C. A., and R. L. Moss. 1976. Facilitation of lordosis in the rat by prostaglandin E2. *J. Endocr.* 71:457–458.

Duellman, W. E., and L. Trueb. 1986. *Biology of Amphibians.* McGraw Hill, New York.

Dufy-Barbe, L., P. Franchimont, B. Dufy, and J. D. Vincent. 1975. Effect of ovariectomy and estradiol benzoate treatment on LH-RH induced LH and FSH release in the female rabbit. *Hormones Res.* 6:372–379.

Duggal, C. L. 1978. Copulatory behaviour of male *Panagrellus redivivus. Nematologica* 24:257–268.

Dukelow, W. R., and J. Erwin. 1986. *Comparative Primate Biology*, vol. 3, *Reproduction and Development.* A. R. Liss, New York.

Dunbar, R.I.M. 1978. Sexual behaviour and social relationships among gelada baboons. *Anim. Behav*. 26:167–178.

Dunn, P. O., and J. T. Lifjeld. 1994. Can extra-pair copulations be used to predict extra-pair paternity in birds? *Anim. Behav*. 47:983–985.

Dunn, P. O., R. J. Robertson, D. Michaud-Freeman, and P. T. Boag. 1994b. Extra-pair paternity in tree swallows: Why do females mate with more than one male? *Behav. Ecol. Sociobiol*. 35:273–281.

Dunn, P. O., L. A. Whittingham, J. T. Lifjeld, R. J. Robertson, and P. T. Boag. 1994a. Effects of breeding density, synchrony, and experience on extrapair paternity in tree swallows. *Behav. Ecol*. 5(2):123–129

Eady, P. E. 1994a. Sperm transfer and storage in relation to sperm competition in *Callosobruchus maculatus. Behav. Ecol. Sociobiol*. 35:123–129.

Eady, P. E. 1994b. Intraspecific variation in sperm precedence in the bruchid beetle *Callosobruchus maculatus. Ecol. Ent*. 19:11–16.

Eady, P. E. 1995. Why do male *Callosobruchus maculatus* beetles inseminate so many sperm? *Behav. Ecol. Sociobiol*. 36:25–32.

Eaton, G. G., A. Slob, and J. A. Resko. 1973. Cycles of mating behaviour, oestrogen and progesterone in the thick-tailed bushbaby (*Galago crassicaudatus crassicaudatus*) under laboratory conditions. *Anim. Behav*. 21:309–315.

Eberhard, W. G. 1974. The natural history and behaviour of the wasp *Trigonopsis cameronii* Kohl (Sphecidae). *Trans R. Ent Soc. Lond*. 125(3):295–328.

Eberhard, W. G. 1975. The ecology and behavior of a subsocial pentatomid bug and two scelenid wasps: Strategy and counterstrategy in a host and its parasites. *Smithson. Contrib. Zool*. 205:1–39.

Eberhard, W. G. 1979. Rates of egg production by tropical spiders in the field. *Biotropica* 11:292–300.

Eberhard, W. G. 1980. Evolutionary consequences of intracellular organelle competition. *Quart. Rev. Biol*. 55:231–249.

Eberhard, W. G. 1985. *Sexual Selection and Animal Genitalia*. Harvard University Press, Cambridge, Mass.

Eberhard, W. G. 1990a. Evolution in bacterial plasmids and levels of selection. *Quart. Rev. Biol*. 65:3–22.

Eberhard, W. G. 1990b. Animal genitalia and female choice. *Am. Sci*. 78:134–141.

Eberhard, W. G. 1990c. Inadvertent machismo? *Trends Ecol. Evol*. 5:263.

Eberhard, W. G. 1991a. Copulatory courtship in insects. *Biol. Rev*. 66:1–31.

Eberhard, W. G. 1991b. Artificial insemination: Can appropriate stimulation improve success rates? *Med. Hypoth*. 36:152–154.

Eberhard, W. G. 1992a. Species isolation, genital mechanics, and the evolution of species-specific genitalia in three species of *Macrodactylus* beetles (Coleoptera, Scarabeidae, Melolonthinae). *Evolution* 46:1774–1783.

Eberhard, W. G. 1992b. Notes on the ecology and behaviour of *Physocyclus globosus* (Araneae, Pholcidae). *Bull. Br. Arachnol. Soc*. 9:38–42.

Eberhard, W. G. 1993a. Copulatory courtship and genital mechanics of three species of *Macrodactylus* (Coleoptera, Scarabeidae, Melolonthinae). *Ecol. Ethol. Evol*. 5:19–63.

Eberhard, W. G. 1993b. Copulatory courtship and the morphology of genitalic coupling in seven *Phyllophaga* species (Coleoptera, Melolonthidae). *J. Nat. Hist.* 27:683–717.

Eberhard, W. G. 1993c. Functional significance of some secondary sexual characters in three species of *Macrodactylus* (Coleoptera: Melolonthidae). *Coleop. Bull.* 47:53–60.

Eberhard, W. G. 1993d. Evaluating models of sexual selection: Genitalia as a test case. *Am. Nat.* 142:564–571.

Eberhard, W. G. 1994. Evidence for widespread courtship during copulation in 131 species of insects and spiders, and implications for cryptic female choice. *Evolution* 48:711–733.

Eberhard, W. G. In press. Sexual selection by cryptic female choice in insects and arachnids. In J. Choe and B. Crespi, eds., *Social Competition and Cooperation in Insects and Arachnids.* Cambridge University Press, Cambridge, U.K.

Eberhard, W. G., and C. Cordero. 1995. Sexual selection by cryptic female choice on male seminal products—a new bridge between sexual selection and reproductive physiology. *TREE* 10:493–496.

Eberhard, W. G., and S. Kariko. In press. The "inside and outside" of copulation behavior in *Macrohaltica* beetles (Coleptera, Chrysomelidae). *J. Ethol.*

Eberhard, W. G., and F. Pereira. 1994. Functions of the male genitalic surstyli in the Mediterranean fruit fly, *Ceratitis capitata* (Diptera: Tephritidae). *J. Kans. Ent. Soc.* 66:427–433.

Eberhard, W. G., and F. Pereira. In press. Functional morphology of male genitalic surstyli in the dungflies *Achisepsis diversiformis* and *A. ecalcarata* (Diptera: Sepsidae). *J. Kans. Ent. Soc.*

Eberhard, W. G., and F. Pereira. Submitted. Male copulation behavior and the process of intromission in the Mediterranean fruitfly, *Ceratitis capitata* (Diptera: Tephritidae). *Psyche.*

Eberhard, W. G., and J. Ugalde. In prep. Female *Macrohaltica* beetles preferentially use conspecific stored sperm to fertilize eggs.

Eberhard, W. G., and E. Vargas. In prep. Intrageneric differences in proteins in male reproductive tissues in *Macrodactylus* and *Phyllophaga* beetles (Coleoptera: Scaraberdae: Melolonthinae).

Eberhard, W. G., R. Achoy, M. C. Marin, and J. Ugalde. 1993a. Natural history and behavior of two species of *Macrohaltica* (Coleptera: Chrysomelidae). *Psyche* 100:93–119.

Eberhard, W. G., S. Guzman-Gomez, and K. Catley. 1993b. Correlation between spermathecal morphology and mating systems in spiders. *Biol. J. Linn. Soc.* 50:197–209.

Eberhard, W. G., R. L. Rodriguez, B. Huber, I. Salas, R. D. Briceño, and V. Rodriguez. In prep. Male genitalia size correlates poorly with male body size in spiders and insects.

Edmonds, S., S. R. Zoloth, and N. T. Adler. 1972. Storage of copulatory stimulation in the female rat. *Physiol. Behav.* 8:161–164.

Edwards, R. 1993. Entomological and mammalogical perspectives on genital differentiation. *TREE* 8:406–409.

Ehrlich, A. H., and P. R. Ehrlich. 1978. Reproductive strategies in the butterflies. I. Mating frequency, plugging, and egg number. *J. Kans. Ent. Soc.* 51:666–697.

Eisenberg, J. F., G. M. McKay, and M. R. Jainudeen. 1971. Reproductive behavior of the Asiatic elephant (*Elephas maximus maximus* L.). *Behaviour* 38:193–225.

Eisenberg, J. F., and E. Gould. 1970. The tenrecs: A study in mammalian behavior and evolution. *Smithson. Contrib. Zool.* 27:1–137.

Engelmann, F. 1970. *The Physiology of Insect Reproduction.* Pergamon Press, New York.

England, B. G., W. C. Foote, D. H. Matthews, A. G. Cardozo, and S. Riera. 1969. Ovulation and corpus luteum function in the llama (*Lama glama*). *J. Endocr.* 45: 505–513.

Erskine, M. S., and M. J. Baum. 1982. Effects of paced coital stimulation on termination of estrus and brain indoleamine levels in female rats. *Pharmac. Biochem. & Behav.* 17:857–861.

Estep, D. Q., T. P. Gordon, M. E. Wilson, and M. L. Walker. 1986. Social stimulation and the resumption of copulation in rhesus (*Macaca mulatta*) and stumptail (*Macaca arctoides*) macaques. *Int. J. Primatol.* 7(5):507–517.

Estop, A., K. Garver, K. Cieply, and V. van Kirk. 1990. Sperm chromosome studies in human male donors. In B. D. Bavister, J. Cummins, and E. S. Roldan, eds., *Fertilization in Mammals*, 443. Serono Symposia, Norwell, Mass.

Etman, A.A.M., and G.H.S. Hooper. 1979. Sperm precedence of the last mating in *Spodoptera litura. Ann. Ent. Soc. Am.* 72:119–120.

Evans, E. I. 1933. The transport of spermatozoa in the dog. *Am. J. Physiol.* 105: 287–293.

Evans, H. E., K. M. O'Neill, and R. P. O'Neill. 1986. Nesting site changes and nocturnal clustering in the sand wasp *Bembecinus guinguespinosus* (Hymenoptera: Sphecidae). *J. Kans. Ent. Soc.* 59:280–286.

Evans, L. E., and D. S. McKenna. 1986. Artificial insemination of swine. In D. Morrow, ed., *Current Therapy in Theriogenology.* W. B. Saunders, Philadelphia.

Ewald, P. 1994. *Evolution of Infectious Diseases.* Oxford University Press, New York.

Ewer, R. F. 1968. *Ethology of Mammals.* Plenum, New York.

Ewer, R. F. 1973. *The Carnivores.* Cornell University Press, Ithaca, N.Y.

Feldman-Muhsam, B. 1986. Observations on the mating behaviour of ticks. In J. R. Sauer and J. A. Hair, eds., *Morphology, Physiology, and Behavioral Biology of Ticks*, 217–232. Halsted Press, New York.

Feldman-Muhsam, B., and S. Borut. 1983. On the spermatophore of ixodid ticks. *J. Insect Physiol.* 29:449–457.

Felsenstein, J. 1985. Phylogenies and the comparative method. *Am. Nat.* 125:1–15.

Fenton, M. B. 1984. Sperm competition? The case of vespertilionid and rhonolophid bats. In R. L. Smith, ed., *Sperm Competition and the Evolution of Animal Mating Systems*, 573–587. Academic Press, New York.

Ferrari, F. 1978. Spermatophore placement in the copepod *Euchaeta norvegica* Boeck 1872 from deepwater dumpsite 106. *Proc. Biol. Soc. Wash.* 91(2):509–521.

Ferrari, F., and M. Dojiri. 1987. The calanoid copepod *Euchaeta antarctica* from southern ocean Atlantic sector midwater trawls, with observations on spermatophore dimorphism. *J. Crust. Biol.* 7(3):458–480.

Field, L. H., and G. R. Sandlant. 1983. Aggression and mating behavior in the

Stenopelmatidae (Orthoptera, Ensifera) with reference to New Zealand wetas. In D. T. Gwynne and G. K. Morris, eds., *Orthopteran Mating Systems*, 120–146. Westview Press, Boulder.

Fincke, O. M. 1987. Female monogamy in the damselfly *Ischnura verticalis* Say (Zygoptera: Coenagrionidae). *Odonatologica* 16:129–143.

Fink, G. 1979. Feedback actions of target hormones on hypothalamus and pituitary with special reference to gonadal steroids. *Ann. Rev. Physiol.* 41:571–85.

Fink, G., and S. R. Henderson. 1977a. Site of modulatory action of oestrogen and progesterone on gonadotrophin response to luteinizing hormone releasing factor. *J. Endocr.* 73:165–170.

Fink, G., and S. R. Henderson. 1977b. Steroids and pituitary responsiveness in female, androgenized female and male rats. *J. Endocr.* 73:157–164.

Follett, B. K. 1984. Birds. In G. E. Lamming, ed., *Marshall's Physiology of Reproduction*, vol. 1, 283–350. Churchill Livingstone, New York.

Foster, S. P. 1993. Neural inactivation of sex pheromone production in mated light brown apple moths, *Epiphyas postvittana* (Walker). *J. Insect Physiol.* 39:267–273.

Fowler, G. 1973. Some aspects of the reproductive biology of *Drosophila*: Sperm transfer, sperm storage, and sperm utilization. *Adv. Genet.* 17:293–360.

Fox, C. W. 1993. Multiple mating, lifetime fecundity and female mortality of the bruchid beetle, *Callosobruchus maculatus* (Coleoptera: Bruchidae). *Func. Ecol.* 7:203–208.

Fox, W. 1963. Special tubules for sperm storage in female lizards. *Nature* 198:500–501.

Fraser, J. B. 1987. Courtship and copulatory behavior of the funnel-web spider, *Hololena adnexa* (Araneae, Agelenidae). *J. Arachnol.* 15:257–262.

Friedel, T., and C. Gillott. 1976. Male accessory gland substance of *Melanoplus sanguinipes*: An oviposition stimulant under the control of the corpus allatum. *J. Insect Physiol.* 22:489–495.

Friedel, T., and C. Gillott. 1977. Contribution of male-produced proteins to vitellogenesis in *Melanoplus sanguinipes*. *J. Insect Physiol.* 23:145–151.

Fuchs, M. S., and E. A. Hiss. 1970. The partial purification and separation of the protein components of matrone from *Aedes aegypti*. *J. Insect Physiol.* 16:931–939.

Fuchs, M. S., F. V. Craig, Jr., and D. D. Despommier. 1969. The protein nature of the substance inducing female monogamy in *Aedes aegypti*. *J. Insect Physiol.* 15:701–709.

Furudate, S., and T. Nakano, 1980. Studies on the pheromonal pregnancy block in the mice. I. Strain differences of mice in pheromonal pregnancy block. *Exp. Anim.* 29(3):341–344.

Fuyama, Y. 1983. Species-specificity of paragonial substances as an isolating mechanism in *Drosophila*. *Experientia* 39:190–192.

Gack, C., and K. Peschke. 1994. Spermathecal morphology, sperm transfer and a novel mechanism of sperm displacement in the rove beetle, *Aleochara curtula* (Coleoptera, Staphylinidae). *Zoomorphology* 114:227–237.

Gage, M.J.G. 1991. Risk of sperm competition directly affects ejaculate size in the Mediterranean fruit fly. *Anim. Behav.* 42:1036–1037.

Gage, M.J.G. 1992. Removal of rival sperm during copulation in a beetle, *Tenebrio molitor*. *Anim. Behav.* 44:587–589.

Gage, M.J.G., and R. R. Baker. 1991. Ejaculate size varies with socio-sexual situation in an insect. *Ecol. Ent.* 16:331–337.

Galiano, M. E. 1987a. Description of *Aillutticus*, new genus (Araneae, Salticidae). *Bull. Brit. Arachnol. Soc.* 7:157–164.

Galiano, M. E. 1987b. Descripción de *Hisukattus* nuevo género (Araneae, Salticidae). *Rev. Soc. Ent. Arg.* 44:137–148.

Galiano, M. E., and L. Baert. 1990. Distribution of the Galapagoan salticid species with the description of three *Sitticus* species (Araneae, Salticidae). *Bull. Inst. Roy. Sci. Nat. Belg. Ent.* 60:123–132.

Galun, R., and M. Warburg. 1967. Studies on the reproductive physiology of the tick *Ornithodoros tholozani* (Laboulbene and Megnin): The effect of mating on oogenesis. *Acta Soc. Zool. Bohem.* 31:329–334

Garófalo, C. A. 1980. Reproductive aspects and evolution of social behavior in bees (Hymenoptera, Apoidea). *Rev. Brasil. Genet.* 3:139–152.

Gartlan, J. S. 1969. Sexual and maternal behaviour of the vervet monkey, *Cercopithecus aethiops*. *J. Reprod. Fert.* 6:137–150.

Gering, R. L. 1953. Structure and function of the genitalia in some American agelenid spiders. *Smithson. Misc. Coll.* 121(4):1–84.

Germond, J. E., and A. Aeschlimann. 1977. Influence of copulation on vitellogenesis and egg-laying in *Ornithodoros moubata* (Ixodoidea: Argasidae). In R. G. Adiyodi and R. G. Adiyodi, eds., *Advances in Invertebrate Reproduction*, 308–318. Wiley, New York.

Giebultowicz, J. M., A. K. Raina, and R. L. Ridgway. 1991a. Role of the spermatheca in the termination of sex pheromone production in mated gypsy moth females. *Proc. Conf. Insect Chem. Ecol. Tabor 1990*, 101–104. Academia Prague and SPB Acad. Publ., The Hague, 1991.

Giebultowicz, J. M., A. K. Raina, E. C. Uebel, and R. L. Ridgway. 1991b. Two-step regulation of sex-pheromone decline in mated gypsy moth females. *Archiv. Insect Biochem. Physiol.* 16:95–105.

Giglioni, M.E.C., and G. F. Mason. 1966. The mating plug in anopheline mosquitoes. *Proc. Roy. Ent. Soc. Lond.* (A) 41:123–129.

Gilbert, D. G. 1981. Ejaculate esterase 6 and initial sperm use by female *Drosophila melanogaster*. *J. Insect Physiol.* 27:641–650.

Gilbert, D. G., R. C. Richmond, and K. B. Sheehan. 1981a. Studies of esterase 6 in *Drosophila melanogaster*. V. Progeny production and sperm use in females inseminated by males having active or null alleles. *Evolution* 35:21–37.

Gilbert, D. G., R. C. Richmond, and K. B. Sheehan. 1981b. Studies of esterase 6 in *Drosophila melanogaster*. VII. Remating times of females inseminated by males having active or null alleles. *Behav. Genetics* 11:195–208.

Gilbert, L. 1976. Postmating female odor in *Heliconius* butterflies: A male-contributed antiaphrodisiac? *Science* 193:419–420.

Gilburn, A. S., S. P. Foster, and T. H. Day. 1993. Genetic correlation between a female mating preference and the preferred male character in seaweed flies (*Coelopa frigida*). *Evolution* 47:1788–1795.

Gillies, M. T. 1956. A new character for the recognition of nulliparous females of *Anopheles gambiae*. *Bull. Wld. Hlth. Org*. 15:451–459.

Gillingham, J. C., C. C. Carpenter, B. J. Brecke, and J. B. Murphy. 1977. Courtship and copulatory behavior of the Mexican milk snake, *Lampropeltis triangulum sinaloae* (Colubridae). *Southw. Nat*. 22:187–194.

Gillott, C. 1988. Arthropoda—Insecta. In K. G. Adiyodi and R. G. Adiyodi, eds., *Reproductive Biology of Invertebrates*, 319–471. Wiley, New York.

Gillott, C., and P. A. Langley. 1981. The control of receptivity and ovulation in the tsetse fly, *Glossina morsitans*. *Physiol. Entomol*. 6:269–281.

Gilman, D. P., L. F. Mercer, and J. C. Hitt. 1979. Influence of female copulatory behavior on the induction of pseudopregnancy in the female rat. *Physiol. Behav*. 22:675–678.

Ginsberg, J. R., and U. W. Huck. 1989. Sperm competition in mammals. *TREE* 4:74–79.

Ginsberg, J. R., and D. I. Rubenstein. 1990. Sperm competition and variation in zebra mating behavior. *Behav. Ecol. Sociobiol*. 26:427–434.

Gladney, W. J., and C. C. Dawkins. 1973. Experimental interspecific mating of *Amblyomma maculatum* and *A. americanum*. *Ann. Ent. Soc. Am*. 66(5):1093–1097.

Glick, B. B. 1980. Ontogenetic and psychobiological aspects of the mating activities of male *Macaca radiata*. In D. G. Lindberg, ed., *The Macaques: Studies in Ecology, Behavior and Evolution*, 345–369. Van Nostrand Reinhold, New York.

Godfray, H.C.J. 1994. *Parasitoids*. Princeton University Press, Princeton.

Gomendio, M., and E.R.S. Roldan. 1991. Sperm competition influences sperm size in mammals. *Proc. Roy. Soc. Lond. B* 243:181–185.

Gomendio, M., and E.R.S. Roldan. 1993. Mechanisms of sperm competition: Linking physiology and behavioural ecology. *TREE* 8:95–100.

Goodall, J. 1986. *The Chimpanzees of Gombe: Patterns of Behavior*. Harvard University Press, Cambridge, Mass.

Gosling, L. M. 1986. Selective abortion of entire litters in the coypu: Adaptive control of offspring production in relation to quality and sex. *Am. Nat*. 127(6): 772–795.

Goulson, D. 1993. Variation in the genitalia of the butterfly *Maniola jurtina* (Lepidoptera: Satyrinae). *Zool. J. Linn. Soc*. 107:65–71.

Gowaty, D. A. 1994. Architects of sperm competition. *TREE* 9:160–162.

Grandison, A.G.C., and S. Ashe. 1983. The distribution, behavioural ecology and breeding strategy of the pygmy toad, *Mertensophryne micranotis* (Lov.). *Bull, Br. Mus. Nat. Hist. (Zool.)* 45(2):85–93.

Grant, B. 1983. On the relationship between average copulation duration and insemination reaction in the genus *Drosophila*. *Evolution* 37:854–856.

Grasshoff, M. 1973. Konstruktions- und Funktionsanalyse an Kopulationsorganen einiger Radnetzspinnen. *Senckenberg. Naturf. Ges. Frankfurt Main Ber*. 24:129–151.

Gray, G. D., M. Zerylnick, H. N. Davis, and D. A. Dewsbury. 1974. Effects of variations in male copulatory behavior on ovulation and implantation in prairie voles, *Microtus ochrogaster*. *Hormones Behav*. 5:389–396.

Gregory, G. E. 1965. The formation and fate of the spermatophore in the African migratory locust, *Locusta migratoria migratorioides* Reiche and Fairmaire. *Trans. Roy. Ent. Soc. Lond.* 117:33–66.

Gregory, P. G., and D. J. Howard. 1994. A post-insemination barrier to fertilization isolates two closely related ground crickets. *Evolution* 48:705–710.

Griffiths, D. A., and J. Boczek. 1977. Spermatophores of some acaroid mites (Astigmata: Acarina). *Int. J. Ins. Morph. Embryol.* 6:231–238.

Griswold, C. 1990. A revision and phylogenetic analysis of the spider subfamily Phyxelidinae (Araneae, Amaurobiidae). *Bull. Ann. Mus. Nat. Hist.* 196:1–206.

Griswold, C. 1991. A revision and phylogenetic analysis of the spider genus *Machadonia* Lehtinen (Araneae, Lycosidae). *Ent. Scand.* 22:305–351.

Gromko, M. H., and T. A. Markow. 1993. Courtship and remating in field populations of *Drosophila. Anim. Behav.* 45:253–262.

Gromko, M. H., M.E.A. Newport, and M. G. Kortier. 1984. Sperm dependence of female receptivity to remating in *Drosophila melanogaster. Evolution* 38:1273–1282.

Gruelich, W. W. 1934. Artificially induced ovulation in the cat (*Felis domestica*). *Anat. Rec.* 58:217–223.

Gwadz, R. W. 1972. Neuro-hormonal regulation of sexual receptivity in female *Aedes aegypti. J. Insect Physiol.* 18:259–266.

Gwadz, R. W., G. B. Craig, Jr., and W. A. Hickey. 1971. Female sexual behavior as the mechanism rendering *Aedes aegypti* refractory to insemination. *Biol. Bull.* 140:201–214.

Gwynne, D. T. 1983. Male nutritional investment and the evolution of sexual differences in Tettigoniidae and other orthoptera. In D. T. Gwynne and G. K. Morris, eds., *Orthopteran Mating Systems*, 337–366. Westview Press, Boulder.

Gwynne, D. 1984a. Sexual selection and sexual differences in Mormon crickets (Orthoptera: Tettigoniidae *Anabrus simplex*). *Evolution* 38:1011–1022.

Gwynne, D. T. 1984b. Courtship feeding increases female reproductive success in bushcrickets. *Nature* 307:361–363.

Gwynne, D. 1984c. Male mating effort, confidence of paternity, and insect sperm competition. In R. L. Smith, ed., *Sperm Competition and the Evolution of Animal Mating Systems*, 117–150. Academic Press, New York.

Gwynne, D. T. 1986. Courtship feeding in katydids (Orthoptera: Tettigoniidae): Investment in offspring or in obtaining fertilizations? *Am. Nat.* 128:342–352.

Gwynne, D. T. 1988a. Courtship feeding and the fitness of female katydids (Orthoptera: Tettigoniidae). *Evolution* 42(3):545–555.

Gwynne, D. T. 1988b. Courtship feeding in katydids benefits the mating male's offspring. *Behav. Ecol. Sociobiol.* 23:373–377.

Gwynne, D. 1990. Operational sex ratios and the control of sexual selection: A test with katydids (Orthoptera: Tettigoniidae). *Am. Nat.* 136:474–484.

Gwynne, D. T. 1993. Food quality controls sexual selection in mormon crickets by altering male mating investment. *Ecology* 74(5):1406–1413.

Gwynne, D. T., and W. D. Brown. 1994. Mate feeding, offspring investment, and sexual differences in katydids (Orthoptera: Tettigoniidae). *Behav. Ecol.* 5(3): 267–272.

Gwynne, D. T., and L. W. Simmons. 1990. Experimental reversal of courtship roles in an insect. *Nature* 346:172–174.

Haacker, U. 1969a. An attractive secretion in the mating behaviour of a millipede. *Z. Tierpsychol.* 26:988–990.

Haacker, U. 1969b. Spermaübertragung von *Glomeris* (Diplopoda). *Naturwissenschaften* 56:467.

Haacker, U. 1970. Das Paarungsverhalten von *Rhinocricus padbergi* Verh. (Diplopoda, Spirobolida). *Rev. Comp. Anim.* 4:35–39.

Haacker, U. 1974. Patterns of communication in courtship and mating behaviour of millipedes (Diplopoda). *Symp. Zool. Soc. Lond.* 32:317–328.

Haacker, U., and S. Fuchs. 1970. Das Paarungsverhalten von *Cylindroiulus punctatus* Leach. *Z. Tierpsychol.* 27:641–648.

Hafernik, J. E., and R. W. Garrison. 1986. Mating success and survival rate in a population of damselflies: Results at variance with theory? *Amer. Nat.* 128:353–365.

Hagedorn, H. H. 1980. Ecdysone, a gonadal hormone in insects. In W. H. Clark, Jr., and T. S. Adams, eds., *Advances in Invertebrate Reproduction*, 97–107. Elsevier North Holland.

Hall, J. C. 1994. The mating of a fly. *Science* 264:1702–1714.

Halliday, T. 1987. Physiological constraints on sexual selection. In J. W. Bradbury and M. Andersson, eds., *Sexual Selection: Testing the Alternatives*. Wiley, New York.

Halliday, T. R., and P. A. Verrell. 1984. Sperm competition in amphibians. In R. L. Smith, ed., *Sperm Competition and the Evolution of Animal Mating Systems*, 487–508. Academic Press, New York.

Halpert, A. P., W. R. Garstka, and D. Crews. 1982. Sperm transport and storage and its relation to the annual sexual cycle of the female red-sided garter snake, *Thamnophis sirtalis parietalis*. *J. Morphol.* 174:149–159.

Hamilton, W. D. 1967. Extraordinary sex ratios. *Science* 156:477–488.

Hancock, R. G., W. A. Foster, and W. L. Yee. 1990. Courtship behavior of the mosquito *Sabethes cyaneus* (Diptera: Culicidae). *J. Insect Behav.* 3:401–416.

Hansson, A. 1947. The physiology of reproduction in mink (*Mustela vison*, Schreb.) with special reference to delayed implantation. *Acta Zool.* 28:1–136.

Harbo, J. R. 1979. The rate of depletion of spermatozoa in the queen honeybee spermatheca. *J. Apicul. Res.* 18(3):204–207.

Harcourt, A. H., and J. Gardiner. 1994. Sexual selection and genital anatomy of male primates. *Proc. Roy. Soc. Lond. B* 255:47–53.

Hardy, D. F., and J. F. DeBold. 1972. Effects of coital stimulation upon behavior of the female rat. *J. Comp. Physiol. Psychol.* 78:400–408.

Harrison, R. M., and R. W. Lewis. 1986. The male reproductive tract and its fluids. *Comp. Primate Biol.* 3:101–148.

Harshman, L. G., and T. Prout. 1994. Sperm displacement without sperm transfer in *Drosophila melanogaster*. *Evolution* 48:758–766.

Hartmann, R., and W. Loher. 1974. Control of sexual behaviour pattern "secondary defence" in the female grasshopper, *Chorthippus curtipennis*. *J. Insect Physiol.* 20:1713–1728.

Hartley, I. R., N. B. Davies, B. J. Hatchwell, A. Desrochers, D. Nebel, and T. Burke. 1995. The polygynandrous mating system of the alpine accentor, *Prunella collaris*. II. Multiple paternity and parental effort. *Anim. Behav.* 49:789–803.

Hartnoll, R. G. 1969. Mating in the Brachyura. *Crustaceana* 16:161–181.

Harvey, P. H., and M. D. Pagel. 1991. *The Comparative Method in Evolutionary Biology*. Oxford University Press, Oxford.

Hass, B. 1990. A quantitative study of insemination and gamete efficiency in different species of the *Rhabditis strongyloides* group (Nematoda). *Invert. Reprod. Devel.* 18:205–208

Heady, S. E. 1993. Factors affecting female sexual receptivity in the planthopper, *Prokelisia dolus*. *Physiol. Ent.* 18:263–270.

Heg, D., B. J. Ens, T. Burke, L. Jenkins, and J. P. Kruijt. 1993. Why does the typically monogamous oystercatcher (*Haematopus ostralegus*) engage in extrapair copulations? *Behaviour* 126(3–4):247–289.

Heming-van Battum, K. E., and B. S. Heming. 1986. Structure, function and evolution of the reproductive system in females of *Hebrus pusillus* and *H. ruficeps* (Hemiptera, Gerromorpha, Hebridae). *J. Morphol.* 190:121–167.

Heming-van Battum, K. E., and B. S. Heming. 1989. Structure, function and evolutionary significance of the reproductive system in males of *Hebrus ruficeps* and *H. pusillus* (Heteroptera, Gerromorpha, Hebridae). *J. Morphol.* 202:281–323.

Hendricky, A. G., R. S. Thompson, D. L. Hess, and S. Prahalada. 1978. Artificial insemination and a note on pregnancy detection in the nonhuman primate. *Symp. Zool. Soc. Lond.* 43:219–240.

Henry, C. S. 1994. Singing and cryptic speciation in insects. *TREE* 9:388–392.

Herman-Giddens, M. E., A. D. Sandler, and N. E. Friedman. 1988. Sexual precocity in girls. *Amer. J. Diseases Children* 142:431–433.

Herre, A. 1993. Population structure and the evolution of virulence in nematode parasites of fig wasps. *Science* 259:1442–1444.

Hewitt, G. M., P. Mason, and R. A. Nichols. 1989. Sperm precedence and homogamy across a hybrid zone in the alpine grasshopper *Podisma pedestris*. *Heredity* 62:343–353.

Higgins, L. 1989. Effect of insemination on the morphology of the internal female genitalia of the spider *Nephila clavipes* (Araneae: Araneidae). *Ann. Ent. Soc. Am.* 82:748–753.

Hihara, F. 1981. Effects of the male accessory gland secretion on oviposition and remating in females of *Drosophila melanogaster*. *Zool. Mag. Tokyo* 90:307–316.

Hill, G. E. 1994. Trait elaboration via adaptive male choice: Sexual conflict in the evolution of signals of male quality. *Ethol. Ecol. Evol.* 6:351–370.

Hinton, H. E. 1974. Symposium on reproduction of arthropods of medical and veterinary importance. III. Accessory functions of seminal fluid. *J. Med. Ent.* 11:19–25.

Hinton, H. E. 1981a. *Biology of Insect Eggs*, vol. 1. Pergamon Press, New York.

Hinton, H. E. 1981b. *Biology of Insect Eggs*, vol. 2. Pergamon Press, New York.

Hjelle, J. T. 1990. Anatomy and morphology. In G. Polis, ed., *The Biology of Scorpions*, 9–63. Stanford University Press, Stanford.

Holler, C., H. Bargen, S. B. Vinson, and D. Witt. 1994. Evidence for the external use of juvenile hormone for host marking and regulation in a parasitic wasp, *Dendrocerus carpenteri*. *J. Insect Physiol.* 40(4):317–322.

Holmes, H. B. 1974. Patterns of sperm competition in *Nasonia vitripennis*. *Can. J. Gen. Cytol.* 16:789–795.

Holt, G. G., and D. T. North. 1970. Effects of gamma irradiation on the mechanism of sperm transfer in *Trichoplusia ni*. *J. Insect Physiol.* 16:2211–2222.

Hong, L. K. 1984. Survival of the fastest: On the origin of premature ejaculation. *J. Sex Res.* 20(2):109–122.

Hope, W. D. 1974. Nematoda. In A. Giese and J. Pearse, eds., *Reproduction of Marine Invertebrates*, vol. 1. Academic Press, New York.

Hopkins, C.C.E. 1978. The male genital system, and spermatophore production and function in *Euchaeta norvegica* Boeck (Copepoda: Calanoida). *J. Exp. Mar. Biol. Ecol.* 35:197–231.

Hopkins, C.C.E., and D. Machin. 1977. Patterns of spermatophore distribution and placement in *Euchaeta norvegica* (Copepoda: Calanoida). *J. Mar. Biol. Assoc. U.K.* 57:113–131.

Hoppe, P. C. 1975. Genetic and endocrine studies of the pregnancy-blocking pheromone of mice. *J. Reprod. Fert.* 45:109–115.

Hoshi, K., T. Aita, K. Yanagida, and A. Sato. 1990. Variation in the cholesterol/phospholipid ratio in human spermatozoa, and its relationship to capacitation. In B. D. Bavister, J. Cummins, and E.R.S. Roldan, eds., *Fertilization in Mammals*, 428. Serono Symposia, Norwell, Mass.

Houck, L. D., S. G. Tilley, and S. J. Arnold. 1985. Sperm competition in a plethodontid salamander: Preliminary results. *J. Herpetol.* 19:420–423.

Howard, D. J., and P. G. Gregory. 1993. Post-insemination signalling systems and reinforcement. *Phil. Trans. Roy. Soc. Lond. B* 338:1–6.

Howard, R. D., and A. G. Kluge. 1985. Proximate mechanisms of sexual selection in wood frogs. *Evolution* 39:260–277.

Howard, D. J., G. L. Waring, C. A. Tibbets, and P. G. Gregory. 1993. Survival of hybrids in a mosaic hybrid zone. *Evolution* 47:789–800.

Howell, J. F., R. B. Hutt, and W. B. Hill. 1978. Codling moth: mating behavior in the laboratory. *Ann. Ent. Soc. Am.* 71:891–895.

Hrdy, S. 1977. Infanticide as a primate reproductive strategy. *Am. Sci.* 65:40–49.

Hrdy, S. B. 1981. *The Woman that Never Evolved*. Harvard University Press, Cambridge, Mass.

Huber, B. A. 1993a. Genital mechanics and sexual selection in the spider *Nesticus cellulanus* (Araneae: Nesticidae). *Can. J. Zool.* 71:2437–2447.

Huber, B. 1993b. Female choice and spider genitalia. *Boll. Acc. Gioenia Sci. Nat.* 26:209–214.

Huber, B. A. 1994a. Genital morphology, copulatory mechanism and reproductive biology in *Psilochorus simoni* (Berland, 1911) (Pholcidae; Araneae). *Netherlands J. Zool.* 44(1–2):85–99.

Huber, B. 1994b. Copulatory mechanics of the funnel-web spiders *Histopona torpida* and *Textrix denticulata* (Agelenidae, Araneae). *Acta Zool.* (Stockholm) 75:379–384.

Huber, B. A. 1995. The retrolateral tibial apophysis in spiders—shaped by sexual selection? *Zool. J. Linn. Soc.* 113:151–163.

Huber, B. A. In press. Genital morphology and copulatory mechanics in *Anyphaena accentuata* (Anyphaenidae) and *Clubiona pallidula* (Clubionidae: Araneae). *J. Zool, Lond.*

Huck, U. W. 1982. Pregnancy block in laboratory mice as a function of male social status. *J. Reprod. Fert.* 66:181–184.

Huck, U. W., and R. D. Lisk. 1985. Determinants of mating success in the golden hamster (*Mesocricetus auratus*). II. Pregnancy initiation. *J. Comp. Psychol.* 99: 231–239.

Huck, U. W., and R. D. Lisk. 1986. Mating-induced inhibition of receptivity in the female golden hamster. *Behav. Neur. Biol.* 45:107–119.

Huck, U. W., R. D. Lisk, J. C. Allison, and C. G. Van Dongen. 1986a. Determinants of mating success in the golden hamster (*Mesocricetus auratus*): Social dominance and mating tactics under seminatural conditions. *Anim. Behav.* 34:971–989.

Huck, U. W., R. D. Lisk, and C. Thierjung. 1986b. Stimulus requirements for pregnancy initiation in the golden hamster (*Mesocricetus auratus*). Change with time of mating during the receptive period. *J. Reprod. Fert.* 76:449–458.

Huck, U. W., R. P. Quinn, and R. D. Lisk. 1985. Determinants of mating success in the golden hamster (*Mesocricetus auratus*). IV. Sperm competition. *Behav. Ecol. Sociobiol.* 17:239–252.

Huck, U. W., R. L. Soltis, and C. B. Coopersmith. 1982. Infanticide in male laboratory mice: Effects of social status, prior sexual experience, and basis for discrimination between related and unrelated young. *Anim. Behav.* 30:1158–1165.

Hughes, A. 1981. Differential male mating success in the white spotted sawyer *Monochamus scutellatus* (Coleoptera: Cerambycidae). *Ann. Ent. Soc. Am.* 74: 180–184.

Huignard, J. 1969. Action stimulatrice du spermatophore sur l'ovogenèse chez *Acanthoscelides obtectus* Say (Insecte Coleoptere). *C. R. Acad. Sci. Paris* 268(D):2938–2940.

Huignard, J. 1970. Analyse expérimentale de certains stimuli externes influençant l'ovogenèse chez *Acanthoscelides obtectus* Say (Coleoptere, Bruchidae). *Collog. Internat. Cent. Nat. Rech. Sci. (Fr.)* 189:357–380.

Huignard, J. 1971. Utilisation d'une substance paralysante, le parathion, pour l'étude de la migration des spermatozoides vers la spermatheque chez *Acanthoscelides obtectus* Say (Coleoptera Bruchidae). *C. R. Acad. Sci. Paris* 273(D): 2557–2559.

Huignard, J. 1983. Transfer and fate of male secretions deposited in the spermatophore of females of *Acanthoscelides obtectus* Say (Coleoptera Bruchidae). *J. Insect Physiol.* 29:55–63.

Huignard, J., A. Quesneau-Thierry, and M. Barbier. 1977. Isolement, action biologique et evolution des substances paragoniales contenues dans le spermatophore d'*Acanthoscelides obtectus* (Coleoptere). *J. Insect Physiol.* 23:351–357.

Humphries, D. A. 1967. The mating behaviour of the hen flea *Ceratophyllus gallinae* (Schrank) (Siphonaptera: Insecta). *Anim. Behav.* 15:82–90.

Hunter, F. M., M. Petrie, M. Otronen, T. Birkhead, and A. P. Moller. 1993. Why do females copulate repeatedly with one male? *TREE* 8:21–26.

Hunter, R.H.F. 1975. Transport, migration and survival of spermatozoa in the female genital tract: Species with intra-uterine deposition of semen. In E.S.E. Hafez and C. G. Thibault, eds., *The Biology of Spermatozoa*, 145–155. Karger, New York.

Huys, R., and G. A. Boxshall. 1991. *Copepod Evolution*. The Ray Society, London.

Ikeda, H. 1974. Multiple copulation: An abnormal mating behaviour which deleteriously affects fitness in *Drosophila mercatorum. Mem. Ehime Univ., Sci.*, ser. B (Biol.) 7:18–28.

Immelmann, K., and C. Beer. 1989. *A Dictionary of Ethology*. Harvard University Press, Cambridge, Mass.

Imms, A. D. 1964. *A General Textbook of Entomology*. Revised. O. W. Richards and R. G. Davies. Methuen, London.

Inoue, M., F. Mitsunaga, H. Ohsawa, A. Takenaka, Y. Sugiyama, A. G. Soumah, and O. Takenaka. 1992. Paternity testing in captive Japanese macaques (*Macaca fuscata*) using DNA fingerprinting. In R. D. Martin, A. F. Dixson and E. J. Wickings, eds. 1992. *Paternity in Primates: Genetic Tests and Theories*, 131–140. Karger, Basel.

Itgi, N. B., V. K. Biradar, and S. B. Mathad. 1982. Primary stimulus for oviposition and egg production in the house cricket *Gryllodes sigillatus* (Walker). *A. Angew. Entomol.* 94:35–41.

Itô, Y., and M. Yamagishi. 1989. Sperm competition in the melon fly, *Dacus cucurbitae* (Diptera: Tephritidae): Effects of sequential matings with normal and virgin or non-virgin sterile males. *Appl. Ent. Zool.* 24:466–477.

Iwasa, Y., and A. Pomiankowski. 1994. The evolution of mate preferences for multiple sexual ornaments. *Evolution* 48:853–867.

Jackson, R. R. 1978. An analysis of alternative mating tactics of the spider *Phidippus johnsoni* (Araneae, Salticidae). *J. Arachnol.* 5:185–230.

Jackson, R. R. 1980. The mating strategy of *Phidippus johnsoni* (Araneae, Salticidae). II. Sperm competition and the function of copulation. *J. Arachnol.* 8:217–240.

Jackson, R. R. 1981. Relationship between reproductive security and intersexual selection in a jumping spider *Phidippus johnsoni* (Araneae: Salticidae). *Evolution* 35(3):601–604.

Jackson, R. R. 1992. Conditional strategies and interpopulation variation in the behaviour of jumping spiders. *N.Z.J. Zool.* 19:99–111.

Jaenson, T.G.T. 1979. Mating behaviour of males of *Glossina pallidipes* Austen (Diptera: Glossinidae). *Bull. Ent. Res.* 69:573–588.

Jaenson, T.G.T. 1980. Mating behaviour of females of *Glossina pallidipes* Austen (Diptera: Glossinidae). *Bull. Ent. Res.* 70:49–60.

Jamieson, B. G. 1987. *The Ultrastructure and Phylogeny of Insect Spermatozoa*. Cambridge University Press, New York.

Jaszczak, S., and E.S.E. Hafez. 1975. Physiopathology of sperm transport in the human female. In E.S.E. Hafez and C. Thibault, eds., *The Biology of Spermatozoa*, 250–256. Karger, New York.

Jennett, F. J. 1985. Morphometric patterns among microtine rodents. I. Sexual selection suggested by relative scent gland development in representative voles (*Microtus*). In D. Durall, D. Muller-Schwarze, and M. Silverstein, eds., *Chemical Signals in Vertebrates 4*, 541–550. Plenum, New York.

Jennions, M. D., P.R.Y. Backwell, and N. I. Passmore. 1992. Breeding behaviour of the African frog, *Chiromantis xerampelina*: Multiple spawning and polyandry. *Anim. Behav.* 44:1091–1100.

Jewett, D. A., and W. R. Dukelow. 1978. Evaluation of caged mating behavior in *Macaca fascicularis*. *J. Med. Primatol.* 7:228–236.

Jöchle, W. 1975. Current research in coitus-induced ovulation: A review. *J. Reprod. Fert.* (suppl.) 22:165–207.

Johnson, F. M., and S. Bealle. 1968. Isozyme variability in species of the genus *Drosophila*. V. Ejaculatory bulb esterases in *Drosophila* phylogeny. *Biochem. Gen.* 2:1–18.

Johnson, R. D., R. L. Kitchell, and H. Gilanpour. 1986. Rapidly and slowly adapting mechanoreceptors in the glans penis of the cat. *Physiol. Behav.* 37:69–78.

Joly, D., M. L. Cariou, and D. Lachaise. 1991. Can female polyandry explain within-ejaculate sperm dimorphism? In B. Bacetti, ed., *Comparative Spermatology 20 Years After*, 771–773. Raven Press, New York.

Jones, J. C., and R. E. Wheeler. 1965. Studies on spermathecal filling in *Aedes aegypti* (Linnaeus). I. Description. *Biol. Bull.* 129:134–150.

Jones, M.D.R., and S. J. Gubbins. 1979. Modification of female circadian flight-activity by a male accessory gland pheromone in the mosquito, *Culex pipiens quinquefasciatus*. *Physiol. Ent.* 4:345–351.

Jones, R. E. 1984. *Human Reproduction and Sexual Behavior*. Prentice Hall, Englewood Cliffs, N.J.

Jones, T. P. 1966. Sex attraction and copulation in *Pelodera teres*. *Nematologica* 12:518–522.

Judson, C. L. 1967. Feeding and oviposition behavior in the mosquito *Aedes aegypti* (L.). I. Preliminary studies of physiological control mechanisms. *Biol. Bull.* 133:369–377.

Kalb, J. M., A. J. DiBenedetto, and M. F. Wolfner. 1993. Probing the function of *Drosophila melanogaster* accessory glands by directed cell ablation. *Proc. Natl. Acad. Sci. USA* 90:8093–8097.

Kanagawa, H., E.S.E. Hafez, M. M. Nawar, and S. Jaszczak. 1972. Patterns of sexual behavior and anatomy of copulatory organs in macaques. *Z. Tierpsychol.* 31:449–460.

Karpenko, C. P., and D. T. North. 1973. Ovipositional response elicited by normal, irradiated, F_1 male progeny, or castrated male *Trichoplusia ni* (Lepidoptera: Noctuidae). *Ann. Ent. Soc. Am.* 66:1278–1280.

Karr, T. L. 1991. Intracellular sperm/egg interactions in *Drosophila*: A three-dimensional structural analysis of a paternal product in the developing egg. *Mech. Devel.* 34:101–112.

Kasule, F. K. 1986. Repetitive mating and female fitness in *Dysdercus cardinali* (Hemiptera: Pyrrhocoridae). *Zool. J. Linn. Soc.* 88:191–199.

Katz, D. F., and E. Z. Drobnis. 1990. Analysis and interpretation of the forces generated by spermatozoa. In B. D. Bavister, J. Cummins, and E.R.S. Roldan, eds., *Fertilization in Mammals*, 126–137. Serono Symposia, Norwell, Mass.

Kaufmann, J. H. 1965. A three-year study of mating behavior in a freeranging band of rhesus monkeys. *Ecology* 46(4):500–512.

Kaulenas, M. S. 1992. *Insect Accessory Reproductive Structures*. Springer-Verlag, New York.

Keller, L. 1995. All's fair when love is war. *Nature* 373:190–191.

Keller, L., and H. K. Reeve. 1995. Why do females mate with multiple males? The sexually selected sperm hypothesis. *Adv. Stud. Behav.* 24:291–315.

Kelly, R. W. 1981. Prostaglandin synthesis in the male and female reproductive tract. *J. Reprod. Fert.* 62:293–304.

Kemp, D. H., B. F. Stone, and K. C. Binnington. 1982. Tick attachment and feeding: Role of the mouthparts, feeding apparatus, salivary gland secretions and the host response. In F. D. Obenchain and R. Galun, eds., *Physiology of Ticks*, 119–166. Pergamon Press, New York.

Kenagy, G. J., and S. C. Trombulak. 1986. Size and function of mammalian testes in relation to body size. *J. Mammal.* 67:1–22.

Kenney, A. M., T. G. Hartung, H. N. Davis, Jr., G. D. Gray, M. Zerylnick, and D. A. Dewsbury. 1978. Male copulatory behavior and the induction of ovulation in female voles: A quest for species specificity. *Hormones Behav.* 11:123–130.

Kerr, W. E., R. Zucchi, J. Nakadaira, and J. E. Butolo. 1962. Reproduction in the social bees (Hymenoptera: Apidae). *J. N.Y. Ent. Soc.* 70:265–276.

Keverne, E. B., and A. E. Rosser. 1985. The evolutionary significance of the olfactory block to pregnancy. In D. Duval, D. Muller-Schwarze, and R. M. Silverstein, eds., *Chemical Signals in Vertebrates* 4, 433–440. Plenum, New York.

Khalifa, A. 1949. The mechanism of insemination and the mode of action of the spermatophore in *Gryllus domesticus*. *Quart. J. Micro. Sci.* 90:281–292.

Khalil, G. M., and N. M. Shanbaky. 1975. The subgenus *Persicargas* (Ixodoidea: Argasidae: *Argas*) 21. The effect of some factors in the process of mating on egg development and oviposition in *A.* (*P.*) *arboreus* Kaiser, Hoogstraal & Kohls. *J. Med. Ent.* 12:47–51.

Khalil, G. M., D. E. Sonenshine, O. A. Sallam, and P. J. Homsher. 1983. Mating regulation and reproductive isolation in the ticks *Hyalomma dromedarii* and *H. anatolicum excavatum* (Acari: Ixodoidea: Ixodidae). *J. Med. Ent.* 20:136–145.

Kimura, K., and Y. Tsubaki. 1985. Egg weight variation associated with female age in *Pieris rapae crucivora* Boisduval (Lepidoptera: Pieridae). *Appl. Ent. Zool.* 20:500–501.

Kimura, T., K. Yasuyama, and T. Yamaguchi. 1988. Effect of some drugs and the bioactive substance derived from the spermatophore on the mechanical activity of the spermathecal duct in the female cricket. *Zool. Sci.* 5:1197.

Kingan, T. G., A. K. Raina, and P. Thomas-Laemont. 1993a. Control of reproductive behavior of female moths by factors in male seminal fluids. In R. D. Lumsden and J. L. Vaughn, eds., *Pest Management: Biologically Based Technologies*, 117–120. ACS Conference Proceedings Series. American Chemical Society, New York.

Kingan, T. G., P. A. Thomas-Laemont, and A. K. Raina. 1993b. Male accessory gland factors elicit change from "virgin" to "mated" behaviour in the female corn earworm moth *Helicoverpa zea*. *J. Exp. Biol.* 183:61–76.

Kirkendall, L. R. 1990. Sperm is a limiting resource in the pseudogamous bark beetle Ips acuminatus (Scolytidae). *Oikos* 57:80–87.

Kirkpatrick, M. 1987. The evolutionary forces acting on female mating preferences in polygynous animals. In J. W. Bradbury and M. B. Andersson, eds., *Sexual Selection: Testing the Alternatives*, 67–82. Wiley, Chichester, U.K.

Kirkpatrick, M., and M. J. Ryan. 1991. The evolution of mating preferences and the paradox of the lek. *Nature* 350:33–38.

Klauber, L. M. 1972. *Rattlesnakes*. Vol. 1. University of California Press, Berkeley.

Klepal, W., H. Barnes, and M. Barnes. 1977. Studies on the reproduction of cirripedes. VI. Passage of the spermatozoa into the oviducal sac and closure of pores. *J. Exp. Mar. Biol. Ecol.* 27:289–304.

Klostermeyer, L. E., and A. W. Anderson. 1976. Anatomy, histology, and post-larval development of the reproductive systems of the sugarbeet root maggot. *Ann. Ent. Soc. Am.* 69(4):625–631.

Klowden, M. J. 1979. Blood intake by *Aedes aegypti* not regulated by insemination. *J. Insect Physiol.* 25:349–351.

Klowden, M. J., and G. M. Chambers. 1991. Male accessory gland substances activate egg development in nutritionally starved *Aedes aegypti* mosquitoes. *J. Insect Physiol.* 37:721–726.

Knowlton, N., and S. R. Greenwell. 1984. Male sperm competition avoidance mechanisms: The influence of female interests. In R. L. Smith, ed., *Sperm Competition and the Evolution of Animal Mating Systems*, 62–84. Academic Press, New York.

Kodric-Brown, A. 1992. Male dominance can enhance mating success in guppies. *Anim. Behav.* 44:165–167.

Kodric-Brown, A., and J. H. Brown. 1984. Truth in advertising: The kinds of traits favored by sexual selection. *Am. Nat.* 124:309–323.

Koehler, J. K. 1965. An electron microscope study of the dimorphic spermatozoa of *Asplanchna* (Rotifera). *Zeit. Zellforsch.* 67:57–76.

Koeniger, G. 1986. Reproduction and mating behavior. In T. Rinderer, ed., *Bee Genetics and Breeding*, 255–279. Academic Press, New York.

Koeniger, G. 1991. Diversity in *Apis* mating systems. In D. R. Smith, ed., *Diversity in the Genus Apis*, 189–212. Westview Press, Boulder.

Koga, T., Y. Henmi, and M. Murai. 1993. Sperm competition and the assurance of underground copulation in the sand-bubbler crab *Scopimera globosa* (Brachyura: Ocypodidae). *J. Crust. Biol.* 13:134–137.

Kon, M., A. Oe, and H. Numata. 1993. Intra- and interspecific copulations in the two congeneric green stink bugs, *Nezara antennata* and *N. viridula* (Heteroptera, Pentatomidae), with reference to postcopulatory changes in the spermatheca. *J. Ethol.* 11:83–89.

Kopf, G. S. 1990. The zona pellucida-induced acrosome reaction: A model for sperm signal transduction. In B. D. Bavister, J. Cummins, and E.R.S. Roldan, eds., *Fertilization in Mammals*, 253–264. Serono Symposia, Norwell, Mass.

Koprowski, J. L. 1992. Removal of copulatory plugs by female tree squirrels. *J. Mammal.* 73(3):572–576.

Kotrba, M. 1990. Sperm transfer by spermatophore in an acalyptrate fly (Diptera: Diopsidae). *Ent. Gener.* 15:181–183.

Kraak, S.B.M., and E. P. Van Den Bergh. 1992. Do female fish assess paternal quality by means of test eggs? *Anim. Behav.* 43:865–867.

Krieger, F., and E. Krieger-Loibl. 1958. Beiträge zum Verhalten von *Ischnura elegans* und *Ischnura pumilio* (Odonata). *Zeitschr. Tierpsychol.* 15:82–93.

Kroger, W. S., and S. C. Freed. 1950. Psychosomatic aspects of sterility. *Am. J. Obst. Gynecol.* 59:867–874.

Kuba, H., and Y. Itô. 1993. Remating inhibition in the melon fly, *Bactrocera* (=*Dacus*) *cucurbitae* (Diptera: Tephritidae): Copulation with spermless males inhibits female remating. *J. Ethol.* 11:23–28.

Kuester, J., A. Paul, and J. Arnemann. 1992. Paternity determination by oligonucleotide DNA fingerprinting in Barbary macaques (*Macaca sylvanus*). In R. D. Martin, A. F. Dixson, and E. J. Wickings, eds., *Paternity in Primates: Genetic Tests and Theories*, 141–154. Karger, Basel.

Kumar, A., and G. U. Kurup, 1985. Sexual behavior of the lion-tailed macaque, *Macaca silenus*. In P. G. Heltne, ed., *The Lion-Tailed Macaque: Status and Conservation*, 109–148. Liss, New York.

Kunze, L. 1957. Die funktionsanatomischen Grundlagen der Kopulation der Zwergzikaden, untersucht an *Euscelis plebejus* (Fall.) und einigen Typhlocybinen. *Deut. Ent. Zeitschr.* 6:322–387.

Kurland, J. A. 1973. A natural history of kra macaques (*Macaca fascicularis* Raffles, 1821) at the Kutai Reserve, Kalimantan Timur, Indonesia. *Primates* 14(2–3):245–262.

Kuster, J. E., and K. G. Davey. 1983. The effect of allatectomy or neurosecretory cell ablation on protein synthesis in the spermathecae of *Rhodnius prolixus*. *Int. J. Invert. Reprod.* 6:189–195.

Labov, J. B. 1981. Pregnancy blocking in rodents: Adaptive advantages for females. *Am. Nat.* 118:361–371.

Lachmann, A. 1994. Struktur und Evolution der Fortpflanzungsorgane und Fortpflanzungssystem von fünf *Coproica*—Arten (Diptera, Sphaeroceridae). Doctoral thesis, Freie Universität, Berlin.

Lafon-Cazal, M., D. Gallois, J. Lehouelleur, and J. Bockaert. 1987. Stimulatory effects of male accessory-gland extracts on the myogenicity and the adenylate cyclase activity of the oviduct of *Locusta migratoria*. *J. Insect Physiol.* 33:909–915.

Laidlaw, Jr., H. H. 1944. Artificial insemination of the queen bee (*Apis mellifera* L.): Morphological basis and results. *J. Morph.* 74:429–465.

Lake, P. E. 1975. Gamete production and the fertile period with particular reference to domesticated birds. *Symp. Zool. Soc. Lond.* 35:225–244.

Lambert, R. D., and C. Tremblay. 1978. Effet du coit sur la montée des spermatozoides et la fécondation chez le lapin. *Rev. Can. Biol.* 37:1–4.

LaMunyon, C. W. 1994. Paternity in naturally-occurring *Utetheisa ornatrix* (Lepidoptera: Arctiidae) as estimated using enzyme polymorphism. *Behav. Ecol. Sociobiol.* 34:403–408.

LaMunyon, C. W. 1995. Paternity in naturally occurring *Utetheisa ornatrix* (Lepidoptera: Arctiidae) as estimated using enzyme polymorphism. *Behav. Ecol. Sociobiol.* 34:403–408.

LaMunyon, C. W., and T. Eisner. 1993. Postcopulatory sexual selection in an arctiid moth (*Utetheisa ornatrix*). *Proc. Nat. Acad. Sci. USA* 90:4689–4692.

LaMunyon, C. W., and T. Eisner. 1994. Spermatophore size as determinant of paternity in an arctiid moth (*Utetheisa ornatrix*). *Proc. Nat. Acad. Sci. USA* 91: 7081–7084.

Land, R. B., and T. E. McGill. 1967. The effects of the mating pattern of the mouse on the formation of corpora lutea. *J. Reprod. Fert.* 13:121–125.

Landa, V. 1960. Origin, development and function of the spermatophore in cockchafer (*Melolontha melolontha* L.). *Casopis Ceskoslovenske Spolecnosti Ent.* 57:297–316.

Lange, A. B. 1984. The transfer of prostaglandin-synthesizing activity during mating in *Locusta migratoria. Insect Biochem.* 14:551–556.

Lange, A. B., and B. G. Loughton. 1985. An oviposition-stimulating factor in the male accessory reproductive gland of the locust, *Locusta migratoria. Gen. Comp. Endocrinol.* 57:208–215.

Langtimm, C. A., and D. A. Dewsbury. 1991. Phylogeny and evolution of rodent copulatory behaviour. *Anim. Behav.* 41:217–225.

Lanier, D. L., D. Q. Estep, and D. A. Dewsbury. 1979. Role of prolonged copulatory behavior in facilitating reproductive success in a competitive mating situation in laboratory rats. *J. Comp. Physiol. Psychol.* 93:781–792.

Larsen, R. E. 1986. Semen collection from the boar. In D. Morrow, ed., *Current Therapy in Theriogenology*, 969–972. W. B. Saunders, Philadelphia.

Lea, A. O., and J. D. Edman. 1972. Sexual behavior of mosquitoes. 3. Age dependence of insemination of *Culex nigripalpus* and *C. pipiens quinquefasciatus* in nature. *Ann. Ent. Soc. Am.* 65:290–293.

Lea, A. O., and D. G. Evans. 1972. Sexual behavior of mosquitoes. 1. Age dependence of copulation and insemination in the *Culex pipiens* complex and *Aedes taeniorhynchus* in the laboratory. *Ann. Ent. Soc. Am.* 65:285–289.

Leahy, M. G. 1967. Nonspecificity of the male factor enhancing egg-laying in Diptera. *J. Insect Physiol.* 13:1283–1292.

Leahy, M. G. 1970. Effect of the male accessory gland secretion on mosquitos and fruit flies. *Collog. Internat.* 189:297–309.

Leahy, M. G. 1973. Oviposition of virgin *Schistocerca gregaria* (Forskal) (Orthoptera: Acrididae) after implant of the male accessory gland complex. *J. Ent.* [A]48:69–78.

Leahy, M. G., and G. B. Craig, Jr. 1965. Accessory gland substance as a stimulant for oviposition in *Aedes aegypti* and *A. albopictus. Mosquito News* 25:448–452.

Leahy, M. G., and R. Galun. 1972. Effect of mating on oogenesis and oviposition in the tick *Argas persicus* (Oken). *Parasitol.* 65:167–178.

Le Cato, III, G. L., and R. L. Pienkowski. 1972. Alfalfa weevil oviposition: Influence of sperm stored in the spermatheca. *Ann. Ent. Soc. Am.* 65:979–980.

Leckie, P. A., J. G. Watson, and S. Chaykin. 1973. An improved method for the artificial insemination of the mouse (*Mus musculus*). *Biol. Reprod.* 9:420–425.

Lee, D. C. 1974. Rhodacaridae (Acari: Mesostigmata) from near Adelaide, Australia. III: Behaviour and development. *Acarologia* 16:21–44.

Lee, J. C., and E. A. Bernays. 1990. Food tastes and toxic effects: Associative learning by the polyphagous grasshopper *Schistocerca americana* (Drury) (Orthoptera: Acrididae). *Anim. Behav.* 39:163–173.

Leegwater-van der Linden, M. E., and E.P.M. Tiggelman. 1984. Multiple mating and inseminating potential of *Glossina pallidipes*. *Ent. Exp. Appl.* 35:283–294.

Leigh-Sharpe, W. H. 1920. The comparative morphology of the secondary sexual characters of elasmobranch fishes. I. The claspers, clasper siphons, and clasper glands. *J. Morph.* 34:245–265.

Leigh-Sharpe, W. H. 1922. The comparative morphology of the secondary sexual characters of elasmobranch fishes. *J. Morph.* 36:191–197, 198–219, 221–243.

Leong, J. M., and J. E. Hafernik, Jr. 1992. Hybridization between two damselfy species (Odonata: Coenagrionidae): Morphometric and genitalic differentiation. *Ann. Ent. Soc. Am.* 85(6):662–670.

Leopold, R. A. 1976. The role of male accessory glands in insect reproduction. *Ann. Rev. Ent.* 21:199–221.

Leopold, R. A., and M. E. Degrugillier. 1973. Sperm penetration of housefly eggs: Evidence for involvement of a female accessory secretion. *Science* 181:555–557.

Leopold, R. A., S. Meola, and M. E. Degrugillier. 1978. The egg fertilization site within the house fly, *Musca domestica* (L.) (Diptera: Muscidae). *Int. J. Insect Morphol. Embryol.* 7:111–120.

Leopold, R. A., A. C. Terranova, and E. M. Swilley. 1971a. Mating refusal in *Musca domestica*: Effects of repeated mating and decerebration upon frequency and duration of copulation. *J. Exp. Zool.* 176:353–360.

Leopold, R. A., A. C. Terranova, B. J. Thorson, and M. E. Degrugillier. 1971b. The biosynthesis of the male housefly accessory secretion and its fate in the mated female. *J. Insect Physiol.* 17:987–1003.

Letsinger, J. T., and M. H. Gromko. 1985. The role of sperm numbers in sperm competition and female remating in *Drosophila melanogaster*. *Genetica* 66:195–202.

Leuthold, W. 1977. *African Ungulates*. Springer-Verlag, New York.

Levi, H. W. 1964a. American spiders of the genus *Episinus* (Araneae: Theridiidae). *Bull. Mus. Comp. Zool.* 131:1–25.

Levi, H. W. 1964b. The spider genera *Stemmops*, *Chrosiothes*, and the new genus *Cabello* from America. *Psyche* 71:73–92.

Levi, H. W. 1964c. The spider genus *Helvibus* (Araneae, Theridiidae). *Trans. Am. Microscop. Soc.* 83:133–142.

Levi, H. W. 1964d. American spiders of the genus *Phoroncidia* (Araneae: Theridiidae). *Bull. Mus. Comp. Zool.* 131:65–86.

Levi, H. W. 1964e. The spider genus *Thymoites* in America (Araneae: Theridiidae). *Bull. Mus. Comp. Zool.* 130:445–471.

Levi, H.W. 1967a. Cosmopolitan and pantropical species of theridiid spiders (Araneae: Theridiidae). *Pac. Ins.* 9:175–186.

Levi, H. W. 1967b. The theridiid spider fauna of Chile. *Bull. Mus. Comp. Zool.* 136:1–20.

Levi, H. W. 1974a. The orb-weaver genera *Araniella* and *Nuctenea* (Araneae: Araneidae). *Bull. Mus. Comp. Zool.* 146:291–316.

Levi, H. W. 1974b. The orb-weaver genus *Zygiella* (Araneae: Araneidae). *Bull. Mus. Comp. Zool.* 146:267–290.

Levi, H. W. 1975. The American orb-weaver genera *Larinia*, *Cercidia* and *Man-*

gora north of Mexico (Araneae, Araneidae). *Bull. Mus. Comp. Zool.* 147:101–135.

Levi, H. W. 1976. The orb-weaver genera *Verrucosa*, *Acanthepeira*, *Wagneriana*, *Acacesia*, *Wixia*, *Scoloderus* and *Alpaida* north of Mexico (Araneae: Araneidae). *Bull. Mus. Comp. Zool.* 147:351–391.

Levi, H. W. 1977a. The American orb-weaver genera *Cyclosa*, *Metazygia*, and *Eustala* north of Mexico (Araneae: Araneidae). *Bull. Mus. Comp. Zool.* 148:61–127.

Levi, H. W. 1977b. The orb-weaver genera *Metepeira*, *Kaira* and *Aculepeira* in America north of Mexico (Araneae: Araneidae). *Bull. Mus. Comp. Zool.* 148: 185–238.

Levi, H. W. 1978. The American orb-weaver genera *Colphepeira*, *Micrathena* and *Gasteracantha* north of Mexico (Araneae: Araneidae). *Bull. Mus. Comp. Zool.* 148:417–442.

Levi, H. W. 1982. The spider genera *Psechrus* and *Fecenia* (Arenaea: Psechridae). *Pac. Ins.* 24:114–138.

Levi, H. W. 1986. The neotropical orb-weaver genera *Chrysometa* and *Homalometa* (Araneae: Tetragnathidae). *Bull. Mus. Comp. Zool.* 151:91–215.

Levi, H. W. 1991. The neotropical orb-weaver genera *Edricus* and *Wagneriana* (Araneae: Araneidae). *Bull. Mus. Comp. Zool.* 152:363–415.

Levi, H. W., Y. D. Lubin, and M. H. Robinson. 1982. Two new *Achaearanea* species from Papua New Guinea with notes on other theridiid spiders (Araneae: Theridiidae). *Pac. Ins.* 24:105–113.

Levine, L. 1967. Sexual selection in mice. IV. Experimental demonstration of selective fertilization. *Am. Nat.* 101:289–294.

Lew, A. C., and H. J. Ball. 1980. Effect of copulation time on spermatozoan transfer of *Diabrotica virgifera* (Coleoptera: Chrysomelidae). *Ann. Ent. Soc. Am.* 73: 360–361.

Lewis, C. T., and J. N. Pollock 1975. Engagement of the phallosome in blowflies. *J. Ent.* (A) 49(2):137–147.

Lewis, S. M., and S. Austad. 1990. Sources of intraspecific variation in sperm precedence in red flour beetles. *Am. Nat.* 135:351–359.

Lewis, S. M., and O. T. Wang. 1991. Reproductive ecology of two species of *Photinus* fireflies (Coleoptera: Lampyridae). *Psyche* 98:293–307.

Lewontin, R. C. 1974. *The Genetic Basis of Evolutionary Change*. Columbia University Press, New York.

Licht, L. E. 1969. Unusual aspects of anuran sexual behavior as seen in the red-legged frog, *Rana aurora aurora*. *Can. J. Zool.* 47:505–509.

Licht, P. 1984. Reptiles. In G. E. Lamming, ed., *Marshall's Physiology of Reproduction*, vol. 1, 206–282. Churchill Livingstone, New York.

Liebherr, J. K. 1992. Phylogeny and revision of the *Platynus degallieri* species group (Coleoptera: Carabidae: Platynini). *Bull. Am. Mus. Nat. Hist.* 214:1–115.

Lifjeld, J. T., and R. J. Robertson. 1992. Female control of extra-pair fertilization in tree swallows. *Behav. Ecol. Sociobiol.* 31:89–96.

Lifjeld, J. T., P. O. Dunn, R. J. Robertson, and P. T. Boag. 1993. Extra-pair paternity in monogamous tree swallows. *Anim. Behav.* 45:213–229.

Liley, N. R. 1966. Ethological isolating mechanisms in four sympatric species of poeciliid fishes. *Behaviour* (suppl.) 13:1–197.

Linley, J. R. 1981. Emptying of the spermatophore and spermathecal filling in *Culicoides melleus* (Coq.) (Diptera: Ceratopogonidae). *Can. J. Zool.* 59:347–356.

Linley, J. R., and G. M. Adams. 1972. A study of the mating behaviour of *Culicoides melleus* (Coquillett) (Diptera: Ceratopogonidae). *Trans. Roy. Ent. Soc. Lond.* 124:81–121.

Linley, J. R., and G. M. Adams. 1974. Sexual receptivity in *Culicoides melleus* (Diptera: Ceratopogonidae). *Trans. Roy. Ent. Soc. Lond.* 126:279–303.

Linley, J. R., and M. J. Hinds. 1974. Male potency in *Culicoides melleus* (Coq.) (Diptera, Ceratopogonidae). *Bull. Ent. Res.* 64:23–128.

Linley, J. R., and M. J. Hinds. 1975a. Quantity of the male ejaculate influenced by female unreceptivity in the fly, *Culicoides melleus. J. Insect Physiol.* 21:281–285.

Linley, J. R., and M. J. Hinds. 1975b. Sperm loss at copulation in *Culicoides melleus. J. Ent.* (A)50:37–41.

Linley, J. R., and M. S. Mook. 1975. Behavioural interaction between sexually experienced *Culicoides melleus* (Coquillett) (Diptera Ceratopogonidae). *Behaviour* 54:7–110.

Linley, J. R., and K. R. Simmons. 1981. Sperm motility and spermathecal filling in lower Diptera. *Intl. J. Invert. Reprod.* 4:137–146.

Linley, J. R., and K. R. Simmons. 1983. Quantitative aspects of sperm transfer in *Simulium decorum* (Diptera: Simuliidae). *J. Insect Physiol.* 29:581–584.

Lofts, B. 1987. Testicular function. In D. O. Norris and R. E. Jones, eds., *Hormones and Reproduction in Fishes, Amphibians, and Reptiles*, 283–326. Plenum Press, New York.

Lofts, B., and R. K. Murton. 1973. Reproduction in birds. In D. S. Farner and J. R. King, eds., *Avian Biology*, vol. 3, 1–109. Academic Press, New York.

Loher, W. 1979. The influence of prostaglandin E2 on oviposition in *Teleogryllus commodus. Ent. Exp. Appl.* 25:107–119.

Loher, W. 1981. The effect of mating on female sexual behavior of *Teleogryllus commodus* Walker. *Behav. Ecol. Sociobiol.* 9:219–225.

Loher, W. 1984. Behavioral and physiological changes in cricket females after mating. *Adv. in Invert. Reprod.* 3:189–201.

Loher, W., and K. Edson. 1973. The effect of mating on egg production and release in the cricket *Teleogryllus commodus. Ent. Exp. Appl.* 16:483–490.

Loher, W., I. Ganjian, I. Kubo, D. Stanley-Samuelson, and S. S. Tobe. 1981. Prostaglandins: Their role in egg-laying of the cricket *Teleogryllus commodus. Proc. Nat. Acad. Sci. USA* 78:7835–7838.

Loher, W., T. Weber, and F. Huber. 1993. The effect of mating on phonotactic behaviour in *Gryllus bimaculatus* (De Geer). *Physiol. Ent.* 18:57–66.

Loibl, E. 1958. Zur Ethologie und Biologie der deutschen Lestiden (Odonata). *Z. Tierpsychol.* 15:54–81.

Longo, F. J. 1987. *Fertilization*. Chapman and Hall, New York.

Lopez, A. 1987. Glandular aspects of sexual biology. In W. Nentwig, ed., *Ecophysiology of Spiders*, 121–132. Springer-Verlag, New York.

Lopez-Leon, M. D., M. C. Pardo, J. Cabrero, and J.P.M. Camacho. 1994. Dynamics of sperm storage in the grasshopper *Eyprepocnemis plorans. Physiol. Ent.* 19:46–50.

Lorch, P. D., G. S. Wilkinson, and P. R. Reillo. 1993. Copulation duration and sperm precedence in the stalk-eyed fly *Cyrtodiopsis whitei* (Diptera: Diopsidae) *Behav. Ecol. Sociobiol.* 32:303–311.

Lorkovic, Z. 1952. L'accouplement artificiel chez les Lépidoptères et son application dans les recherches sur la fonction de l'appareil génital des insectes. *Physiol. Comp. Oecol.* 3:313–319.

Lounibos, L. P. 1994. Variable egg development among *Anopheles* (*Nyssorhynchus*): Control by mating? *Physiol. Ent.* 19:51–57.

Lovecky, D. V., D. Q. Estep, and D. A. Dewsbury. 1979. Copulatory behaviour of cotton mice (*Peromyscus gossypinus*) and their reciprocal hybrids with white-footed mice (*P. leucopus*). *Anim. Behav.* 27:371–375.

Lum, P.T.M. 1961. The reproductive system of some Florida mosquitoes. II. The male accessory glands and their role. *Ann. Ent. Soc. Am.* 54:430–433.

Lyons, S. A., and L. L. Getz. 1993. Reproductive activation of virgin female prairie voles (*Microtus ochrogaster*) by paired and unpaired males. *Behav. Proc.* 29: 191–200.

MacBean, I. T., and P. A. Parsons. 1966. The genotypic control of the duration of copulation in *Drosophila melanogaster*. *Experientia* 22:101–102.

Madsen, T., R. Shine, J. Loman, and T. Hakansson. 1992. Why do female adders copulate so frequently? *Nature* 355:440–441.

Mallory, F. F., and R. J. Brooks. 1978. Infanticide and other reproductive strategies in the collared lemming, *Dicrostonyx groenlandicus*. *Nature* 273:144–146.

Mallory, F. F., and F. V. Clulow. 1977. Evidence of pregnancy failure in the wild meadow vole, *Microtus pennsylvanicus*. *Can. J. Zool.* 55(1):1–17.

Mane, S. D., L. Tompkins, and R. C. Richmond. 1983. Male esterase 6 catalyzes the synthesis of a sex pheromone in *Drosophila melanogaster* females. *Science* 222:419–421.

Mange, A. P. 1970. Possible nonrandom utilization of X- and Y-bearing sperm in *Drosophila melanogaster*. *Genetics* 65:95–106.

Mann, T. 1984. *Spermatophores*. Springer-Verlag, New York.

Mann, T., and C. Lutwak-Mann. 1981. *Male Reproductive Function and Semen*. Springer-Verlag, New York.

Manning, A. 1967. The control of sexual receptivity in female *Drosophila*. *Anim. Behav.* 15:239–250.

Manson, J. H. 1994. Male agression: A cost of female choice in Cayo Santiago rhesus macaques. *Anim. Behav.* 48:473–475.

Marchlewska-Koj, A. 1983. Pregnancy blocking by pheromones. In J. G. Vandenbergh, ed., *Pheromones and Reproduction in Mammals*, 151–174. Academic Press, New York.

Markow, T. A. 1985. A comparative investigation of the mating system of *Drosophila hydei*. *Anim. Behav.* 33:775–781.

Markow, T. A., and P. F. Ankney. 1988. Insemination reaction in *Drosophila*: Found in species whose males contribute material to oocytes before fertilization. *Evolution* 42:1097–1101.

Marlow, B. J. 1961. Reproductive behaviour of the marsupial mouse, *Antechinus flavipes* (Waterhouse) (Marsupialia) and the development of the pouch young. *Aust. J. Zool.* 9:209–218.

Marquez, C. 1994. Historia natural de *Anolis aquaticus* Taylor 1956 (Sauria, Polychridae) en la quebrada La Palma, Puriscál, San José, Costa Rica. Master's thesis, University of Costa Rica.

Marraro, C. H., and J. R. Nursall. 1983. The reproductive periodicity and behaviour of *Ophioblennius atlanticus* (Pisces: Blenniidae) at Barbados. *Can. J. Zool.* 61:317–325.

Martin, R. H. 1990. Analysis of human sperm chromosome complements. In B. D. Bavister, J. Cummins, and E.R.S. Roldan, eds., *Fertilization in Mammals*, 365–372. Serono Symposia, Norwell, Mass.

Massoud, Z., and J.-M. Betsch. 1972. Étude sur les insectes collemboles, II: Les caractères sexuels secondaires des antennes des symphplèones. *Rev. Ecol. Biol. Sol* 9:55–97.

Masters, W. H., and V. E. Johnson. 1966. *The Human Sexual Response*. Little, Brown, New York.

Mastroianni, Jr., L. 1958. Clinical concepts of infertility. In J. T. Velardo, ed., *Essentials of Human Reproduction*, 171–189. Oxford University Press, New York.

Masumoto, T. 1991. Males' visits to females' webs and female mating receptivity in the spider, *Agelena limbata* (Araneae: Agelenidae). *J. Ethol.* 9:1–7.

Masumoto, T. 1993. The effect of the copulatory plug in the funnel-web spider, *Agelena limbata* (Araneae: Agelenidae). *J. Arachnol.* 21:55–59.

Matthews, M., and N. T. Adler. 1977. Facilitative and inhibitory influences of reproductive behavior on sperm transport in rats. *J. Comp. Physiol. Psychol.* 91: 727–741.

Maynard Smith, J. 1956. Fertility, mating behaviour and sexual selection in *Drosophila subobscura. J. Genet.* 54:261–279.

Maynard Smith, J. 1976. Sexual selection and the handicap principle. *J. theor. Biol.* 57:239–242.

Maynard Smith, J. 1987. Sexual selection—a classification of models. In J. W. Bradbury and M. B. Andersson, eds., *Sexual Selection: Testing the Alternatives*, 9–20. Wiley, New York.

Maynard Smith, J. 1991. Theories of sexual selection. *TREE* 6:146–151.

Mayr, E. 1982. *The Growth of Biological Thought*. Belknap Press of Harvard University Press, Cambridge.

Mazomenos, B., J. L. Nation, W. J. Coleman, K. C. Dennis, and R. Esponda. 1977. Reproduction in Caribbean fruit flies. Comparisons between a laboratory strain and a wild strain. *Fla. Ent.* 60:139–144.

McAlpine, D. K. 1988. Studies in upside-down flies (Diptera: Neurochaetidae). Part II. Biology, adaptations, and specific mating mechanisms. *Proc. Linn. Soc. N.S.W.* 110:59–82.

McClintock, M. K. 1984. Group mating in the domestic rat as a context for sexual selection: Consequences for the analysis of sexual behavior and neuroendocrine responses. *Adv. Stud. Behav.*

McClintock, M. K., J. J. Anisko, and N. T. Adler. 1982a. Group mating among Norway rats. II. The social dynamics of copulation: Competition, cooperation, and mate choice. *Anim. Behav.* 30:410–425.

McClintock, M. K., J. P. Toner, N. T. Adler, and J. J. Anisko. 1982b. Postejaculatory quiescence in female and male rats: Consequences for sperm transport during group mating. *J. Comp. Physiol. Psychol.* 96:268–277.

McComb, K. 1987. Roaring by red deer stags advances the date of oestrus in hinds. *Nature* 330:648–649.

McGill, T. E. 1962. Sexual behavior in three inbred strains of mice. *Behaviour* 19:341–350.

McGill, T. E. 1977. Reproductive isolation, behavioral genetics, and functions of sexual behavior in rodents. In J. S. Rosenblatt and B. R. Komisaruk, eds., *Reproductive Behavior and Evolution*, 78–109. Plenum Press, New York.

McGill, T. E., and T. W. Ransom. 1968. Genotypic change affecting conclusions regarding the mode of inheritance of elements of behaviour. *Anim. Behav.* 16:88–91.

McGill, T. E., D. M. Corwin, and D. T. Harrison. 1968. Copulatory plug does not induce luteal activity in the mouse *Mus musculus*. *J. Reprod. Fert.* 15:149–151.

McKinney, F., K. M. Cheng, and D. J. Bruggers. 1984. Sperm competition in apparently monogamous birds. In R. L. Smith, ed., *Sperm Competition and the Evolution of Animal Mating Systems*, 523–545. Academic Press, New York.

McKinney, F., W. R. Siegfried, I. J. Ball, and P.G.H. Frost. 1978. Behavioral specializations for river life in the African black duck (*Anas sparsa*) (Eyton). *Z. Tierpsychol.* 48:349–400.

McLain, D. K. 1980. Female choice and the adaptive significance of prolonged copulation in *Nezara viridula* (Hemiptera: Pentatomidae). *Psyche* 87:325–336.

McLain, D. K. 1981. Sperm precedence and prolonged copulation in the southern green stink bug, *Nezara viridula*. *J. Georgia Ent. Soc.* 16:70–77.

McLain, D. K., and N. B. Marsh. 1990. Male copulatory success: Heritability and relationship to mate fecundity in the southern green stinkbug, *Nezara viridula* (Hemiptera: Pentatomidae). *Heredity* 64:161–167.

McLain, D. K., D. L. Lanier, and N. B. Marsh. 1990. Effects of female size, male size, and number of copulations on fecundity, fertility and longevity of *Nezara viridula* (Hemiptera: Pentatomidae). *Ann. Ent. Soc. Am.* 83:1130–1136.

McVey, M. E. 1988. The opportunity for sexual selection in a territorial dragonfly, *Erythemis simplicicollis*. In T. H. Clutton-Brock, ed., *Reproductive Success—Studies of Individual Variation in Contrasting Breeding Systems*, 44–58. University of Chicago Press, Chicago.

McVey, M. E., and B. J. Smittle. 1984. Sperm precedence in the dragonfly *Erythemis simplicicollis*. *J. Insect Physiol.* 30(8):619–628.

Menard, N., W. Scheffrahn, D. Vallet, C. Zidane, and C. Reber. 1992. Application of blood protein electrophoresis and DNA fingerprinting to the analysis of paternity and social characteristics of wild barbary macaques. In R. D. Martin, A. F. Dixson, and E. J. Wickings, eds., *Paternity in Primates: Genetic Tests and Theories*, 155–174. Karger, Basel.

Mendonça, M. T., and D. Crews. 1990. Mating-induced ovarian recrudescence in the red-sided garter snake. *J. Comp. Physiol.* A 166:629–632.

Merle, J. 1968. Fonctionnement ovarien et réceptivité sexuelle de *Drosophila melanogaster* après implantation de fragments de l'appareil gènital male. *J. Insect Physiol.* 14:1159–1168.

Metter, D. E. 1964. On breeding and sperm retention in *Ascaphus. Copeia* 1964:710–711.

Michael, R. P., and G. S. Saayman. 1967. Individual differences in the sexual behaviour of male rhesus monkeys (*Macaca mulatta*) under laboratory conditions. *Anim. Behav.* 15:460–466.

Michael, R. P., and D. Zumpe. 1971. Patterns of reproductive behavior. In E.S.E. Hafez, ed., *Comparative Reproduction of Nonhuman Primates*, 205–242. Thomas, Springfield.

Mikkola, K. 1992. Evidence for lock-and-key mechanisms in the internal genitalia of the *Apamea* moths (Lepidoptera, Noctuidae). *Syst. Ent.* 17:145–153.

Miller, P. L. 1984. The structure of the genitalia and the volumes of sperm stored in male and female *Nesciothemis farinosa* (Foerster) and *Orthetrum chrysostigma* (Burmeister) (Anisoptera: Libellulidae). *Odonatologica* 13:415–428.

Miller, P. L. 1990. Mechanisms of sperm removal and sperm transfer in *Orthetrum coerulescens* (Fabricius) (Odonata: Libellulidae). *Physiol. Ent.* 15:199–209.

Milledge, F. 1993. Further remarks on the taxonomy and relationships of the Linyphiidae, based on the epigynal duct conformation and other characters (Araneae). *Bull. Br. Arachnol. Soc.* 9:145–156.

Milligan, S. R. 1980. Pheromones and rodent reproductive physiology. *Symp. Zool. Soc. Lond.* 45:251–275.

Milligan, S. R. 1982. Induced ovulation in mammals. *Oxford Reviews of Reprod. Biol.* 4:1–46.

Missakian, E. A., L. R. Del Rio, and R. E. Myers. 1969. Reproductive behavior of captive male rhesus monkeys (*Macaca mulatta*). *Comm. Behav. Biol.* 4:231–235.

Mitchell, G. D. 1979. *Behavioral Sex Differences in Nonhuman Primates*. Van Nostrand-Reinhold, New York.

Mock, D. W., and M. Fujioka. 1990. Monogamy and long-term pair bonding in vertebrates. *TREE* 5:39–43.

Møller, A. P. 1987a. Copulation behaviour in the goshawk, *Accipiter gentilis*. *Anim. Behav.* 35:755–763.

Møller, A. P. 1987b. Behavioural aspects of sperm competition in swallows (*Hirundo rustica*). *Behaviour* 100:92–104.

Møller, A. P. 1987c. House sparrow, *Passer domesticus*, communal displays. *Anim. Behav.* 35:203–210.

Møller, A. 1988. Female choice selects for male sexual tail ornaments in the monogamous swallow. *Nature* 332:640–642.

Møller, A. P. 1991a. Sperm competition and sperm counts: An evaluation of the empirical evidence. In B. Bacetti, ed., *Comparative Spermatology 20 Years After*, 775–777. Raven Press, New York.

Møller, A. P. 1991b. Sperm competition, sperm depletion, paternal care and relative testes size in birds. *Am. Nat.* 137:882–906.

Møller, A. P., and Pomiankowski, A. 1993. Why have birds got multiple sexual ornaments? *Behav. Ecol. Sociobiol.* 32:167–176.

Monsma, S. A., and M. F. Wolfner. 1988. Structure and expression of a *Drosophila* male accessory gland gene whose product resembles a peptide pheromone precursor. *Genes Dev.* 2:1063–1073.

Monsma, S. C., H. A. Harada, and M. F. Wolfner. 1990. Synthesis of two *Drosophila* male accessory gland proteins and their fate after transfer to the female during mating. *Develop. Biol.* 142:465–475.

Montgomerie, R., and R. Thornhill. 1989. Female advertisement in birds: A means of inciting male-male competition? *Ethology* 81:209–220.

Moore, A. J. 1994. Genetic evidence for the "good genes" process of sexual selection. *Behav. Ecol. Sociobiol.* 35:235–241.

Moorhouse, D. E. 1969. The attachment of some ixodid ticks to their natural hosts. *Proc. 2nd Int. Congr. Acarol.* 1967:319–327.

Morali, G., and C. Beyer. 1992. Motor aspects of masculine sexual behavior in rats and rabbits. *Adv. Stud. Behav.* 21:201–219.

Morón, M. A. 1986. *El Género Phyllophaga en México, Morfología, Distribución, y Sistemática Supraespecífica.* Pub. 20, Instituto de Ecología, Mexico, D.F.

Morrison, P. E., K. Venkatesh, and B. Thompson. 1982. The role of male accessory-gland substance on female reproduction with some observations of spermatogenesis in the stable fly. *J. Insect Physiol.* 28:607–614.

Morrow, D., ed., 1986. *Current Therapy in Theriogenology.* Saunders, Philadelphia.

Morton, D. B., and T. D. Glover. 1974a. Sperm transport in the female rabbit: The role of the cervix. *J. Reprod. Fert.* 38:131–138.

Morton, D. B., and T. D. Glover. 1974b. Sperm transport in the female rabbit: The effect of inseminate volume and sperm density. *J. Reprod. Fert.* 38:139–146.

Mulcahy, D. L. 1975. Preface. In D. Mulcahy, ed., *Gamete Competition in Plants and Animals.* American Elsevier, New York.

Mulder, R. A., and M.J.L. Magrath. 1994. Timing of prenuptial molt as a sexually selected indicator of male quality in superb fairy-wrens (*Malurus cyaneus*). *Behav. Ecol.* 5(4): 7853.

Mulder, R. A., P. O. Dunn, A. Cockburn, K. A. Lazenby-Cohen, and M. J. Howell. 1994. Helpers liberate female fairy-wrens from constraints on extra-pair mate choice. *Proc. Roy. Soc. Lond. B* 255:223–229.

Muma, M. H. 1966. Mating behaviour in the solpugid genus *Eremobates* Banks. *Anim. Behav.* 14:346–350.

Munne, S., A. Estop, J. Santalo, and V. Catala. 1990. In vitro ageing of mouse sperm produces structural chromosome abnormalities. In B. D. Bavister, J. Cummins, and E.R.S. Roldan, eds., *Fertilization in Mammals*, 450. Serono Symposia, Norwell, Mass.

Murai, M., S. Goshima, and Y. Henmi. 1987. Analysis of the mating system of the fiddler crab, *Uca lactea. Anim. Behav.* 35:1134–1342.

Murtaugh, M. P., and D. L. Denlinger. 1982. Prostaglandins E and F_2 in the house cricket and other insects. *Insect Biochem.* 12:599–603.

Murtaugh, M. P., and D. L. Denlinger. 1985. Physiological regulation of long-term oviposition in the house cricket, *Acheta domesticus. J. Insect Physiol.* 31:611–617.

Murtaugh, M. P., and D. L. Denlinger. 1987. Regulation of long-term oviposition in the house cricket, *Acheta domesticus*: Roles of prostaglandin and factors associated with sperm. *Arch. Insect Biochem. Physiol.* 6:59–72.

Murvosh, C., R. L. Fye, and G. C. Labrecque. 1964. Studies on the mating behavior of the house fly, *Musca domestica* L. *Ohio J. Sci.* 64:264–271.

Nadler, R. D., and L. A. Rosenblum. 1969. Sexual behavior of male bonnet monkeys in the laboratory. *Brain Behav. Evol.* 2:482–497.

Nadler, R. D., and L. A. Rosenblum. 1973. Sexual behavior of male pigtail macaques in the laboratory. *Brain Behav. Evol.* 7:18–33.

Nakagawa, S., G. J. Farias, D. Suda, R. T. Cunningham, and D. L. Chambers. 1971. Reproduction of the Mediterranean fruit fly: Frequency of mating in the laboratory. *Ann. Ent. Soc. Am.* 64:949–950.

Nakamura, K., and K. Sekiguchi. 1980. Mating behavior and oviposition in the pycnogonid *Propallene longicaps*. *Mar. Ecol. Prog. Ser.* 2:163–168.

Nalepa, C. 1988. Cost of parental care in the woodroach *Cryptocercus punctulatus* Scudder (Dictyoptera: Cryptocercidae). *Behav. Ecol. Sociobiol.* 23:135–140.

Negro, J. J., J. A. Donazar, and F. Hiraldo. 1992. Copulatory behaviour in a colony of lesser kestrels: Sperm competition and mixed reproduction strategies. *Anim. Behav.* 43:921–930.

Nelson, D. R., T. S. Adams, and J. G. Pomonis. 1969. Initial studies on the extraction of the active substance inducing monocoitic behavior in house flies, black blow flies, and screw-worm flies. *J. Econ. Ent.* 62:634–639.

Nilakhe, S. S., and E. J. Villavaso. 1979. Measuring sperm competition in the boll weevil by the use of females whose spermathecae have been surgically removed. *Ann. Ent. Soc. Am.* 72:500–502.

Norris, M. J. 1933. Contributions toward the study of insect fertility. II. Experiments on the factors influencing fertility in *Ephestia kuhniella* A. (Lepidoptera, Phycitidae). *Proc. Zool. Soc. Lond.* 4:903–934.

Oakeshott, J. G., T. M. Boyce, R. J. Russell, and M. J. Healy. 1995. Molecular insights into the evolution of an enzyme; esterase-6 in *Drosophila*. *TREE* 10:103–110.

Obara, Y. 1982. Mate refusal hormone in the cabbage white butterfly? *Naturwissenschaften* 69:551–552.

Oberhauser, K. S. 1989. Effects of spermatophores on male and female monarch butterfly reproductive success. *Behav. Ecol. Sociobiol.* 25:237–246.

Oberhauser, K. S. 1992. Rate of ejaculate breakdown and intermating intervals in monarch butterflies. *Behav. Ecol. Sociobiol.* 31:367–373.

O'Connell, S. M., and G. Cowlishaw. 1994. Infanticide avoidance, sperm competition and male choice: The function of copulation calls in female baboons. *Anim. Behav.* 48:687–694.

Oglesby, J. M., D. L. Lanier, and D. A. Dewsbury. 1981. The role of prolonged copulatory behavior in facilitating reproductive success in male Syrian golden hamsters (*Mesocricetus auratus*) in a competitive mating situation. *Behav. Ecol. Sociobiol.* 8:47–54.

O'Hanlon, J. K., and B. D. Sachs. 1986. Fertility of mating in rats (*Rattus norvegicus*): Contributions of androgen-dependent morphology and actions of the penis. *J. Comp. Psychol.* 100:178–187.

Okelo, O. 1979. Mechanisms of sperm release from the receptaculum seminis of *Schistocerca vaga* Scudder (Orthoptera: Acrididae). *Int. J. Invert. Reprod.* 1:121–131.

Oldfield, G. N., and I. M. Newell. 1973. The role of the spermatophore in the reproductive biology of protogynes of *Aculus cornutus* (Acarina: Eriophyidae). *Ann. Ent. Soc. Am.* 66(1):160–163.

Olds-Clarke, P. 1990. Variation in the quality of sperm motility and its relationship to capacitation. In B. D. Bavister, J. Cummins, and E.R.S. Roldan, eds., *Fertilization in Mammals*, 91–99. Serono Symposia, Norwell, Mass.

Oliver, J. H. 1982. Tick reproduction: Sperm development and cytogenetics. In F. D. Obenchain and R. Galun, eds., *Physiology of Ticks*, 245–275. Pergamon Press, New York.

Oliver, J. H. 1986. Induction of oogenesis and oviposition in ticks. In J. R. Sauer and J. A. Hair, eds., *Morphology, Physiology and Behavioral Biology of Ticks*, 233–247. Wiley, New York.

Oliver, J. H., R. W. Murphy, and F. D. Obenchain. 1975. Reproduction in ticks (Acari: Ixodoidea). 4. Effects of mechanical and chemical stimulation on oocyte development in *Amblyomma americanum*. *J. Parasitol.* 61:782–784.

Oliver, J. H., M. J. Pound, and R. H. Andrews. 1984. Induction of egg maturation and oviposition in the tick *Ornithodoros parkeri* (Acari: Argasidae). *J. Parasitol.* 70:337–342.

Olsson, M., A. Gullberg, and H. Tegelstrom. 1994. Sperm competition in the sand lizard *Lacerta agilis*. *Anim. Behav.* 48:193–200.

O'Meara, G. F., and D. G. Evans. 1977. Autogeny in saltmarsh mosquitoes induced by a substance from the male accessory gland. *Nature* 267:342–344.

Omura, S. 1938. Structure and function of the female genital system of *Bombyx mori* with special reference to the mechanism of fertilization. *J. Fac. Agric. Hokkaido Univ.* 40:111–128.

Ono, T., M. T. Siva-Jothy, and A. Kato. 1989. Removal and subsequent ingestion of rivals' semen during copulation in a tree cricket. *Physiol. Ent.* 14:195–202.

Opell, B. D. 1979. Revision of the genera and tropical american species of the spider family Uloboridae. *Bull. Mus. Comp. Zool.* 148:443–549.

Opell, B. D., and J. A. Beatty. 1976. The nearctic Hahniidae (Arachnida: Araneae). *Bull. Mus. Comp. Zool.* 147:393–433.

Orshan, L., and M. P. Pener. 1991. Effects of the corpora allata on sexual receptivity and completion of oocyte development in females of the cricket, *Gryllus bimaculatus*. *Physiol. Ent.* 16:231–242.

Otronen, M. 1984a. The effect of differences in body size on the male territorial system of the fly *Dryomyza anilis*. *Anim. Behav.* 32:882–890.

Otronen, M. 1984b. Male contests for territories and females in the fly *Dryomyza anilis*. *Anim. Behav.* 32:891–898.

Otronen, M. 1989. Female mating behaviour and multiple matings in the fly *Dryomyza anilis*. *Behav.* 111:77–97.

Otronen, M. 1990. Mating behavior and sperm competition in the fly, *Dryomyza anilis*. *Behav. Ecol. Sociobiol.* 26:349–356.

Otronen, M. 1993a. Size-related male movement and its effect on spatial and temporal male distribution at female oviposition sites in the fly *Dryomyza anilis*. *Anim. Behav.* 46:731–740.

Otronen, M. 1993b. Size assortative mating in the yellow dung fly *Scatophaga stercoraria*. *Behaviour* 126(1–2):63–76.

Otronen, M. 1994a. Repeated copulation as a strategy to maximize fertilization in the fly, *Dryomyza anilis* (Dryomyzidae). *Behav. Ecol.* 5(1):51–56.

Otronen, M. 1994b. Fertilisation success in the fly *Dryomyza anilis* (Dryomyzidae): Effects of male size and the mating situation. *Behav. Ecol. Sociobiol.* 35: 33–38.

Otronen, M., and M. T. Siva-Jothy. 1991. The effect of postcopulatory male behaviour on ejaculate distribution within the female sperm storage organs of the fly, *Dryomyza anilis* (Diptera: Dryomyzidae). *Behav. Ecol. Sociobiol.* 29:33–37.

Overstreet, J. W., and D. F. Katz. 1977. Sperm transport and selection in the female genital tract. In M. H. Johnson, ed., *Development in Mammals*, vol. 2, 31–63. North-Holland, New York.

Packer, C., and A. E. Pusey. 1983. Adaptations of female lions to infanticide by incoming males. *Am. Nat.* 121:716–728.

Page, Jr., R. E. 1986. Sperm utilization in social insects. *Ann. Rev. Ent.* 31:297–320.

Palombit, R. A. 1994. Extra-pair copulations in a monogamous ape. *Anim. Behav.* 47:721–723.

Palti, Z., and M. Freund. 1972. Spontaneous contractions of the human ovary *in vitro*. *J. Reprod. Fertil.* 28:113–115.

Pappas, P. J., and J. H. Oliver. 1972. Reproduction in ticks (Acari: Ixodoidea). 2. Analysis of the stimulus for rapid and complete feeding of female *Dermacenter variabilis* (Say). *J. Med. Ent.* 9:47–50.

Parker, G. A. 1970a. Sperm competition and its evolutionary effect on copula duration in the fly *Scatophaga stercoraria*. *J. Insect Physiol.* 16:1301–1328.

Parker, G. A. 1970b. The reproduction behaviour and nature of sexual selection in *Scatophaga stercoriaria* L. (Diptera: Scatophagidae). II. The fertilization rate and spatial and temporal relationships of each sex around the site of mating and oviposition. *J. Anim. Ecol.* 39:205–228.

Parker, G. A. 1970c. Sperm competition and its evolutionary consequences. *Biol. Rev.* 45:525–567.

Parker, G. A. 1972. Reproductive behavior of *Sepsis cynipsea* (L.) (Diptera: Sepsidae). I. Preliminary analysis of the reproductive strategy and its associated behaviour patterns. *Behaviour* 41:172–206.

Parker, G. A. 1979. Sexual selection and sexual conflict. In M. S. Blum and N. Blum, eds., *Sexual Selection and Reproduction Competition in Insects*, 123–166. Academic Press, New York.

Parker, G. A. 1984. Sperm competition and the evolution of animal mating strategies. In R. L. Smith, ed., *Sperm Competition and the Evolution of Animal Mating Systems*, 2–60. Academic Press, New York.

Parker, G. A. 1990. Sperm competition games: Raffles and roles. *Proc. Roy. Soc. Lond. B*. 242:120–126.

Parker, G. A. 1992. Snakes and female sexuality. *Nature* 355:395–396.

Parker, G. A., and L. W. Simmons. 1991. A model of constant random sperm displacement during mating: Evidence from *Scatophaga*. *Proc. Roy. Soc. Lond. Ser. B* 246:107–115.

Parker, G. A., and J. L. Smith. 1975. Sperm competition and the evolution of the precopulatory passive phase behaviour in *Locusta migratoria migratorioides*. *J. Ent.* [A]49:155–171.

Parker, G. A., R. R. Baker, and V.G.F. Smith. 1972. The origin and evolution of gamete dimorphism and the male-female phenomenon. *J. theor. Biol.* 36:529–533.

Parker, G. A., L. W. Simmons, and H. Kirk. 1990. Analyzing sperm competition data: Simple models for predicting-mechanisms. *Behav. Ecol. Sociobiol.* 27:55–65.

Parks, J. E., and E. Ehrenwald. 1990. Cholesterol efflux from mammalian sperm and its potential role in capacitation. In B. D. Bavister, J. Cummins, and E.R.S. Roldan, eds., *Fertilization in Mammals*, 155–168. Serono Symposia, Norwell, Mass.

Parrish, J. J., and R. H. Foote. 1985. Fertility differences among male rabbits determined by heterospermic insemination of fluorochrome-labeled spermatozoa. *Biol. Reprod.* 33:940–949.

Paterson, H.E.H. 1978. More evidence against speciation by reinforcement. *S. Afr. J. Sci.* 74:369–371.

Paterson, H.E.H. 1982. Perspective on speciation by reinforcement. *S. Afr. J. Sci.* 78:53–57.

Paul, A. J. 1984. Mating frequency and viability of stored sperm in the tanner crab *Chionoecetes bairdi* (Decapoda, Majidae). *J. Crust. Biol.* 4:375–381.

Peretti, A. V. 1992. The spermatophore of *Bothriurus bonariensis* (C. L. Koch) (Scorpiones, Bothriuridae): Morphology and functioning. *Bol. Soc. Biol.* (Concepcion, Chile) 63:157–167.

Peretti, A. V. In press. Analisis del comportamiento de transferencia espermática de *Bothriurus flavidus* Kraepelin (Scorpiones, Bothriuridae). *Rev. Bras. Biol.*

Perreault, S. D. 1989. Regulation of sperm nuclear reactivation during fertilization. In B. D. Bavister, J. Cummins, and E.R.S. Roldan, eds., *Fertilization in Mammals*, 285–296. Serono Symposia, USA, Norwell, Mass.

Petersson, E. 1989. Mating in swarming caddis flies (Trichoptera: Leptoceridae). Ph.D. thesis, Uppsala University.

Petrie, M. 1992. Copulation behaviour in birds: Why do females copulate more than once with the same male? *Anim. Behav.* 44:790–792.

Petrie, M., M. Hall, T. Halliday, H. Budgey, and C. Pierpoint. 1992. Multiple mating in a lekking bird: Why do peahens mate with more than one male and with the same male more than once? *Behav. Ecol. Sociobiol.* 31:349–358.

Pickford, R., and C. Gillott. 1971. Insemination in the migratory grasshopper, *Melanoplus sanguinipes. Can. J. Zool.* 49:1583–1588.

Pickford, R., A. B. Ewen, and C. Gillott. 1969. Male accessory gland substance: An egg-laying stimulant in *Melanoplus sanguinipes* (F.) (Orthoptera: Acrididae). *Can. J. Zool.* 47:1199–1203.

Pierce, Jr., J. D., B. Ferguson, A. L. Salo, D. K. Sawrey, L. E. Shapiro, S. A. Taylor, and D. A. Dewsbury. 1990. Patterns of sperm allocation across successive ejaculates in four species of voles (*Microtus*). *J. Reprod. Fert.* 88:141–149.

Pitkjanen, I. G. 1959. Some data on the transport of semen in the genital tract of the sow. *Animal Breeding Abstracts* 27:212.

Pitnick, S. 1991. Male size influences mate fecundity and remating interval in *Drosophila melanogaster. Anim. Behav.* 41:735-745.

Pitnick, S. 1993. Operational sex ratios and sperm limitation in populations of *Drosophila pachea. Behav. Ecol. Sociobiol.* 33:383–391.

Pitnick, S., and T. A. Markow. 1994. Male gametic strategies: Sperm production and the allocation of ejaculate among successive mates by the sperm-limited fly *Drosophila pachea* and its relatives. *Am. Nat.* 143:785–819.

Pitnick, S., T. A. Markow, and M. F. Riedy. 1991. Transfer of ejaculate and incorporation of male-derived substances by females in the *nannoptera* species group (Diptera: Drosophilidae). *Evolution* 45:774–780.

Platnick, N. I., and M. U. Shadab. 1974. A revision of the *tranquillus* and *speciousus* groups of the spider genus *Trachelas* (Araneae, Clubionidae) in North and Central America. *Am. Mus. Novitat.* 255:1–34.

Platnick, N. I., and M. U. Shadab. 1975. A revision of the spider genus *Gnaphosa* (Araneae, Gnaphosidae) in America. *Bull. Am. Mus. Nat. Hist.* 155:1–66.

Platnick, N. I., and M. U. Shadab. 1976. A revision of the spider genera *Lygromma* and *Neozimiris* (Araneae, Gnaphosidae). *Am. Mus. Novitat.* 2598:1–23.

Poccia, D. 1989. Reactivation and remodeling of the sperm nucleus following fertilization. In H. Schatten and G. Schatten, eds., *The Molecular Biology of Fertilization*, 115–135. Academic Press, New York.

Poinar, Jr., G. O., and E. A. Herre. 1991. Speciation and adaptive radiation in the fig wasp nematode, *Parasitodiplogaster* (Diplogasteridae: Rhabditida) in Panama. *Rev. Nematol.* 14:361–374.

Polak, M., and R. Trivers. 1994. The science of symmetry in biology. *TREE* 9:122–124.

Polis, G. A., and W. D. Sissom. 1990. Life history. In G. A. Polis, ed., *The Biology of Scorpions*, 161–223. Stanford University Press, Stanford.

Pollard, J. W., C. Plante, W. A. King, P. J. Hansen, and S. S. Suarez. 1990. Sperm fertilizing capacity is maintained by binding to oviductal epithelial cells. In B. D. Bavister, J. Cummins, and E.R.S. Roldan, eds., *Fertilization in Mammals*, 433. Serono Symposia USA, Norwell, Mass.

Pomiankowski, A., and L. Sheridan. 1994. Linked sexiness and choosiness. *TREE* 9:242–244.

Porter, A. H., and A. M. Shapiro. 1990. Lock-and-key hypothesis: Lack of mechanical isolation in a butterfly (Lepidoptera: Pieridae) hybrid zone. *Ann. Ent. Soc. Am.* 83:107–114.

Potapov, M. A., G. G. Nazarova, and V. I. Evsikov. 1993. Female choice, male social rank and mating success. Abstr. *23rd Int. Congr. Ethol.*:189.

Prenter, J., R. Elwood, and S. Colgan. 1994. The influence of prey size and female reproductive state on the courtship of the autumn spider, *Metellina segmentata*: A field experiment. *Anim. Behav.* 47:449–456.

Price, D., and H. G. Williams-Ashman. 1961. The accessory reproductive glands of mammals. In W. C. Young, ed., *Sex and Internal Secretions*, 366–448. Williams and Wilkins, Baltimore.

Price, R. M., and E. R. Meyer. 1979. An amplexus call made by the male American toad, *Bufo americanus americanus* (Amphibia, Anura, Bufonidae). *J. Herpetol.* 13(4):506–509.

Proctor, H. C., R. L. Baker, and D. T. Gwynne. In press. Mating behaviour and spermatophore morphology: A comparative test of the female-choice hypothesis. *Can. J. Zool.*

Profet, M. 1993. Menstruation as a defense against pathogens transported by sperm. *Quart. Rev. Biol.* 68(3):335.

Proshold, F. I., and L. E. LaChance. 1974. Analysis of sterility in hybrids from interspecific crosses between *Heliothis virescens* and *H. subflexa*. *Ann. Ent. Soc. Am.* 67:445–449.

Prout, T., and J. Bundgaard. 1977. The population genetics of sperm displacement. *Genetics* 85:95–121.

Queller, D. C. 1987. Sexual selection in flowering plants. In J. W. Bradbury and M. B. Andersson, eds., *Sexual Selection: Testing the Alternatives*, 165–179. Wiley, New York.

Queller, D. 1994. Male-female conflict and parent-offspring conflict. *Am. Nat.* 144 (suppl.):584–599.

Raabe, M. 1986. Insect reproduction: Regulation of successive steps. *Adv. Insect Physiol.* 19:29–154.

Rabb, G. B., and M. S. Rabb. 1960. On the mating and egg-laying behavior of the Surinam toad, *Pipa pipa*. *Copeia* 4:271–276.

Rabb, G. B., and M. S. Rabb. 1963. On the behavior and breeding biology of the African pipid frog *Hymenochirus boettgeri*. *Sonder. Z. Tierpsychol.* 20(2):215–241.

Radwan J., and W. Witalinski. 1991. Sperm competition. *Nature* 352:671–672.

Raina, A. K. 1989. Male-induced termination of sex pheromone production and receptivity in mated females of *Heliothis zea*. *J. Insect Physiol.* 35(11):821–826.

Raina, A. K., and E. A. Stadelbacher. 1990. Pheromone titer and calling in *Heliothis virescens* (Lepidoptera: Noctuidae): Effect of mating with normal and sterile backcross males. *Ann. Ent. Soc. Am.* 83:987–990.

Ramalingam, S., and G. B. Craig, Jr. 1976. Functions of the male accessory gland secretions of *Aedes mosquitoes* (Diptera: Culicidae): Transplantation studies. *Can. Ent.* 108:955–960.

Ramalingam, S., and G. B. Craig, Jr. 1978. Fine structure of the male accessory glands in *Aedes triseriatus*. *J. Insect Physiol.* 24:251–259.

Ramaswamy, S. G., and N. E. Cohen. 1992. Ecdysone: An inhibitor of receptivity in the moth, *Heliothis virescens*? *Naturwissenschaften* 79:29–31.

Rao, K. T., and M.V.S. Rao. 1990. Incubation and hatching of the Indian star tortoise *Geochelone elegans* in captivity. *J. Bombay Nat. Hist. Soc.* 87:461–462.

Raulston, J. R., J. W. Snow, H. M. Graham, and P. D. Lingren. 1975. Tobacco budworm: Effect of prior mating and sperm content on the mating behavior of females. *Ann. Ent. Soc. Am.* 68:701–704.

Real, L. 1991. Search theory and mate choice. I. Models of single-sex discrimination. *Am. Nat.* 136:376–404.

Rehfeld, V. K., and W. Sudhaus. 1985. Comparative studies on sexual behaviour of two sibling species of *Rhabditis* (Nematoda). *Zool. Jb. Syst.* 112:435–454.

Reichard, U. 1995. Extra-pair copulation in a monogamous gibbon (*Hylobates lar*). *Ethology* 100:99–112.

Reinhold, K., and K.-G. Heller. 1993. The ultimate function of nuptial feeding in the bushcricket *Poecilimon veluchianus* (Orthoptera: Tettigoniidae: Phaneropterinae). *Behav. Ecol. Sociobiol.* 32:55–60.

Retnakaran, A. 1974. The mechanism of sperm precedence in the spruce budworm, *Choristoneura fumiferana* (Lepidoptera: Tortricidae). *Can. Ent.* 106:1189–1194.

Retnakaran, A., and J. Percy. 1985. Fertilization and special modes of reproduction. In G. A. Kerkut and L. Gilbert, eds., *Comprehensive Insect Physiological Biochemistry and Pharmacology*, vol. 1, 231–293. Pergamon Press, New York.

Richards, O. W. 1924. The mating habits of certain species of *Micropteryx*. *Ent. Mon. Mag.* 60:31–34.

Richards, O. W. 1927. Sexual selection and allied problems in the insects. *Biol. Rev.* 2:298–360

Richmond, R. C., and A. Senior. 1981. Esterase-6 of *Drosophila melanogaster*: Kinetics of transfer to females, decay in females and male recovery. *J. Insect Physiol.* 27:849–853.

Ridley, M. 1983. *The Explanation of Organic Diversity*. Clarendon Press, Oxford.

Ridley, M. 1988. Mating frequency and fecundity in insects. *Biol. Rev.* 63:509–549.

Ridley, M. 1989a. The timing and frequency of mating in insects. *Anim. Behav.* 37:535–545.

Ridley, M. 1989b. The incidence of sperm displacement in insects: Four conjectures, one corroboration. *Biol. J. Linn. Soc.* 38:349–367.

Riemann, J. G., and B. J. Thorson. 1969. Effect of male accessory material on oviposition and mating by female house flies. *Ann. Ent. Soc. Am.* 62:828–834.

Riemann, J. G., and B. J. Thorson. 1974. Viability and use of sperm after irradiation of the large milkweed bug. *Ann. Entomol. Soc. Am.*. 67(6):871–876.

Riemann, J. G., D. J. Moen, and B. J. Thorson. 1967. Female monogamy and its control in houseflies. *J. Insect Physiol.* 13:407–418.

Robertson, H. M. 1982. Male activity during *Drosophila* copulation. *Drosophila Inf. Serv.* 58:129.

Robertson, H. M., and H.E.H. Paterson. 1982. Mate recognition and mechanical isolation in *Enallagma* damselflies (Odonata: Coenagrionidae). *Evolution* 36: 243–250.

Robinson, G. G. 1942. The mechanism of insemination in an argasid tick, *Ornithodorus moubata* Murray. *Parasitology* 34:195–198.

Robinson, M. H. 1982. Courtship and mating behavior in spiders. *Ann. Rev. Ent.* 27:1–20.

Robinson, M. H., and B. Robinson. 1980. Comparative studies of the courtship and mating behavior of tropical araneid spiders. *Pac. Ins. Monogr.* 36:1–218.

Robinson, T. J. 1975. Contraception and sperm transport in domestic animals. In E.S.E. Hafez and C. G. Thibault, eds., *The Biology of Spermatozoa*, 202–213. Karger, New York.

Rodriguez, R. L. In prep. Copulation behavior of two species of *Ozophora* bugs (Hemiptera: Lygaeidae).

Rodriguez, R. L., and W. G. Eberhard. 1994. Male courtship before and during copulation in two species of *Xyonisius* bugs (Hemiptera, Hygacidae). *J. Kans. Ent. Soc.* 67:37–45.

Rodriguez, R. L., and W. G. Eberhard. In prep. Inter- and intra-specific divergence in copulation behavior in *Pseudoxychila* beetles (Coleoptera: Cicindellidae).

Rodriguez, V. 1994a. Fuentes de variación en la precedencia de espermatozoides de *Chelymorpha alternans* Boheman 1854 (Coleoptera: Chrysomelidae: Cassidinae). Master's thesis, University of Costa Rica.

Rodriguez, V. 1994b. Function of the spermathecal muscle in *Chelymorpha alternans* Boheman (Coleoptera: Chrysomelidae: Cassidinae). *Physiol. Ent.* 19:198–202.

Rodriguez, V. 1994c. Sexual behavior in *Omaspides convexicollis* Spaeth and *O. bistriata* Boheman. (Coleoptera: Chrysomelidae: Cassidinae), with notes on maternal care of eggs and young. *Coleop. Bull.* 48:140–144.

Rodriguez, V., W. G. Eberhard, and D. Windsor. Submitted. Longer genitalia confer a sexually selected advantage in a beetle. *Proc. Roy. Soc. Lond. B.*

Roig-Alsina, A. 1993. The evolution of the apoid endophallus, its phylogenetic implications, and functional significance of the genital capsule (Hymenoptera, Apoidea). *Boll. Zool.* 60:169–183.

Rojas, J. C., E. A. Malo, A. Gutierrez-Martinez, and R. N. Ondarza. 1990. Mating behavior of *Triatoma mazzottii* Usinger (Hemiptera: Reduviidae) under laboratory conditions. *Ann. Ent. Soc. Am.* 83:598–602.

Roldan, E.R.S. 1990. Physiological stimulators of the acrosome reaction. In B. D. Bavister, J. Cummins, and E. S. Roldan, eds., *Fertilization in Mammals*, 197–204. Serono Symposia, Norwell, Mass.

Roldan, E.R.S., and R.A.P. Harrison. 1990. Molecular mechanisms leading to exocytosis during the sperm acrosome reaction. In B. D. Bavister, J. Cummins, and E.R.S. Roldan, eds., *Fertilization in Mammals*, 179–196. Serono Symposia, Norwell, MA.

Roldan, E.R.S., M. Gomendio, and A. D. Vitullo. 1992. The evolution of eutherian spermatozoa and underlying selective forces: Female selection and sperm competition. *Biol. Rev.* 67:551–593.

Rood, J. P. 1972. Ecological and behavioral comparisons of three genera of Argentine cavies. *Anim. Behav. Monogr.* 5:1–83.

Roth, L. M. 1962. Hypersexual activity induced in females of the cockroach *Nauphoeta cinerea*. *Science* 138:1267–1268.

Roth, L. M., and B. Stay. 1961. Oocyte development in *Diploptera punctata* (Eschscholtz) (Blattaria). *J. Insect Physiol.* 7:186–202.

Rowe, L. G., G. Arnqvist, A. Sih, and J. J. Krupa. 1994. Sexual conflict and the evolutionary ecology of mating patterns: Water striders as a model system. *TREE* 9:289–293.

Rowell, T. E. 1988. Beyond the one-male group. *Behaviour* 104:189–201.

Rowlands, I. W. 1958. Insemination by intraperitoneal injection. *Proc. Soc. Stud. Fertil.* 10:150–157.

Rubenstein, D. I. 1989. Sperm competition in the water strider, *Gerris remigis*. *Anim. Behav.* 38:631–636.

Ruknudin, A., and V. V. Raghavan. 1988. Initiation, maintenance and energy metabolism of sperm motility in the bed bug *Cimex hemipterus*. *J. Ins. Physiol.* 34:137–142.

Rutowski, R. L., G. W. Gilchrist, and B. Terkanian. 1987. Female butterflies mated with recently mated males show reduced reproductive output. *Behav. Ecol. Sociobiol.* 20:319–322.

Ryan, M. J. 1985. *The Túngara frog: A study in sexual selection and communication*. University of Chicago Press, Chicago.

Ryan, M. J. 1990. Sexual selection, sensory systems and sensory exploitation. *Oxford Surveys in Evol. Biol.* 7:157–195.

Ryan, M. J., and A. Keddy-Hector. 1992. Directional patterns of female mate choice and the role of sensory bias. *Am. Nat.* 139:S4-S35.

Sachs, B. D., and R. J. Barfield. 1976. Functional analysis of masculine copulatory behavior in the rat. *Adv. Stud. Behav.* 7:91–154.

Sahli, R., J. E. Germond, and P. A. Diehl. 1985. *Ornithodoros moubata*: Spermataleosis and secretory activity of the sperm. *Expt. Parasit.* 60:383–395.

Sakai, M., Y. Taoda, K. Mori, M. Fujino, and C. Ohta. 1991. Copulation sequence and mating termination in the male cricket *Gryllus bimaculatus* DeGeer. *J. Insect Physiol.* 37:599–615.

Sakaluk, S. K. 1984. Male crickets feed females to ensure complete sperm transfer. *Science* 223:609–610.

Sakaluk, S. K. 1985. Spermatophore size and its role in the reproductive behaviour of the cricket, *Gryllodes supplicans* (Orthoptera: Gryllidae). *Can. J. Zool.* 63: 1652–1656.

Saling, P. M., D. O. Bunch, P. LeGuen, and L. Leyton. 1990. ZP-3 induced acrosomal exocytosis: A new model for triggering. In B. D. Bavister, J. Cummins, and E.R.S. Roldan, eds., *Fertilization in Mammals*, 239–252. Serono Symposia, Norwell, Mass.

Salmon, M. 1984. The courtship, aggression and mating system of a "primitive" fiddler crab (*Uca vocans*: Ocypodidae). *Trans. Zool. Soc. Lond.* 37:1–50.

Salthe, S. N., and J. S. Mecham. 1974. Reproductive and courtship patterns. In B. Lofts, ed., *Physiology of the Amphibia*, 309–521. Academic Press, New York.

Sandor, T. 1980. Steroids in invertebrates. In W. H. Clark, Jr., and T. S. Adams, eds., *Advances in Invertebrate Reproduction*, 81–91. Elsevier North Holland, New York.

Sasaki, T., and O. Iwahashi. 1995. Sexual cannibalism in an orb-weaving spider *Argiope aemula*. *Anim. Behav.* 49:1119–1121.

Saul, S. H., and S. D. McCombs. 1993. Dynamics of sperm use in the Mediterranean fruit fly (Diptera: Tephritidae): Reproductive fitness of multiple-mated females and sequentially mated males. *Ann. Ent. Soc. Am.* 86(2):198–202.

Saumiton-Laprade, P., J. Cuguen, and P. Vernet. 1994. Cytoplasmic male sterility in plants: Molecular evidence and the nucleocytoplasmatic conflict. *TREE* 9:431–435.

Saunders, D. S., and C.W.H. Dodd. 1972. Mating, insemination, and ovulation in the tsetse fly, *Glossina morsitans*. *J. Insect Physiol.* 18:187–198.

Schadler, M. H. 1981. Postimplantation abortion in pine voles (*Microtus pinetorum*) induced by strange males and pheromones of strange males. *Biol. Reprod.* 25:295–297.

Schaller, G. 1972. *The Serengeti Lion*. University of Chicago Press, Chicago.

Schaller, G. B., H. Jinchu, P. Wenshi, and Z. Jing. 1985. *The Giant Pandas of Wolong*. University of Chicago Press, Chicago.

Scheller, R. H., J. F. Jackson, L. B. McAllister, B. S. Rothman, E. Mayeri, and

R. Axel. 1983. A single gene encodes multiple neuropeptides mediating a stereotyped behavior. *Cell* 32:7–22.

Schenkel, R. 1965. On sociology and behaviour in impala (*Aepyceros melampus* Lichtenstein). *E. Afr. Wildl. J.* 4:99–114.

Schlager, G. 1960. Sperm precedence in the fertilization of eggs in *Tribolium castaneum*. *Ann. Ent. Soc. Am.* 53:557–560.

Schoenmakers, H.J.N. 1980. The possible role of steroids in vitellogenesis in the starfish *Asterias rubens*. In W. H. Clark, Jr., and T. S. Adams, eds., *Advances in Invertebrate Reproduction*, 127–149. Elsevier North Holland, New York.

Schuh, R. T., and J. A. Slater. 1995. *True Bugs of the World*. Comstock Publishing Associates, Ithaca, N.Y.

Schult, J., and U. Sellenschlo. 1983. Morphologie und Funktion der Genitalstrukturen bei *Nephila* (Arachn., Aran., Araneidae). *Mittl. Hamb. Zool. Mus. Inst.* 80:221–230.

Schwagmeyer, P. L., and D. W. Foltz. 1990. Factors affecting the outcome of sperm competition in thirteen-lined ground squirrels. *Anim. Behav.* 39:156–162.

Schwartz, J. M., G. F. McCracken, and G. M. Burghardt. 1989. Multiple paternity in wild populations of the garter snake, *Thamnophis sirtalis*. *Behav. Ecol. Sociobiol.* 25:269–273.

Scott, D. 1986. Inhibition of female *Drosophila melanogaster* remating by a seminal fluid protein (esterase 6). *Evolution* 40:1084–1091.

Scott, D., and R. C. Richmond. 1985. An effect of male fertility on the attractiveness and oviposition rates of mated *Drosophila melanogaster* females. *Anim. Behav.* 33:817–824.

Scott, D., and R. C. Richmond. 1990. Sperm loss by remating *Drosophila melanogaster* females. *J. Insect Physiol.* 36:451–456.

Scott, D., and E. Williams. 1993. Sperm displacement after remating in *Drosophila melanogaster*. *J. Insect Physiol.* 39:201–206.

Scott, D., R. C. Richmond, and D. A. Carlson. 1988. Pheromones exchanged during mating: A mechanism for mate assessment in *Drosophila*. *Anim. Behav.* 36: 1164–1173.

Searcy, W. A. 1982. The evolutionary effects of mate selection. *Ann. Rev. Ecol. Syst.* 13:57–85.

Seo, S. T., R. I. Vargas, J. E. Gilmore, R. S. Kurashima, and M. S. Fujimoto. 1990. Sperm transfer in normal and gamma-irradiated, laboratory-reared Mediterranean fruit flies (Diptera: Tephritidae). *J. Econ. Ent.* 83(5):1949–1953.

Setty, B.N.Y., and T. R. Ramaiah. 1979. Isolation and identification of prostaglandins from the reproductive organs of male silkmoth, *Bombyx mori* L. *Insect Biochem.* 9:613–617.

Setty, B.N.Y., and T. R. Ramaiah. 1980. Effect of prostaglandins and inhibitors of prostaglandin biosynthesis on oviposition in the silkmoth *Bombyx mori*. *Ind. J. Exp. Biol.* 18:539–541.

Sever, Z., and H. Mendelssohn. 1988. Copulation as a possible mechanism to maintain monogamy in porcupines, *Hystrix indica*. *Anim. Behav.* 36:1541–1558.

Shapiro, A. M., and A. H. Porter. 1989. The lock-and-key hypothesis: Evolutionary and biosystematic interpretation of insect genitalia. *Ann. Rev. Ent.* 34:231–245.

Sharman, G. B., J. H. Calaby, and W. E. Poole. 1966. Patterns of reproduction in female diprotodont marsupials. In R. I. Rowlands, ed., *Comparative Biology of Reproduction in Mammals*, 205–232. Symp. Zool. Soc. Lond. 15.

Sheldon, B. C. 1994. Sperm competition in the chaffinch: The role of the female. *Anim. Behav.* 47:163–173.

Sheldon, B. C., and T. Burke. 1994. Copulation behavior and paternity in the chaffinch. *Behav. Ecol. Sociobiol.* 34:149–156.

Shirk, P. D., G. Bhaskaran, and H. Roller. 1980. The transfer of juvenile hormone from male to female during mating in the Cecropia silkmoth. *Experientia* 36:682–683.

Shively, C., and D. G. Smith. 1985. Social status and reproductive success of male *Macaca fascicularis*. *Am. J. Primatol.* 9:129–135.

Shively, C., S. Clarke, N. King, S. Schapiro, and G. Mitchell. 1982. Patterns of sexual behavior in male macaques. *Am. J. Primatol.* 2:373–384.

Shuster, S. M. 1989. Female sexual receptivity associated with molting and differences in copulatory behavior among the male morphs in *Paracerceis sculpta* (Crustacea: Isopoda). *Biol. Bull.* 117:331–337.

Shuster, S. M. 1991. The ecology of breeding females and the evolution of polygyny in *Paracerceis sculpta*, a marine isopod crustacean. In R. T. Bauer and J. W. Martin, eds., *Crustacean Sexual Biology*, 91–110. Columbia University Press, New York.

Silberglied, R. E., J. Shepard, and J. Dickinson. 1984. Eunuchs: The role of apyrone sperm in Lepidoptera? *Am. Nat.* 123:255–265.

Simmons, L. W. 1986. Female choice in the field cricket *Gryllus bimaculatus* (De Geer). *Anim. Behav.* 34:1463–1470.

Simmons, L. W. 1987a. Sperm competition as a mechanism of female choice in the field cricket, *Gryllus bimaculatus*. *Behav. Ecol. Sociobiol.* 21:197–202.

Simmons, L. W. 1987b. Heritability of a male character chosen by females of the field cricket, *Gryllus bimaculatus*. *Behav. Ecol. Sociobiol.* 21:129–133.

Simmons, L. W. 1988. The contribution of multiple mating and spermatophore consumption to the lifetime reproductive success of female field crickets (*Gryllus bimaculatus*). *Ecol. Ent.* 13:57–69.

Simmons, L. W. 1990. Nuptial feeding in tettigoniids: Male costs and the rates of fecundity increase. *Behav. Ecol. Sociobiol.* 27:43–47.

Simmons, L. W., and W. J. Bailey. 1990. Resource influenced sex roles of zaprochiline tettigoniids. *Evolution* 44:1853–1868.

Simmons, L. W., and D. T. Gwynne. 1991. The refractory period of female katydids (Orthoptera: Tettigoniidae): Sexual conflict over the remating interval? *Behav. Ecol.* 2:276–282.

Simmons, L. W., and G. A. Parker. 1992. Individual variation in sperm competition success of yellow dung flies, *Scatophaga stercoraria*. *Evolution* 46:366–375.

Simmons, L. W., and P. I. Ward. 1991. The heritability of sexually dimorphic traits in the yellow dung fly *Scathophaga stercoraria* (L.). *J. Evol. Biol.* 4:593–601.

Simpson, J. 1970. The male genitalia of *Apis dorsata* (F.) (Hymenoptera: Apidae). *Proc. Roy. Ent. Soc. Lond.* [A]45:169–171.

Siva-Jothy, M. T. 1987a. The structure and function of the female sperm-storage organs in libellulid dragonflies. *J. Insect Physiol.* 33:559–567.

Siva-Jothy, M. T. 1987b. Variation in copulation duration and the resultant degree of sperm removal in *Orthetrum cancellatum* (L.) (Libellulidae: Odonata). *Behav. Ecol. Sociobiol.* 20:147–151.

Siva-Jothy, M. T., and Y. Tsubaki. 1989a. Variation in copulation duration in *Mnais pruinosa pruinosa* Selys (Odonata: Calopterygidae). 1. Alternative mate-securing tactics and sperm precedence. *Behav. Ecol. Sociobiol.* 24:39–45.

Siva-Jothy, M. T., and Y. Tsubaki. 1989b. Variation in copulation duration in *Mnais pruinosa pruinosa* Selys (Odonata: Calopterygidae). 2. Causal factors. *Behav. Ecol. Sociobiol.* 25:261–267.

Sivinski, J. 1984. Effect of sexual experience on male mating success in a lek forming tephritid *Anastrepha suspensa* (Loew). *Fla. Ent.* 67(1):126–130

Sivinski, J., and T. Burk. 1989. Reproductive and mating behaviour. In A. S. Robinson and G. Hooper, eds., *Fruit Flies, Their Biology, Natural Enemies and Control*, vol. 3A, 343–351. Elsevier, New York.

Sivinski, J., and Smittle, B. 1987. Male transfer of materials to mates in the Caribbean fruit fly, *Anastrepha suspensa* (Diptera: Tephritidae). *Fla. Ent.* 70:233–238.

Sivinski, J., T. Burk, and J. C. Webb. 1984. Acoustic courtship signals in the Caribbean fruit fly, *Anastrepha suspensa* (Loew). *Anim. Behav*. 32:1011–1016.

Slip, D. J., and R. Shine. 1988. The reproductive biology and mating system of diamond pythons, *Morelia spilota* (Serpentes, Boidae). *Herpetologica* 44:396–404.

Sluys, R. 1989. Sperm resorption in triclads (Platyhelminthes, Tricladida). *Invert. Reprod. and Dev*. 15:89–95.

Small, M. F. 1993. *Female Choices—Sexual Behavior of Female Primates*. Cornell University Press, Ithaca, N.Y.

Smith, P. H., L. Barton Browne, and A.C.M. van Gerwen. 1988. Sperm storage and utilisation and egg fertility in the sheep blowfly, *Lucilia cuprina*. *J. Insect Physiol.* 34:125–129.

Smith, P. H., L. Barton Browne, and A.C.M. van Gerwen. 1989. Causes and correlates of loss and recovery of sexual receptivity in *Lucilia cuprina* females after their first mating. *J. Insect Behav*. 2:325.

Smith, P. H., C. Gillott, L. Barton Browne, and A.C.M. van Gerwen. 1990. The mating-induced refractoriness of *Lucilia cuprina* females: Manipulating the male contribution. *Physiol. Ent.* 15:469–481.

Smith, R. L. 1979. Repeated copulation and sperm precedence: Paternity assurance for a male brooding water bug. *Science* 205:1029–1031.

Smith, R. L. 1984. *Sperm Competition and the Evolution of Animal Mating Systems.* Academic Press, New York.

Smith, T. T. 1990. Which sperm in the hamster oviduct will fertilize the eggs? In B. D. Bavister, J. Cummins, and E.R.S. Roldan, eds., *Fertilization in Mammals*, 435. Serono Symposia, Norwell, Mass.

Smuts, B. B. 1992. Male aggression against women: An evolutionary perspective. *J. Hum. Nat.* 3:1–44.

Smuts, G. L., J. Hanks, and I. J. Whyte. 1978. Reproduction and social organization of lions from the Kruger National Park. *Carnivore* 1:17–28.

Snedden, W. A. 1990. Determinants of male mating success in the temperate crayfish *Orconectes rusticus*: Chela size and sperm competition. *Behaviour* 115:100–113.

Snow, J. W., R. J. Jones, D. T. North, and G. G. Holt. 1972. Effects of irradiation on ability of adult male corn earworms to transfer sperm, and field attractiveness of females mated to irradiated males. *J. Econ. Ent.* 65:906–908.

Solinas, M., and G. Nuzzaci. 1984. Functional anatomy of *Dacus oleae* Gmel. female genitalia in relation to insemination and fertilization processes. *Entomologica* (Bari) 19:135–165.

Somers, J.-A., H. H. Shorey, and L. K. Gaston. 1977. Reproductive biology and behavior of *Rhabditis pellio* (Schneider) (Rhabditida: Rhabditidae). *J. Nematol.* 9:143–148.

Sorenson, L. G. 1994. Forced extra-pair copulation and mate guarding in the white-cheeked pintail: Timing and trade-offs in an asynchronously breeding duck. *Anim. Behav.* 48:519–533.

Spencer, J. L., G. L. Bush, Jr., J. E. Keller, and J. R. Miller. 1992. Modification of female onion fly, *Delia antiqua* (Meigen), reproductive behavior by male paragonial gland extracts (Diptera: Anthomyiidae). *J. Insect Behav.* 5:689–697.

Spielman, A. 1964. The mechanics of copulation in *Aedes aegypti. Biol. Bull.* 127:324–344.

Spielman, A., M. G. Leahy, Sr., and V. Skaff. 1967. Seminal loss in repeatedly mated female *Aedes aegypti. Biol. Bull.* 132:404–412.

Spielman, A., M. G. Leahy, Sr., and V. Skaff. 1969. Failure of effective insemination of young female *Aedes aegypti* mosquitoes. *J. Insect Physiol.* 15:1471–1479.

Stacey, N. E. 1984. Control of the timing of ovulation by exogenous and endogenous factors. In G. W. Potts and R. J. Woottow, eds., *Fish Reproduction: Strategies and Tactics*, 207–222. Academic Press, New York.

Stalker, H. D. 1976. Enzymes and reproduction in natural populations of *Drosophila Euronotus. Genetics* 84:375–384.

Stamps, J. A. 1975. Courtship patterns, estrus periods and reproductive condition in a lizard, *Anolis aeneus. Physiol. Behav.* 14:531–535.

Stanley-Samuelson, D. W., and W. Loher. 1983. Arachidonic and other long-chain polyunsaturated fatty acids in spermatophores and spermathecae of *Teleogryllus commodus*: Significance in prostaglandin-mediated reproductive behaviour. *J. Insect Physiol.* 29:41–45.

Stanley-Samuelson, D. W., R. A. Jurenka, G. J. Blomquist, and W. Loher. 1987. Sexual transfer of prostaglandin precursor in the field cricket, *Teleogryllus commodus. Physiol. Ent.* 12:347–354.

Starr, C. 1984. Sperm competition, kinship, and sociality in the aculeate Hymenoptera. In R. L. Smith, ed., *Sperm Competition and the Evolution of Animal Mating Systems*, 427–464. Academic Press, New York.

Stehn, R. A., and M. E. Richmond. 1975. Male-induced pregnancy termination in the prairie vole, *Microtus ochrogaster. Science* 187:1211–1213.

Stenseth, N. C., L. R. Kirkendall, and N. Moran. 1985. On the evolution of pseudogamy. *Evolution* 39:294–307.

Stephenson, A. G., and R. I. Bertin. 1983. Male competition, female choice, and sexual selection in plants. In L. Real, ed., *Pollination Biology*, 109–149. Academic Press, New York.

Sternberg, S., B. A. Peleg, and R. Galun. 1973. Effect of irradiation on mating competitiveness of the male tick *Argas persicus* (Oken). *J. Med. Ent.* 10:137–142.

Stille, B., T. Madsen, and M. Niklasson. 1986. Multiple paternity in the adder, *Vipera berus*. *Oikos* 47:173–175.

Strahl, S. D. 1988. The social organization and behavior of the hoatzin *Opisthocomus hoazin* in central Venezuela. *Ibis* 130:483–502.

Stumm-Zollinger, E., and P. S. Chen. 1988. Gene expression in male accessory glands of interspecific hybrids of *Drosophila*. *J. Insect Physiol.* 34:59–74.

Suarez, S. S., M. Drost, K. Redfern, and W. Gottlieb. 1990. Sperm motility in the oviduct. In B. D. Bavister, J. Cummins, and E.R.S. Roldan, eds., *Fertilization in Mammals*, 111–124. Serono Symposia, Norwell, Mass.

Sugawara, T. 1979. Stretch reception in the bursa copulatrix of the butterfly, *Pieris rapae crucivora*, and its role in behaviour. *J. Comp. Physiol.* 130:191–199.

Sugawara, T. 1993. Oviposition behaviour of the cricket *Teleogryllus commodus*: Mechanosensory cells in the genital chamber and their role in the switch-over of steps. *J. Insect Physiol.* 39:335–346.

Sugawara, T., and W. Loher. 1986. Oviposition behaviour of the cricket *Teleogryllus commodus*: Observation of external and internal events. *J. Insect Physiol.* 32:179–188.

Sussman, R. W., and P. A. Garber. 1987. A new interpretation of the social organization and mating system of the *Callitrichidae*. *Inter. J. Primatol.* 8(1):73–93.

Suter, R. B. 1990. Courtship and the assessment of virginity by male bowl and doily spiders. *Anim. Behav.* 39:307–313.

Suter, R. B., and V. S. Parkhill. 1990. Fitness consequences of prolonged copulation in the bowl and doily spider. *Behav. Ecol. Sociobiol.* 26:369–373.

Suter, R. B., and G. Renkes. 1984. The courtship of *Frontinella pyramitela* (Araneae, Linyphiidae): Patterns, vibrations and functions. *J. Arachnol.* 12:37–54.

Svard, L., and J. N. McNeil. 1994. Female benefit, male risk: Polyandry in the true armyworm *Pseudaletia unipuncta*. *Behav. Ecol. Sociobiol.* 35:319–326.

Svard, L., and C. Wiklund. 1991. The effects of ejaculate mass on female reproductive output in the European swallowtail butterfly, *Papilio machaon* (L.) (Lepidoptera: Papilionidae). *J. Insect Behav.* 4:33–42.

Svare, B., and M. Mann. 1981. Infanticide: Genetic, developmental and hormonal influences in mice. *Physiol. Behav.* 27:921–927.

Swailes, G. E. 1971. Reproductive behavior and effects of the male accessory gland substance in the cabbage maggot, *Hylemya brassicae*. *Ann. Ent. Soc. Am.* 64:176–179.

Tadler, A. 1993. Genitalia fitting, mating behaviour and possible hybridization in millipedes of the genus *Craspedosoma* (Diplopoda, Chordeumatida, Craspedosomatidae). *Acta Zool.* 74:215–225.

Tait, N. N., and D. A. Briscoe. 1990. Sexual head structures in the Onychophora: Unique modifications for sperm transfer. *J. Nat. Hist.* 24:1517–1527.

Talbot, P., G. DiCarlantonio, P. Zao, J. Penkala, and L. T. Haimo. 1985. Motile cells lacking hyaluronidase can penetrate the hamster oocyte cumulus complex. *Dev. Biol.* 108:387–398.

Tang-Martinez, Z., L. Mueller, and G. T. Taylor. 1993. Individual odours and mating success in the golden hamster, *Mesocricetus auratus*. *Anim. Behav*. 45:1141–1151.

Tatchell, R. J. 1962. Studies on the male accessory reproductive glands and the spermatophore of the tick, *Argas persicus*. *Oken. Parasitol*. 52:133–142.

Taub, D. M. 1980. Female choice and mating strategies among wild Barbary macaques (*Macaca sylvanus* L.). In D. G. Lindburg, ed., *The Macaques: Studies in Ecology, Behavior and Evolution*, 287–344. Van Nostrand-Reinhold, New York.

Taub, D. M. 1982. Sexual behavior of wild barbary macaque males. *Am. J. Primatol*. 2:109–113.

Tavolga, M. C., and F. S. Essapian. 1957. Behavior of the bottlenosed dolphin (*Tursiops truncatus*): Mating, pregnancy, parturition and mother-infant behavior. *Zoologica* 42:11–31.

Taylor, B. J. 1989. Sexually dimorphic neurons in the terminalia of *Drosophila melanogaster*. I. Development of sensory neurons in the genital disc during metamorphosis. *J. Neurogenet*. 5:173–192.

Taylor, M. H. 1986. Environmental and endocrine influences on reproduction of *Fundalus heteroclitus*. *Am. Zool*. 26:159–171.

Taylor, P. D., and G. C. Williams. 1982. The lek paradox is not resolved. *Theor. Pop. Biol*. 22:392–409.

Taylor, O. R. 1967. Relationship of multiple mating to fertility in *Atteva punctella* (Lepidoptera: Yponomeutidae). *Ann. Ent. Soc. Am*. 60:583–590.

Tennessen, K. J. 1982. Review of reproductive isolating barriers in Odonata. *Adv. Odonatol*. 1:251–265.

Terranova, A. C., R. A. Leopold, M. E. Degrugillier, and J. R. Johnson. 1972. Electrophoresis of the male accessory secretion and its fate in the mated female. *J. Insect Physiol*. 18:1573–1591.

Thibout, E. 1975. Analyse des causes de l'inhibition de la réceptivité sexuelle et de l'influence d'une eventuelle séconde copulation sur la réproduction chez la teigne du poireau, *Acrolepia assectella* (Lepidoptera: Plutellidae). *Ent. Exp. Appl*. 18:105–116.

Thibout, E. 1976. Consequences de l'ablation de la spermathèque ou de la bourse copulatrice sur la réproduction de la teigne du poireau, *Acrolepiopsis* (*Acrolepia*) *assectella* Zell. (Lepidoptera). *C. R. Acad. Sci. Paris* 282(D):2199–2202.

Thibout, E. 1977. La migration spermatique chez *Acrolepiopsis* (*Acrolepia*) *assectella* Zell. (*Lep. Plutellidae*): Role de la motilité des spermatozoides et de la musculature de l'appareil génital femelle. *Ann. Soc. Ent. Fr*. (N.S.) 13(2):381–389.

Thomas, D. W., M. B. Fenton, and R.M.R. Barclay. 1979. Social behavior of the little brown bat, *Myotis lucifugus*. *Behav. Ecol. Sociobiol*. 6:129–136.

Thomas, R. H., and D. W. Zeh. 1984. Sperm transfer and utilization strategies in arachnids: Ecological and morphological restraints. In R. L. Smith, ed., *Sperm Competition and the Evolution of Animal Mating Systems*, 180–222. Academic Press, New York.

Thor, D. H., and W. J. Carr. 1979. Sex and aggression: Competitive mating strategy in the male rat. *Behav. Neur. Biol*. 26:261–265.

Thornhill, R. 1976. Sexual selection and nuptial feeding behavior in *Bittacus apicalis* (Insecta: Mecoptera). *Am. Nat*. 110:529–548.

Thornhill, R. 1980a. Rape in *Panorpa* scorpionflies and a general rape hypothesis. *Anim. Behav*. 28:52–59.

Thornhill, R. 1980b. Mate choice in *Hylobittacus apicalis* (Insecta: Mecoptera) and its relation to some models of female choice. *Evolution* 34:519–538.

Thornhill, R. 1983. Cryptic female choice and its implications in the scorpionfly *Harpobittacus nigriceps*. *Am. Nat*. 122:765–788.

Thornhill, R. 1984. Alternative female choice tactics in the scorpionfly *Hylobittacus apicalis* (Mecoptera) and their implications. *Am. Zool*. 24:367–383.

Thornhill, R., and J. Alcock. 1983. *The Evolution of Insect Mating Systems*. Harvard University Press, Cambridge, Mass.

Thornhill, R., and D. T. Gwynne. 1986. The evolution of sexual differences in insects. *Am. Sci*. 74:382–389.

Thornhill, R., and K. P. Sauer. 1991. The notal organ of the scorpionfly (*Panorpa vulgaris*): An adaptation to coerce mating duration. *Behav. Ecol*. 2(2):156–164.

Thornhill, R., S. W. Gangestad, and R. Comer. 1995. Human female orgasm and mate fluctuating asymmetry. *Anim. Behav*. 50:1601–1615.

Thresher, R. E. 1984. Basslets (Pseudochromoids). In *Reproduction in Reef Fishes*, 100–333. T.F.H. Publications, Neptune City, N.J.

Tobe, S. S., and W. Loher. 1983. Properties of the prostaglandin synthetase complex in the cricket *Teleogryllus commodus*. *Insect Biochem*. 13:137–141.

Tompkins, L., and J. C. Hall. 1981. The different effects on courtship of volatile compounds from mated and virgin *Drosophila* females. *J. Insect Physiol*. 27:17–21.

Tompkins, L., and J. C. Hall. 1983. Identification of brain sites controlling female receptivity in mosaics of *Drosophila melanogaster*. *Genetics* 103:179–195.

Toro, H. 1985. Ajuste mecànico para la còpula de *Callonychium chilense* (Hymenoptera, Andrenidae). *Rev. Chil. Ent*. 12:153–158.

Toro, H. 1989. Ajuste genital en la còpula de Thynninae (Hymenoptera, Tiphiidae). *Acta Ent. Chil*. 15:123–130.

Toro, H. and E. de la Hoz. 1976. Factores mecànicos en la aislaciòn reproductiva de *Apoidea* (Hymenoptera). *Rev. Soc. Ent. Arg*. 35:193–202.

Tortosa, F. S., and T. Redondo. 1992. Frequent copulations despite low sperm competition in white storks (*Ciconia ciconia*). *Behaviour* 121(3–4):288–315.

Trail, P. W., and E. S. Adams. 1989. Active mate choice at cock-of-the-rock leks: Tactics of sampling and comparison. *Behav. Ecol. Sociobiol*. 25:283–292.

Troisi, A., and M. Carosi. 1994. Male dominance rank influences orgasmic response in Japanese macaque females. *Ethol. Ecol. Evol*. 6:449–450.

Tschinkel, W. R. 1987. Relationship between ovariole number and spermathecal sperm count in ant queens: A new allometry. *Ann. Ent. Soc. Am*. 80:208–211.

Tschinkel, W. R., and S. D. Porter. 1988. Efficiency of sperm use in queens of the fire ant, *Solenopsis invicta* (Hymenoptera: Formicidae). *Ann. Ent. Soc. Am*. 81: 777–781.

Tschudi-Rein, K., and G. Benz. 1990. Mechanisms of sperm transfer in female *Pieris brassicae* (Lepidoptera: Pieridae). *Ann. Ent. Soc. Am*. 83:1158–1164.

Tsubaki, Y., and Y. Sokei. 1988. Prolonged mating in the melon fly, *Dacus cucurbitae* (Diptera: Tephritidae): Competition for fertilization by sperm-loading. *Res. Pop. Ecol*. 30:343–352.

Tutin, C.E.G. 1979. Mating patterns and reproductive strategies in a community of wild chimpanzees (*Pan troglodytes schweinfurthii*). *Behav. Ecol. Sociobiol.* 6:29–38.

Tyler, S. J. 1972. The behaviour and social organization of the New Forest ponies. *Anim. Behav. Monogr.* 5:85–196.

Ueno, H. 1994. Intraspecific variation of P2 value in a coccinellid beetle *Harmonia axyridis*. *J. Ethol.* 12:169–174.

Uhl, G. 1992. Sperm storage and repeated egg production in female *Pholcus phalangioides* Fuesslin (Araneae). *Bull. Soc. Neuschatel. Sci. Nat.* 116:245–252.

Uhl, G. 1993. Mating behaviour and female sperm storage in *Pholcus phalangioides* (Fuesslin) (Araneae). *Mem. Queensland Mus.* 33:667–674.

Uhl, G. 1994. Genital morphology and sperm storage in *Pholcus phalangioides* (Fuesslin, 1775) (Pholeidae, Araneae). *Acta Zool.* (Stockholm) 75:1–12.

Uhl, G. 1994c. Reproduktionsbiologie von Zitterspinnen (*Pholcus phalangioides*: Pholcidae; Araneae). Ph.D. thesis. Albert-Ludwigs-Universität, Freiburg.

Uhl, G., B. A. Huber, and W. Rose. 1994. Male pedipalp morphology and copulatory mechanism in *Pholcus phalangioides* (Fuesslin, 1775) (Araneae, Pholcidae). *Bull. Br. Arachnol. Soc.* 10:1–10.

VanDemark, N. L., and R. L. Hays. 1952. Uterine motility responses to mating. *Am. J. Physiol.* 170:518–521.

Vandenbergh, J. G. 1987. Regulation of puberty and its consequences on population dynamics of mice. *Amer. Zool.* 27:891–898.

Vandenbergh, J. G., and D. M. Coppola. 1986. The physiology and ecology of puberty modulation by primer pheromones. *Adv. Stud. Behav.* 16:71–83.

van der Assem, J., and J. Visser. 1976. Aspects of sexual receptivity in the female *Nasonia vitripennis* (Hymenoptera: Pteromalidae). *Biol. Behav.* 1:37–56.

van der Assem, J., and J. H. Werren. 1994. A comparison of the courtship and mating behavior of three species of *Nasonia* (Hymenoptera: Pteromalidae). *J. Insect Behav.* 7:53–66.

van Helsdingen, P. J. 1965. Sexual behaviour of *Lepthyphantes leprosus* (Ohlert) (Araneida, Linyphiidae), with notes on the function of the genital organs. *Zool. Mededelingen* 41:15–42.

Vardell, H. H., and J. H. Brower. 1978. Sperm precedence in *Tribolium confusum* (Coleoptera: Tenebrionidae). *J. Kans. Ent. Soc.* 51:187–190.

Venier, L. A., P. O. Dunn, J. T. Lifjeld, and R. J. Robertson. 1993. Behavioural patterns of extra-pair copulation in tree swallows. *Anim. Behav.* 45:412–415.

Venkatesh, K., and P. E. Morrison. 1980. Some aspects of oogenesis in the stable fly *Stomoxys calcitrans* (Diptera: Muscidae). *J. Insect Physiol.* 26:711–715.

Ventura, W. P., and M. Freund. 1973. Evidence for a new class of uterine stimulants in rat semen and male accessory gland secretions. *J. Reprod. Fert.* 33:507–511.

Verrell, P. A. 1992. Primate penile morphologies and social systems: Further evidence for an association. *Folia Primat.* 59:114–120.

Verrell, P. A. 1994. Males may choose larger females as mates in the salamander *Desmognathus fuscus*. *Anim. Behav.* 47:1465–1467.

Villavaso, E. J. 1975a. Functions of the spermathecal muscle of the boll weevil, *Anthonomus grandis*. *J. Insect Physiol.* 21:1275–1278.

Villavaso, E. J. 1975b. The role of the spermathecal gland of the boll weevil, *Anthonomus grandis*. *J. Insect Physiol.* 21:1457–1462.

von Euler, U. S. 1936. On the specific vaso-dilating and plain muscle stimulating substances from accessory genital glands in man and certain animals (prostaglandin and vesiglandin). *J. Physiol.* 88:213–234.

von Helversen, D., and O. von Helversen. 1991. Pre-mating sperm removal in the bushcricket *Metaplastes ornatus* Ramme 1931 (Orthoptera, Tettigonoidea, Phaneropteridae). *Behav. Ecol. Sociobiol.* 28:391–396.

Voss, R. 1979. Male accessory glands and the evolution of copulatory plugs in rodents. *Occ. Pap. Mus. Zool. Univ. Mich.* 689:1–27.

Waage, J. K. 1979. Dual function of the damselfly penis: Sperm removal and transfer. *Science* 203:916–918.

Waage, J. K. 1984. Sperm competition and the evolution of odonate mating systems. In R. L. Smith, ed., *Sperm Competition and the Evolution of Animal Mating Systems*, 251–290. Academic Press, New York.

Waage, J. K. 1986. Evidence for widespread sperm displacement ability among Zygoptera (Odonata) and the means for predicting its presence. *Biol. J. Linn. Soc.* 28:285–300.

Waddy, S. L., and D. E. Aiken. 1991. Mating and insemination in the American lobster, *Homarus americanus*. In R. T. Bauer and J. W. Martin, eds., *Crustacean Sexual Biology*, 126–144. Columbia University Press, New York.

Wagner, R. H. 1991. The use of extrapair copulations for mate appraisal by razorbills, *Alca torda*. *Behav. Ecol.* 2(3):198–203.

Wagner, W. E., Jr., A.-M. Murray, and W. H. Cade. 1995. Phenotypic variation in the mating preferences of female crickets. *Anim. Behav.* 49:1269–1281.

Walker, W. 1980. Sperm utilization strategies in nonsocial insects. *Am. Nat.* 115: 780–799.

Wall, R., and P. A. Langley. 1993. The mating behaviour of tsetse flies (*Glossina*): A review. *Physiol. Ent.* 18:211–218.

Walton, W. 1960. Copulation and natural insemination. In A. S. Parkes, ed., *Marshall's Physiology of Reproduction*, vol. 2, 130–160. Longmans, London.

Ward, P. I. 1993. Females influence sperm storage and use in the yellow dung fly *Scathophaga stercoraria* (L.). *Behav. Ecol. Sociobiol.* 32:313–319.

Ward, P. I., and L. W. Simmons. 1991. Copula duration and testes size in the yellow dung fly, *Scathophaga stercoraria* (L.): The effects of diet, body size, and mating history. *Behav. Ecol. Sociobiol.* 29:77–85.

Ward, S., and J. S. Carrel. 1979. Fertilization and sperm competition in the nematode *Coenorhabditis elegans*. *Dev. Biol.* 73:304–321.

Ware, A. D., and B. D. Opell. 1989. A test of the mechanical isolation hypothesis in two similar spider species. *J. Arachnol.* 17:149–162.

Warren, R. J., R. W. Vogelsang, R. L. Kirkpatrick, and P. F. Scanlon. 1978. Reproductive behaviour of captive white-tailed deer. *Anim. Behav.* 26:179–183.

Wasser, S. K., and D. P. Barash. 1983. Reproductive suppression among female mammals: Implications for biomedicine and sexual selection theory. *Quart. Rev. Biol.* 58:513–538.

Wasser, S. K., and D. Y. Isenberg. 1986. Reproductive failure among women: Pathology or adaptation? *J. Psychosom. Ob. Gyn.* 5:153–175.

Watson, P. J. 1990. Female-enhanced male competition determines the first mate and principal sire in the spider *Linyphia litigiosa* (Linyphiidae). *Behav. Ecol. Sociobiol.* 26:77–90.

Watson, P. J. 1991. Multiple paternity as genetic bet-hedging in female Sierra dome spiders, *Linyphia litigiosa* (Linyphiidae). *Anim. Behav.* 41:343–360.

Watson, P. J., and J.R.B. Lighton. 1994. Sexual selection and the energetics of copulatory courtship in the Sierra dome spider, *Linyphia litigiosa. Anim. Behav.* 48:615–626.

Watson, P. J., and R. Thornhill. 1994. Fluctuating asymmetry and sexual selection. *TREE* 9:21–25.

Wcislo, W. T. 1990. Parasitic and courtship behavior of *Phalacrotophora halictorum* (Diptera: Phoridae) at a nesting site of *Lasioglossum figueresi* (Hymenoptera: Halictidae). *Rev. Biol. Trop.* 38:205–209.

Wcislo, W. T., and S. L. Buchmann. In press. Mating behavior in the bees *Dieunomia heteropoda* and *Nomia tetrazonata*, with a review of courtship in Nomiinae (Hymenoptera: Halictidae). *J. Nat. Hist.* 24.

Wcislo, W. T., R. L. Minckley, and H. C. Spangler. 1992. Pre-copulatory courtship behavior in a solitary bee, *Nomia triangulifera* Vachal (Hymenoptera: Halictidae). *Apidologie* 23:431–442.

Webb, J. C., J. Sivinski, and C. Litzkow. 1984. Acoustical behavior and sexual success in the Caribbean fruit fly, *Anastrepha suspensa* (Loew) (Diptera: Tephritidae). *Environ. Ent.* 13:650–656.

Webber, H. H. 1977. Gasteropoda: Prosobranchia. In A. C. Giese and J. S. Pearse, eds., *Reproduction of Marine Invertebrates*, vol. 4, 1–97. Academic Press, New York.

Webster, A. B., R. G. Gartshore, and R. J. Brooks. 1981. Infanticide in the meadow vole, *Microtus pennsylvanicus*: Significance in relation to social system and population cycling. *Behav. Neur. Biol.* 31:342–347.

Webster, R. P., and R. T. Cardé. 1984. The effects of mating, exogenous juvenile hormone and a juvenile hormone analogue on pheromone titre, calling and oviposition in the omnivorous leafroller moth (*Platynota stultana*). *J. Insect Physiol.* 30(2):113–118.

Wedell, N. 1991. Sperm competition selects for nuptial feeding in a bushcricket. *Evolution* 45:1975–1978.

Wedell, N. 1992. Protandry and mate assessment in the wartbiter *Decticus verrucivorus* (Orthoptera: Tettigoniidae). *Behav. Ecol. Sociobiol.* 31:301–308.

Wedell, N. 1993. Spermatophore size in bushcrickets: Comparative evidence for nuptial gifts as a sperm protection device. *Evolution* 47(4):1203–1212.

Wedell, N., and A. Arak. 1989. The wartbiter spermatophore and its effect on female reproductive output (Orthoptera: Tettigoniidae, *Decticus verrucivorus. Behav. Ecol. Sociobiol.* 24:117–125.

Weigensberg, I., and D. J. Fairbairn. 1994. Conflicts of interest between the sexes: A study of mating interactions in a semiaquatic bug. *Anim. Behav.* 48:893–901.

Wells, M. J., and J. Wells. 1977. Cephalopoda: Octopoda. In A. C. Giese and J. S. Pearse, eds., *Reproduction of Marine Invertebrates*, 291–317. Academic Press, New York.

West-Eberhard, M. J. 1983. Sexual selection, social competition, and speciation. *Quart. Rev. Biol.* 58(2):155–183.

West-Eberhard, M. J. 1984. Sexual selection, competitive communication and species-specific signals in insects. In T. Lewis, ed., *Insect Communication*, 283–324. Academic Press, New York.

West-Eberhard, M. J. 1991. Sexual selection and social behavior. In M. H. Robinson and L. Tiger, eds., *Man and Beast Revisited*, 159–172. Smithsonian Institution Press, Washington, D.C.

Westneat, D. F. 1993. Temporal patterns of within-pair copulations, male mate-guarding, and extra-pair events in eastern red-winged blackbirds (*Agelaius phoeniceus*). *Behaviour* 124:267–290.

Westneat, D. F. 1995. Paternity and paternal behaviour in the red-winged blackbird, *Agelaius phoeniceus. Anim. Behav.* 49:21–35.

Westneat, D. F., P. W. Sherman, and M. L. Morton. 1990. The ecology and evolution of extra-pair copulations in birds. *Curr. Ornithol.* 7:331–369.

Wetton, J. H., and D. T. Parkin. 1991. An association between fertility and cuckoldry in the house sparrow, *Passer domesticus. Proc. Roy. Soc. B* 245:227–233.

Weygoldt, P. 1969. *The Biology of Pseudoscorpions*. Harvard University Press, Cambridge, Mass.

Whalen, M., and T. G. Wilson. 1986. Variation and genomic localization of genes encoding *Drosophila melanogaster* male accessory gland proteins separated by sodium dodecyl sulfate-polyacrylamide gel electrophoresis. *Genetics* 114:77–92.

Wharton, R. A. 1987. Biology of the diurnal *Metasolpuga picta* (Kraepelin) (Solifugae, Solpugidae) compared with that of nocturnal species. *J. Arachnol.* 14:363–383.

Wheeler, D., A. Wong, and J.M.C. Ribeiro. 1993. Scaling of feeding and reproductive structures in the mosquito *Aedes aegypti* L. (Diptera: Culicidae). *J. Kans. Ent. Soc.* 66:121–124.

Wheeler, M. R. 1947. IV. The insemination reaction in intraspecific matings of *Drosophila. Univ. Texas Publ.* 4720:78–115.

Whitman, D. W., and W. Loher. 1984. Morphology of male sex organs and insemination in the grasshopper *Taeniopoda eques* (Burmeister). *J. Morphol.* 179:1–12.

Whittier, J. M., and D. Crews. 1989. Mating increases plasma levels of prostaglandin F_{2a} in female garter snakes. *Prostaglandins* 37:359–366.

Whittier, J. M., and M. Tokarz. 1992. Physiological regulation of sexual behavior in female reptiles. In C. Gans and D. Crews, eds., *Biology of the Reptilia*, vol. 18, *Physiology E. Hormones, Brain, and Behavior*, 24–69. University of Chicago Press, Chicago.

Whittier, J. M., R. T. Mason, and D. Crews. 1987. Plasma steroid hormone levels of female red-sided garter snakes, *Thamnophis sirtalis parietalis*: Relationship to mating and gestation. *Gen. Comp. Endocrinol.* 67:33–43.

Wiewandt, T. A. 1971. Breeding biology of the Mexican leaf-frog. *Fauna* 2:29–34.

Wigglesworth, V. B. 1965. *The Principles of Insect Physiology*. Methuen, London.

Wilcox, R. S., and J. R. Spence. 1986. The mating system in two hybridizing species of water striders (Gerridae). I. Ripple signal functions. *Behav. Ecol. Sociobiol.* 19:79–85.

Wildt, D. E. 1990. Potential application of IVF technology for species conservation. In B. D. Bavister, J. Cummins, and E.R.S. Roldan, eds., *Fertilization in Mammals*, 349–364. Serono Symposia, Norwell, Mass.

Wilkes, A. 1965. Sperm transfer and utilization by the arrhenotokous wasp *Dahlbominus fuscipennis* (Zett.) (Hymenoptera: Eulophidae). *Can. Ent.* 97:647–657.

Wilkes, A. 1966. Sperm utilization following multiple insemination in the wasp *Dahlbominus fuscipennis. Can. J. Genet. Cytol.* 8:451–461.

Wilkinson, R., and Birkhead, T. 1995. Copulation behaviour in the vasa parrots *Coracopsis vasa* and *C. nigra. Ibis* 137:117–119.

Will, M. W., and S. K. Sakaluk. 1994. Courtship feeding in decorated crickets: Is the spermatophylax a sham? *Anim. Behav.* 48:1309–1315.

Willey, M. B. 1992. Mating behavior and male-male competition in two New Zealand spiders, *Ixeuticus martius* and *Ixeuticus robustus* (Araneae: Desidae). *Am. Arachnol.* 46:13.

Williams, G. C. 1992. *Natural Selection: Domains, Levels, and Challenges.* Oxford University Press, New York.

Williams, G. C., and R. M. Nesse. 1991. The dawn of Darwinian medicine. *Quart. Rev. Biol.* 66:1–22.

Williams, R. W., N.K.B. Hagan, A. Berger, and D. D. Despommier. 1978. An improved assay technique for matrone, a mosquito pheromone, and its application in ultrafiltration experiments. *J. Insect Physiol.* 24:127–132.

Willson, M. F. 1979. Sexual selection in plants. *Am. Nat.* 113:777–790.

Willson, M. F. 1990. Sexual selection in plants and animals. *TREE* 5:210–214.

Willson, M. F. 1994. Sexual selection in plants: Perspective and overview. *Am. Nat.* 144 (suppl.):S13-S39.

Willson, M. F., and N. Burley. 1983. *Mate Choice in Plants.* Princeton University Press, Princeton.

Wilson, G.D.F. 1991. Functional morphology and evolution of isopod genitalia. In R. T. Bauer and J. W. Martin, eds., *Crustacean Sexual Biology*, 228–245. Columbia University Press, New York.

Wiman, F. H. 1979a. Hybridization and the detection of hybrids in the fairy shrimp genus *Streptocephalus. Am. Midl. Nat.* 102:149–156.

Wiman, F. H. 1979b. Mating patterns and speciation in the fairy shrimp genus *Streptocephalus. Evolution* 33:172–181.

Wirtz, P. 1978. The behaviour of the Mediterranean *Tripterygion* species (Pisces, Blennioidei). *Z. Tierpsychol.* 48:142–174.

Wishart, G. J. 1987. Regulation of the length of the fertile period in the domestic fowl by numbers of oviductal spermatozoa as reflected by those trapped in laid eggs. *J. Reprod. Fertil.* 80:493–498.

Witalinski, W., E. Szlendak, and J. Boczek. 1990. Anatomy and ultrastructure of the reproductive systems of *Acarus siro* (Acari: Acaridae). *Exp. App. Acarol.* 10:1–31.

Witter, M. S., I. C. Cuthill, and R.H.C. Bonser. 1994. Experimental investigations of mass-dependent predation risk in the European starling, *Sturnus vulgaris. Anim. Behav.* 48:201–222.

Wojcik, D. P. 1969. Mating behavior of 8 stored-product beetles (Coleoptera: Dermestidae, Tenebrionidae, Cucujidae, and Curculionidae). *Fla. Ent.* 52:171–197.

Wright, S. J., C. Simerly, H. Schatten, and G. Schatten. 1989. Nuclear-cytoskeletal interactions during mouse fertilization and development. In B. D. Bavister, J. Cummins, and E.R.S. Roldan, eds., *Fertilization in Mammals*, 269–283. Serono Symposia, USA, Norwell, Mass.

Wu, C.-I. 1983. Virility deficiency and the sex-ratio trait in *Drosophila pseudoobscura*. I. Sperm displacement and sexual selection. *Genetics* 105:651–662.

Yamagishi, M., and Y. Tsubaki. 1990. Copulation duration and sperm transfer in the melon fly, *Dacus cucurbitae* Coquillett (Diptera: Tephritidae). *Appl. Ent. Zool.* 25(4):517–519.

Yamagishi, M., Y. Ito, and Y. Tsubaki. 1992. Sperm competition in the melon fly, *Bactrocera cucurbitae* (Diptera: Tephritidae): Effects of sperm "longevity" on sperm precedence. *J. Insect Behav.* 5:599–608.

Yamanaka, H. S., and A. L. Soderwall. 1960. Transport of spermatozoa through the female genital tract of hamsters. *Fertil. Steril.* 11:470–474.

Yamaoka, K., and T. Hirao. 1977. Stimulation of virginal oviposition by male factor and its effect on spontaneous nervous activity in *Bombyx mori*. *J. Insect Physiol.* 23:57–63.

Yamazaki, K., G. K. Beauchamp, and C. J. Wysocki. 1983. Recognition of H-2 types in relation to the blocking of pregnancy in mice. *Science* 221:186–188.

Yanagimachi, R., and M. C. Chang. 1963. Sperm ascent through the oviduct of the hamster and rabbit in relation to the time of ovulation. *J. Reprod. Fert.* 6:413–420.

Yano, I., R. A. Kanna, R. N. Oyama, and J. A. Wyban. 1988. Mating behaviour in the penaeid shrimp *Penaeus vannamei*. *Mar. Biol. (Biol.)* 97:171–175.

Yasui, Y. 1988. Sperm competition of *Macrocheles muscaedomesticae* (Scopoli) (Acarina: Mesostigmata: Macrochelidae), with special reference to precopulatory mate guarding behavior. *J. Ethol.* 6:83–90.

Yasuyama, K., T. Kimura, and T. Yamaguchi. 1988. Ultrastructural and immunohistological investigation of the spermathecal duct of the female cricket, *Gryllus bimaculatus*. *Zool. Sci.* 5:1196.

Yen, S.S.C., B. L. Lasley, C. F. Wang, H. Leblanc, and T. M. Siler. 1975. The operating characteristics of the hypothalamic-pituitary system during the menstrual cycle and observations of biological action of somatostatin. *Recent Progr. Horm. Res.* 31:321–357.

Yokoi, N. 1990. The sperm removal behavior of the yellow spotted longicorn beetle *Psacothea hilaris* (Coleoptera: Cerambycidae). *Appl. Ent. Zool.* 25:383–388.

Young, A.D.M., and A.E.R. Downe. 1982. Renewal of sexual receptivity in mated female mosquitoes, *Aedes aegypti*. *Physiol. Ent.* 7:476–477.

Young, A.D.M., and A.E.R. Downe. 1983. Influence of mating on sexual receptivity and oviposition in the mosquito, *Culex tarsalis*. *Physiol. Ent.* 8:213–217.

Young, A.D.M., and A.E.R. Downe. 1987. Male accessory gland substances and the control of sexual receptivity in female *Culex tarsalis*. *Physiol. Ent.* 12:233–239.

Young, R. J., and W. C. Starke. 1990. Characterization of activated motility of rabbit spermatozoa. In B. D. Bavister, J. Cummins, and E.R.S. Roldan, eds., *Fertilization in Mammals*, 437. Serono Symposia, Norwell, Mass.

Yuval, B., and A. Spielman. 1990. Sperm precedence in the deer tick *Ixodes dammini*. *Physiol. Ent.* 15:123–128.

Zahavi, A. 1988. Mate guarding in the Arabian babbler, a group-living songbird. *Proc. Int. Orn. Congr.* 20:420–427.

Zarrow, M. X., and J. H. Clark. 1968. Ovulation following vaginal stimulation in a spontaneous ovulator and its implications. *J. Endocr.* 40:343–352.

Zavaleta, D., and F. Ogasawara. 1987. A review of the mechanism of the release of spermatozoa from storage tubules in the fowl and turkey oviduct. *World's Poultry Sci. J.* 43:132–139.

Zeh, J., and D. W. Zeh. 1994. Last male sperm procedence breaks down when females mate with three males. *Proc. Ray. Soc. Lond. B* 257:287–292.

Zeh, D. W., and J. A. Zeh. In prep. Patchy habitats, promiscuous females and male variability in a beetle-riding pseudoscorpion.

Zimmering, S., and G. L. Fowler. 1968. Progeny: Sperm ratios and nonfunctional sperm in *Drosophila melanogaster*. *Genet. Res. Camb.* 12:359–363.

Zimmering, S., J. M. Barnabo, J. Femino, and G. L. Fowler. 1970. Progeny: Sperm ratios and segregation-distorter in *Drosophila melanogaster*. *Genetica* 41:61–64.

Zuk, M., J. D. Ligon, and R. Thornhill. 1992. Effects of experimental manipulation of male secondary sex characters on female mate preference in red jungle fowl. *Anim. Behav.* 44:999–1006.

Zumpe, D., and R. P. Michael. 1983. A comparison of the behavior of *Macaca fascicularis* and *Macaca mulatta* in relation to the menstrual cycle. *Am. J. Primatol.* 4:55–72.

Zunino, M. 1988. La evolución de los aparatos copuladores: Comentarios a W. G. Eberhard, "Sexual selection and animal genitalia." *Elytron* 1:105–107.

Subject Index

Taxonomic Index

William G. Eberhard is a member of the scientific staff of the Smithsonian Tropical Research Institute, and Professor of Biology at the Escuela de Biologia of the Universidad de Costa Rica. He is the author of *Sexual Selection and Animal Genitalia.*